GIS Applications in Agriculture

Volume Two

Nutrient Management for Energy Efficiency

GIS Applications in Agriculture

Series Editor

Francis J. Pierce

Washington State University, Prosser

GIS Applications in Agriculture, *edited by Francis J. Pierce and David E. Clay*

GIS Applications in Agriculture, Volume Two: Nutrient Management for Energy Efficiency, *edited by David E. Clay and John F. Shanahan*

GIS Applications in Agriculture, Volume Three: Invasive Species, *edited by Sharon A. Clay*

GIS Applications in Agriculture

Volume Two

Nutrient Management for Energy Efficiency

Edited by

David E. Clay

John F. Shanahan

GIS APPLICATIONS IN AGRICULTURE SERIES

CRC Press
Taylor & Francis Group
Boca Raton London New York

CRC Press is an imprint of the
Taylor & Francis Group, an **informa** business

CRC Press
Taylor & Francis Group
6000 Broken Sound Parkway NW, Suite 300
Boca Raton, FL 33487-2742

Printed in the United States of America on acid-free paper
10 9 8 7 6 5 4 3 2 1

International Standard Book Number: 978-1-4200-9270-7 (Hardback)

Library of Congress Cataloging-in-Publication Data

GIS applications in agriculture. Volume two, Nutrient management for energy efficiency / editors: David E. Clay and John F. Shanahan.
p. cm. -- (GIS applications in agriculture)
Includes bibliographical references and index.
ISBN 978-1-4200-9270-7 (alk. paper)
1. Agriculture--Remote sensing. 2. Geographic information systems. 3. Agricultural mapping. 4. Agriculture--Data processing. 5. Plants--Nutrition. I. Clay, David (David E.) II. Shanahan, John Francis, 1955- III. Title: Nutrient management for energy efficiency. IV. Series: GIS applications in agriculture series.

S494.5.R4G57 2011
631.8'1--dc22 2010034458

Visit the Taylor & Francis Web site at
http://www.taylorandfrancis.com

and the CRC Press Web site at
http://www.crcpress.com

Contents

Series Preface

From its initial use in the 1960s by cartographers who wanted to adopt computer techniques in mapmaking, geographic information systems (GIS) have become an essential and efficient toolkit in all aspects of agriculture from farm management and resource conservation to a broad range of agribusiness applications. Recognizing that few examples from agriculture were used in teaching or demonstrating GIS, my good friend Max Crandall envisioned a book dedicated to applications of GIS in agriculture that could provide learning opportunities to scientists, educators, students, consultants, and farmers in either formal or informal settings. Max formed a small group, including Pierre C. Robert, Harold Reitz Jr., Matthew Yen, and myself, that shared numerous ideas about such a book over phone conversations. With the untimely death of Pierre C. Robert in December 2003, interest in the book diminished until 2006 when CRC Press, through the efforts of one of its editors, John Sulzycki, agreed to publish the book entitled *GIS Applications in Agriculture* edited by Francis J. Pierce and David E. Clay, which appeared in February 2007. With a long-term commitment from CRC Press to Max Crandall's vision, this book became a book series on GIS Applications in Agriculture for which I am series editor.

GIS Applications in Agriculture: Nutrient Management for Energy Efficiency edited by David E. Clay and John F. Shanahan is the second volume in this book series. This book includes 24 chapters on various topics dealing with geospatial aspects of tillage, and nutrient, water, and energy use in agriculture. Like the first volume, detailed applications are provided in many chapters with datasets and color figures on a separate CD for readers to use in teaching and learning GIS or directly applying them to situations they face in agriculture. I am grateful to David E. Clay for agreeing to develop this volume and to his coeditor, John Shanahan, for the hard work that went into organizing and editing it, and to the chapter authors for their excellent contributions to what I believe are interesting and useful applications of GIS in agriculture. I would also like to thank Randy Brehm, the CRC Press editor for this volume, and all those at CRC Press who made this volume possible.

As series editor, it is my responsibility to seek new book ideas and capable editors to create additional books in this series on topics of importance to agriculture that provide relevant applications of GIS in agriculture. I invite those who have ideas for new volumes in the GIS Applications in Agriculture series to contact myself or CRC Press to discuss publishing opportunities.

Francis J. Pierce
Series Editor

Preface

The concept for this book evolved from numerous scientific discussions and grower education meetings, including a soil carbon workshop jointly sponsored by South Dakota State University and the South Dakota Corn Utilization Council, and a United States North Central Sustainable Agriculture (NC-SARE) conference, leading to a white paper providing research recommendations. At these meetings, the scientific community expressed the need for the development of techniques that could simultaneously increase soil carbon storage and reduce agriculture's energy footprint, while producers from around the world mentioned the need to reduce their fertilizer input costs. A review of activities showed that many different approaches were being tested. The use of site-specific or precision agriculture has arisen as a common theme for addressing the multitude of concerns and issues raised.

Agriculture is being changed by three fundamental forces: the expanding capacity of personal computers, molecular biology revolution, and developments in information technology like geographical information systems (GIS). Through precision farming, all three technologies can be packaged and delivered to producers. The combined impact is likely to lead to the greatest intellectual transition that has ever occurred in agriculture. Who would have thought 50 years ago that a large percentage of tractors traversing our vast fields today would be under the control of autosteer systems rather than being driven by farmer operators? Who would have thought that the farming community would be using a single line of machinery powerful and efficient enough to farm 4000 ac? Who would have thought that the major defensive mechanisms in our battle with pests would come in our seed bags? We are entering a new era in production agronomics, an era dominated by site-specific spatial management of farming inputs. This is an era during which the agricultural foundations will undergo revolutionary changes.

For centuries, agronomy was dominated by the biological sciences and the ability to work hard. We are now witnessing an era when creativity and mathematics are assuming equal importance. Throughout the chapters of this book there is evidence that attests to how complex mathematical and spatial modeling approaches can serve as the basis for much of our present and, certainly, future management. It is mathematics that will enable producers to make full use of the technological advances made during the 21st century. This book focuses on the use of mathematics and creativity to develop nutrient management practices that will help producers improve their profitability and energy efficiency. It highlights successes and discusses the nuts and bolts associated with implementing the proposed techniques. The topics discussed in this book include calculating energy efficiency, devising techniques for overcoming yield-limiting factors, collecting and analyzing soil information, using remote sensing for improving management decisions, developing an economically optimum site-specific corn plant population equation based on experiment containing many field sites, assessing and implementing site-specific carbon and water management systems, analyzing energy efficiency of compost and manures, and estimating soil productivity and energy efficiency using online data sources. For color figures, please refer to the accompanying CD-ROM.

Funding for organizing this book was obtained from the US-USDA-NC-SARE program. The NC SARE mission is to "strengthen rural communities, increase farmer/rancher profitability, and improve the environment by supporting research and education." NC SARE contact information is http://www.sare.org/ncrsare/phone (402) 472-7081.

For MATLAB® and Simulink® product information, please contact:

The MathWorks, Inc.
3 Apple Hill Drive
Natick, MA, 01760-2098 USA
Tel: 508-647-7000
Fax: 508-647-7001
E-mail: info@mathworks.com
Web: www.mathworks.com

Editors

Dr. David E. Clay is the director of the South Dakota Drought Tolerance Center, Brookings, South Dakota, professor of plant sciences, and fellow of the American Society of Agronomy, Madison, Wisconsin. Currently, he is serving as senior editor for the *Agronomy Journal* and as associate editor for the *Journal of Plant Nutrition*, the *International Journal of Agriculture*, and *Precision Agriculture*. He has edited a number of books that include *South Dakota Best Management Manual* (SDSU), *Site-Specific Farming Guidelines Manual* (International Plant Nutrition Institute), *GIS Applications in Agriculture* (CRC Press), and *Soil Science: A Step-by-Step Field Analysis* (Soil Science Society of American). He has also authored the new book, *Mathematics and Science for Improved Agronomic Decisions*, which is being published by the International Plant Nutrition Institute, Norcross, Georgia. He has published over 120 refereed journal papers. Dr. Clay has received the SDSU Ag/Bio College Deans Award for Excellence (1994) and Teamwork (2004), SDSU Gamma Delta Research Award (1996), SDSU F.O. Butler Award for Excellence in Research (2004), SDSU Sigma Xi Presidents Award for Service (2006), USDA-ARS Collaboration Award (2006), and the PrecisionAG Award of Excellence in education and research (PrecisionAG Institute, 2009). Dr. Clay's research goal is to develop and test sustainable agricultural management systems that enhance environmental quality, maintain rural economies, and enable energy self-sufficiency.

Dr. John F. Shanahan is currently an agronomy research manager with Pioneer Hi-Bred International of Johnston, Iowa. He was a professor in the Department of Soil and Crop Sciences at Colorado State University from 1982 to 1998 and a research agronomist with USDA-ARS from 1998 to 2010 in Lincoln, Nebraska. He received his BS in agronomy from the University of Nebraska in 1977 and his MS (1979) and PhD (1982) in crop breeding and physiology from Colorado State University. His major fields of interest are agronomy, soil science, remote sensing, and precision agriculture, with the goal of providing growers with innovative management solutions for improving agricultural productivity and sustainability. Dr. Shanahan has reported results of his research in more than 70 peer-reviewed publications, book chapters, and conference proceedings, written more than 60 extension publications (including bulletins, newsletter articles, and popular press reports), and delivered more than 250 extension presentations providing information to crop producers, consultants, extension educators, government personnel, and the public. He has served on the editorial boards for *Agronomy Journal* and *Journal of Precision Agriculture*. He is a member of the American Society of Agronomy, Crop Science Society of America, and Soil Science Society of America and a fellow of the American Society of Agronomy.

Contributors

Viacheslav I. Adamchuk
Biological Systems Engineering Department
University of Nebraska-Lincoln
Lincoln, Nebraska

Mahdi Al-Kaisi
Department of Agronomy
Iowa State University
Ames, Iowa

Telmo Jorge Carneiro Amado
Soil Department Federal University of Santa Maria, Santa Maria
Rio Grande do Sul, Brazil

Daryl B. Arnall
Department of Plant and Soil Sciences
Oklahoma State University
Stillwater, Oklahoma

A.S. Birr
Minnesota Department of Agriculture
Pesticide and Fertilizer Management Division
Rochester, Minnesota

Yacine M. Bouroubi
Horticulture Research and Development Centre
Agriculture and Agri-Food Canada
Saint-Jean-sur-Richelieu, Quebec, Canada

Kevin F. Bronson
USDA-ARS Arid Land Agricultural Research Center
Maricopa, Arizona

Gregg Carlson
Plant Science Department
South Dakota State University
Brookings, South Dakota

David E. Clay
Plant Science Department
South Dakota State University
Brookings, South Dakota

Sharon A. Clay
Plant Science Department
South Dakota State University
Brookings, South Dakota

D.A. Crouse
Department of Soil Science
North Carolina State University
Raleigh, North Carolina

A. Ehni
Wells County North Dakota Soil Conservation District
Fessenden, North Dakota

R.E. Engel
Department of Land Resources and Environmental Sciences
Montana State University
Bozeman, Montana

David W. Franzen
Soil Science Program
School of Natural Resource Sciences
North Dakota State University
Fargo, North Dakota

Christopher J. Graham
Department of Crop and Soil Sciences
Cornell University
Ithaca, New York

Robert D. Grisso
Biological Systems Engineering Department
Virginia Polytechnic Institute and State University
Blacksburg, Virginia

Serge Guillaume
Cemagref, UMR ITAP
Montpellier, France

J.L. Havlin
Department of Soil Science
North Carolina State University
Raleigh, North Carolina

C.B. Hedley
Landcare Research
Palmerston North, New Zealand

R. Kyle Heimerl
Plant Science Department
South Dakota State University
Brookings, South Dakota

D.W. Kelley
Department of Geography
University of St. Thomas
St. Paul, Minnesota

Newell R. Kitchen
USDA-ARS
Columbia, Missouri

Michael F. Kocher
Biological Systems Engineering Department
University of Nebraska-Lincoln
Lincoln, Nebraska

David A. Laird
Department of Agronomy
Iowa State University
Ames, Iowa

Rattan Lal
Carbon Management and Sequestration Center
School of Environment and Natural Resources
The Ohio State University
Columbus, Ohio

Adam J. Liska
Department of Biological Systems Engineering
University of Nebraska
Lincoln, Nebraska

Daniel S. Long
USDA-ARS Columbia Plateau Conservation Research Center
Pendleton, Oregon

Douglas D. Malo
Plant Science Department
South Dakota State University
Brookings, South Dakota

Jeffrey J. Melkonian
Department of Crop and Soil Sciences
Cornell University
Ithaca, New York

Umakant Mishra
Energy Biosciences Institute
University of California Berkeley
Berkeley, California

and

Carbon Management and Sequestration Center
School of Environment and Natural Resources
The Ohio State University
Columbus, Ohio

Robert W. Mullen
School of Environment and Natural Resources
The Ohio State University
Columbus, Ohio

Nsalambi V. Nkongolo
Department of Agriculture and Environmental Sciences
Lincoln University of Missouri
Jefferson City, Missouri

John Nowatzki
Agricultural and Biosystems Engineering Department
North Dakota State University
Fargo, North Dakota

Bernard Panneton
Horticulture Research and Development Centre
Agriculture and Agri-Food Canada
Saint-Jean-sur-Richelieu, Quebec, Canada

F. Mamani Pati
UAC Carmen Pampa
Bolivian Catholic University
Coroico, La Paz-Bolivia

Richard K. Perrin
Department of Agricultural Economics
University of Nebraska
Lincoln, Nebraska

Shafiqur Rahman
Department of Agricultural and Biosystems Engineering
North Dakota State University
Fargo, North Dakota

Kurtis D. Reitsma
Plant Science Department
South Dakota State University
Brookings, South Dakota

M.P. Russelle
Plant Science Research Unit
USDA-Agricultural Research Service
St. Paul, Minnesota

Antônio Luis Santi
CESNORS/Federal University of Santa Maria, Frederico Westphalen
Rio Grande do Sul, Brazil

Peter C. Scharf
Plant Science Department
University of Missouri
Columbia, Missouri

Joseph Schefers
Monsanto Company
Brookings, South Dakota

Thomas E. Schumacher
Plant Science Department
South Dakota State University
Brookings, South Dakota

H. Smeltekop
UAC Carmen Pampa
Bolivian Catholic University
Coroico, La Paz-Bolivia

D.G. Tiffany
Department of Applied Economics
University of Minnesota
St. Paul, Minnesota

Nicolas Tremblay
Horticulture Research and Development Centre
Agriculture and Agri-Food Canada
Saint-Jean-sur-Richelieu, Quebec, Canada

Harold M. van Es
Department of Crop and Soil Sciences
Cornell University
Ithaca, New York

Philippe Vigneault
Horticulture Research and Development Centre
Agriculture and Agri-Food Canada
Saint-Jean-sur-Richelieu, Quebec, Canada

R.J. Wiederholt
Carrington Research Extension Center
North Dakota State University
Carrington, North Dakota

I.J. Yule
New Zealand Centre for Precision Agriculture
Massey University
Palmerston North, New Zealand

Xiaodong Zhang
Earth Systems Science and Policy
University of North Dakota
Grand Forks, North Dakota

1 Energy and Climate Implications for Agricultural Nutrient Use Efficiency

Adam J. Liska and Richard K. Perrin

CONTENTS

1.1 EXECUTIVE SUMMARY

Energy and climate change are beginning to dominate the global political agenda and will drive policy formation that will shape the future of agriculture. Energy issues threaten national security and economic stability, as well as access to low-cost nutrient inputs for agriculture. Climate change has the potential to cause serious disruption to agricultural productivity. Paradoxically, nutrient use in agriculture to increase crop yields has the potential to negatively impact climate. This chapter will discuss recent and future energy and climate trends, the relationships between agricultural nutrient use efficiency and biofuels, and how global land limitations will shape agriculture in the future. Comparative gross energy yield and nitrogen use efficiency for ethanol production from crop residue, switchgrass, grain sorghum, sweet sorghum, and corn grain is presented, showing small differences in nitrogen use efficiency, but large differences in gross energy yields. In addition to considering the need to increase crop productivity to meet the demands of a growing population and bioenergy, agricultural nutrient use efficiency must be reconsidered with respect to the important energy and climate challenges shaping agriculture today.

1.2 ENERGY AND CLIMATE TRENDS

Nutrient application in agriculture is essential to maintain a sufficient food supply for a growing global population and to meet an increased demand for bioenergy. The fixation of atmospheric nitrogen (N) by the Haber–Bosch process has enabled higher crop yields necessary to support the growth of global population by roughly three billion people in the twentieth century, or almost half of humanity.[20] Energy is required to process, deliver, and apply nutrients to land, which is costly and contributes to climate change. Increased nutrient use efficiency is essential to increase crop productivity and energy efficiency of bioenergy production in a sustainable manner while limiting negative environmental impacts and reducing costs.[8]

Energy issues are beginning to dominate the global political agenda. First, there is growing concern that global production of easily accessible oil is nearing its peak rate.[13,56] Global oil production is dominated by giant oil fields, with the 500 largest fields contributing over 60% of production.[27] In 2008, 580 of the 651 largest oil fields globally were reported to have passed their peak production rate and are now producing an average of about 6% less oil per year.[27,29] In accordance with reported declining trends, an independent analysis from Uppsala University in Sweden found that global oil production will decline from 84 million barrels per day (mb/day) (including natural gas liquids) in 2007 to roughly 76 mb/day by 2030.[2] In contradiction to these findings, official analysis from the International Energy Agency (IEA) optimistically anticipates that petroleum production will continue to increase through 2030, reaching a level 20% higher than current levels.[30] The U.S. Energy Information Administration (EIA) also anticipates an increase in production over this period of about 15%.[15] Contrary to these assertions, the Swedish study states that historic trends of reduced field productivity will continue in the future at the same rate, which means lower productivity than other estimates that are more likely to be politically influenced. These conflicting expectations add uncertainty and volatility to world energy markets that are already vulnerable to political and economic vagaries.

Increases in oil demand that exceed rates of supply increase will cause oil prices to climb. By 2030, both the IEA and EIA project oil to reach about $190 per barrel in nominal dollars ($115–$130 in 2008 dollars). Some suggest, however, that the recent oil price spike in 2008 to $147 per barrel (compared to roughly $80 per barrel in December 2009) has stimulated increased conservation and adaptation which may keep oil prices relatively lower in the near term due to reduced demand.[45] The current recession has also reduced demand for oil. Nonetheless, oil prices and the trend in total cost of U.S. crude oil imports are likely to continue to increase (Figure 1.1). In 2007 with oil at $70 per barrel, the U.S. trade deficit in petroleum products was $293 billion, or 36% of the total trade deficit of $819 billion.[62] Increasing production of nonconventional sources of petroleum such as oil (tar) sands from Canada will also help maintain petroleum supply,[29] while production from current major fields is declining. Oil sands could contribute as much as 20% of U.S. gasoline supply by 2020.[41]

Global growth in population and the world economy have required greater energy use to sustain improving living standards. With increasingly narrow margins between energy supply and demand, analysis suggests that disruption of the oil supply and

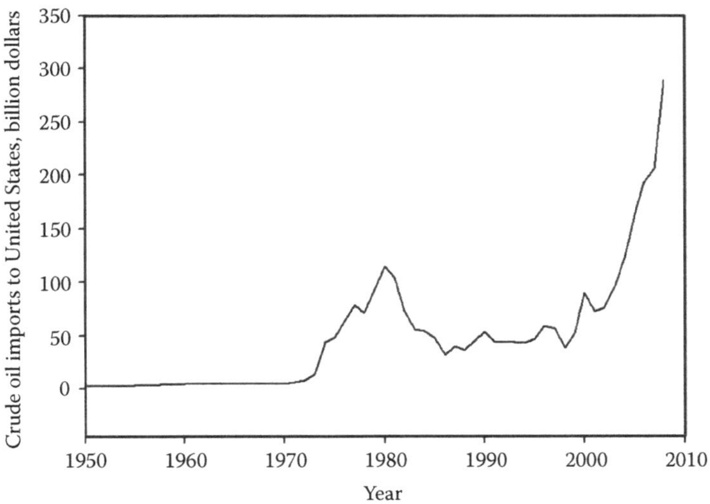

FIGURE 1.1 Inflation-adjusted annual value of crude oil imports into the United States in billions from 1950 to 2008 (in 2000 dollars). (From EIA, *Annual Energy Review 2008*, U.S. Department of Energy, Washington, DC, 2009.)

accompanying oil price spikes can have significant negative impacts on the global economy.[26] While the interrelationships between the business cycle and petroleum price are complicated and not easily resolved,[36] Brown[6] reports that 9 of the past 10 U.S. recessions since 1945 were preceded by significant oil price spikes. The relationships between oil price and the health of the economy suggest that current high and unstable oil prices could have broad implications for economic stability.

The high oil prices of 2008 led to the transfer of nearly $1 trillion to members of OPEC, the Organization of the Petroleum Exporting Countries.[17] National oil companies in OPEC and other countries, such as Saudi Aramco (Saudi Arabia), the National Iranian Oil Company (Iran), Petrochina (China), Petrobras (Brazil), and Gazprom (Russia), control approximately 90% of the world's oil reserves and 75% of global oil production—similar numbers apply for natural gas.[69]

In conjunction with these trends, growing military intervention to ensure access to foreign oil has amplified the threat of international conflict. There is growing consensus in the economic and military communities that oil played a large part in the U.S.-led invasion of Iraq in 2003.[4,26,44,50] Iraq has the third largest oil reserves globally at 115 billion barrels (~9% of global crude oil reserves), ranking below only Saudi Arabia and Iran.[16] The full monetary cost of the Iraq war is projected by Stiglitz and Bilmes[58] to range between $2.7 and $5 trillion and the conflict has resulted in between 90,000 and 800,000 violent deaths of Iraqi civilians and more than 4000 U.S. military deaths since 2003.[63]

As a corollary to the invasions of Iraq, ongoing U.S. military activities in Afghanistan are also motivated, in part, by proposed pipeline installation for transportation of oil and gas from Central Asia to the Indian Ocean.[24,53,61] In that region, Kazakhstan contains three of the world's 10 largest giant oil fields (newly discovered), and they are now Chevron's leading source of petroleum, which is exported via pipelines heading west through Georgia.[38,39,56] In U.S. Congressional testimony in 2006, Steven Mann, Principal Deputy Assistant Secretary for South and Central Asian Affairs at the State Department, clearly outlined U.S. intentions concerning energy in the Afghanistan region:

> "Since the independence of the new Caspian states 15 years ago, the United States has been in the forefront of oil and gas development in the region, and our efforts are paying off." "With the completion of the first phase of the East-West Energy Corridor [through Georgia], we must now press on with the second phase of supporting new energy routes out of Central Asia." "The United States and the countries of the broader region share an interest in the free movement of energy, people, goods, and information from the Kazakh steppes to the Indian Ocean. We want not only to support economic development along a north-south axis, but also afford Afghanistan access to a wider world, thus becoming a bridge, not a barrier."[61]

In addition to these two costly military ventures with significant relationships to energy resources, costly military security for global oil transportation includes protection of unstable maritime oil transit routes, which has been recently estimated to cost between $104 and $138 billion per year.[11,12,19]

Serious energy-related security issues also arise from the fact that international terrorism has been supported by tens of millions of dollars from the sale of Middle Eastern oil.[7,47] Some of this money is thought to have been used to support the September 11, 2001 "9/11" terrorist attacks on the United States,[47] although it is recognized that relatively little money was actually required to carry out most terrorist attacks in the last 10 years.[67]

These various challenges for the petroleum economy consisting of limited and fragile supply, wealth transfer, contribution to national deficits, costly military operations, and terrorism are all serious national security and economic issues. These issues have stimulated support for the development of alternative energy sources.[14,71] Biomass resources are of particular importance because they can be converted to liquid fuels to substitute for petroleum which can be used in existing infrastructure with limited modification. The U.S. *Energy Independence and Security Act of 2007* (EISA) mandates that 36 Bgal of biofuels be produced annually by 2022, of which 15 Bgal/year are to be grain ethanol, 16 Bgal/year are to be cellulosic ethanol, and 5 Bgal are to be other advanced renewable fuels. Furthermore, the U.S. Air Force, the world's single largest consumer of petroleum, recently announced a plan to substitute 50% of their fuel use with alternative fuels, with particular emphasis on biofuels.[3] Continued expansion of the biofuel industry will place greater demands on agricultural productivity and efficiency.

While limited energy supply is increasingly problematic, the impacts of atmospheric emissions from fossil-based energy sources on global climate change are becoming well established.[31] Anthropogenic climate change from greenhouse gas (GHG) emissions from burning fossil fuels will impact agriculture in a number of ways. Some of the changes that will have negative impacts on agriculture include higher average night-time temperatures,[51] greater frequency of heat waves, heavy rainfall events, destructive storms, and regional droughts.[35] In addition, rising sea level has the potential to submerge coastal agricultural regions, decreasing the availability of fertile agricultural land. Finally, rising temperature is contributing to the global disappearance of glaciers that threatens the water supply of regions dependent on glacial meltwater for irrigation to support agricultural productivity.[49,65]

The sum of these challenges for agriculture may be quite significant in the near-term future. Commodity prices will rise, and these price increases will stimulate adoption of more efficient production techniques as well as expansion of agricultural systems. Nutrient efficiency research will provide new information that will permit these adjustments to be made, which otherwise would not be possible with lower food and biofuel prices.

1.3 AGRICULTURAL NUTRIENT USE EFFICIENCY AND BIOFUELS

Energy price increases raise both the cost of field operations and the prices of crop inputs. Most N fertilizers are manufactured from natural gas or petroleum, so their costs will obviously rise with oil prices. Even though potash and phosphate are not manufactured from fossil energy, substantial amounts of energy are required in their extraction and processing, and their prices have also recently followed oil prices (Figure 1.2a). The primary nutrient applied in the United States is N (Figure 1.2b), and its price is most closely related to oil price. Therefore, sustained oil prices above $100 per barrel will lead to fertilizer prices substantially higher than previously experienced, and drive investment in practices to limit fertilizer expenditures and increase nutrient use efficiency.

Improved management has contributed to recent gains in nutrient use efficiency for corn in the United States to the extent that on average 37% of applied N is now taken up by the crop. From 1980 to 2000, N application remained relatively constant at 146 kg/ha[1] (Figure 1.2c) while the partial factor productivity (kg grain yield per kg of N applied) increased by 36% (from 42 to 57 kg/kg[1]).[8] Increasing agricultural nutrient use efficiency will also reduce the negative climate impact of crop production.

GHG emissions from agriculture are a large positive source of global warming potential. In 2005, nonfossil fuel emissions from global agriculture contributed 10%–12% of total of anthropogenic GHG emissions, with methane (CH_4) contributing a little over half of emissions and nitrous oxide (N_2O) emissions contributing the majority of the remainder.[57] These agricultural emissions contributed roughly 50% of global methane emissions and 60% of nitrous oxide emissions. Methane is produced when organic materials decompose in oxygen-deprived conditions, with significant sources from digestion in ruminant livestock, manure, and flooded rice. Nitrous oxide, on the other hand, is produced by the microbial transformation of N in soil and manure. Global agricultural N_2O emissions are projected to increase by 35%–60% by 2030 due to increased N fertilizer use and increased animal manure production. In the United States, however, synthetic N applications are projected to remain relatively constant, and increases in N_2O emissions are projected mainly from manure.[57] Large and uncertain net fluxes of carbon dioxide from global agriculture are not thought to contribute much to net GHG emissions from agriculture overall.

In addition to biogenic GHG emissions, additional GHG emissions from production and application of cropping inputs must also be considered. After totaling net biogenic and fossil fuel inputs for a corn grain cropping systems in U.S. Corn Belt, N_2O

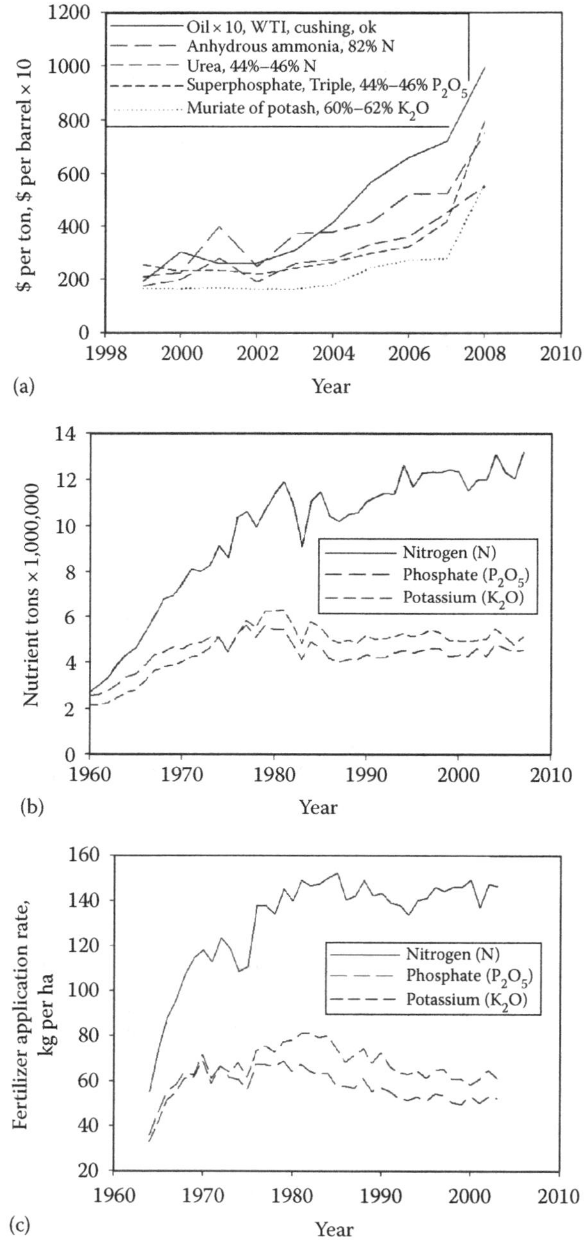

FIGURE 1.2 Average annual fertilizer and oil prices (a), total fertilizer consumption in the United States (b), and average fertilizer application rates per hectare for corn (c). (From EIA, Official Energy Statistics from the U.S. Government. Accessed at http://tonto.eia.doe.gov, November 2009; ERS [Economic Research Service], USDA, U.S. Fertilizer Use and Price. Accessed at http://www.ers.usda.gov/Data/FertilizerUse/, November 2009; USDA [United States Department of Agriculture], *Agricultural Prices Supplement*, August 2009. Accessed at http://usda.mannlib.cornell.edu/usda/current/AgriPricSu/AgriPricSu-08-05-2009.pdf, November 2009.)

TABLE 1.1
Energy Inputs and Direct GHG Emissions from an Average Corn Cropping System in the Midwest, Assuming Soil Carbon Dynamics Are Neutral

	Energy Inputs, %	GHG Emissions, %
Nitrogen fertilizer	39.7	15.0
N_2O emissions from N fertilizer[a]: denitrification, volatilization, leaching, runoff	—	35.5
N_2O emissions from crop and biomass N: crop residue, roots, manure	—	12.7
Phosphorus fertilizer	2.2	3.2
Potassium fertilizer	3.7	1.5
Lime	0.2	6.3
Herbicides	11.5	5.2
Insecticides	0.5	0.2
Seed	1.2	0.6
Gasoline	3.8	1.7
Diesel	16.2	8.7
LPG	8.3	3.4
Natural gas	4.8	1.7
Electricity	6.0	3.1
Farm machinery	2.0	0.9
Total	**100.0**	**100.0**

Source: Liska, A.J. et al., *J. Ind. Ecol.*, 13, 58, 2009.
[a] Of synthetic N fertilizer applied, 1.33% is lost as N_2O.
Emissions from "fertilizer[s]" are from fossil fuel use in upstream production.

emissions from fertilizer N inputs (based on IPCC emission factors) were found to be roughly 36% of net GHG emissions[32,42] (Table 1.1). Applications of N are also a significant fraction of energy inputs for corn grain production accounting for roughly 40% of agricultural energy inputs. By improving N use efficiency, energy inputs and emissions can be reduced. Evidence suggests that N_2O emissions can be reduced by 20% by increasing crop N use efficiency via crop management, by 10% via use of either soil N tests or fertilizer timing, and by 5% with the use of either nitrification and urease inhibitors or N fertilizer placement.[9] Different types of N fertilizer may also have the potential to reduce N_2O emissions. Furthermore, efficient N use can be improved by adjusting application rates using GIS-based precision estimates of crop needs.

Crops accumulate their biomass carbon from atmospheric carbon dioxide and have the potential to be an energy source that does not contribute to net growth in atmospheric GHGs. If bioenergy is not to significantly contribute to GHG emissions, efficient collection and conversion to liquid forms is required for minimal use of fossil fuels. Ethanol production from corn grain in the United States and from sugarcane in Brazil are models for increased utilization of agricultural resources to meet societies' energy needs while reducing GHG emissions relative to the gasoline they replace (although indirect GHG contributions of both gasoline and biofuel production are not yet accurately accounted for in current analytical methods).[41] Significant research and development efforts are also underway to better utilize biomass resources for transportation fuels via production of cellulosic ethanol, and other so-called "second-generation" biofuels, although these systems are not yet profitable.[43]

Biofuel production is a "system of systems"[34] composed of distributed independent complex systems for crop production, biomass transportation, biorefining, co-product use, fossil fuel production and delivery, fertilizer and chemical inputs, and end-use vehicle systems. The efficiency of the overall biofuel production system can be analyzed using life cycle assessment (LCA), which is used to compare biofuel performance with fossil fuel systems. In addition, LCA enables the environmental impacts of a production system to be analyzed to identify areas for improvements in efficiency.[25]

The life cycle efficiency of biofuel production is primarily analyzed in terms of energy efficiency and net GHG emissions.[42,43] Crop production contributes approximately 50% of positive life cycle emissions.[42] In the past and now, research into the life cycle energy efficiency of biofuel production systems (particularly corn-ethanol) has been marked by conflicting results, but greater consensus from recent research shows positive energy and GHG benefits are derived from biofuel production and use.[1,5,23,40,42] The EISA legislation now requires that biofuels must reduce life cycle GHG emissions, including indirectly caused emissions, compared to fossil fuels. Comparisons of the indirect GHG emissions resulting from gasoline production and corn-ethanol production are still primitive.[41] EISA requires that corn-ethanol must reduce emissions by 20% compared to gasoline, but the precise methodologies for these life cycle calculations are still under development. Other recent state legislation, such as in California, will restrict market access if biofuels do not meet life cycle GHG emissions reduction targets.

Recent improvements in biorefinery energy efficiency have greatly influenced the life cycle energy efficiency of corn-ethanol production.[42] In 2001, survey data reported that energy inputs for the biorefinery were 13.9 MJ/L[1] (primarily coal and natural gas) and comprised 67% of life cycle energy inputs.[23] Since 2001, the U.S. corn-ethanol industry has significantly expanded with new more efficient production capacity, composed primarily of natural gas powered dry mill biorefineries (Figure 1.3). Survey data from 2006 documents the increasing average energy efficiency of the industry, with biorefinery energy inputs reduced to 7.7 MJ/L[1], contributing 56% to life cycle energy inputs.[42,52] Use of recent data suggest that corn-ethanol has a net energy return of 1.6 units of energy per unit of energy invested.[42] Furthermore, compared to gasoline, corn-ethanol has been shown to reduce direct GHG emissions by approximately 47% on average;[5,40] this estimate, however, does not include emissions from indirect land use change.[55]

Other agricultural biofuel production systems are under development, but most other systems suffer from relatively lower energy yield per hectare and lower energy yield per unit of nutrient applied compared with ethanol produced from corn and sorghum grain. For example, in the United States, soybean-biodiesel produces on average 15% of the biofuel volume per hectare compared with corn-ethanol.[43] Correcting for energy density differences, soybean-biodiesel yields 23% of the gross energy yield of corn-ethanol on average. Cellulosic ethanol produced from crop residues and perennial grasses is currently under development, and only a very limited production

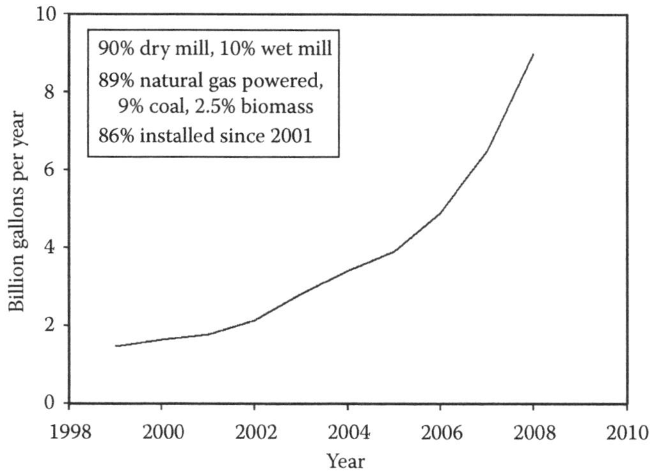

FIGURE 1.3 Increasing ethanol production capacity utilizing corn grain in the United States. Total installed capacity in 2009 is 12.5 Bgal/year. Industry statistics provided by Cooper. (From Cooper G., Personal communication, Renewable Fuels Association, Washington, DC, 2009.)

capacity is installed. Ethanol from sweet sorghum provides an alternative cropping system that shows potential to be competitive with corn grain.[70]

State-level production statistics and field trial data provide a comparison among these selected ethanol production systems in Nebraska (Table 1.2). Gross energy yield per hectare as ethanol was found to range from 10 to 95 GJ/ha[1], with use of both corn grain and residue being the most productive system. Despite this wide range in productivity per area, the N use efficiency only varies from 0.47 to 0.70, with an average of 0.586 GJ ethanol per kg N applied (Figure 1.4). As a stand-alone component, residue is the least productive system, but had the highest efficiency, while the most productive system (corn grain plus residue) only had a slightly higher efficiency than the average. Ethanol from grain sorghum and switchgrass had N use efficiencies below average for the five systems. Of these systems, sweet sorghum stood out as having the third highest energy yield, but field trials found no significant response to N applications over a 2 year trial period.[70] This response is suggested because of a more gradual rate of nutrient uptake in sweet sorghum, and N uptake later in the season compared with the grain crops. This is significant because it indicates that with appropriate crop rotations, a relatively high yielding system could exist with limited N fertilizer inputs. Sweet sorghum should be researched further in the future to explore the lower limits of N applications for high-yield biofuel systems.

Nutrient use efficiency is just one important aspect in defining the life cycle GHG emissions and energy efficiency of biofuel production systems. Another important consideration for nutrient use and GHG emissions is changes in soil carbon.[37,68,70] Removing crop residue for biofuels also removes soil phosphorus (P) and potassium (K); one metric ton of corn residue harvested removes 8 kg of N, 0.79 kg of P, and 6.74 kg of K.[28] If ethanol production from crop residue and energy crops such as switchgrass is pursued in the future, appropriate nutrient replacement will be vital to maintain crop productivity.[66]

TABLE 1.2
Nutrient Use Efficiency of Selected Cropping Systems for Ethanol Production in Nebraska

	Nitrogen Rate, kg/ha	Biomass Yield, Mg/ha	Biofuel Conversion Efficiency, L/Mg	Gross Energy Yield, GJ/ha	Nutrient Use Efficiency,[a] (GJ/kg) N
Corn residue, 20% removal	14.4	1.62	294	10.0	0.70
Switchgrass	78.1	7.10	294	44.0	0.56
Grain sorghum	96.0	4.29	501	45.3	0.47
Sweet sorghum	0	3.54	665	49.7	—
Corn grain	144.0	8.09	501	85.4	0.59
Corn grain + residue, 20%	158.4	9.70	—	95.4	0.63

[a] Partial factor productivity. All yields are on a dry matter basis. Corn grain and residue yields[46] (2004–2006) and conversion of grain to ethanol and ethanol energy density (21.1 MJ/L) were previously reported.[42] Nitrogen rates for corn were previously reported[21] and 20% residue removal is allocated N applied for 10% of above ground biomass. Switchgrass yields in NE were also previously reported,[54] and it is assumed that N will be applied at the recommended rate of 11 kg N/Mg biomass yield.[66] Conversion yield for switchgrass and residue is 70 gal/ton based on Iogen technology from a DOE-funded facility (http://www.energy.gov/news/4827.htm). Grain sorghum yields[46] (2004–2006) and N application rates[21] were for Nebraska. Sweet sorghum theoretical yield is based on field studies and estimated from 80% juice extracted and brix reading with the Simon cv. in 2008.[70] The corn plus residue system assumes additional N rate for the 20% of N removed.

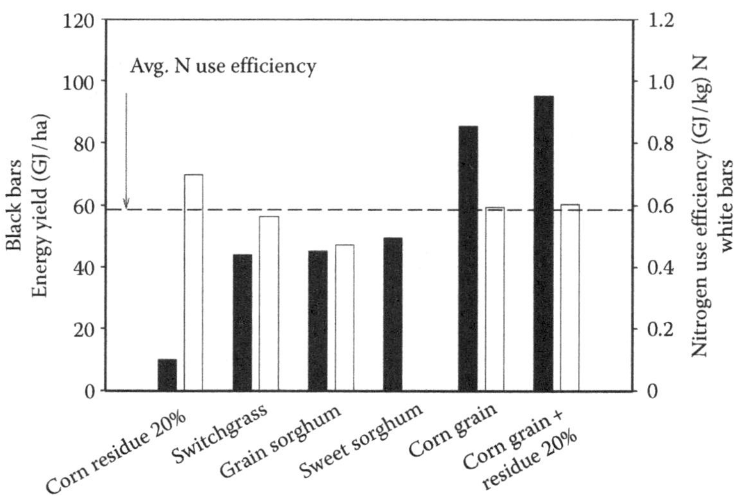

FIGURE 1.4 Bioenergy productivity versus nutrient use efficiency: ethanol gross energy yield (GJ/ha, black bars) and nitrogen use efficiency (GJ/kg N applied, white bars) for selected cropping systems for ethanol production in Nebraska. Average N use efficiency (partial factor productivity) for the five systems reporting N application is shown with a dashed line.

1.4 LAND LIMITATIONS AND GLOBAL AGRICULTURAL PRODUCTION

Pressure for increased agricultural production over the next 40 years will come from three sources: world population growth, per capita income growth, and demand for biofuels. Population growth alone is projected to require a one-third increase in crop production, and increased demand for livestock products, made possible by higher incomes, is expected to further increase the required production to approximately 50% above current levels by 2050 (Figure 1.5a). In addition, biofuel production will result in yet additional demands on agricultural resources. Yet, there is little potential for increasing the critical agricultural resources necessary to provide this additional biomass, making it crucial that more efficient production techniques be developed and adopted.

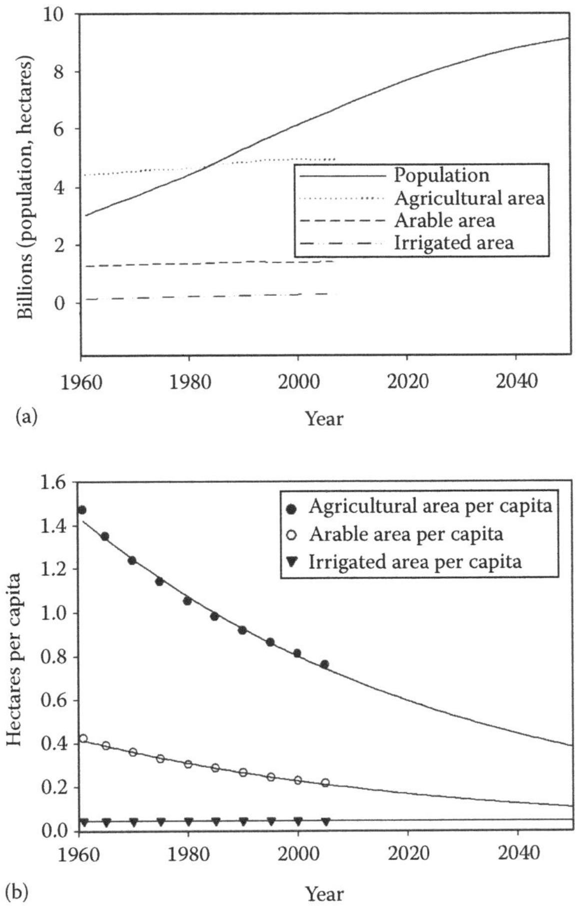

FIGURE 1.5 Trends and projections for global population and agriculture land resources (a) and agricultural area per capita (b). (From FAO [Food and Agriculture Organization of the United Nations], FAOSTAT. Available at http://faostat.fao.org. Accessed November 24, 2009; UN [United Nations], *World Population Prospects: The 2008 Revision*. Available at http://esa.un.org/unpp. Accessed November 24, 2009.)

Crops for biofuel production were grown on about 36 million hectares (Mha) around the globe, or 2.5% of the world's arable land in 2008.[60] Comparable figures for current biofuel area from another recent report are 41 Mha (2.8%).[48] Projections of future growth in land use for biofuel production are difficult to make, because of uncertainty about the relative prices of food and energy and uncertainty about policies that either encourage or discourage biofuel production. The UNEP[60] reports projections to 60–80 Mha, or even 166 Mha, by 2020, which are equivalent to 4%–11% of the current stock of arable land, or 1%–3% of total agricultural land.

Recent and potential increases of biofuel crops in Brazil and Indonesia are dramatic. Cropping area devoted to sugarcane in Brazil has increased from 7 to 9 Mha between 2007 and 2008, now constituting about 15% of the 60 Mha of arable land in Brazil.[60] This upward trend will continue, though the government intends to limit the expansion into sensitive ecosystems. Soybeans, also used in part for biodiesel, occupied about 23 Mha in 2005, and are expected to occupy the majority of an additional 60 Mha that will likely be converted from savannah to crop land. The Indonesian government intends that the current 6 Mha of oil palm be augmented by another 18–20 Mha,[33] with about two-thirds of this to be planted on land currently covered by rainforests.

It is clear that world agricultural land resources will not increase much, based on the experience of the last 10 years (Figure 1.5a). Of the approximately 5 Bha of total agricultural land, only about 1.4 Bha, or 28%, is arable. Irrigated land has increased only about a half a percent per year, arable land less than 0.2% per year, while total agricultural land has actually declined.

Population has of course increased during this time, resulting in the very dramatic decreases in land per capita (Figure 1.5b). If the per capita land resource base continues to decline along the trend of the last 50 years, a great deal of pressure will be placed on agricultural research to achieve the kind of productivity improvement that will be needed.

Water is another critical component of the agricultural resource base, one that will more likely decline rather than increase with population. Irrigated land, as noted above, has increased very little over the past decade. Aquifers are being depleted, snowpack's and glaciers are declining, and climate change may reduce rainfall in many regions, while contributing to higher rainfall, and more extreme rainfall events in other areas.[35] It is clear that in the case of agricultural water, efficiency will have to increase if there is to be any chance of providing the needed production increases.

As opposed to land, global per capita consumption of fertilizer materials has increased over the last 50 years, though irregularly (Figure 1.6a). The increased fertilizer use over the past decade has helped make possible a slight per capita increase in agricultural production over the past two decades. With this increase in usage, worldwide average N efficiency has begun to decline, while phosphorus and potassium efficiencies have stabilized after realizing improvements over the previous 2 or 3 decades (Figure 1.6b). Figures reported here are measured in terms of fertilizer materials rather than in fertilizer elements; therefore, any change in the mix of materials over this period may slightly distort the trends in fertilizer elements.

It appears that quantities of land and water allocated to crop production will not increase much in the future, whereas crop nutrients and energy inputs will be

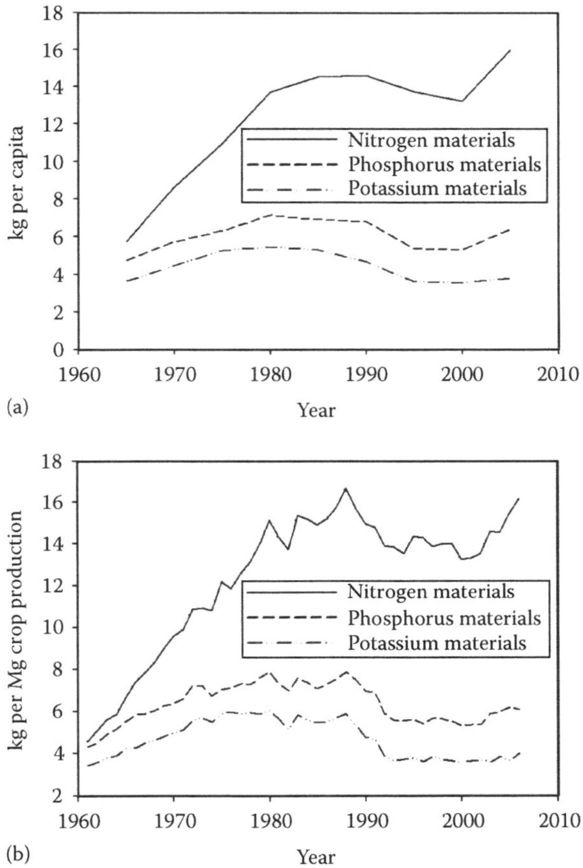

FIGURE 1.6 Global fertilizer materials consumption per capita (a) and fertilizer materials per ton of crop production (b) derived from Refs. [22,59]. (From FAO [Food and Agriculture Organization of the United Nations], FAOSTAT. Available at http://faostat.fao.org. Accessed November 24, 2009; UN [United Nations], *World Population Prospects: The 2008 Revision*. Available at http://esa.un.org/unpp. Accessed November 24, 2009.)

more elastic in supply. But increases in land, energy inputs, and nutrients will entail increases in GHG emissions, making it critical that the required increase in output be achieved with maximum possible efficiency of input use.

Land use change is a significant source of GHG emissions and a driver of climate change, being responsible for approximately 20% of global anthropogenic GHG emissions in the 1990s.[31] These emissions result from the burning of forest and savanna biomass when those lands are converted to cultivated crop production, and from loss of soil carbon stocks when the land is cultivated. It is therefore likely that climate change policies will provide further barriers to the expansion of arable land, adding another source of urgency for improving efficiencies in crop production.[41] Similarly, the GHG emissions related to fertilizer use, mentioned above, may lead to higher costs of nutrients or restrictions on the use of nutrients, adding incentives for generating improved efficiencies in nutrient use.

1.5 CONCLUSIONS

Multiple lines of evidence point to a global oil economy that is increasing unstable, which has broad consequences for economic growth, international military activity, and the future costs of agricultural production. Energy and climate issues are stimulating the production of biofuels from agricultural products which has numerous implications for nutrient use. Climate change has multiple potentially serious impacts on agriculture as well. Growing demand for agricultural products, in conjunction with volatile weather and more costly cropping inputs will lead to greater pressure in the future to increase yields, while minimizing nutrient inputs. These trends will provide significant incentive to heavily invest in better management of nutrient applications in the future and to develop the science necessary to keep improving nutrient use efficiency while meeting new demands on agricultural production from both biofuels and population growth.

REFERENCES

1. Anex R. and Lifset R. Post script to the corn ethanol debate: Reaching consensus? *Journal of Industrial Ecology* 13, 996, 2009.
2. Aleklett K., Höök M., Jakobsson K., Lardelli M., Snowden S., and Söderbergh B. The peak of the oil age—Analyzing the world oil production reference scenario in world energy outlook 2008. *Energy Policy*, 38, 1398, 2010.
3. Aviation Week. USAF Launches Major Biofuel Initiative. http://www.aviationweek.com/aw/generic/story_generic.jsp?channel=aerospacedaily&id=news/BIOF013009.xml, 2009.
4. Bacevich A. *The Limits of Power: The End of American Exceptionalism*. Metropolitan Books, New York, 2008.
5. Bremer V.R., Liska A.J., Klopfenstein T.J., Erickson G.E., Yang H.S., Walters D.T., and Cassman K.G. Emissions savings in the corn-ethanol life cycle from feeding co-products to livestock. *Journal of Environmental Quality* 39, 472, 2010.
6. Brown S.P.A. Oil and the U.S. Macroeconomy, Federal Reserve. Bank of Dallas, June 2004. *Presented at the WEAI Annual Meeting*, July 2, 2004, Vancouver BC, 2004.
7. Burr J.M. and Collins R.O. *Alms for Jihad: Charity and Terrorism in the Islamic World*. Cambridge University Press, Cambridge, 2006.
8. Cassman K.G., Dobermann A., and Walters D.T. Agroecosystems, nitrogen-use efficiency, and nitrogen management. *AMBIO* 31, 132, 2002.
9. CAST (Council for Agricultural Science and Technology). *Climate Change and Greenhouse Gas Mitigation: Challenges and Opportunities for Agriculture*. Task Force Report 141, Ames, IA, 2004.
10. Cooper G. Personal communication. Renewable Fuels Association, Washington, DC, 2009.
11. Copulos M.R. *The Hidden Costs of Imported Oil: An Update*. National Defense Council Foundation, Arlington, VA, 2007.
12. Dancs A., Orisich M., and Smith S. *The Military Cost of Securing Energy*. National Priorities Project. www.nationalpriorites.org, 2008.
13. Deffeyes K. *Hubbert's Peak: The Impending World Oil Shortage*. Princeton University Press, Princeton, Oxford, U.K., 2001.
14. Deutch J. and Schlesinger J.R. *National Security Consequences of U.S. Oil Dependency*. Council on Foreign Relations. Washington, DC, 2006.
15. EIA (Energy Information Administration). *International Energy Outlook 2009*. U.S. Department of Energy. Washington, DC, 2009.
16. EIA. *Annual Energy Review 2008*. U.S. Department of Energy, Washington, DC, 2009.

17. EIA. *OPEC Revenues Fact Sheet*. U.S. Department of Energy, 2009. Available at http://www.eia.doe.gov/emeu/cabs/OPEC_Revenues/pdf.pdf
18. EIA. Official Energy Statistics from the U.S. Government. Accessed at http://tonto.eia.doe.gov, November 2009.
19. EIA. *World Oil Transit Chokepoints*. U.S. Department of Energy, 2008. Available at http://www.eia.doe.gov/cabs/World_Oil_Transit_Chokepoints/pdf.pdf
20. Erisman J.W., Sutton M.A., Galloway J., Klimont Z., and Winiwarter W. How a century of ammonia synthesis changed the world. *Nature Geoscience* 1, 636, 2008.
21. ERS (Economic Research Service). USDA, U.S. Fertilizer Use and Price. Accessed at http://www.ers.usda.gov/Data/FertilizerUse/, November 2009.
22. FAO (Food and Agriculture Organization of the United Nations). FAOSTAT. Available at http://faostat.fao.org. Accessed November 24, 2009.
23. Farrell A.E., Plevin R.J., Turner B.T., Jones A.D., O'Hare M., and Kammen D.M. Ethanol can contribute to energy and environmental goals. *Science* 311, 506, 2006.
24. Foster J. *A Pipeline through a Troubled Land: Afghanistan, Canada, and the New Great Energy Game*. Foreign Policy Series, vol. 3, no. 1. Canadian Centre for Policy Alternatives, Ottawa, Ontario, Canada, 2008.
25. Graedel T. and Allenby B. *Industrial Ecology and Sustainable Engineering*. Prentice-Hall, Englewood Cliffs, NJ, 2010.
26. Greenspan A. *The Age of Turbulence: Adventures in a New World*. Penguin, New York, 2008.
27. Höök H., Hirsch R., and Aleklett K. Giant oil field decline rates and their influence on world oil production. *Energy Policy* 37, 2262, 2009.
28. Hoskinson R.L., Karlen D.L., Birrell S.J., Radtke C.W., and Wilhelm W.W. Engineering, nutrient removal, and feedstock conversion evaluations of four corn stover harvest scenarios. *Biomass & Bioenergy* 31, 126, 2007.
29. IEA (International Energy Agency). *World Energy Outlook 2008*, Paris, France, 2008.
30. IEA. *World Energy Outlook 2009*, Paris, France, 2009.
31. IPCC (Intergovernmental Panel on Climate Change). *Climate Change 2007: The Physical Basis*. Cambridge University Press, Cambridge, U.K., 2007.
32. IPCC. Eggleston H.S., Buendia L., Miwa K., Ngara T. and Tanabe K. 2006. IPCC Guidelines for National Greenhouse Gas Inventories. Prepared by the National Greenhouse Gas Inventories Programme. IGES, Hayama, Japan, 2006.
33. Jakarta Globe CPO Biodiesel Production Plan May Lift Regions, December 2, 2009. http://thejakartaglobe.com/business/cpo-biodiesel-production-plan-may-lift-regions/345113
34. Jamshidi M. *System of Systems Engineering*. Wiley, New York, 2008.
35. Karl T.R., Melillo J.M., and Peterson T.C. (eds.). *Climate Change Impacts in the United States*. Cambridge University Press, Cambridge, U.K., 2009.
36. Kilian L. The economic effects of energy price shocks. *Journal of Economic Literature* 46, 871, 2008.
37. Lal R. and Stewart B.A. (eds.). *Soil Quality and Biofuel Production*. CRC Press, Boca Raton, FL, 2010.
38. LeVine S. and Bush J. Kazakh oil: A war of nerves. *Business Week*, September 22, 2008.
39. LeVine S. *The Oil and the Glory: The Pursuit of Empire and Fortune on the Caspian Sea*. Random House, New York, 2007.
40. Liska A.J. and Cassman K.G. Response to Plevin: Implications for life cycle emissions regulations. *Journal of Industrial Ecology* 13, 508, 2009.
41. Liska A.J. and Perrin R.K. Indirect land use emissions in the life cycle of biofuels: Regulations vs. science. *Biofuels, Bioproducts, and Biorefining* 3, 318, 2009.
42. Liska A.J., Yang H.S., Bremer V.R., Klopfenstein T.J., Walters D.T., Erickson G.E., and Cassman K.G. Improvements in life cycle energy efficiency and greenhouse gas emissions of corn-ethanol. *Journal of Industrial Ecology* 13, 58, 2009.

43. Liska A.J. and Cassman K.G. Towards standardization of life-cycle metrics for biofuels: Greenhouse gas emissions mitigation and net energy yield. *Journal of Biobased Materials and Bioenergy* 2, 187–203, 2008.
44. Moran D. and Russell J.A. (eds.). *Energy Security and Global Politics: The Militarization of Resource Management*. Routledge, New York, 2008.
45. Morse E.L. Low and behold: Making the most of cheap oil. *Foreign Affairs*, September/October, 2009.
46. NASS (National Agricultural Statistics Service), USDA, Quick Stats, State and County Data. Accessed at http://www.nass.usda.gov/, November 2009.
47. NCTA (National Commission on Terrorist Attacks). *The 9/11 Commission Report: Final Report of the National Commission on Terrorist Attacks upon the United States*, Authorized Edition. W.W. Norton, New York, 2004.
48. OECD/FAO. *OECD-FAO Agricultural Outlook 2008–2017*, Paris, Rome, 2008. Available at http://www.agri-outlook.org
49. Orlove B., Wiegandt E., and Luckman B.H. (eds.). *Darkening Peaks: Glacier Retreat, Science, and Society*. University of California Press, Berkeley, CA, 2008.
50. Pelletiere S. *America's Oil Wars*. Praeger Publishers, Westport, CT, 2004.
51. Peng S., Huang J., Sheehy J.E., Laza R.C., Visperas R.M., Zhong X., Centeno G.S. et al. Rice yields decline with higher night temperature from global warming. *Proceedings of the National Academy of Sciences* 101, 9971, 2004.
52. Perrin R.K., Fretes N., and Sesmero J.P. Efficiency in Midwest US corn ethanol plants: A plant survey. *Energy Policy* 37, 1309, 2009.
53. Rashid A. *Taliban: Militant Islam, Oil and Fundamentalism in Central Asia*. Yale University Press, New Haven, CT, 2000.
54. Schmer M.R., Vogel K.P., Mitchell R.B., and Perrin R.K. Net energy of cellulosic ethanol from switchgrass. *Proceedings of the National Academy of Sciences, USA* 105, 464, 2008.
55. Searchinger T., Heimlich R., Houghton R.A., Dong F., Elobeid A., Fabiosa J. et al. Use of US croplands for biofuels increases greenhouse gases through emissions from land-use change. *Science* 319, 1238, 2008.
56. Simmons M. *Twilight in the Desert: The Coming Saudi Oil Shock and the World Economy*. Wiley, Hoboken, NJ, 2005.
57. Smith P., Martino D., Cai Z., Gwary D., Janzen H., Kumar P., McCarl B. et al. Agriculture. In Metz B., Davidson O.R., Bosch P.R., Dave R., Meyer L.A. (ed.). *Climate Change 2007: Mitigation*. Contribution of Working Group III to the Fourth Assessment Report of the Intergovernmental Panel on Climate Change. Cambridge University Press, Cambridge, U.K., 2007.
58. Stiglitz J.E. and Bilmes L.J. *The Three Trillion Dollar War: The True Cost of the Iraq Conflict*. W.W. Norton, New York, 2008.
59. UN (United Nations). *World Population Prospects: The 2008 Revision*. Available at http://esa.un.org/unpp. Accessed November 24, 2009, 2008.
60. UNEP (United Nations Environmental Programme). *Towards Sustainable Production and Use of Resources: Assessing Biofuels*, Paris, France, 2009.
61. U.S. Congress, House of Representatives. *Assessing Energy and Security Issues in Central Asia*. Hearing before the Committee of International Relations, July 25, 2006.
62. U.S. Congressional Research Service. *The U.S. Trade Deficit, the Dollar, and the Price of Oil*. U.S. Congressional Research Service, Washington, DC, 2008.
63. U.S. Congressional Research Service. *Iraqi Civilian Deaths Estimates*. U.S. Congressional Research Service, Washington, DC, 2008.
64. USDA (United States Department of Agriculture). *Agricultural Prices Supplement*, August 2009. Accessed at http://usda.mannlib.cornell.edu/usda/current/AgriPricSu/AgriPricSu-08-05-2009.pdf, November 2009.

65. USGS (United States Geological Survey). Krimmel R.M. *Glaciers of North America—Glaciers of the Conterminous United States, Glaciers of the Western United States*. U.S. Geological Survey Professional Paper 1386-J-2, 2002.
66. Vogel K., Brejda J., Walters D.T., and Buxton D.R. Switchgrass biomass production in the Midwest: Harvest and nitrogen management. *Agronomy Journal* 94, 413, 2002.
67. Warde I. *The Price of Fear: The Truth behind the Financial War on Terror*. University of California Press, Berkeley, CA, 2007.
68. Wilhelm W.W., Johnson J.M.F., Karlen D.L., and Lightle D.T. Corn stover to sustain soil organic carbon further constrains biomass supply. *Agronomy Journal* 99, 1665, 2007.
69. World Bank. *A Citizen's Guide to National Oil Companies: Part A*. Technical Report, Washington, DC, 2008.
70. Wortmann C.S., Liska A.J., Ferguson R.B., Klein R.N., Lyon D.J., and Dweikat I. Dryland performance of sweet sorghum and grain crops for biofuel in Nebraska. *Agronomy Journal* 102, 319, 2010.
71. Zubrin R. *Energy Victory: Winning the War on Terror by Breaking Free of Oil*. Prometheus Books, Amherst, NY, 2009.

2 Nutrient Management for Improved Energy Efficiency

F. Mamani Pati, David E. Clay, and Gregg Carlson

CONTENTS

2.1 EXECUTIVE SUMMARY

To meet future food, fiber, and energy requirements, it is anticipated that agricultural yields will need to be doubled by 2050. Meeting this goal is complicated by increasing cost and availability of fertilizers, anticipated shortages in critical resources (land, P fertilizer, and liquid fuels), and urbanization that is reducing arable lands. Doubling food production with diminishing resources will require wide-scale investments in production agriculture, the development of new genomics that increase energy and production efficiency, and the adoption of precision management techniques that increase productivity as well as resource use efficiency. This chapter provides a framework for assessing energy efficiency and an example using the readily available life-cycle assessment (LCA) model, biofuel energy simulator (BESS), to calculate energy returns at two landscape positions.

2.2 INTRODUCTION

Landscape positions influence water availability, which in turn has numerous impacts on the soil biological processes and the ability to profitability produce a crop. For example, Clay et al.[1] reported that in a poorly drained area located in a South Dakota field, there was a net loss of 95 kg N/ha, while in adjacent tile drained areas there was a net gain of 98 kg N/ha. In a different report from this same field, Clay et al.[2] found that water stress reduced yields by 50%–60% in higher summit/shoulder areas. These differences were attributed to differential amounts of water across the landscape impacting crop growth, denitrification, aerobic N mineralization, and symbiotic relationships between yield-limiting factors.[3,4] Precision farming can be used to develop management practices that overcome landscape position differences. It is likely that landscape variability also impact energy efficiency and net energy yields. Once the energy efficiency is understood, management practices designed to increase gains can be implemented. This chapter demonstrates how the BESS LCA model (http://www.bess.unl.edu/) can be used to calculate energy gains at two landscape positions.

2.2.1 Precision Farming and Energy Efficiency

Precision farming is an integrated agricultural management system that incorporates state-of-the-art agronomic knowledge, information from multiple sources, and the global positioning system, geographical information system, yield monitor, variable rate, and remote sensing technologies. Precision farming allows producers to make management decisions about discrete areas of the field, with the goal of optimizing the crop response based on the production potential and constraints of the specific region. The techniques of precision farming are compatible with providing good stewardship of the land for future generations, preserving the land's potential for multiple uses, and implement techniques that increase agricultural energy efficiency. In the past, most precision farming assessments have concentrated on calculating economic returns and have not considered net energy yields.

Many modern agricultural inputs require large amounts of input energy (Table 2.1). Maximizing the efficiency of production inputs, through the adoption of

TABLE 2.1
Conversion Factors Used to Change Corn Production Inputs into Energy

	Agricultural Energy Input							
	BESS Model[5]		Pimentel et al.[6]		Dias De Oliveira et al.[7]		Shapouri et al.[8]	
	MJ/kg	BTU/lb	MJ/kg	BTU/lb	MJ/kg	BTU/lb	MJ/kg	BTU/lb
Nitrogen (N)	51.2	22,030	66.94	28,803	57	24,526	43.0	18,505
Phosphorus (P_2O_5)	7.21	3,102	17.37	7,474	7.03	3,025	4.76	2,048
Potassium (K_2O)	11.3	4,862	13.65	5,873	6.85	2,947	8.71	3,748
Herbicides	356	153,180	418.4	180,029	267	114,695	216	93,117
Seed	9.7	4,174	103.6	44,577	103	44,319	3.94	1,695

TABLE 2.2
Commonly Used Energy-Conversion Factors

Unit 1	Unit 2
1 J	0.239 cal
1 BTU	1.055 kJ
2.47 ac	1 ha
2.21 lb	1 kg
1 gal	3.79 L
1 U.S. bu corn	56 lb
1 U.S. bu wheat	60 lb
1 U.S. bu soybean	60 lb
1 U.S. gal gasoline	115,000 BTU
1 U.S. gal ethanol	76,000 BTU
During production of ethanol, 1 bu corn produces approximately	2.7–2.85 gal ethanol
During production of ethanol, 1 bu corn produces approximately	18 lb CO_2
During production of ethanol, 1 bu corn produces approximately	18 lb of DDGS

precision agriculture, will allow the targeting of more resources to zones where they are needed and a reduction of resource to areas where not required. Precision farming can be used to improve fertilizer, seeding, and irrigation rates as well as the better targeting of insecticides and herbicides toward pests. This book focuses on using precision farming techniques for improving energy efficiency through improved nutrient management. Nutrients considered in this book include N, P, K, lime, water, and carbon. Nutrients can either be applied in returned crop residues, fertilizers, or manures. To compare findings and energy efficiency from different studies requires the ability to convert units from one form to another, clearly identified boundary conditions, and the ability to conduct an LCA (Table 2.2).

2.2.2 Life-Cycle Assessment

An LCA is a compilation of the inputs, outputs, and the potential environmental impacts through the life cycle of the product.[9] In an LCA, a "cradle-to-grave" or "cradle-to-cradle" assessment is conducted to ensure that improvements or problems do not move up or down the supply or process chain. One common approach for conducting an LCA is to convert all inputs and outputs to energy following ISO 14040-14043 guidelines. LCA can be conducted either by using existing models or by developing new models using programs such as PRé SimaPro (http://www.pre.nl/simapro/).

The life-cycle analysis considered in this chapter is highlighted in Figure 2.1. In this production scenario, agricultural inputs are used to produce corn, which in turn produces ethanol and distiller's dried grains with solubles (DDGS). In this scenario, manure is not reapplied to the land (Figure 2.1).

The corn crop is produced from inputs such as fertilizers, herbicides, diesel for tractors, combines, and planters, seeds, and pesticides. The grain is then delivered to the ethanol plant which produces ethanol and DDGS. At the ethanol plant, corn

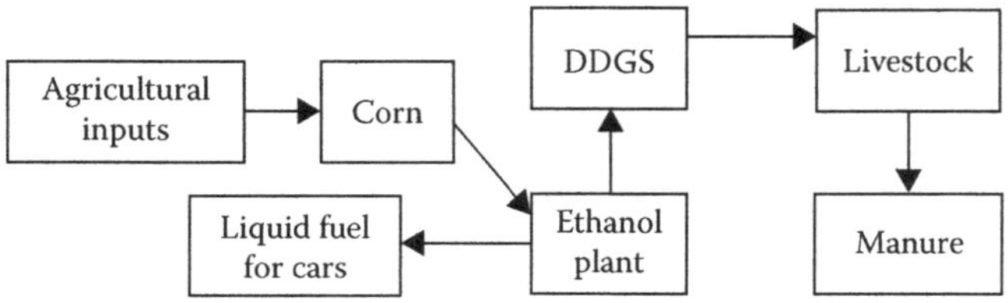

FIGURE 2.1 Production system considered in the life-cycle analysis model.

starch is converted to ethanol and CO_2. Currently, 1 kg of corn grain produces approximately 0.48 L of ethanol, 0.32 kg of CO_2, and 0.32 kg of DDGS.

Different papers have used different factors to convert agricultural inputs into energy (Table 2.1). Based on these differences, some studies have reported that corn production consumes energy,[6] while others have reported that corn production produces energy.[8] In addition, across the United States, different states have different energy efficiencies.[8,10,11] Regional variability is related to management practices and environmental conditions having unique input requirements that maximize profitability.[12–20]

2.3 METHODS: BASICS IN ENERGY CALCULATIONS

Two common approaches to report energy efficiency are net energy yields and the output/input ratio.[8] The energy gain is the difference between output and input energies, whereas the output/input energy ratio is a unit-less value that tends to increase with reductions in inputs. To optimize the amount of energy gained per unit area, an optimization analysis is conducted. At the optimum value, the last unit of input energy added equals the amount of output energy. If the ratio is used as the guide to control management recommendations, then in many situations decisions that do not maximize energy gains will be accepted. For example, Mamani Pati et al.[14] reported that two corn plant populations had identical output/input ratio, yet the 149,000 plants/ha had an energy gain of 65.7 GJ/(ha year) while the 74,500 plants/ha had an energy gain of 58.9 GJ/(ha year).

When calculating energy gains, the boundary conditions must be clearly identified. Different boundary conditions can produce different results. The BESS model version 2008.3.1 model used in this example was designed to calculate the energy gains and efficiency, greenhouse gas (GHG) emissions, and natural resource requirements of corn to ethanol biofuel production systems. Different scenarios are available for different production systems. The BESS model includes four components: crop production, ethanol biorefinery, cattle feedlot, and anaerobic biodigester (optional). Field data obtained from Clay et al.[2] was used in the following example.

2.4 STEP-BY-STEP GUIDE TO CALCULATE ENERGY GAINS

Download the BESS model version 2008.3.1 at http://www.bess.unl.edu.

Step 1. Install the software by running the BESS 2008.3.1 setup.exe file and read the directions.

Step 2. Click the BESS icon on the desktop to start the software.

TABLE 2.3
Agricultural Energy Input for Lower and Upper Landscape Position

		2001		2002	
Input Parameters		**Lower Backslope**	**Upper Backslope**	**Lower Backslope**	**Upper Backslope**
Productivity	Corn yield (Mg/ha)	9.60	7.22	12.55	8.47
	Soil C sequestration (Mg C/ha)	0	0	0	0
Material inputs	Nitrogen (kg N/ha)	179.34	179.34	151.31	151.31
	Manure (kg N/ha)	0	0	0	0
	Phosphorus (kg P_2O_5/ha)	78.24	78.24	66.24	66.24
	Potassium (kg K_2O/ha)	0	0	0	0
	Lime (kg/ha)	0	0	0	0
	Herbicides (kg/ha)	5.25	5.25	5.25	5.25
	Insecticides (kg/ha)	0	0	0	0
	Seed (kg/ha)	25.11	25.11	25.11	25.25
	Irrigation water (cm)	0	0	0	0

Step 3. Click on start the model to begin a simulation with BESS.

Step 4. Click on open a scenario. BESS model provides eight default simulation scenarios that can be used to initiate simulation. These default scenarios can be easily altered to adapt the user's needs for local information. In this analysis, select default simulation scenario 2, USA Midwest average (based on 12 states). A window shows four subpages: crop production, ethanol biorefinery, cattle feedlot, and biodigestor (optional) subpages of the input: operation settings.

Step 5. Fill in crop production data for the lower backslope in 2001. Select fuel consumption by field operation and no-tillage (Table 2.3).

Step 6. Modify the production values in scenario 2 with landscape-specific data from Clay et al.[2] These data are summarized and converted to the appropriate units above. For this analysis, select itemized energy and GHG co-product credit. Co-products from ethanol biorefinery provide a life-cycle energy savings by displacing corn and urea in cattle diets.

Step 7. Click the button. Compute to run the simulation. When a run is completed, this page shows four operational pages: (1) input: operation settings, (2) output: individual scenarios, (3) output: scenario comparison, and (4) summary report.

Step 8. Click on summary report to see the output data. The summary report shows all input parameter values and output results as shown in Table 2.3.

Step 9. Repeat the analysis for the other landscape positions.

Step 10. Parameters included energy inputs, outputs, net energy ratio (NER), and energy gains for create the four simulation runs. Our results are given in Table 2.4.

TABLE 2.4
Output Data for Life Cycle Analysis Parameters

Year	Landscape Position	Input *N*, kg/ha	Input *P*, kg/ha	Corn Yield, Mg/ha	Energy Input, GJ/ha	Energy Output, GJ/ha	Net Energy Ratio	Net Energy Yield (Gain), GJ/ha
2001	Lower backslope	179.3	78.2	9.60	53.8	96.8	1.8	44.4
	Summit/ shoulder	179.3	78.2	7.22	43.4	73.9	1.7	31.5
2002	Lower backslope	151.3	66.2	12.55	65.2	124.4	1.9	61.4
	Summit/ shoulder	151.3	66.2	8.47	45.3	81.6	1.8	39.2

2.5 RESULTS AND DISCUSSION

The BESS program generates a summary report that includes simulation results, input settings, and internal model parameters. Total energy inputs is the sum of energy used to: (1) produce and transport N, P, and seed to the field, (2) conduct disking and planting, (3) produce the ethanol and distillers grain, and (4) produce the facilities and equipment (depreciable capital). Total energy output is the energy contained in the fuel plus energy contained in the co-products and is defined as, energy output (GJ/ha) = energy output in ethanol (GJ/ha) + energy output credit for co-product (GJ/ha). NER is equal to the ratio between energy output and energy input. Net energy value or energy gain is the difference between the energy output in ethanol included credit for co-product and the energy needed to produce ethanol (feedstock) or simply is the difference between output and input energies.

Findings from the above analysis showed that the energy gain was higher in the lower backslope than the summit/shoulder area. These results were attributed to higher yields in the lower backslope position. It may be possible to further increase the energy efficiency by developing higher yielding hybrids and varieties, improving the efficiency of the ethanol plant, adopting management practices that reduce N and P production requirements, applying only the N required to optimize the energy gain, subtracting the amount of nutrients contained in the manure from the fertilizer recommendations, reducing transportation requirements, and considering landscape variability when applying agricultural inputs.[12,20–22]

Across the midcontinent area of the United States, similar energy gains were reported for switchgrass (*Panicum virgatum*) (60 GJ/[ha year])[23] and soybean (*Glycine max*) (50 GJ/[ha year]),[24] whereas higher values (135 GJ/[ha year]) were reported for alfalfa (*Medicago sativa* L.).[24] In Nebraska, Rathke et al.[20] reported that tillage and rotations also influenced energy efficiency. Across years, they reported that energy input requirements were lower in no-tillage (7.34 GJ/ha) than chisel plow (7.83 GJ/ha) and that energy inputs were lower for soybean. Lower input requirements were attributed to soybeans having the ability to fix atmospheric N so that N fertilizer is not needed. Rathke et al.[20] also reported that continuous corn (98.0 GJ/ha) had higher energy gains than continuous soybean (58.0 GJ/ha). It is important to point out that care must be used when comparing studies because different boundary conditions may be used.

In the above example, manure was not applied to the land. However, significant improvements in energy gains can be achieved through the land application of manure. Higher gains with manure result from reduction in the amount of fertilizers needed to optimize yields. The ethanol manufacturing process converts the starch in corn to ethanol, with the remaining nutrients become concentrated in the DDGS component, resulting in approximately tripling in the nutrient concentrations, when compared to corn grain. If the DDGS is used as a feed, then nutrients within the DDGS are used to meet the livestock daily requirements. Livestock utilize only 10%–30% of these nutrients, excreting the remainder as manure.[25] These calculations suggest that 70% of the nutrients originally contained within the grain could be reapplied to the land in the manure. To maximize the efficiency of these manure-based nutrients, storage and application approaches should be used that minimize losses.[26] For example, Reiman et al.[26] showed that the net impact of manure placement on total N was that deep manure injection (45 cm) had 31, 59, and 44 more kilograms of soil inorganic N/ha than shallow injection (15 cm) 12, 18, and 30 months after application, respectively. These results were attributed to deep injection reducing the loss of nitrate-N.

2.5.1 Wisdom of Ethanol Production

A worldwide debate concerning agricultural intensification, ethanol production, and energy efficiency is being conducted. Many people believe that agricultural intensification and ethanol production are linked and therefore by adopting policies that promote ethanol production will (1) accelerate the deforestation of the tropical rain forest; (2) lead to water shortages, food costs, and widespread starvation; and (3) lead to higher commodity prices, which reduce the amount of land dedicated for supporting wildlife. In addition, many people believe that ethanol production should not be supported because it consumes more energy than it produces. The energy efficiency of biofuel production has some basis in truth. For example, 30 years ago the development of ethanol from corn would have required more inputs than outputs. However, just like other industries, agriculture and the biofuel industries have made major improvements in becoming more efficient. For example, 30 years ago most fields were plowed, disked, and cultivated multiple times, whereas today many tillage operations have been eliminated. In addition, many of the fertilizer inputs (N, P, and K) are the same as 30 years ago even though yields have increased significantly.

An often used argument against ethanol production is that its production consumes water. Keeney and Muller[27] reported that each liter of ethanol produced consumes between 3.5 and 6 L of water. This water use must be compared with 19 L of water used per day by swine and 75 L used per day by beef or dairy. Water consumption must also be compared with other energy sources. For example, operating a 2400 W (2400 J/s · 3600 s = 8,640,000 J) fan heater for 1 h consumes 0.01 L of water if wind is the energy source, 0.26 L if solar is the energy source, 4.5 L if coal is the energy source, or 5.5 L if nuclear power is the energy source.[28] If ethanol is the energy source then approximately 2.04 L of water (8,640,000 J · [1 BTU/1,055 J] · [gal ethanol/76,000] ·

[5 gal water/gal ethanol] · [3.79 L/gal]) is consumed in the direct production of ethanol. It should also be noted that some places in the world are short of water and other places have surpluses. In places with large surpluses, water utilization is a nonissue.

The primary benefits of biofuel production have been income for producers, jobs for rural communities, decreases in gasoline prices, and reduction in the use of fossil fuels. Building an ethanol plant provides a one-time boost to the local community and salaries for employees. The economic opportunities created by an ethanol plant can be substantial.[29] Sneller and Durantee[30] reported that at Plainview, Nebraska, the development of a 25 Mgal ethanol plant resulted in the creation of 33 new jobs, $30 million being paid to local farmers, and $128,772 paid in property taxes. This income is needed to help reverse the trend of declining rural populations.[30–34] For example, recent population increases in Groton, South Dakota, were attributed to the James Valley Ethanol plant.[30] At the local level, higher commodity prices help increase profitability. For example, in Iowa in 2008, Duffy[35] estimated that production costs for corn following corn was $604/ac, which was an increase of 18% from 2007. The break even point for this system was $4.17/bu, and therefore returning to corn selling for less than $2.00/bu would likely bankrupt many producers.

At the regional level, ethanol production improves regional infrastructure, such as rail transport of biomass from farms to processing plants, and increases tax revenues shared by local, regional, and national governments. At the national and international levels, ethanol production in 2004: (1) added $14 billion to the nation's gross domestic product,[31] (2) helped create 147,206 jobs in all sectors of the economy,[31] (3) resulted in $2.5 billion of tax revenue that was distributed to federal, state, and local governments,[31] and (4) provided wealth needed for genetic improvements. In addition ethanol production reduces the price for gasoline from between 29 and 40 cents/gal.[34]

Mass balance dictates that if grain is diverted to produce ethanol, then less is available for food. However, only a portion of the grain used to produce ethanol is diverted from the food chain. End products of the distillation process are almost equally distributed between ethanol (2.8 gal/bu), CO_2, and DDGS. When the ethanol processing is complete, approximately one-third of the grain is exported as the livestock feed additive, DDGS, which has greater protein and nutrient content than raw corn.

The bottom line is that no matter what your feelings are about ethanol production, most agronomists and environmentalist agree that agricultural production and energy efficiency must increase. Due to a growing world population and a shrinking natural resource base (oil, arable land, and mined fertilizer), meeting food, fiber, and energy needs will become increasingly difficult. Pimentel and Giampietro[36] estimated that for each new person added, approximately 1 ac of farm land is lost to urbanization and highway construction. By 2050, an increasing world population could reduce the per capita arable farmland from 1.8 ac in 1994 to 0.6 ac. In addition, the production of NH_3-based fertilizer requires energy (natural gas), and both oil and phosphate rock are nonrenewable resources. It is estimated that peak P production will be reached between 2020 and 2040,[37] while oil production is expected to peak between 2010 and 2020. Phosphorus is especially critical because it is a required element in food production and cannot be synthesized. Global warming further complicates the system and will likely result in a further need to increase energy efficiency. Meeting demands from a growing world population will require infrastructure improvements and investments in production agriculture that will increase energy efficiency.

2.6 SUMMARY

The example showed that energy gains were 30%–40% lower in corn grown in the summit/shoulder position than the lower backslope positions. Lower energy gains in summit/shoulder positions were attributed to lower yields and fertilizer inputs that were uniformly distributed across the landscape. Although not tested, it is likely that precision farming could be used to increase the energy gains and fertilizer efficiency in summit/shoulder areas.

2.7 SELECTED MODELS AVAILABLE FOR LIFE-CYCLE ANALYSIS

There are numerous life-cycle models that have been used to assess agricultural activities. This chapter used the BESS model version 2008.3.1. This model was designed to calculate the energy efficiency, GHG emissions, and natural resource requirements of corn to ethanol biofuel production systems. The BESS model includes four components: crop production, ethanol biorefinery, cattle feedlot, and anaerobic biodigester (optional). The BESS model is available to the public for download at http://www.bess.unl.edu. One of the most widely used model is the GREET model (greenhouse gases regulated emissions and energy use in transportation). This model has the capacity to evaluate and compare the environmental impacts of a wide range of renewable and conventional transportation fuels and motor vehicle fleets. The model includes detailed information on corn farming and chemical manufacturing. The model and its documents are posted at http://www.transportation.anl.gov/modeling_simulation/GREET/. A third model is the Erg Biofuels Analysis Meta-Model (EBAMM) structured to provide a relatively simple, transparent tool that can be used to compare biofuel production processes. More information about this model is available at http://rael.berkeley.edu/ebamm/.

ACKNOWLEDGMENTS

Support for this chapter was provided in part by South Dakota Soybean Research and Promotion Council, South Dakota Corn Utilization Council, NASA, SARE, the South Dakota Agricultural Experiment Station, USDA-CSREES, and SD 2010 initiative.

REFERENCES

1. Clay, D.E., Chang, J., Clay, S.A., Ellsbury, M., Carlson, C.G., Malo, D.D., Woodson, D. et al. Field scale variability of nitrogen and delta 15-N in soil and plants. *Communications in Soil Science and Plant Analysis* 28, 1513, 1997.
2. Clay, D.E., Carlson, C.G., Clay, S.A., Chang, J., and Malo, D.D. Soil organic C maintenance in a corn (*Zea mays* L.) and soybean (*Glycine max* L.) as influenced by elevation zone. *Journal of Soil and Water Conservation* 60, 342, 2005.
3. Clay, D.E., Clay, S.A., Liu, Z., and Reese, C. Spatial variability of ^{13}C isotope discrimination in corn. *Communications in Soil Science and Plant Analysis* 32, 1813, 2001.

4. Kim, K., Clay, D.E., Carlson, C.G., Clay, S.A., and Trooien, T. Do synergestic relationships between nitrogen and water influence the ability of corn to use nitrogen derived from fertilizer and soil? *Agronomy Journal* 100, 551, 2008.
5. Liska, A.J., Yang, H.S., Bremer, V., Walters, D.T., Erickson, G., Klopfenstein, T., Kenney, D. et al. *BESS: Biofuel Energy Systems Simulator; Life-Cycle Energy and Emissions Analysis Model for Corn-Ethanol Biofuel* (*Version 2008.3.1. www.bess.unl.edu*). University of Nebraska, Lincoln, 2008.
6. Pimentel, D., Patzek, T., and Cecil, G. Ethanol production: Energy, economic and environmental losses. *Reviews of Environmental Contamination and Toxicology* 189, 25, 2007.
7. Dias de Oliveira, M.E., Vaughan, B.E., and Rykiel, Jr. E.J. Ethanol as fuel: Energy, carbon dioxide balances, and ecological footprint. *BioScience* 55, 593, 2005.
8. Shapouri, H., Duffield, J.A., and Wang, M. The energy balance of corn ethanol: An update. *Agricultural Economic Report 813*, U.S. Department of Agriculture, Washington, DC, 2002, 14 pp. Available at http://www.transportation.anl.gov/pdfs/AF/265.pdf
9. Guinee, J.B. *Handbook on Life Cycle Assessment*. Kluwer Academic, London, 2002.
10. Liska, A.J., Yang, H.S., Bremer, V.R., Klopfenstein, T.J., Walters, D.T., Erickson, G.E., and Cassman, K.G. Improvement in life cycle energy efficiency and greenhouse gas emissions of corn-ethanol. *Journal Industry Ecology* 13, 58, 2009.
11. Swanton, C.J., Murphy, S.D., Hume, D.J., and Clements, D.R. Recent improvements in the energy efficiency of agriculture: Case studies from Ontario, Canada. *Agricultural Systems* 52, 399, 1996.
12. Borin, M., Menini, C., and Sartori, L. Effect of tillage systems on energy and carbon balance in north eastern Italy. *Soil and Tillage Research* 40, 209, 1997.
13. Uhlin, H. Why energy productivity is increasing: An I–O analysis of Swedish agriculture. *Agricultural Systems* 56, 443, 1998.
14. Mamani Pati, F., Clay, D.E., Carlson, C.G., and Clay, S.A. Production, profitability, and energy life cycle audits can produce contrary results for corn (*Zea mays*) used in ethanol production. *Journal of Plant Nutrition* 2009, submitted.
15. Ceccon, P., Coiutti, C., and Giovanardi, R. Energy balances of four farming systems in north-eastern Italy. *Italian Journal of Agronomy* 6, 73, 2002.
16. Haciseferogullari, H., Acaroglu, M., and Gezer, I. Determination of the energy balance of the sugar beet plant. *Energy Sources* 25, 15, 2003.
17. Zentner, R.P., McConkey, B.G., Stumborg, M.A., Campbell, C.A., and Selles, F. Energy performance of conservation tillage management for spring wheat production in the brown soil zone. *Canadian Journal of Plant Science* 78, 553, 1998.
18. Zentner, R.P., Lafond, G.P., Derksen, D.A., Nagy, C.N., Wall, D.D., and May, W.E. Effect of tillage method and crop rotation on non-renewable energy use efficiency for a thin black chernozem in the Canadian prairies. *Soil and Tillage Research* 77, 125, 2004.
19. Hadi, S.H. Energy efficiency and ecological sustainability in conventional and integrated potato production system. In Ubertini, L. (ed.). *Proceedings of the IASTED Conference on Advanced Technology in the Environmental Field*, Lanzarote, Canary Islands, Spain, 2006.
20. Rathke, G.W., Wienhold, B.J., Welhelm, W.W., and Diepenbrock, W. Tillage and rotation effect on corn-soybean energy balances in eastern Nebraska. *Soil and Tillage Research* 97, 60, 2007.
21. Harrison, J.D. and Smith, D.R. *Nutrient Concentrations in Manure Storage Facilities*. Extension Utah State University, Logan, 2004. Available at http://extension.usu.edu/files/factsheets/AG-AWM-02-1.pdf
22. Kuesters, J. and Lammel, J. Investigations of the energy efficiency of the production of winter wheat and sugar beet in Europe. *European Journal of Agronomy* 11, 35, 1999.

23. Schmer, M.R., Vogel, K.P., Mitchell, R.B., and Perrin, R.K. Net energy of cellulosic ethanol from switchgrass. *Proceedings of the National Academy of Sciences of the United States of America* 105, 464–469, 2008.
24. Russelle, M.P., Birr, A.S., and Tiffany, D.G. Estimated net energy yields in a biomass fuelsheds [abstract]. *ASA–CSSA–SSSA Annual Meeting Abstracts*. CD-ROM. Paper No. 167-2, 2006.
25. James, R., Eastridge, M.L., Brown, L.C., Elder, K.H., Foster, S.S., Hoorman, J., Joyce, M.J. et al. *Ohio Livestock Manure Management Guide*. The Ohio State University Bulletin 604-06, 2006. Available at http://ohioline.osu.edu/b604/index.html
26. Reiman, M., Clay, D.E., Carlson, C.G., Humburg, D.E., Reicks, G., Clay, D.W., and Clay, S.A. Deep manure placement impact on soil N and P concentrations, corn (*Zea mays*) and soybean (*Glycine max*) yields, and water infiltration. *Environmental Science and Health Part B* 44, 1–10, 2009.
27. Keeney, D. and Muller, M. *Water Use by Ethanol Plants: Potential Challenges*. Institute for Agriculture and Trade Policy, Minneapolis, MN, 2006. Available at http://www.agobservatory.org/library.cfm?refid=89449
28. Wareham, S. and Green, J. Nuclear power and water scarcity. *Science Alert*, 2007. Available at http://www.sciencealert.com.au/opinions/20072910-16508.html
29. Janssen, L., Klein, N., Taylor, G., Opoku, E., and Halbeck, M. *Conservation Reserve Program in South Dakota: Major Findings Form 2007 Survey of South Dakota CRC*. SDSU Economics Department Research Report 2008-1, 2007. Available at http://econ.sdstate.edu/Research/aboutresearch.htm
30. Sneller, T. and Durantee, D. Issue brief: Economic impacts of ethanol production. Ethanol across American, 2006. Available at http://www.ethanolacrossamerica.net/CFDC_EconImpact.pdf
31. Johansen, H.E. The small town in urbanized society. In Brown, D.L. et al. (eds.). *The Demography of Rural Life*. Cornell University Press, Ithaca, NY, 1993.
32. Coffman, J. and Athan, G. Do small towns have a future? *The Future of Small Towns*. Minnesota Public Radio, May 9, 2005. Web site: http://www.luc.edu/depts/sociology/johnson/p99webn.html
33. Cantrell, R. *Rural Depopulation: A Closer Look at Nebraska's Counties and Communities*. Rural Initiative, University of Nebraska, Lincoln, NE, 2005. Available at http://ruralinitiative.nebraska.edu/includes/downloads/ruraldepopulation.pdf
34. Du, X. and Hayes, D.J. *The Impact of Ethanol Production on U.S. Regional Prices and the Profitability of U.S. Oil Refinery Industry*. Center for Agriculture and Rural Development, Iowa State University. Working Paper 08-wp 467, 2008. Available at http://www.card.iastate.edu/publications/DBS/PDFFiles/08wp467.pdf
35. Duffy, M. *Estimated Cost of Crop Production in Iowa—2008*. Iowa State University, Ames, Iowa, FM1712, 2008. Available at http://www.extension.iastate.edu/agdm/crops/pdf/a1-20.pdf
36. Pimentel, D. and Giampietro, M. 1994. *Food, Land, Population and the U.S. Economy-Full Report*. Available at http://dieoff.org/page55.htm
37. White, S. and Cordell, D. Peak phosphorus: The sequel to peak oil. *Sustainable Phosphorus Futures*, 2008. Available at http://phosphorusfutures.net/index.php?option=com_content&task=view&id=16&Itemid=30

3 Using Precision Farming to Overcome Yield-Limiting Factors in Southern Brazil Oxisols: A Case Study

Telmo Jorge Carneiro Amado
and Antônio Luis Santi

CONTENTS

3.1 EXECUTIVE SUMMARY

Many agronomists have a limited understanding of Brazilian farming systems. It may be possible to increase the agronomy efficiency of these systems by adopting precision farming (PF) techniques. The purpose of this paper is to identify yield-limiting factors (YLF) in Southern Brazilian farming systems, using PF as a tool to improve soil and crop management. It is estimated that in Rio Grande do Sul State close to 1 Mha will be under PF by 2009/2010. Adoption of PF in this part of Brazil has increased exponentially over the past 7 years. Topics addressed in this chapter include (1) soil sampling for soil nutrients status evaluation; (2) soil nutrient variability in the field; (3) spatial and temporal variability of soil nutrient concentration in PF; (4) soil buffering capacity; (5) variable rate lime, phosphorus, and potassium applications; (6) real-time N applications using optical spectroscopy; (7) use of yield maps as a tool to improve soil management; and (8) water availability impacts on temporal yield variability.

3.2 INTRODUCTION

3.2.1 Evolution of Soil Management in South Brazil

Natural ecosystems in Brazil range widely from temperate native grasslands to tropical rainforest. The expansion of agricultural activities into these biomes has occurred at different times. In Southern Brazil (Rio Grande do Sul, Santa Catarina, and Parana States) during the 1960s, wheat (*Triticum aestivum* L.) and soybean (*Glycine max* L., Merr.) crops replaced native grassland and araucaria forest (*Araucaria brasiliensis* Rich.), whereas agricultural development in the Cerrado (Tropical Savanna) and the North Brazil frontiers occurred later. Nowadays, the Cerrado region (Central Brazil) is the main agricultural region of the country. The dominant soil order in Brazil is Oxisol (Figure 3.1). These soils are largely used in agriculture under tropical and subtropical climate conditions. The Oxisols generally have good physical attributes and poor chemical characteristics. The chemical characteristics of these soils generally include low basis saturation, high subsoil aluminum (Al) content, and high phosphorus (P) fixation capacity, whereas the physical attribute include deep horizons, high water infiltration rates, and high soil aggregation and good aeration.

FIGURE 3.1 Oxisol profile under agriculture in Southern Brazil. Santo Augusto, RS. (Photo courtesy of Dirceu Gassen, 2006.)

The soil management applied in the first three decades after land use change was poor with intensive plowing and disking operations and crop residue burning that resulted in bare soil exposed to intensive rainfall during several months of the year.[1] The traditional cropping system in Southern Brazil is double cropping (wheat or cover crops as black oat in winter/soybean or corn in summer) resulting in frequent tillage operations during winter and summer seasons. Implementing this management in fields with undulating landscape and an annual precipitation ranging from 1500 to 1700 mm per year has led to extensive soil erosion,[2–5] nutrients losses, and depletion of soil organic matter (SOM). Cassol[6] estimated that after 25 years of agriculture, two-thirds of agricultural lands of Southern Brazil had low SOM, poor water infiltration, soil structural degradation, compaction, and reduced availability of plant water. In this region, soil degradation resulted in the abandonment of many lands[4] and contamination of water resources.[5,7,8] The use of terraces for soil erosion control has not been generally effective. During the 1970s, Mielniczuk[1] estimated that for each kilogram of soybean harvested, approximately 10 kg of soil were lost. Consequences of poor soil management were modest temporal yield increases and the need to replace nutrients lost through erosion with fertilizer inputs. In response, farmers and researchers during the early 1970s gradually improved their soil management practices by (1) reducing tillage intensity; (2) adopting practices where crop residues are not burned; and (3) avoiding winter fallow. The turning point occurred during the 1970s when tillage trials, conducted in Londrina and Ponta Grossa in the State of Parana, showed that no-till produced similar yields as conventional management with minimal soil erosion.[5,9,10] At the same time, early adopters (pioneers) in Southern farmers started adopting no-tillage on their farms. Since the 1990s, no-tillage has been adopted at an annual rate of approximately 1.5 Mha $year^{-1}$, reaching 25 Mha in Brazil (http://www.febrapdp.org.br/arquivos/BREvolucaoPD72a06.pdf). Even nowadays, it is not achieved the plateau (stabilization) due to no-till adoption in Central and North new frontiers.

Brazilian grain crop yields have been gradually increasing over the last decade. These increases are partially attributed to the success in no-till adoption. Nowadays, continuous efforts in soil management improvements are enhancing profits, improving nitrogen and energy use efficiency, reducing fossil fuel use, getting higher yields, and improving water quality. Preliminary work suggests that PF in Brazil can further increase farming and energy efficiency. In this context, the PF could be the new stage of the journey of improvement in soil and plant management in tropical and subtropical environments.

3.2.2 Adoption of Precision Farming in Southern Brazil

Research conducted at São Paulo (ESALQ and UNICAMP), Parana (UFRP), Minas Gerais, and Cerrado (EMBRAPA) pioneered the use of PF in Brazil during the late 1990s. It was estimated that in 1999 there were 60 combines equipped with yield sensors.[11] In 2000, the "Aquarius Project," located in

Não me Toque, Rio Grande do Sul State implemented PF in commercial farms in South Brazil. This study used, as main practices, soil sampling, variable fertilizer rates, and yield maps to increase farm profitability and increase input use efficiency. The partnership of AGCO (provide yield map and SGIS–Fieldstar system), STARA (provide fertilizer appliers, variable rate of subsoiling, planters with variable plant population), YARA (provide N-Sensor, optical spectroscopy and fertilizers), COTRIJAL (farmers cooperative provide fields), and UFSM (Federal University of Santa Maria carried out the research with graduate students) started with two fields consisting of 256 ha. Nowadays, the study has been expanded to 13 fields all located in the northeast of Rio Grande do Sul State. The dominant soil in this region is clay Rhodic Hapludox. The clay content in this soil varies from 400 to 600 g clay kg^{-1} soil. The average precipitation is approximately 1700 mm $year^{-1}$, without dry season i.e., all months with more than 100 mm $month^{-1}$, and the average annual temperature is 19.2°C. The weather is wet subtropic Cfa, in Koeppen classification. Other general soil features are gentle to moderate slopes, well drained, poor in natural soil fertility, especially in plant-available P, and acid soils with Al and manganese (Mn) toxicity. The main cash crops are soybean, wheat, maize, canola, and black oat (cover crop). Typical soybean, maize, and wheat yields are 3000, 7000, and 2500 kg ha^{-1}, respectively. The general cropping system is double cropping (two crops per year).

Aquarius Project has provided reference information needed for the expansion of PF in Southern Brazil, as well as training for agronomists and consultants. Important information provided by the Aquarius Project includes equipment requirements and PF research findings from soybean, wheat, and corn studies. Many farmers start PF through the application of variable rate fertilizers based on geo-referenced soil samples. Grid sampling sizes generally range from 1 to 5 ha. PF services, soil sampling, and variable application of fertilizers and lime are provided by farm cooperatives and private consultants. The farm cooperatives provide PF services even to small holders (30–50 ha) that generally are not able to afford private consultants costs.

3.3 METHODS AND MAIN EQUIPMENTS AND PROCEDURES USED

The results reported in this study were obtained from the "Aquarius Project" and trials on selected commercial fields that have been carried out by UFSM. All these areas were sampled using a regular grid of 100 m × 100 m (Figure 3.2).

The soil samples were collected based on a geo-referenced point located in the center of a grid cell (100 m × 100 m) (Figure 3.3). At each grid point, eight random cores were collected within a radius of 5 m having the point as center. When the cropland was fertilized in row, the cores including row and interrow were sampled.

Soil data were analyzed with soiltec-SGIS software (http://www.agco.com.br/marcasSoilteqSGIS.asp?op=5). This SIG also was used to create yield maps,

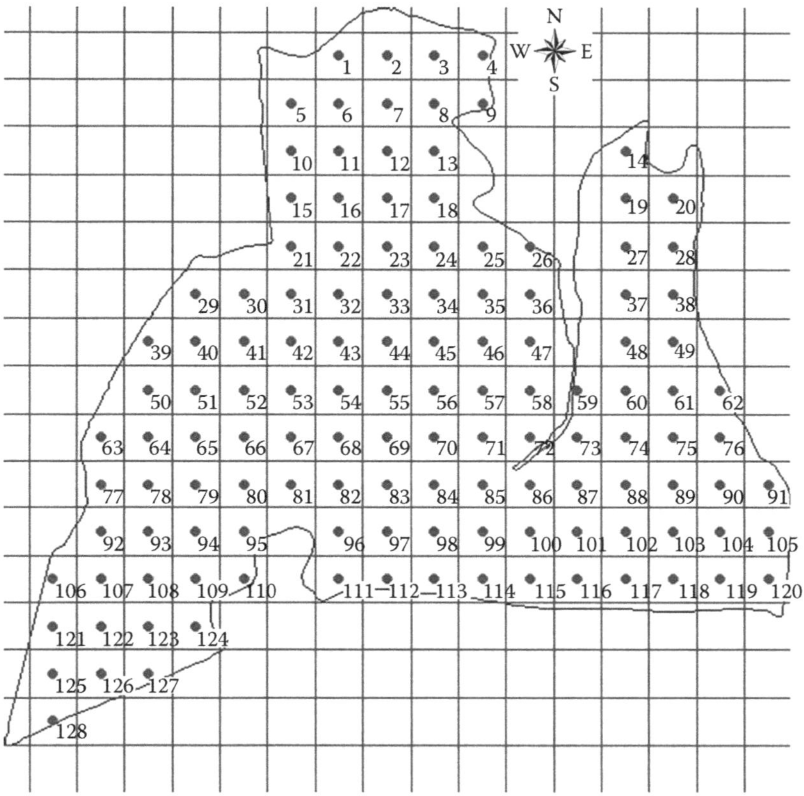

FIGURE 3.2 Soil sampling grid in Lagoa cropland with 126 ha located in Rio Grande do Sul State. (From Aquarius Project, 2009.)

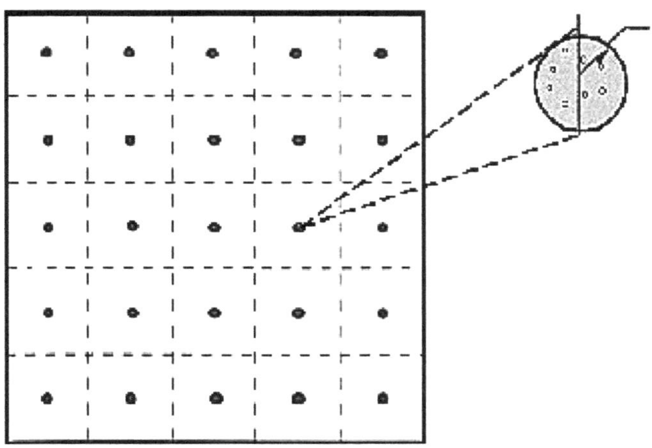

FIGURE 3.3 Protocol of soil sampling eight cores collected in a 5 m radius around the central grid point used in Aquarius Project fields. (Adapted from Rehm, G.W. et al. *Soil Sampling for Variable Rate Fertilizer and Lime Application*, North Central Multistate Report 348, University of Minnesota, St. Paul, MN, 2002.)

fertilizer prescription maps, and other documents for site-specific management. Also, the Campeiro software (http://www.rural.ccr.ufsm.br/pc5/apre_projeto.html) has been used. Campeiro is a Portuguese language PF GIS software developed by UFSM.[13] It is in the seventh version and is used by many Brazilian producers and PF consultants. In this study, relationships among plant and soil attributes were derived using an electronic spreadsheet.

The strategy for variable-rate fertilization (VRF) in the Aquarius Project was to identify fertilizer responsive and nonresponsive field zones.[12] In general, the main strategy was the reallocation of fertilizer from zones with high nutrient contents (above critical level) to lower nutrient contents (below critical level). Each field contained check (adjacent area) where uniform fertilizer was applied following traditional farmer practices. All the other soil and plant management practices were identical in the PF and check areas.

Equipments available for the application of VRF in Brazil and used in the Aquarius Project can be accessed at http://www.stara.com.br/web/index.php?menu=produtos&id=12. For broadcast fertilization Tornado (http://www.stara.com.br/web/index.php?menu=produtos&id=44) and Hércules (http://www.stara.com.br/web/index.php?menu=produtos&id=12) spreaders are commonly selected. Prior to 2009, only broadcast VRF applications were used in the Aquarius Project. Since 2009, producers can also select variable band fertilization and variable plant population rates using the Victoria planter (http://www.stara.com.br/web/index.php?menu=produtos&id=8).

In Aquarius Project, the main fertilizers applications were done by broadcast nutrients with Hércules 7000 oriented by DGPS (Figure 3.4). The width of application was set to 24 m.

Each nutrient was applied in separate operations. The average error between the planned fertilization and applied fertilizer rates was lower than 5% in six of the Aquarius Project fields.[14] Fertilizer application followed the scheme in Figure 3.5.

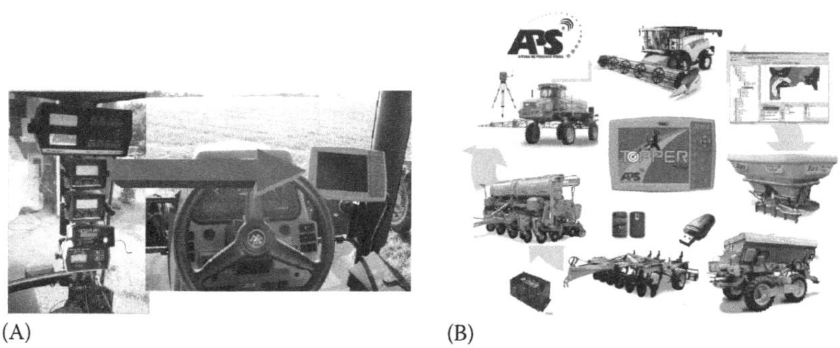

(A) (B)

FIGURE 3.4 Illustration of equipment, datavision, prescription map, and data storage used in variable fertilization rate in Aquarius Project. (From Dellamea, R.B.C., *Eficiência da adubação a taxa variável em áreas manejadas com agricultura de precisão no Rio Grande do Sul*, Mestrado em ciência do solo, Universidade Federal de Santa Maria, RS, Brasil, 2008, 162 pp.)

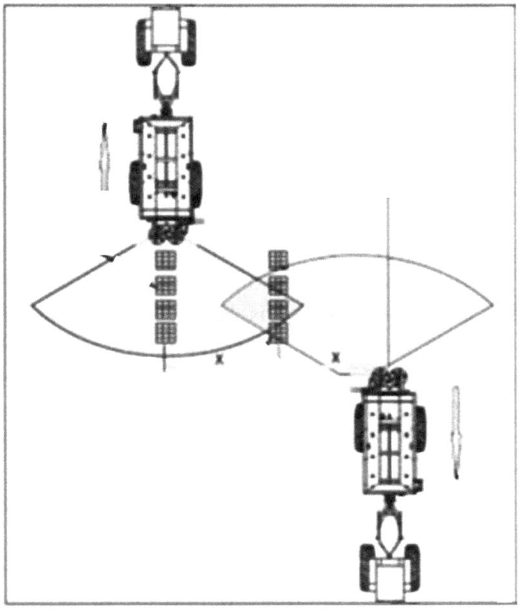

FIGURE 3.5 Variable fertilization with Hércules spreader in Aquarius Project. (From Dellamea, R.B.C., *Eficiência da adubação a taxa variável em áreas manejadas com agricultura de precisão no Rio Grande do Sul*, Mestrado em ciência do solo, Universidade Federal de Santa Maria, RS, Brasil, 2008, 162 pp.)

When implementing the PF practices, farmer's concerns about broadcast P fertilizer efficiency were noted. This concern was due to very slow P movement through Oxisol soil profile. Usually, the broadcast fertilization results in high concentration of P in the 0–0.05 m soil layer. In order to avoid very high soil P concentration in the upper soil layer, the fertilization strategy could be split in banded in row and broadcast application. The first fertilization was set up to 50% of the total P rate as uniform rate applied in row at seeding with traditional farm equipment. The second fertilization was broadcast variable rate application with Hércules spreader.

In the Aquarius Project, the crop yields were registered with combines equipped with yield monitors and Fieldstar system provided by Massey Ferguson (Figure 3.6).

FIGURE 3.6 Combine equipped to PF.

3.4 RESULTS

3.4.1 Soil Variability in Southern Brazil Fields

Soil nutrient variability has vertical and horizontal components,[15,16] which can be increased through the application of fertilizers and use of no-tillage or other conservation tillage practices.[17] For research conducted in long-term no-till field denominate Coxilha located in Palmeira das Missões, Santi[16] showed that average available P content (21.4 ppm) was three times higher in the surface 0–0.05 m than in the 0.10–0.20 m layer (6.8 ppm) and that topsoil K concentrations were 3.8 times higher when comparing the same previous depths. Reactivity and concentration of P and K variability with soil depth can result in fertilization programs that require deep-band fertilizer aiming to stimulate the plant root growth and water-use efficiency.

In order to increase fertilizer efficiency in these high P-fixing soils, the fertilizer is often band applied. Banding the fertilizer in no-tillage systems can result in P concentration in row that may be two to five times higher than in the inter-row.[18,19] Capturing large- and small-scale variability is a challenge in mature no-till fields (Figure 3.7). In mature no-till, a shallow soil sampling depth would result in underfertilization, whereas a deep soil sampling protocol would result in overfertilization. On the other side, interrow soil sampling would result in overfertilization, whereas in row soil sampling in under-fertilization (Figure 3.7). In this case, a high number of cores including row and interrow sampling should be necessary to capture the spatial variability.[12,19] The small-scale variability in no-till is a strong argument against the use of point sampling systems or reduced number of cores to compose a soil sample.[12]

The geostatistic analysis of two Rio Grande do Sul State fields, which have been in no-tillage for 10 years, is shown in Table 3.1. The Trindade do Sul, cropland is a 51.8 ha center pivot irrigated field with high available soil P coefficient of

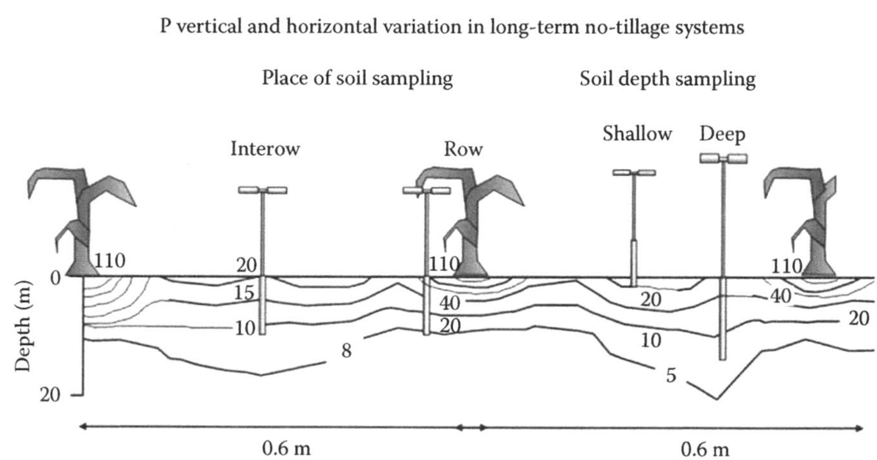

FIGURE 3.7 Conceptual model of vertical and horizontal phosphorus variability in long-term no-till Oxisol corn field. (Adapted from Murrell, 2005; Courtesy of Anghinoni, 2008.)

TABLE 3.1
Statistic and Geostatistic Analysis of Soil Attributes at 0–0.10 m Soil Depth in Two Fields in Rio Grande do Sul State

	Statistic Analysis					Geostatistic Analysis				
Soil Attributes	**Minimum**	**Maximum**	**Average**	**SD**	**CV**	**Nugget**	**Range**	**Sill**	**Model**	**Class**
Trindade do Sul field										
Clay (g dm^3)	64.17	81.28	74.01	4.58	6.19	2.431	510	21.008	Spheric	Strong
pH H_2O (1:1)	4.84	5.93	5.42	0.28	5.22	0.014	408	0.080	Spheric	Strong
P (ppm)	1.00	45.50	14.43	8.98	62.23	43.904	102	80.583	Exponential	Moderate
K (ppm)	133.72	295.11	213.87	39.59	18.51	308.346	408	1567.245	Spheric	Strong
SOM (%)	3.27	4.10	3.64	0.21	5.80	0.014	156	0.045	Spheric	Moderate
Al ($cmol_c$ dm^3)	0.26	1.74	0.88	0.38	43.27	0.060	153	0.193	Spheric	Moderate
Mg ($cmol_c$ dm^3)	1.04	3.10	2.03	0.52	25.73	0.009	220	0.041	Spheric	Strong
Ca ($cmol_c$ dm^3)	2.65	6.89	4.59	0.95	20.62	0.183	459	0.896	Exponential	Strong
CEC ($cmol_c$ dm^3)	5.70	10.30	8.05	1.07	13.24	0.223	204	0.595	Spheric	Moderate
Al saturation (%)	3.00	17.90	10.97	4.37	39.84	13.861	480	54.279	Spheric	Moderate
Basis saturation (%)	44.00	75.60	59.22	7.37	12.44	5.137	204	19.102	Spheric	Moderate

(continued)

TABLE 3.1 (continued)
Statistic and Geostatistic Analysis of Soil Attributes at 0–0.10 m Soil Depth in Two Fields in Rio Grande do Sul State

	Statistic Analysis					Geostatistic Analysis				
Soil Attributes	**Minimum**	**Maximum**	**Average**	**SD**	**CV**	**Nugget**	**Range**	**Sill**	**Model**	**Class**
Palmeira das Missões field										
Clay (g dm^3)	50.05	84.07	61.83	7.73	12.50	7.061	561	59.722	Spheric	Strong
pH H_2O (1:1)	5.75	6.75	6.27	0.25	4.03	0.016	385	0.064	Spheric	Moderate
P (ppm)	11.31	23.75	15.86	3.57	22.48	3.556	255	12.712	Spheric	Moderate
K (ppm)	118.82	278.45	186.89	30.50	15.94	183.346	416	930.037	Spheric	Strong
SOM (%)	2.20	3.40	3.01	0.25	8.40	0.010	459	0.061	Spheric	Strong
Al ($cmol_c$ dm^3)	0	0	0	0	0	0	0	0	—	—
Mg ($cmol_c$ dm^3)	1.11	2.07	1.53	0.20	13.22	0.009	220	0.041	Spheric	Strong
Ca ($cmol_c$ dm^3)	4.68	7.56	6.00	0.68	11.28	0.082	488	0.459	Spheric	Strong
CEC ($cmol_c$ dm^3)	6.40	9.90	8.02	0.80	9.91	0.141	510	0.632	Exponential	Strong
Al saturation (%)	0	0	0	0	0	0	0	0	—	—
Basis saturation (%)	67.00	89.30	81.36	5.02	6.17	5.832	510	25.166	Exponential	Strong

Source: Adapted from Amado, T.J.C. et al., *Revista Brasileira de Ciência do Solo*, 33, 831–834, 2009.
P and K determined by Melich-1 extractor.

variation (CV) and a large difference between the minimum and maximum values.[20] Following the classification of Warrick and Nielsen,[21] i.e., low variability at CV < 12%, medium at 12 ≤ CV ≤ 62%, and high at CV > 62%, the P had high variability (62.2%) (Table 3.1). Also, this nutrient had the lowest range. Calcium (Ca), magnesium (Mg), K, and Al had medium variability. While clay (texture), pH water, and SOM had low variability and high ranges. For comparative purposes, Table 3.1 also shows results from a 58.2 ha Palmeira das Missões irrigated field. This field had better soil fertility management than Trindade do Sul field, which resulted in a lower P CV (22.5%) and slightly higher soil test P level than in the Trindade do Sul. A high CV for P and a range of 120 m was reported previously by Coelho[11] for Brazilian soils.

SOM and pH variability frequently are related to topography and landscapes.[12] In Trindade do Sul, the average slope is 6.0% while in Palmeira das Missões, the slope is 4.5%.[20] Pretest erosion is an important factor impacting pH and SOM spatial variability. In landscape with concave or convex shapes, water erosion has reduced soil organic carbon (SOC) in backslope and shoulder areas, which in turn contributed to increased SOC contents in depressional zones.

3.4.2 Variable-Rate Phosphorus, Potassium, and Lime Applications

In a 21.9 ha field located in Vista Alegre (RS), P and K contour maps were developed (Figure 3.8). These maps show significant variability for both P and K concentration. In this field, the average available soil P was 22.9 ppm (Melich-1), with values ranging from 2.5 (minimum) to 47.3 ppm (maximum) and a CV = 43.6%, while the average soil K content was 226 ppm with values ranging from 84 to 348 ppm with a CV = 29.1%. Areas with P and K concentrations less than 15 and 120 ppm, respectively, were defined as below the critical levels. Thus, 29% and 14% of the field required additional P and K fertilizer, respectively. On the other side, P and K fertilizer could be saved on 8% and 41% of the field, respectively, due to high soil nutrient content.

It should be highlighted that in this field the zones that needed more P were not coincident with the areas requiring additional K. This way, two different fertilization recommendations were needed (Figure 3.9). For P, the triple superphosphate rates ranged from 152 to 326 kg ha^{-1}, while for K rates ranged from 33 to 133 kg ha^{-1}. These rates should be compared with the farm fertilization of 80 kg P_2O_5 ha^{-1} (equivalent to 200 kg ha^{-1} triple superphosphate) and 98 kg K_2O ha^{-1} (equivalent to 169 kg ha^{-1} potassium chloride) used in check as a reference.

Most of Brazilian Oxisols are naturally acid and may contain high concentrations of Al and Mn. In some situations, Al and Mn may be plant toxic. To increase yields and reduce Al and Mn toxicity typically, 2–3 Mg ha^{-1} of lime is applied uniformly in no-tillage fields every 4–5 years. Using variable-rate technology, it may be possible to reduce the total amount of lime applied across mature no-till fields. For example, in a 38 ha Nei Manica's field, the total amount of lime applied was reduced by 31% (Figure 3.10) relative to a uniform rate of 3 Mg ha^{-1}.

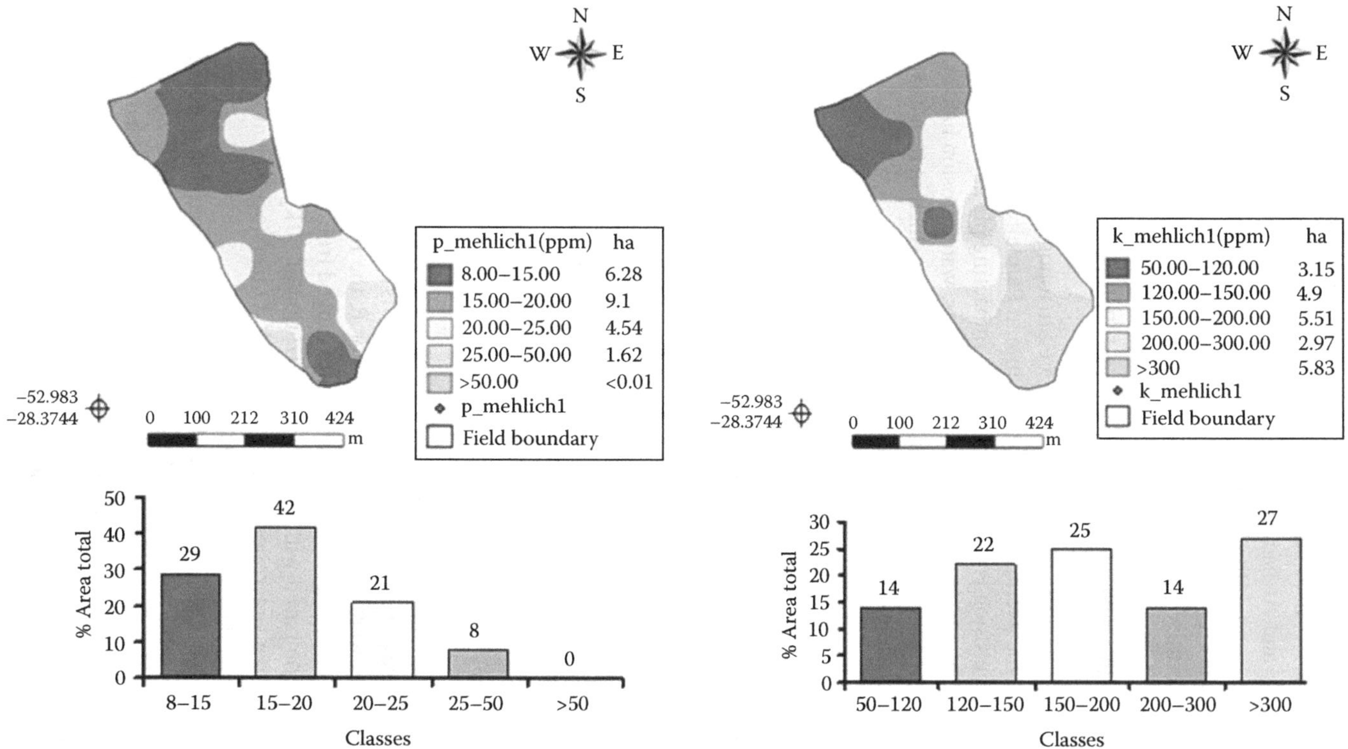

FIGURE 3.8 Phosphorus and potassium spatial variability in soil samples at 0–0.10 m in Juliano Michelini's field with 21.9 ha located in Vista Alegre, RS. (From Aquarius Project, 2009.)

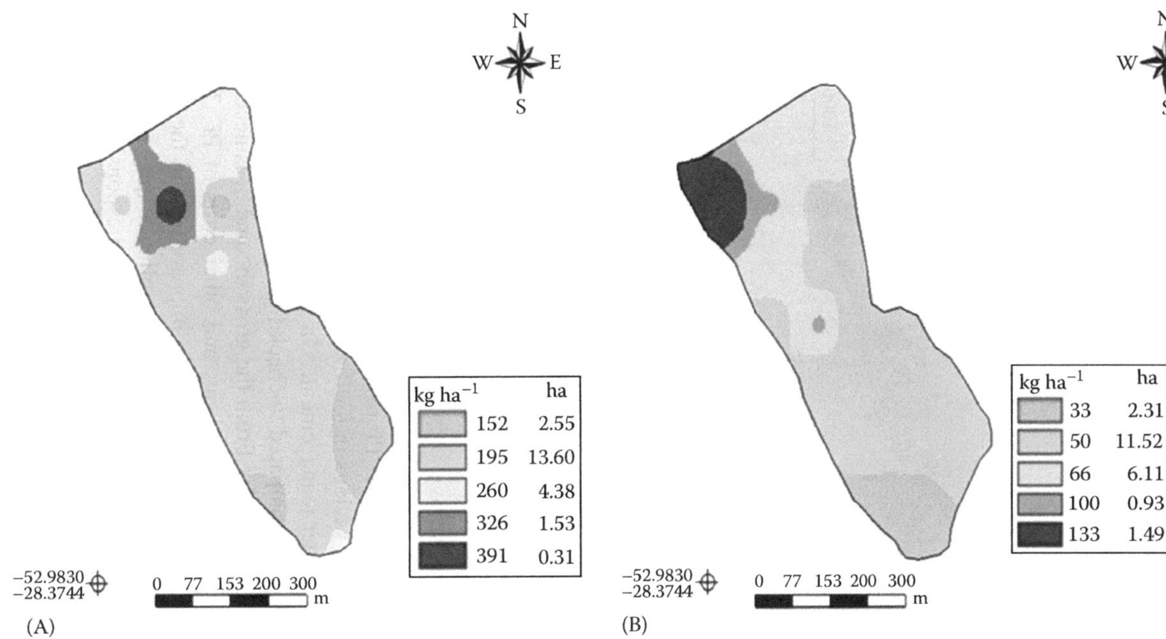

FIGURE 3.9 Variable fertilization recommendation in Juliano Michelini's field. The P fertilization was done with triple superphosphate (A) and K with potassium chloride (B). (From Aquarius Project, 2009.)

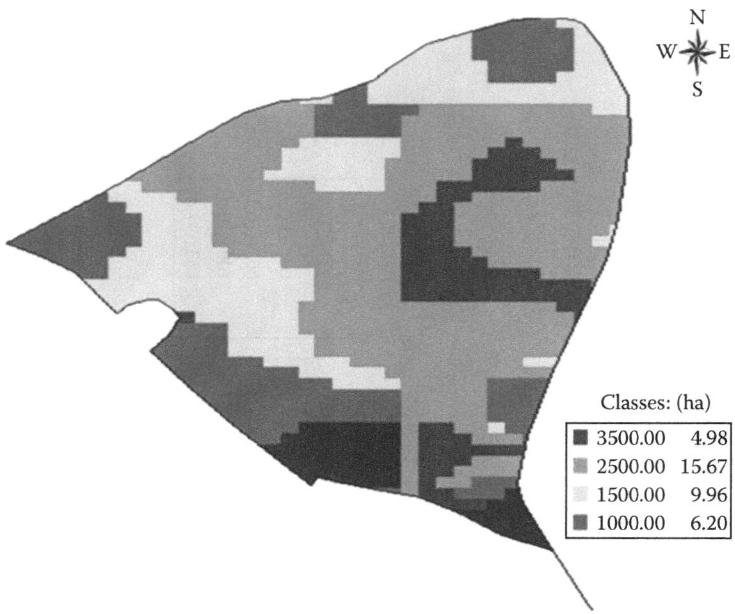

FIGURE 3.10 Variable-rate lime application in Nei Manica's field with 18 ha located in Não me Toque, RS. (From Aquarius Project, 2009.)

3.4.2.1 Yield Response to Variable-Rate Fertilizer

The average corn (*Zea mays* L.) yield obtained in Nei Manica's rainfed field was 8.867 kg ha^{-1} ranging from 4.346 kg ha^{-1} (minimum yield) to 11.520 kg ha^{-1} (maximum) with 13% moisture (Figure 3.11). The yield in the PF treatment was 13% higher than in the adjacent reference field. Over a number of PF fields investigated in the project, PF on average increases corn yield by 14% when compared to check fields.

The higher corn yields under PF resulted in increased gross income (increase in the yield combined with savings in fertilization inputs) of US$ 150 ha^{-1} when compared to traditional farmer management.[14] In soybean fields, PF increased soybean yield on average by 10% compared to check fields, resulting in an average increase in gross income of U$ 51 ha^{-1}.[14] From the gross income, it is necessary to debit the investment in equipments, soil analysis, and other costs of PF. The positive economic results of PF could be explained by improvement in the knowledge of soil nutrient status associated with intensive soil sampling, improvement in nitrogen use efficiency (NUE) by reallocation of fertilizer from nonresponsive zones to responsive ones in VRF, and better fertilizer recommendation.

3.4.3 Temporal Variability in Soil Nutrients

A benefit of PF is a reduction in size of field areas with low nutrient concentrations. When the first soil samples were collected in 2001 in Lagoa field, 54% of the field contained areas deficient in P. In 2003 and 2005, these areas were reduced to 29%

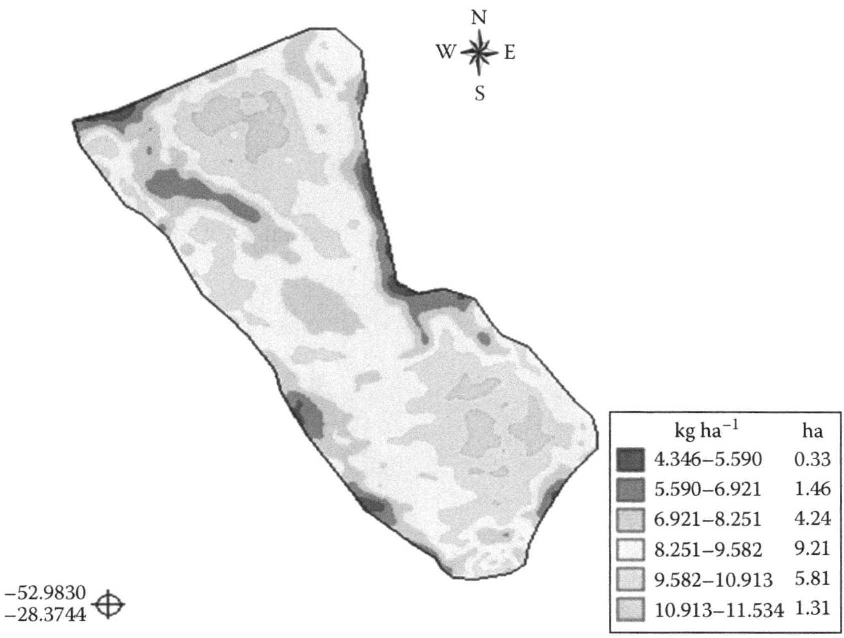

FIGURE 3.11 Corn yield in 2007 in Juliano Michelini's field located in Vista Alegre, RS. (From Aquarius Project, 2009.)

and 8%, respectively. This way, the soil P content was gradually brought to a range considered as ideal. These results suggest that PA was an efficient tool to reduce the size of areas where yield losses due to P deficiencies were occurring (Figure 3.12). After that, the remain soil P variability is not more a YLF, because the soil P content is already above the critical level in the whole field.[11]

Similar results were observed in a field located in Santo Antônio (RS), where after only 2 years, the share of low P areas decreased from 58% to 8% (Figure 3.13).

PF techniques in short term may not eliminate soil nutrient spatial variability. To help assess P and K fertilizer progress, lower and upper soil nutrient desirable levels were identified. For soil P (Melich-1), the target range was 15–30 ppm at 0–0.10 m, whereas for K, a target range between 150 and 300 ppm was established. In two fields, PA helped reduce the field areas outside these values (Figure 3.14).

3.4.4 Phosphorus and Potassium Apparent Soil Buffer Capacity Determined by Nutrient Balance

In Oxisol soils with high buffering capacity, fertilizer applications in excess of the crop removal rates may be needed to meet the needs of the plant to sustain high yields. In a 57 ha Coxilha field located in Palmeira das Missões, Santi[16] found that the apparent P and K buffer capacities in the 0–0.10 m depth were 16 kg P_2O_5 ha^{-1} and 2.9 kg K_2O ha^{-1} for each ppm of available P (Melich-1) and available K, respectively. Contrasting with this result, lower P buffering capacities were previously observed in

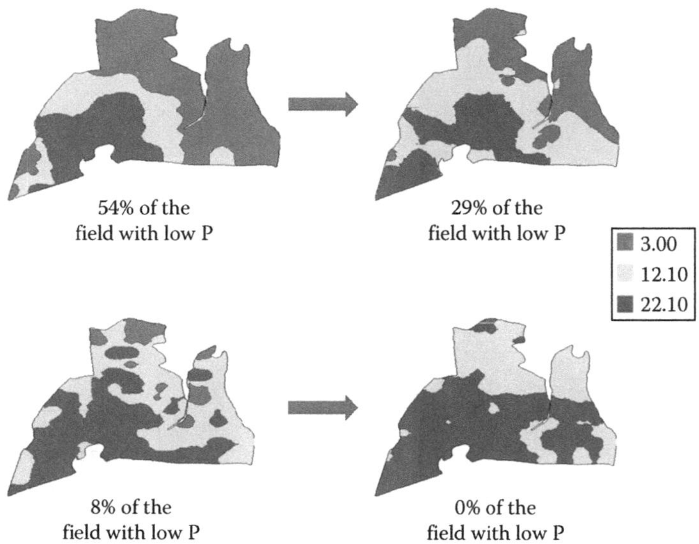

FIGURE 3.12 Lagoa field temporal evolution in soil P content with successive variable-rate fertilizations in Não me Toque, RS. (From Aquarius Project, 2009.)

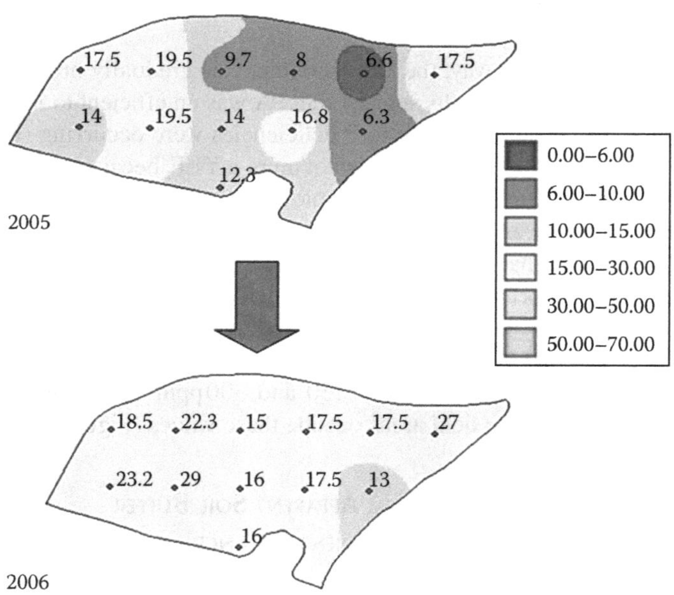

FIGURE 3.13 Temporal evolution of P content in Jairo Kohlrausch's field with 13 ha after two variable-rate fertilizations in Santo Antônio, RS. (From Aquarius Project, 2009.)

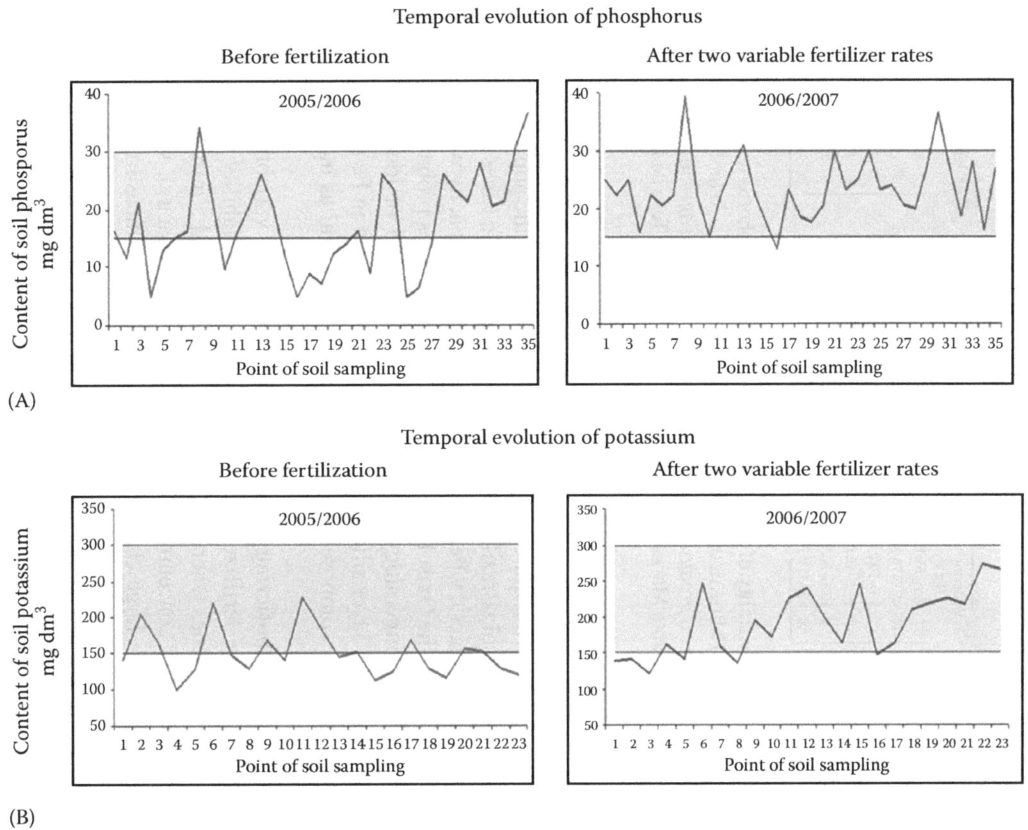

FIGURE 3.14 Temporal evolution of soil P (Jairo Kohlraush field in Santo Antônio, RS) (A) and K (Gilberto Maldaner field in Lagoa Três Cantos, RS) (B) after two variable-rate fertilizations. (From Aquarius Project, 2009.)

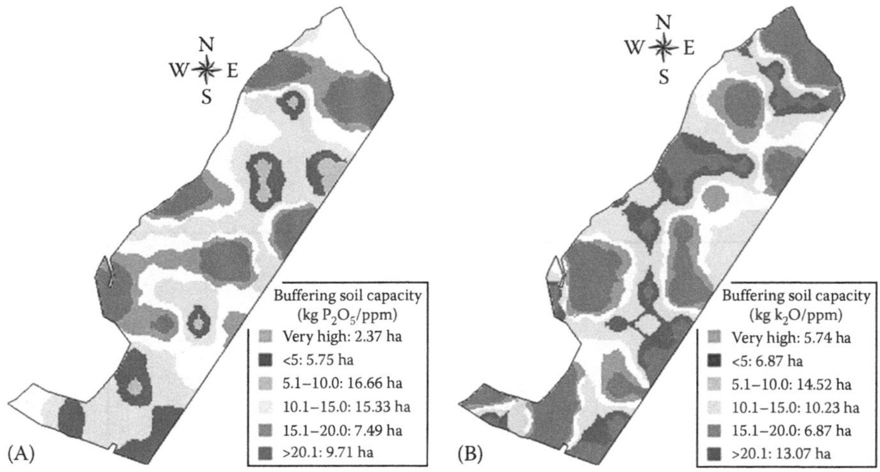

FIGURE 3.15 Spatial variability of apparent buffer capacity of phosphorus (A) and potassium (B) in Colorado cropland, Palmeira das Missões, RS. (From Santi, A.L., *Relações entre indicadores de qualidade do solo e a produtividade das culturas em áreas com agricultura de precisão*, Universidade Federal de Santa Maria, Santa Maria, RS, 2007, 175 pp. [Tese de Doutorado].)

a Mollisol located in Illinois (United States). Franzen[22] found that to increase soil P and K concentrations by 1 ppm in the surface 0.15 m of that soil, 6.5 kg P_2O_5 ha^{-1} and 6.7–8.0 kg K_2O ha^{-1} were required. These results suggest that the apparent P buffering capacity in the Oxisol (Brazil) soil was approximately three times higher than in the Mollisol (United States), whereas the apparent K buffer capacity was half. These discrepancies between the Oxisol and Mollisol buffer capacities probably are related to differences in soil mineralogy. The Oxisol was composed of low-activity clay mineral as kaolinite 1:1 layer silicate (500 g clay kg^{-1} soil) rich in Fe and Al oxides, while the Mollisol was composed of high-activity clay mineral as montmorillonite 2:1 layer silicates.

The soil's P and K buffering capacities contained spatial variability and influenced the impact of the fertilizer on the plant nutrients availability (Figure 3.15). Gray areas represent zones where the addition of fertilizer had a minimal temporal impact on soil nutrient concentrations (very high buffer capacity), whereas green zones were areas where large changes in soil nutrients were measured (low buffer capacity). These results highlight that the buffer capacity is site specific and it should be investigated specially for the high-fixing soil classes in order to optimize decision rules for VRF.

3.4.5 Variable-Rate Nitrogen Fertilization in Real Time Using Optical Spectroscopy

In the 2008–2009 growing season, we started the use of real-time variable-rate N fertilization with YARA N-Sensor ALS (http://www.yara.com/products_services/fertilizers/support_services/support_tools/) in the Aquarius Project. The YARA

FIGURE 3.16 N-Sensor ALS (YARA) with Hércules 7.000 spreader (STARA) during top-dress variable-rate N fertilization in real time. Cruz Alta, RS. (Photo courtesy of Telmo Amado, 2009.)

N-Sensor is being used in large scale in Europe where approximately 800 units were used in spring 2009, mainly in Germany and in the United Kingdom. This sensor provided real-time remote sensing information to Hércules 7.000 fertilizer spreader (STARA), allowing the N fertilization on-the-go as shown in Figure 3.16. The YARA N-Sensor provides vigor index readings that are converted to N fertilizer prescription by an algorithm (Figure 3.16).

In Brazil, the traditional corn N fertilization program is to apply one-third of the total N rate at seeding and the remaining two-third of the N rate in two top-dressed fertilizations. Due to high leaching potentials, pre-seeding N applications are not used. In Rio Grande do Sul State, the traditional corn N fertilization program is a broadcast application of between 80 and 120 kg N ha^{-1} of urea. The N fertilization rate is determined based on SOM content, previous crop (legume or graminea) and yield goal.[23] In 2009, two experiments were carried out in rainfed corn fields using the YARA N-Sensor. The strategy used was to apply uniform rates of 27 kg N ha^{-1} at seeding and 32 kg N ha^{-1} at V-4 (first topdress), followed by a variable-rate application of on average 91 kg N ha^{-1} at V-8 (second topdress) with N-Sensor base rate. The option for using a uniform N rate at V-4 is a relatively weak relationship between N-Sensor readings and corn N status when the plant biomass is still low.

This sensor was tested in the Cruz Alta trials. In this study, there were three replications per treatment (Figure 3.17). In this figure, it is possible to check the variability of the vigor index calculated from N-Sensor reading and the respective N fertilizer prescription given by the N-Sensor at 8-V in each of these replicated strips. In the first left strip (first block), there were the highest vigor index readings (greenish colors) and, as a consequence, the lowest recommended N fertilizer

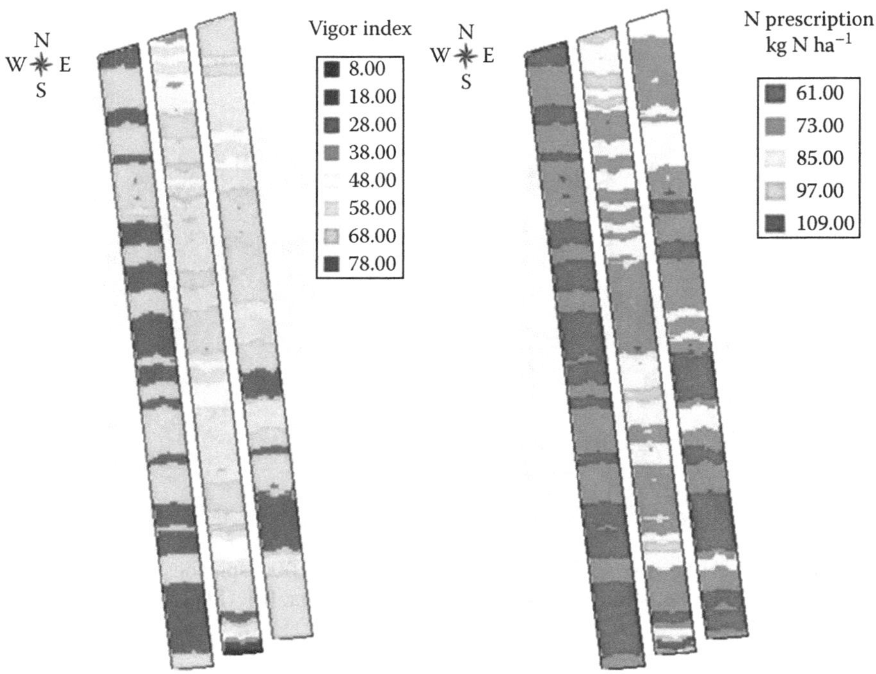

FIGURE 3.17 Spatial variability of vigor index and N prescription by YARA N-Sensor. Cruz Alta, 2009. (Courtesy of Bragagnolo, 2009.)

rates (reddish colors), as can be seen in the prescription map (first left strip of the second block). On the other side, the lowest vigor index readings were obtained in the second strip of the first block and, therefore, highest N fertilizer rates were recommended for this replication (second strip of the second block). The principle of N-Sensor is to estimate the plant N nutrition status through vigor index readings that are closely related to the current N uptake of the crop. This way, where the N nutrition status is satisfactory, the N fertilizer is partially reallocated to other zones where the N status is deficient, increasing the NUE.

In Brazil, the mineral soil N (NO_3 + NH_4) content at the beginning of the growing season usually shows low relationship with N uptake at flowering and corn yield. These results are attributed to intense rainfall through the corn-growing season resulting in risk of mineral N leaching. Rains with 100 mm over a 2-day period are not unusual. The intensive leaching process in tropical and subtropical wet climates explains why mineral soil N is not as useful a parameter to N fertilization programs as it is in North America for instance. In Brazil, SOM is the main soil parameter considered in N fertilizer program. Although, SOM also has some limitations in advance predicting N mineralization, i.e., the process of conversion of organic N to mineral N forms that are uptaken by plants, and as consequence, the soil N credits necessary to adjust the top-dress N fertilization. This way, in wet tropical and subtropical climates, the soil parameters have serious limitations in estimating with accuracy the soil N credits. In this situation, real-time N fertilization with optical

TABLE 3.2
Corn Grain Yield and NUE with Uniform Rates of Topdress Fertilization and Based on the Sensor-Based Nitrogen Rate

Treatment	Grain Yield, kg ha^{-1}	Relative Yield, %	Yield Response, kg ha^{-1}	NUE, kg Grain kg^{-1} of Applied N Topdress
1. 27 kg N ha^{-1} at seeding (control)	5478	63	—	—
2. 80 kg N ha^{-1} [a]	6306	72	828	15.6
3. 120 kg N ha^{-1} [a]	8286	95	2808	30.2
4. 140 kg N ha^{-1} [a]	8460	97	2982	26.4
5. 150 kg N ha^{-1} SBNR[a]	8662	99	3184	25.9
6. 160 kg N ha^{-1} [a]	8726	100	3248	24.4

Source: Aquarius Project (2009).
[a] Topdress fertilization applied at four and eight-leaves corn. First topdress with 32 kg ha^{-1}, second topdress with the remaining.

spectroscopy has high potential to improve the efficiency of fertilization in tropical and subtropical wet environments.

Findings from a field located in Tio Hugo (RS), where an N strip experiment (350 m long × 15 m wide) was carried out, are shown in Table 3.2. All treatments received the same base dressing at seeding (27 kg N ha^{-1}). N rates at topdressing were varied between 0 kg N ha^{-1} (control) and 133 kg N ha^{-1}, applied at V4 and V8 corn stages (six treatments). The data shows significant yield response to N fertilizer. Treatments 2 and 3 are the traditional farm N fertilization over the years according to the common relation of fertilizer cost and corn price. Treatments 4, 5, and 6 were designed to achieve the highest yields and NUE. Results showed that the high traditional farm N fertilization rate (treatment 3) had the highest NUE and achieved 95% of the highest yield. The low traditional farm N fertilization rate (treatment 2) showed lower corn yield and NUE. The treatment 1 had only base dressing at seeding and it showed the lowest yield. The treatment 5 with YARA N-Sensor was among the highest yielding treatments. Increasing the N fertilization rate above the optimum was associated to a decrease of NUE as expected. It should be noted that during the study, a late drought probably reduced the maximum yields. This study was one of the first with real-time variable rate of N fertilization in Brazil. Although the results obtained are not conclusive, they show a potential of this tool to improve N fertilization program. Next year trials with corn and wheat will be carried out in Brazil to confirm the potential of the N-Sensor in improving N fertilization in wet tropical and subtropical climates.

3.4.6 Yield Maps as a Tool to Improve Soil Management

The yield zones of corn fields in the Aquarius Project were defined as low yield = average field yield × 0.95, medium yield = average field yield × 1.05 and high yield > average field yield × 1.05[24] (Figure 3.18). In three fields in 2007/2008,

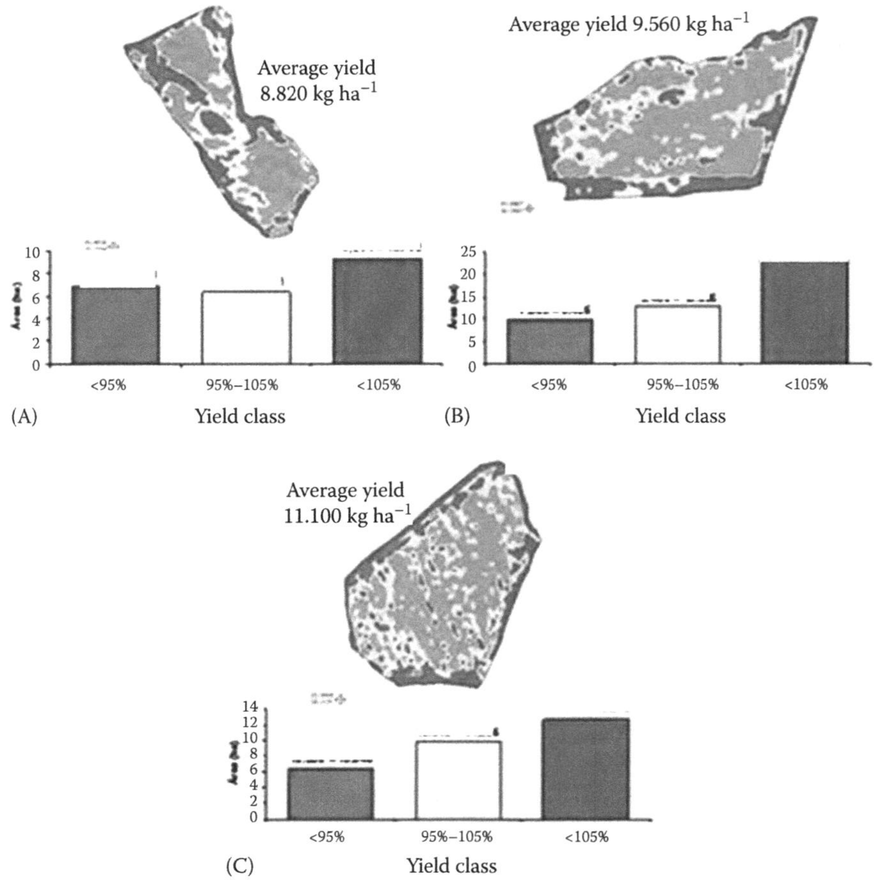

FIGURE 3.18 Spatial variability of low, medium, and high yield in three rainfed corn croplands (A, B, C). (From Aquarius Project, 2009.)

the average corn yields were 8.820, 9.900, and 11.100 kg ha^{-1}. These high yields were attributed to timely rainfall and the good plant and soil management practices adopted by the producer. It should be highlighted that these fields have been managed under continuous no-till for at least 10 years. Among the good soil management practices, it should be highlighted, the crop rotation adopted. Summer crop rotation (soybean/corn) and winter cover crops (black oat [*Avena strigosa* Schreber], common vetch [*Vicia sativa* (L.) Walp.], radish oil [*Raphanus sativus* L.]) were used in these fields to increase biodiversity and improve soil quality. The low yield zones were primarily located on the field borders (Figure 3.18). These border areas are subject to intensive transit of machines and equipments and, therefore, are prone to soil compaction. Also, these areas are close to native forest and natural pastures, which increases the pest attacks. In the entrance and exit of the fields is common errors of yield data register by the combine. These errors were eliminated in order to avoid misinterpretation in the field margins. Ponteli[25] investigating soybean and corn yield maps in Palmeira das Missões (RS) reported that due to erosion, low yield zones were associated to low SOM contents in the ondulating landscape of Southern Brazil.

In order to identify the YLF, two rainfed and two irrigated fields were selected for study. For this analysis, the field was split into cells with the dimension of 17 m × 17 m (rainfed areas) (Figure 3.19) and 13 m × 13 m (irrigated areas) (Figure 3.20) were calculated. The data were interpolated with inverse of square with a distance of 100 m using the software CR-Campeiro5.

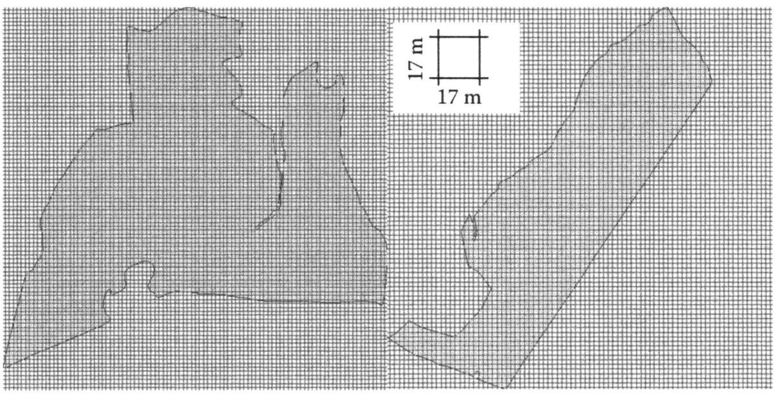

FIGURE 3.19 Lagoa field (Não me Toque) with 132 ha and Coxilha field (Palmeira das Missões) with 57 ha both are rainfed areas. (From Santi, A.L., *Relações entre indicadores de qualidade do solo e a produtividade das culturas em áreas com agricultura de precisão*, Universidade Federal de Santa Maria, Santa Maria, RS, 2007, 175 pp. [Tese de Doutorado].)

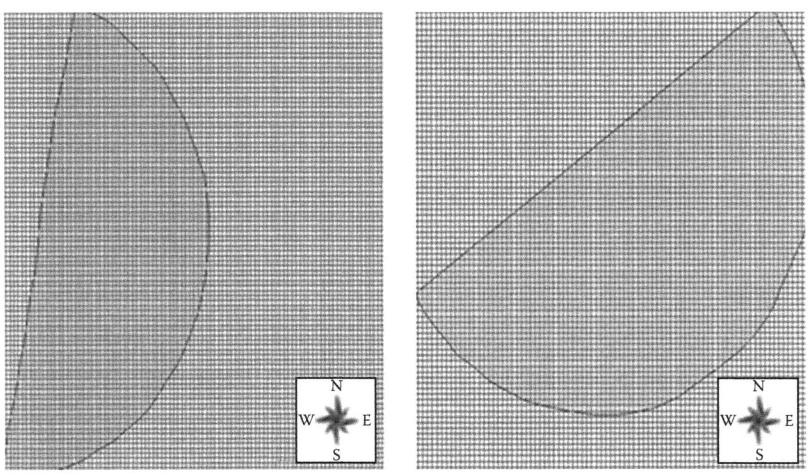

(A) (B)

FIGURE 3.20 Croplands in Trindade do Sul with 57 ha (A) and Palmeira das Missões with 55 ha (B). The two croplands are irrigated with pivot Central. (Courtesy of Lemainski, 2008.)

In Coxilha field, in order to investigate the temporal yield stability, six yield data were analyzed: soybean 2001, corn 2002, wheat 2002, soybean 2003, soybean 2004, and corn 2005. The yields were normalized based on average field yield to each year in order to allow a temporal analysis. In Lagoa fields were registered: maize 2002, wheat 2003, soybean 2003, and maize 2005.

The cumulative normalized yield maps of rainfed fields are presented in Figure 3.21 and the cumulative yield maps of irrigated fields are shown in Figure 3.22. Based on these maps, five points in each yield zone were selected for further investigation of YLF.

Among soil physical attributes, the infiltration has been considered one of the best soil quality indicator because it integrates many other attributes such as soil structure, porosity, soil resistance, bulk density, and soil compaction. Cooper[26] reported that in high-yield fields, the most common limiting factor is plant water availability.

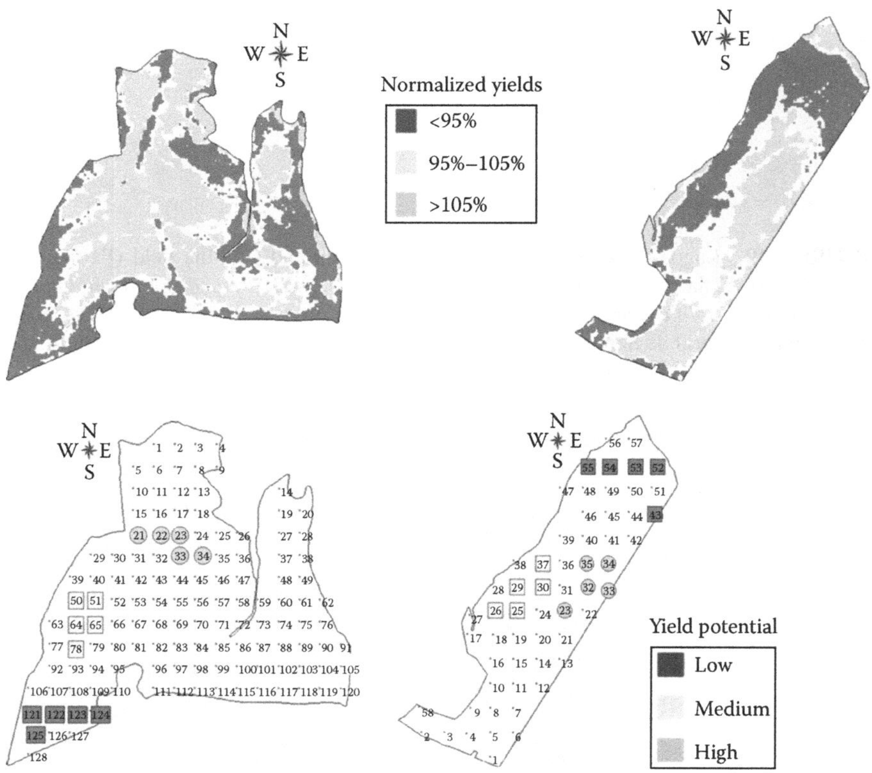

(A) Não–Me–Toque (B) Palmeira das Missões

FIGURE 3.21 Cumulative normalized yields in two rainfed fields—Lagoa and Coxilha croplands. Points select in low, medium, and high yield zones in order to investigate soil physic attributes. Croplands are located in Não me Toque and Palmeira das Missões, RS. (From Santi, A.L., *Relações entre indicadores de qualidade do solo e a produtividade das culturas em áreas com agricultura de precisão*, Universidade Federal de Santa Maria, Santa Maria, RS, 2007, 175 pp. [Tese de Doutorado].)

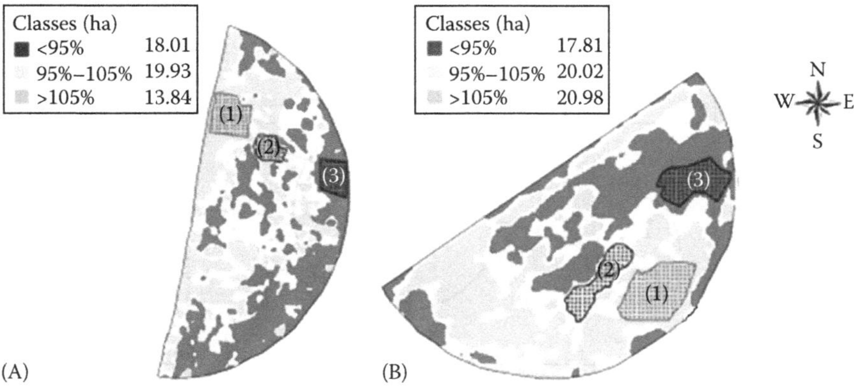

FIGURE 3.22 Cumulative normalized yields in two irrigated fields—Trindade do Sul (A) and Palmeira das Missões (B). Five points were selected in low (red color) represented by number 3, medium (yellow color) represented by number 2, and high (green color) yield zones in order to investigate soil physical attributes. (Courtesy of Lemainksi, 2008.)

In Palmeira das Missões and Lagoa rainfed fields, Santi[16] performed five infiltration tests in each selected grid point using the Cornell Sprinkle Infiltrometer described by Van Es and Schindelbeck[27] (Figure 3.23). Santi[16] reported that in the Palmeira das Missões field, runoff started after 10, 17, and 27 min in low-, medium-, and high-yield zones, respectively. At the Lagoa field, the runoff started at 9, 12, and 16 min in the low-, medium-, and high-yield zones, respectively (data not shown). In Lagoa field (Não me Toque), the test was performed at two different times in March and

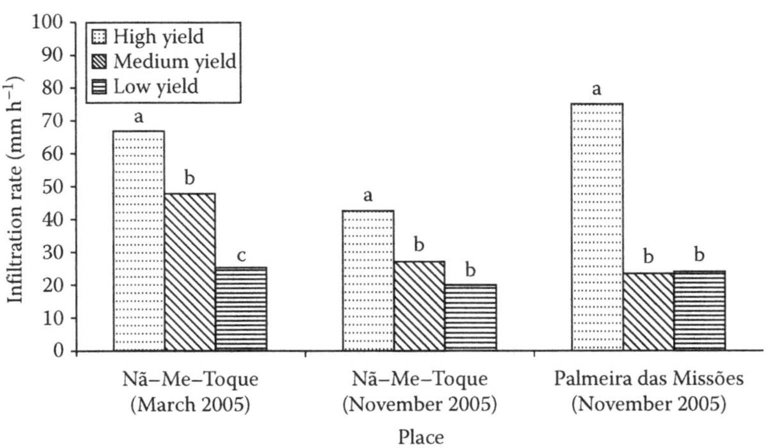

FIGURE 3.23 Infiltration in three yield zones evaluated by Soil Quality Test Kit Guide/USDA-ARS in two rainfed fields. Means followed by the same letter are not different by Tukey test ($p < 0.05$). (From Santi, A.L., *Relações entre indicadores de qualidade do solo e a produtividade das culturas em áreas com agricultura de precisão*, Universidade Federal de Santa Maria, Santa Maria, RS, 2007, 175 pp. [Tese de Doutorado].)

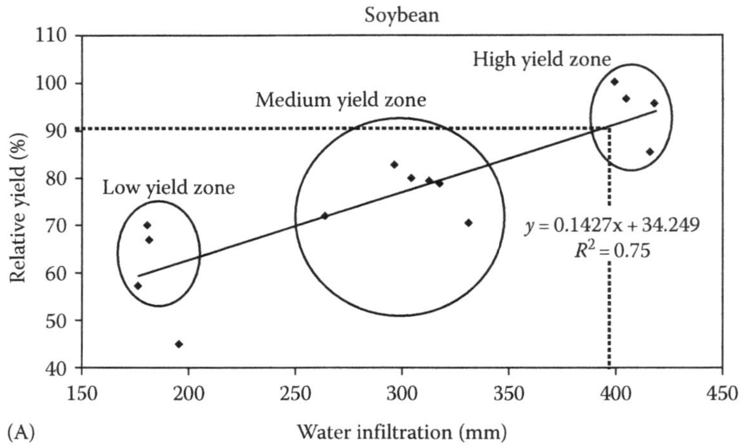

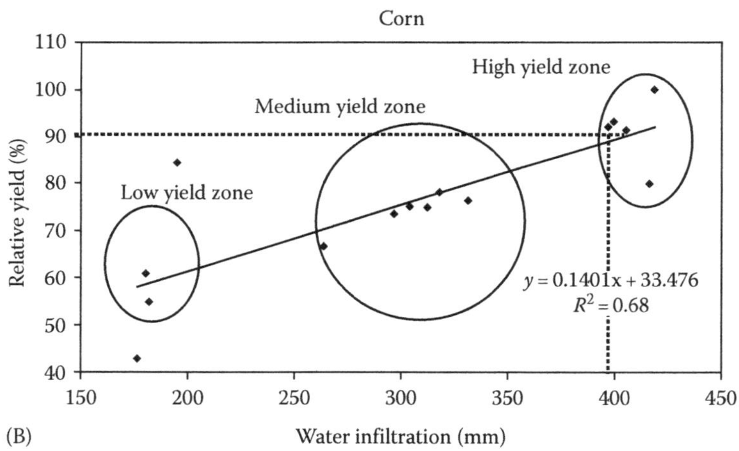

FIGURE 3.24 Relation between infiltration and average relative yield of soybean (3 year average) and corn (2 year average). Palmeira das Missões, RS. (From Santi, A.L., *Relações entre indicadores de qualidade do solo e a produtividade das culturas em áreas com agricultura de precisão*, Universidade Federal de Santa Maria, Santa Maria, RS, 2007, 175 pp. [Tese de Doutorado].)

November 2005. The results show that, in both fields, the high-yield zone had higher infiltration while the low-yield zone had lower infiltration.

Santi[16] established the relation between average relative yields for soybean (3 years) and corn (2 years) with the infiltration. In that research, Santi[16] made five replications of the infiltration to each point (five points in each yield zone) using double rings as proposed by Bouwer.[28] The results are presented in Figure 3.24.

In an irrigated Trindade do Sul field (Figure 3.21A), Amado et al.[20] reported that average black bean (*Phaseolus vulgaris* L.) yields (2005 and 2006) were 1.0 Mg ha^{-1} (low yield zone), 1.2 Mg ha^{-1} (medium yield zone), and 1.5 Mg ha^{-1} (high yield zone), while in Palmeira das Missões, the average corn yields (2003 and 2004) were 6.9 Mg ha^{-1} (low yield zone), 8.7 Mg ha^{-1} (medium yield zone), and 9.7 Mg ha^{-1}

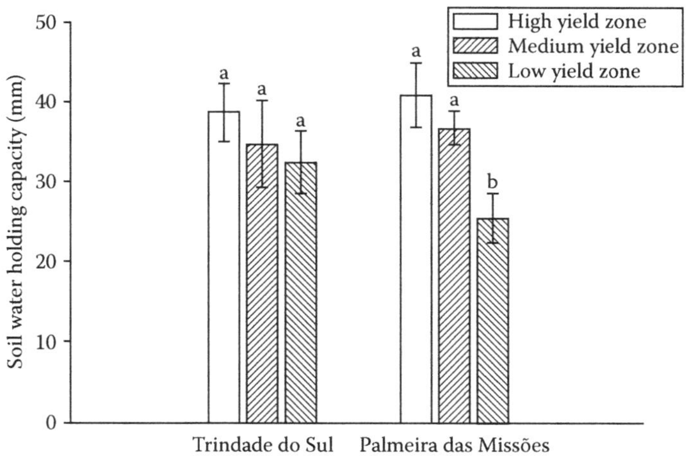

FIGURE 3.25 Relation between soil water-holding capacity and yield zones in two irrigated fields (Trindade do Sul and Palmeira das Missões). Means followed by the same letter are not different by Tukey test ($p < 0.05$). (Adapted from Amado, T.J.C. et al., *Revista Brasileira de Ciência do Solo*, 33, 831, 2009.)

(high yield zone). Plant-available water (6–1500 kPa) and aggregate sizes were also influenced by yield zone. Low-yield zone tended to have lower water holding capacity (Figure 3.25), whereas high-yielding zones had larger aggregates (Figure 3.26).[29,30] The large size aggregates may be associated with water infiltration rates described above and soil quality in general.

3.5 CONCLUSIONS

The large-scale adoption of no-till since the 1990s in Brazil and other South American countries was an important step forward to improve soil management, allowing expansion of the cropped area, erosion control, and increasing crop yields. The rapid adoption of precision agriculture since 2000 associated with no-till represents the next step to sustain high yields and efficient use of inputs in tropical and subtropical climates. VRF based on grid soil sampling was the entrance point to precision agriculture in Brazil. Yield mapping still is in a stage of incipient adoption and its use should grow in next few years.

A series of experiments were conducted to assess the YLF on Southern Brazil Oxisols. Factors that limited yields were soil nutrient contents, nutrient buffering capacity, water infiltration, and plant-available water. Grid soil sampling was an important tool for identifying zones with nutrient concentrations below the critical level. The study also identified that spatial variability of nutrient buffering capacity should be assessed in Oxisols. Variable-rate applications of fertilizers and lime made it possible to reduce the amount applied in field zones where the nutrients already were high or very high, allowing a reallocation to zones with low nutrient contents. In general, under PF, it was possible to reduce the fertilizer and lime inputs in relation to traditional farmer practice. Inspection of yield maps

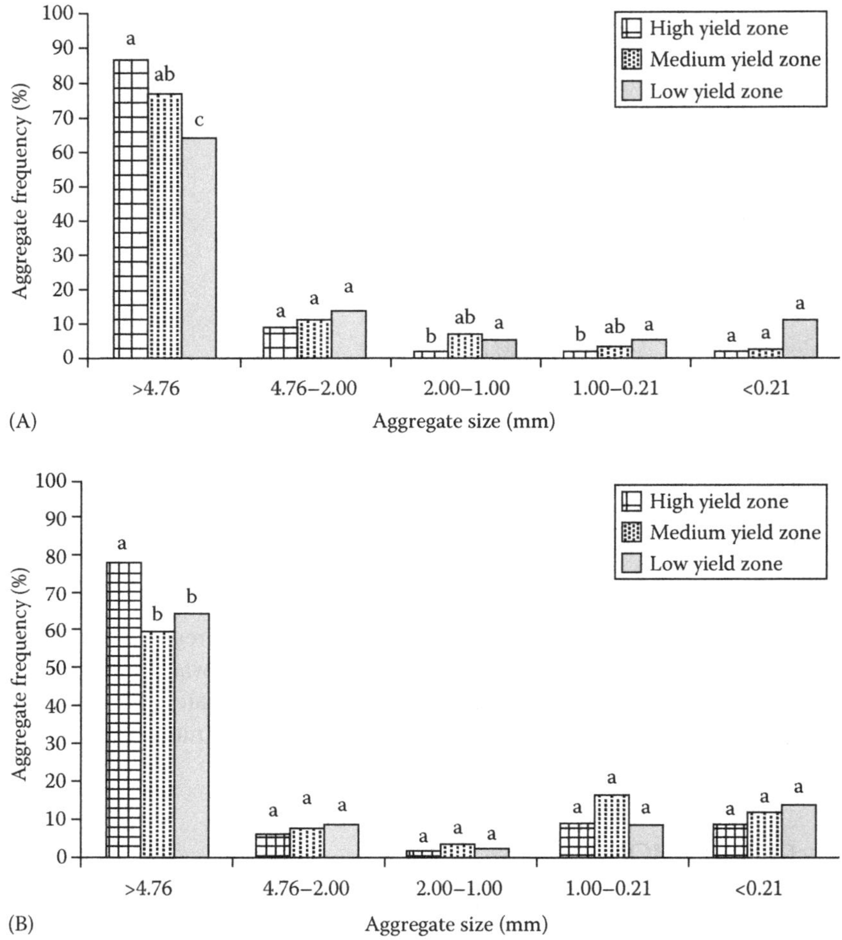

FIGURE 3.26 Water stable aggregates in Palmeira das Missões (A) and Lagoa (B) fields at 0–0.05 m in yield zones. Means followed by the same letter are not different by Tukey test ($p < 0.05$). (From Santi, A.L., *Relações entre indicadores de qualidade do solo e a produtividade das culturas em áreas com agricultura de precisão*, Universidade Federal de Santa Maria, Santa Maria, RS, 2007, 175 pp. [Tese de Doutorado].)

suggested that in eroded zones and field margins, yields were limited by soil compaction, depleted SOM, reduced water infiltration, and reduced soil water-holding capacity.

The site-specific management should have a holistic approach of soil management with evaluation of chemical, physical, and even biological indicators in order to establish the main interventions necessary to improve soil quality and sustain long-term high crop yields. In order to achieve this objective, vigor maps (optical spectroscopy) and yield maps that use the plant as an indicator of soil quality could be an efficient approach to be used complementary to soil fertility maps.

ACKNOWLEDGMENTS

Aquarius Project is supported by Massey Ferguson, STARA Ltda, YARA Ltda, and Cotrijal. Also acknowledgments to farmers who allowed us to carry out this research in their farms, and graduate and undergraduate students responsible for field and laboratory activities. To Dr. Luis I. Prochnow (IPNI, Brazil) for the opportunity, to Dr. David E. Clay (South Dakota State University, SD) and Dr. Joerg Jasper (YARA, Germany) for technical review of the paper.

REFERENCES

1. Mielniczuk, J. Manejo do solo no Rio Grande do Sul: Uma síntese histórica. In *Curso de fertilidade do solo em plantio direto, VI. Passo Fundo, 2003. Resumo de palestras.* Aldeia Norte Editora Ltda., Ibirubá, 2003, pp. 5–14.
2. Cogo, N.P., Drews, C.R., and Gianello, C. Índice de erosividade das chuvas dos municípios de Guaíba, Ijuí, e Passo Fundo, no Estado do Rio Grande do Sul. In *Encontro nacional de pesquisa sobre conservação do solo, II. Passo Fundo, 1978. Anais.* CNPT, Passo Fundo, 1978, pp. 145–152.
3. Gianluppi, D., Scopel, I., and Mielniczuk, J. Alguns prejuízos da erosão do solo no RS. In *Congresso Brasileiro de ciência do solo, XVII, Manaus, 1979. Resumos.* SBCS, Campinas, 1979, 92 pp.
4. Mielnickuk, J. and Schneider, P. Aspectos socio-economicos do Manejo de solos no sul do Brasil. In *Anais do I Simposio de Manejo de Solo e Plantio Direto no Sul Brasil e II Simposio de Conservacao de Solo do Planalto P 3–19.* Passo Fundo, RS, 1984.
5. Bolliger, A. et al. Taking stock of the Brazilian Zero-Till revolution: A review of landmark research and farmer's practice. *Advances in Agronomy* 91, 47, 2006.
6. Cassol, E.A. *Erosão do solo-influência do uso agrícola, do Manejo e preparo do solo.* IPRNR 15. Instituto de Pesquisas de Recursos Naturais Renovaveis. IPRNR, Porto Alegre, 1984, p. 40.
7. Amado, T.J.C. and Reinert, D.J. Zero tillage as a tool for sustainable agriculture in South Brazil. In Benites, J. et al. (eds.). *Conservation Tillage for Sustainable Agriculture. Proceedings of the International Workshop, Harare, Zimbabwe, Annexe III: Background Papers (International)*, June 22–27, GTZ, Eschborn, 1998, pp. 227–238.
8. Pottker, D. *Efeito do tipo de solo, tempo de cultivo e da calagem sobre a mineralização da matéria orgânica em solos do Rio Grande do Sul.* Dissertação de mestrado. Universidade Federal do Rio Grande do Sul, 1977, 128 pp.
9. Derpsch, R. Historical review of zero tillage. In Benites, J. et al. (eds.). *Conservation Tillage for Sustainable Agriculture. Proceedings of the International Workshop, Harare, Zimbabwe, June 22–27, 1998. Annexe III: Background Papers (International).* GTZ, Eschborn, 1998, pp. 205–218.
10. Steiner, K. et al. Promotion of conservation farming by the German development cooperation. In Garcia-Torres, L., Benites, J., and Martinex-Vilela A. (eds.). *Conservation Agriculture: A Worldwide Challenge. Proceedings of the first World Congress on Conservation Agriculture.* Madrid, October 1–5, 2001, vol. 2, XUL Cordoba, 2001, p. 60.
11. Coelho, A.M. Agricultura de precisão: Manejo da variabilidade espacial e temporal dos solos e das culturas. In *Tópicos especiais em ciência do solo.* SBCS, 2003, pp. 249–290.
12. Rehm, G.W. et al. *Soil Sampling for Variable Rate Fertilizer and Lime Application.* North Central Multistate Report 348. University of Minnesota, St. Paul, MN, 2002.
13. Giotto, L., Robaina, A.D., and Sulzbach, L. *A Agricultura de Precisão com o Sistema CR Campeiro5*, 2004. 330 pp.

14. Dellamea, R.B.C. *Eficiência da adubação a taxa variável em áreas manejadas com agricultura de precisão no Rio Grande do Sul.* Mestrado em ciência do solo. Universidade Federal de Santa Maria, RS, Brasil, 2008, 162 pp.
15. Carvalho, M.P., Taketa, E.Y., and Freddi, O.S. Variabilidade espacial de atributos de um solo sob videira em Vitória Brasil (SP). *Revista Brasileira de Ciência do Solo* 27, 695–703, 2003.
16. Santi, A.L. *Relações entre indicadores de qualidade do solo e a produtividade das culturas em áreas com agricultura de precisão.* Universidade Federal de Santa Maria, Santa Maria, RS, 2007, 175 pp. (Tese de Doutorado).
17. Schlindwein, J.A. and Anghinoni, I. Variabilidade espacial de atributos de fertilidade e amostragem de solo no sistema plantio direto. *Revista Brasileira de Ciência do Solo* 24, 85–91, 2000.
18. da Silva, M.A.G. et al. Amostragem e variabilidade nos atributos de fertilidade em um latossolo sob plantio direto em São Miguel do Iguaçu, Estado do Paraná. *Acta Scientiarum* 25, 243–248, 2003.
19. Nicolodi, M., Salet, R.L., and Basso, F.P. É possível fazer uma amostragem representative de solo com trado, no sistema plantio direto? In *Reunião Sul-Brasileira de Ciência do Solo, 3, 2000. Pelotas, RS. Resumos...Rumos da Ciência do Solo para o Desenvolvimento do RS e SC.* SBCS, 2000.
20. Amado, T.J.C. et al. Variabilidade espacial de atributos químicos e físicos de Latossolos e sua relação com os rendimentos de milho e feijão irrigados. *Revista Brasileira de Ciência do Solo* 33, 831–834, 2009.
21. Warrick, A.W. and Nielsen, D.R. Spatial variability of soil physical properties in the field. In Hillel, D. (ed.). *Applications of Soil Physics.* Academic Press, New York, 1980, pp. 319–344.
22. Franzen, D.W. *Summary of 40 Years of Grid Sampling in Illinois.* Fertilizer Conference Proceedings, Illinois, 2007. Available in <http://frec.cropsci.uiuc.edu/2008/report5>, accessed in August 28, 2008.
23. Comissão de química e fertilidade do solo—CQFRS/SC. *Recomendações de adubação e calagem para os estados do Rio Grande do Sul e Santa Catarina.* 3rd edn. Passo Fundo. SBCS, Núcleo Regional Sul, UFRGS, 2004, 400 pp.
24. Molin, J.P. Definição de unidades de manejo a partir de mapas de produtividade. *Engenharia Agrícola* 22, 83–92, 2002.
25. Ponteli, C.B. *Caracterização da variabilidade espacial das características químicas do solo e da produtividade das culturas utilizando as ferramentas da agricultura de precisão.* Dissertação de mestrado em Engenharia Agrícola. Universidade Federal de Santa Maria, RS, Brasil, 2006, 117 pp.
26. Cooper, R.L. Pesquisa sobre produtividade máxima da soja nos EUA. *Informações Agronômicas*—POTAFOS, no. 101, 2003.
27. Van Es, H. and Schindelbeck, R. *Field Procedures and Data Analysis for the Cornell Sprinkle Infiltrometer.* Cornell University, Department of Crop and Soil Sciences. Extension Publication R 03-01, 2003, 8 pp.
28. Bouwer, H. Intake rate: Cylinder infiltrometer. In Klute, A. (ed.). *Methods of Soil Analysis. Physical and Mineralogical Methods.* Part 1. 2nd edn. American Society of Agronomy, Inc.; Soil Science Society of America, Inc. Publishers, Madison, 1986, pp. 825–844.
29. Doran, J.W. and Parkin, T.B. Defining and assessing soil quality. In Doran, J.W. et al. (eds.). *Defining Soil Quality for a Sustainable Environment.* Soil Science Society of America, Madison, 1994, pp. 3–22 (Publication Number 35).
30. Conceição, P.C. et al. Qualidade do solo em sistemas de manejo avaliada pela dinâmica da matéria orgânica e atributos relacionados. *Revista Brasileira de Ciência do Solo* 29, 777–788, 2005.
31. Aquarius Project. Available at www.ufsm.br/projetoaquarius. Accessed 20/12/2009.
32. Murrell, S.T. *Advanced Soil Ferility: Whats New?*, JNPI, 2005.

4 Collecting and Analyzing Soil Spatial Information Using Kriging and Inverse Distance

David W. Franzen

CONTENTS

4.1 EXECUTIVE SUMMARY

Using appropriate soil sampling techniques can increase energy efficiency and reduce the amount of nutrients required to maximize yields. Many crop nutrient needs and rates are determined through the use of soil testing.[1] Soil nutrient recommendations protocols, which relate a soil nutrient extractant or other measurements with a relative supply and availability for specific crops, have been developed for most of the United States and many parts of the world. These procedures are constantly under scientific scrutiny for their continued usefulness and levels of confidence (North American Proficiency Testing Program, Soil Science Society of America, Madison, WI). This chapter presents methods of obtaining representative grid samples and processing them in a manner that will maximize confidence in the resulting nutrient recommendations. There are many geographic areas in which zone sampling has replaced grid sampling; however, some of the principles presented here should also be considered in zone-sampled fields. Data and directions for calculating inverse distance weighting (IDW) factors and semivariances are provided.

4.2 INTRODUCTION

4.2.1 Sampling Patterns

Grid sampling is used and preferred in regions where past fertilization or manure application has been high.[2–5] In many fields, native fertility levels have been masked by past fertilizer and manure applications. Grid sampling is used when there is no apparent logical method of dividing a field into relatively homogeneous areas. A sufficiently dense grid of samples is obtained to reveal fertility patterns within a field. It may be that the field is relatively uniform in fertility level, but this will only be revealed through sampling.

There are several sampling pattern strategies that might be considered for grid sampling.[6,7] These include random (Figure 4.1); regular systematic (Figure 4.2); staggered start systematic, sometimes called triangular or diamond (Figure 4.3); clustered (Figure 4.4); and systematic unaligned (Figure 4.5). A random sampling might be considered in a field with no recent history of fertilization or manure, such as a government set-aside program, breakout field, or an old pasture to be converted to cropland. Regular systematic approach was common grid sampling in the era before global positioning system (GPS) receivers. This approach allowed a sampler to use a tachometer or even "step off" distances to achieve the desired pattern. A staggered start systematic recognized that systematic errors in one direction might be possible, and the start and end of each sampling rank were offset to try to compensate for

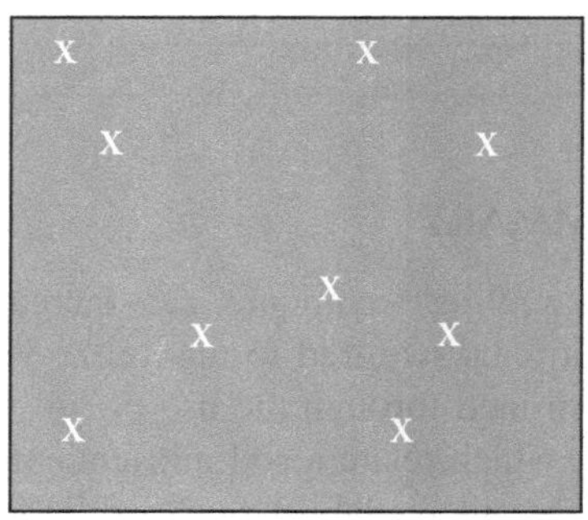

FIGURE 4.1 Example of random sampling.

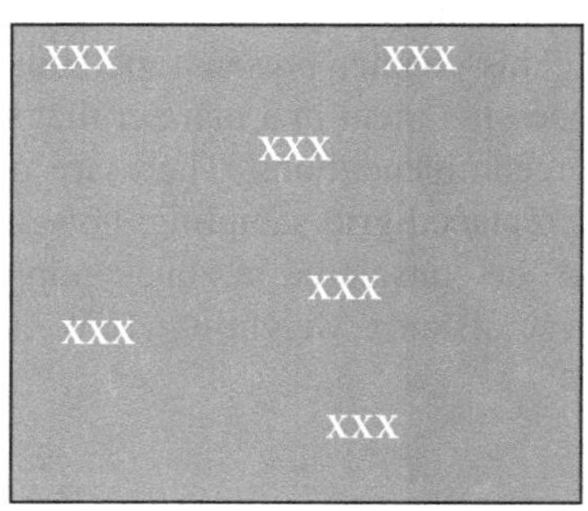

FIGURE 4.2 Example of random cluster sampling.

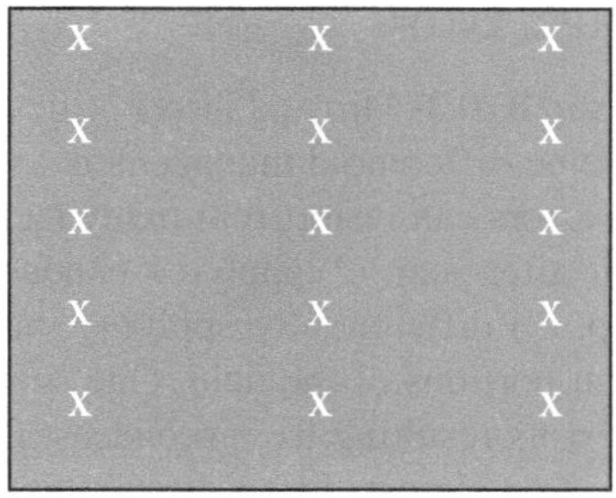

FIGURE 4.3 Example of regular systematic grid sampling.

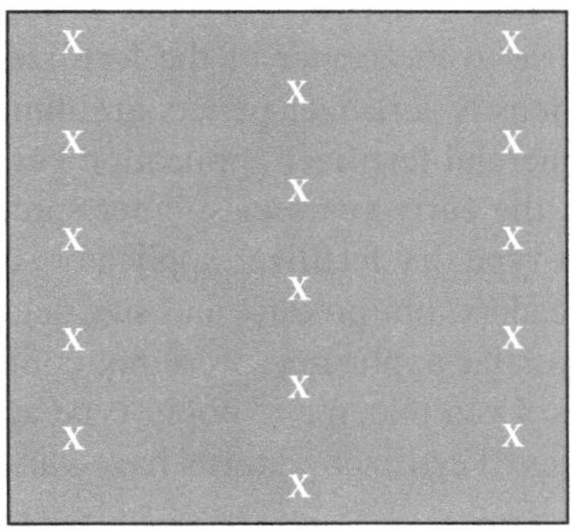

FIGURE 4.4 Example of staggered start (or triangular or diamond) grid sampling.

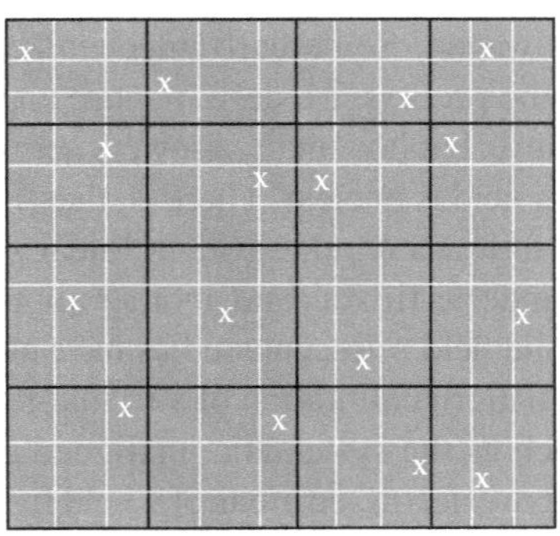

FIGURE 4.5 Example of systematic unaligned grid sampling.

these errors in one direction. The clustered approach is a type of random sample that might help to measure small-scale variability and larger-scale variability by grouping two to three sample core composites around random points a number of locations within the field. The systematic unaligned grid was made practical through a combination of GPS and field software that allows the locating of random grid points within a systematic grid. This approach minimizes the effects of systematic errors in two directions. The systematic unaligned grid is probably the method most used by commercial grid samplers today.

4.2.2 Causes of Sampling Errors

Systematic errors at both small (0.3–1 m) and larger scales (10–30 m) are endemic in fields that have received some differential management and crop removal rates over the past 100 years. These errors can result from many factors including (1) planting in the same direction year after year, (2) applying banded fertilizer either with the seed or on a separate trip, (3) differentially applying manure, and (4) having different rotations in different portions of the field. One source of error generally not considered is systematic errors resulting from broadcast fertilizer applications. Dry fertilizer pellets of one type are almost always different in size or density, so the distances they travel laterally behind a spinner-type applicator are different. Depending on the source of fertilizer, there may be significant fertilizer dust present. Blended fertilizers also always contain pellets and crystals of different sizes. Low-density fertilizer and dust fall close to the center of the fertilizer pattern if applied using spinner spreaders. High-density fertilizer pellets are flung farther out into the pattern. The history of commercial fertilizer application began as early as the 1860s,[8] so residual patterns from the early years of spinner spreader fertilization can still be seen. Pneumatic boom-type dry fertilizer applicators do not have the same density issues as spinner spreaders, but product and size segregation are still possible. Boom-type pneumatic granular applicators were not commonly used in the United States until the late 1980s. Even pneumatic boom-type applicators are not immune to spread pattern anomalies.[9] Fertilizer patterns have varied from about 12–24 m in width over the years.

Even before commercial fertilizers and continuing to present day, manure application was and is even more prone to systematic errors. Dry manures in particular are difficult to manage depending on their condition. In early years, manures were simply shoveled out of a wagon. Some early mechanized manure applicators did little more than simulate the process. Modern manure applicators can be calibrated to more evenly apply manure, but how many growers actually take the effort to calibrate and apply evenly in a field is unclear. There is, therefore, sufficient reason to be concerned when samples follow a regular grid. A linear series of samples might be taken in an area severely underfertilized, and another might be obtained in an area of overfertilization in the same field.[10] Recent studies have emphasized the importance of proper rate and spread pattern calibration of variable-rate applicators.[11,12]

Selecting a system, such as the systematic unaligned grid, minimizes the effect from streaks of under- and overfertilized areas of a field from fertilizer/manure application traffic. Fields have also been consolidated over the years, therefore assuming that the present planting direction has always existed since prairie breakout is unreasonable. The direction of fertilizer application may have been turned 90° from the original direction by the new operators.

Sampling in fields with banded P and K fertilizer applications is an additional challenge due to small-scale variability. Banding more than small amounts of nonmobile fertilizer nutrients leaves a residual level of elevated soil test P and/or K levels in the immediate region of the band.[2] If the bands are related in direction to the row stubble after harvest, it is relatively simple for a sampler to avoid the enriched band under the stubble after the first year. However, after the second year, the sampling strategy is more difficult. Some studies have suggested the use of a sample core transect taken perpendicular to the row direction[2] and increasing the number of sample cores for each sample composite. This strategy would lend itself to grid-point sampling, rather than grid-cell sampling.

4.2.3 Collecting Composite Samples and Grid Density

Once the basic grid strategy has been chosen, it is necessary to consider the protocols for obtaining each individual composite sample. Surface and subsurface composite samples might have different protocols. Under most situations, a composite sample should not consist of a single core. Typically, between 5 and 8 individual samples should be composited for each surface sample. Five might be used when banding is not an issue, while eight or more might be prudent in fields with a history of banding. In many situations, fewer cores are composited from subsurface samples. Many practitioners composite from one to five soil cores for subsoils, when the analyses are intended for nitrates, sulfates, and chlorides. This recommendation depends on the diameter of the sampling core tube and field variability. When first sampling a field, erring on the high side of sample core number for each composite sample is advised.

Grid-point sampling uses the grid point identified by the sampler as the center of a small area, usually not more than 3 m radius, to obtain the additional two to eight soil cores that will represent the grid-point composite. The basis for using a grid point is that it addresses small-scale variability better than grid-cell sampling and is practically faster to obtain the sample cores. In grid-cell sampling, the additional two to eight sample cores are obtained randomly throughout the cell, although some guidelines limit the area to an 80 ft radius around the grid-point location. Most studies examining the strategies have found grid cell to produce better data.[13,14]

A number of studies have examined the effect of grid soil sample density on the ability to represent the "true" status and pattern of nutrients in a field. Wollenhaupt et al.[13] determined a 60 m grid was necessary to represent field P and K patterns. Franzen and Peck[15] found that a 67 m grid, or about 1 sample 0.405 ha^{-1}, was necessary to represent field pH, P, and K levels in two 12.5 ha fields. Ferguson and Hergert[16] recommended that in Nebraska, the optimum grid sampling density is 1 sample 0.405 ha^{-1}. The expense and time required for such an intensive sampling has led many growers to use a less dense sampling protocol. To maximize returns, appropriate protocols should be used. In regions where soil test P or K levels are high, a lower sample density would be expected to provide similar recommendations to the grower. However, in fields where lower nutrient levels are present, the lower density might result in significant underfertilization, whether the recommendations are based on either a buildup maintenance or sufficiency approach.[17]

To illustrate the effect of sample density on soil fertility patterns, Figure 4.6 shows a plot of soil pH values of soil samples taken to a 15.2 cm depth in an Illinois field near Mansfield, about 32 km west of Champaign. The data are part of a long-term sampling project terminated in 2001 but begun in 1961. The entire data set from the project that also includes soil P and K data as well as some years of corn and soybean grid yield data can be found at www.ndsu.soilsci/faculty/franzen/franzen.

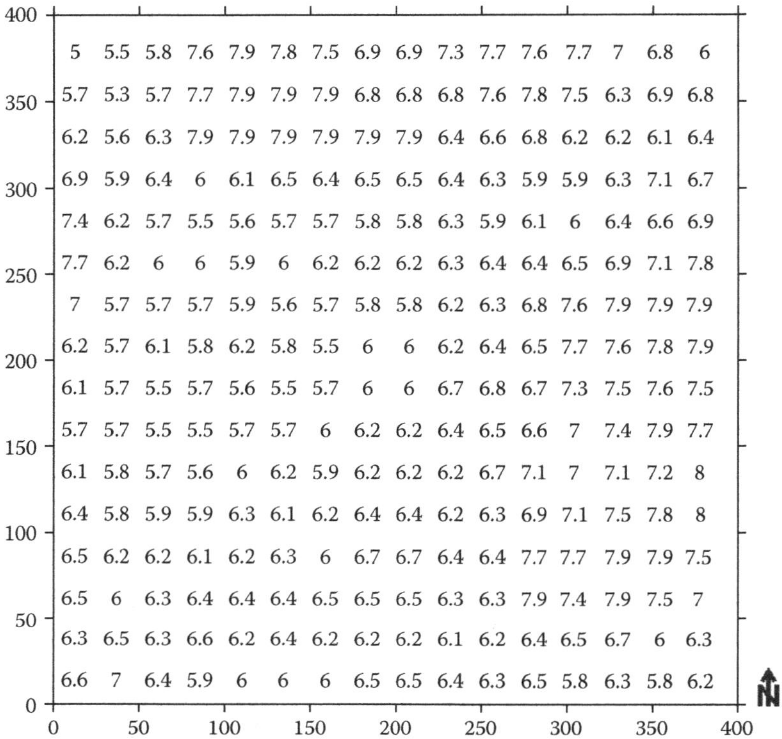

FIGURE 4.6 12.5 ha field at Mansfield, IL, with 24.4 m grid soil pH data displayed.

The soil pH data show that substantial pH variability can exist in production fields (Figure 4.6). Data for this field are provided on the data disk provided with this book. If the 255 data points were mapped using kriging, a map similar to Figure 4.7 would be produced. This map can be created using protocols described previously.[10] The sampling intensity influences the type of map produced. Reducing the number of samples increases error and changes the resulting map (Figure 4.9).

The 24.3 m sampling in Figure 4.6 was conducted in a regular systematic grid. If we select only those samples from Figure 4.6 that represent those that might be taken in a 67 m grid (36 samples/40 ac, or about 1.1 ac/sample (Figure 4.8), the resulting map would be similar to that in Figures 4.9 through 4.11.

As the sample density decreases, the original features of areas of high to low pH from Figure 4.7 change. Areas of relatively similar nutrient values of large relative size may still be recognizable; however, the boundaries between different soil levels change. Some smaller features are lost, while other features are enhanced in size. Selecting the appropriate sampling protocol prior to sampling is difficult. Yield maps, EC surveys, and remote sensing data can assist in these decisions.

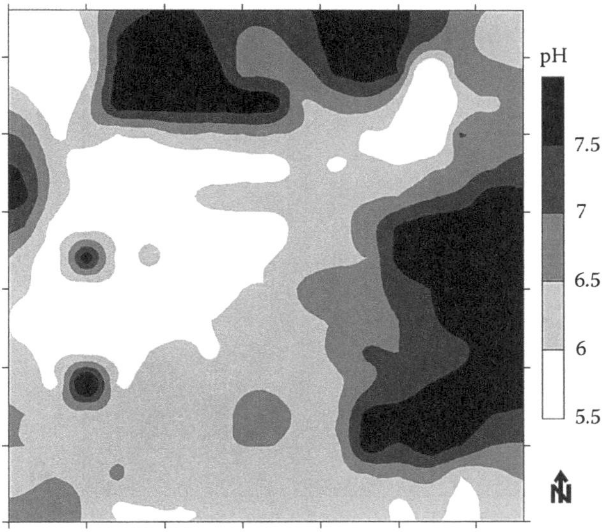

FIGURE 4.7 Contour map using kriging interpolation of soil pH at Mansfield, IL, using the 24.4 m sampling grid values from Figure 4.6.

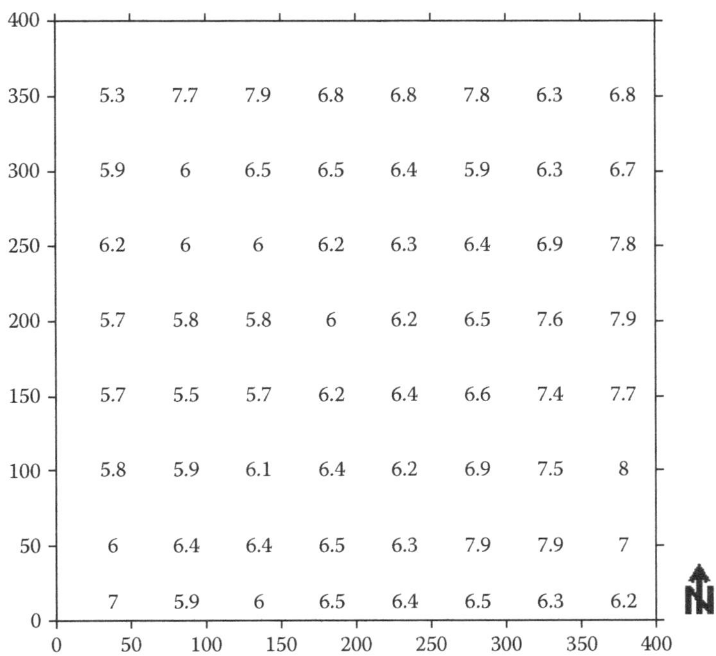

FIGURE 4.8 Plot of Mansfield with samples representing a 5.1 samples ha^{-1} grid.

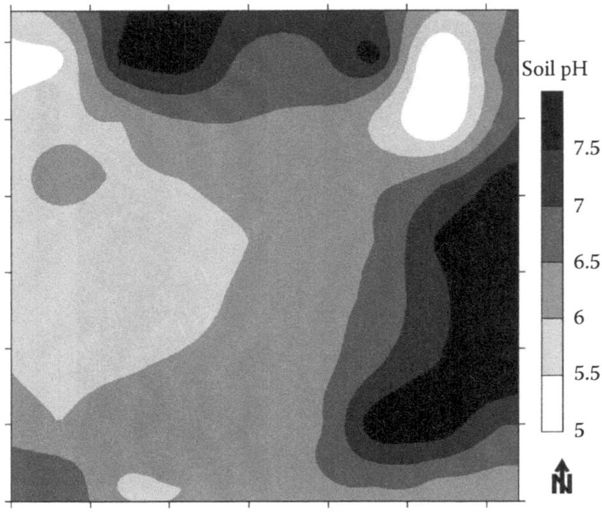

FIGURE 4.9 Mansfield kriged interpolation contour map of soil pH using the 5.1 samples ha^{-1} data from Figure 4.8.

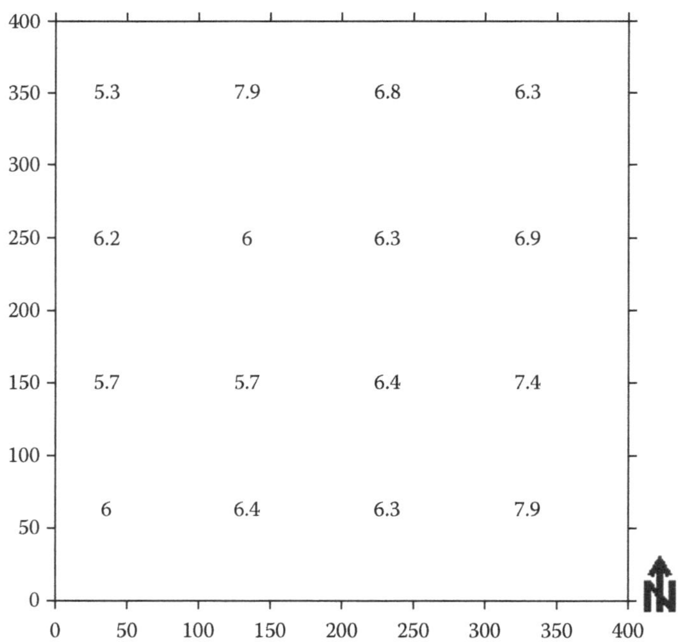

FIGURE 4.10 Mansfield soil pH from Figure 4.6 data set, with only values representing 1.3 samples ha^{-1} plotted.

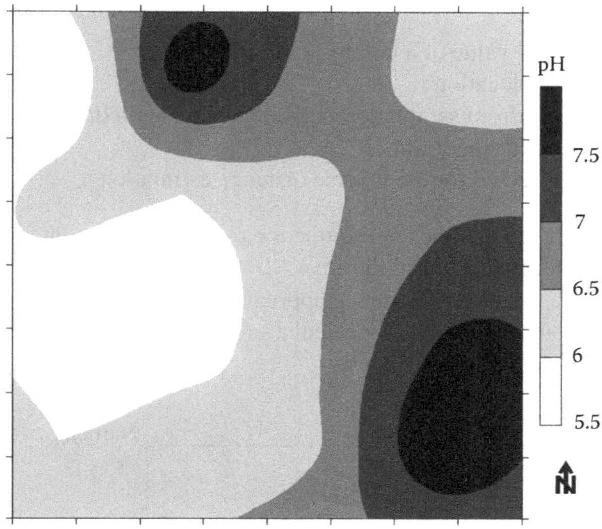

FIGURE 4.11 Mansfield, IL, soil pH kriged contour map of the 1.3 samples ha^{-1} data from Figure 4.10.

4.2.4 Interpolation Techniques

A number of studies have investigated the appropriate data interpolation methods for soil sample data. Interpolation methods used in the literature include inverse distance, kriging, triangulation, and spline. Inverse distance, triangulation, and spline methods are mathematical models that do not possess the inherent ability to estimate the error associated with the estimate. Kriging is a method that intrinsically carries along the data structure to provide a confidence in the estimates produced.

Triangulation is described in several texts, including Isaacs and Srivastava[18] and Davis.[19] In triangulation, the data are related as the corners of triangles. Using easting of the three data points as "a," northing as "b" and a constant as "c," the equation of a plane $z = ax + by + c$ is determined by solving the three linear equations for each of the three triangle corners. Once the equation for the plane is developed, substitution of the coordinates for an unknown point within the triangle will result in the unknown point estimated value. Wollenhaupt et al.[13] tested a version of triangulation, Delaunay triangulation, and found that it fit nutrient level boundaries similar to hand-drawn polygons. The difference between triangulation and Delaunay triangulation is that Delaunay triangulation fits the triangle corners as close to equidistant as possible, which may increase the accuracy of the estimate in some data sets.

Inverse distance to a power estimation is a relatively simple method of estimating a specific value of unsampled locations.[18] The equation for a given estimate using the inverse distance method is

$$z^{*}_{(so)} = \frac{\left(\sum z_i d_{o,i}^{-p}\right)}{\sum d_{o,i}^{-p}}$$

where

z^* is the estimated value of a point not sampled
z_i is the value at a location i
$d_{o,i}$ is the distance from sampled point location to the ith data location s_i
Σ is a symbol signifying "sum of"
p is the power selected for the inverse distance estimation

Figure 4.12 will serve as the data source for a simple illustration for the estimation of the value at X located at 1,2 in Figure 4.12.

To demonstrate the inverse distance approach, it is easiest to set up a table (Table 4.1) and work through the following calculations. Of course, computer software performs these tasks nearly instantaneously.

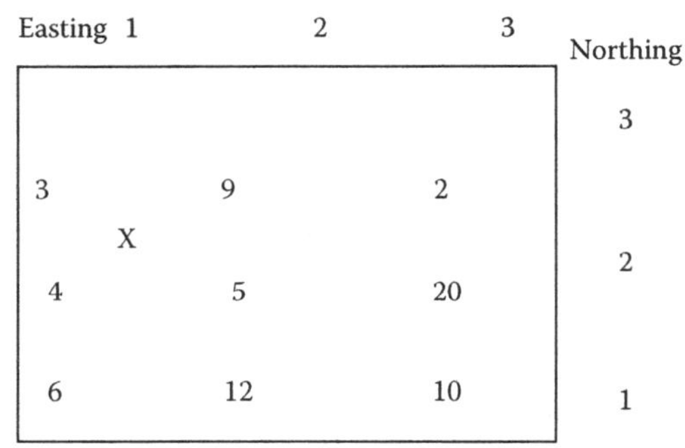

FIGURE 4.12 Illustration data set to be used for determining the value of "X" by inverse distance methods. Assume data points are 1 unit EW/NS and X is centered between four points.

TABLE 4.1
Template to Logically Address Data Input and Calculation for Inverse Distance Interpolation of a Data Set

			$z_i d_{o,i}^{-p}$	
z_i	$d_{o,i}$	$d_{o,i}^{-1}$	First Power	Second Power
3				
9				
4				
5				
2				
20				
6				
12				
10				
Sums				
	Value of unknown			

TABLE 4.2
Distance, Inverse Distance and Inverse Distance Squared for Our Example Data Set and Point 'x'

				$z_i d_{o,i}^{-p}$	
z_i	$d_{o,i}$	$d_{o,i}^{-1}$	$d_{o,i}^{-2}$	First Power	Second Power
3	0.707	1.414	2.0		
9	0.707	1.414	2.0		
4	0.707	1.414	2.0		
5	0.707	1.414	2.0		
2	1.58	0.633	0.40		
20	1.58	0.633	0.40		
6	1.58	0.633	0.40		
12	1.58	0.633	0.40		
10	2.12	0.472	0.22		
Sums					

After listing the z_i's, the distance from each z_i to the unknown is listed. Once this computation is made, we also calculate the inverse of this distance to the first ($d_{o,i}^{-1}$) and second ($d_{o,i}^{-2}$) powers and place the values in Table 4.2.

The next step is to calculate $z_i d_{o,i}^{-p}$ for each z_i. Multiplying the values in the appropriate columns together, we get the following values under the first power and second power columns (Table 4.3).

The next step is to sum the distances to a power for each z_i and also the $z_i d_{o,i}^{-p}$ for each power (Table 4.4).

TABLE 4.3
Values from Table 4.2, with Distance × Distance to a Power Calculated

				$z_i d_{o,i}^{-p}$	
z_i	$d_{o,i}$	$d_{o,i}^{-1}$	$d_{o,i}^{-2}$	First Power	Second Power
3	0.707	1.414	2.0	4.242	6.0
9	0.707	1.414	2.0	12.726	18.0
4	0.707	1.414	2.0	5.656	8.0
5	0.707	1.414	2.0	7.07	10.0
2	1.58	0.633	0.40	1.266	0.8
20	1.58	0.633	0.40	12.66	8.0
6	1.58	0.633	0.40	3.80	2.4
12	1.58	0.633	0.40	7.596	4.8
10	2.12	0.472	0.22	4.72	2.2
Sums					

TABLE 4.4
Values from Table 4.3, with Sums

				$z_i d_{o,i}^{-p}$	
z_i	$d_{o,i}$	$d_{o,i}^{-1}$	$d_{o,i}^{-2}$	First Power	Second Power
3	0.707	1.414	2.0	4.242	6.0
9	0.707	1.414	2.0	12.726	18.0
4	0.707	1.414	2.0	5.656	8.0
5	0.707	1.414	2.0	7.07	10.0
2	1.58	0.633	0.40	1.266	0.8
20	1.58	0.633	0.40	12.66	8.0
6	1.58	0.633	0.40	3.80	2.4
12	1.58	0.633	0.40	7.596	4.8
10	2.12	0.472	0.22	4.72	2.2
Sums		8.66	9.82	59.736	60.2

The resulting estimate varies depending on what power is selected to represent the equation. Choosing inverse distance to the first power results in 59.736/8.66 or **6.9**. Choosing inverse distance to the second power results in 60.2/9.82 or **6.13**.

From the exercise tables, note that the weight of closest values increase with increasing power. Inverse distance squared (IDS) is often chosen because inverse distance to the zero power and first power tend to place too much weight on points farther away, while the third and fourth powers and higher tend to progressively ignore all but the closest points. Intuitively, one would expect points closer to be more related to unsampled points than points farther away. This is one reason many practitioners use IDS.

Spline methods are more rigorous than either triangulation or inverse distance. The equations are sufficiently rigorous that most researchers tend to go the extra step and utilize kriging to determine estimates rather than use splines. Spline methods are described in Wahba,[20] and briefly outlined in Wollenhaupt et al.[7]

Kriging is the general term for geostatistics that use a covariate with distance (semivariance) to construct a matrix of linear equations used to determine the mathematical weights for values and distances between them. The term "kriging" is named for a South African gold mining geologist, Daniel Gerhardus Krige. The system of kriging was first presented by a Canadian geologist Matheron in an article in Economic Geology in 1963. The system can be used to generate the error and confidence in each point and some researchers also map confidences as well as estimates of values for an area. A good general reference on kriging is presented in Isaacs and Srivastava,[18] along with the basic kriging equations. For more rigorous examination of the kriging techniques, see Journel and Huijbregts[21] and Cressie.[22] Many researchers find that understanding and viewing a semivariogram is important in understanding the spatial relationships of soil property data sets. A semivariogram is the graph of the inverse covariance between distance (h) and the semivariance value for that distance.

When the semivariance (γ) is plotted against its distance (h), the resulting points may be fit using a variety of models. The selection of the model is somewhat

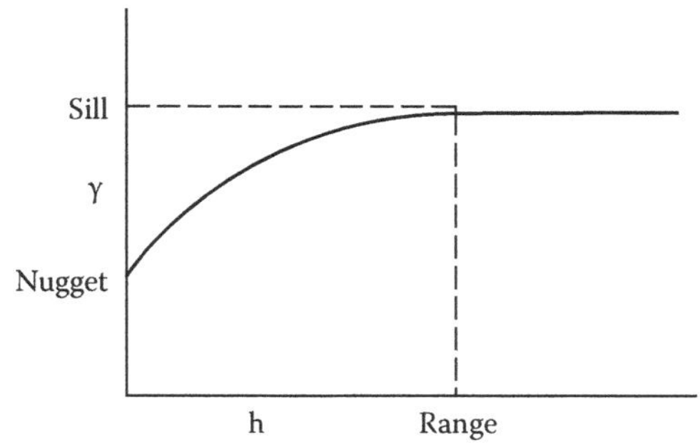

FIGURE 4.13 Idealized variogram versus distance relationship showing nugget, range, and sill.

subjective. The person fitting the data often has the choice of what range of data to fit and the type of equation with which to fit the data. The illustration in Figure 4.13 shows an approximation for a spherical curve. These data cross the *y*-axis at a point designated "nugget," which represents the uncertainty beyond spatial relationships in the data. The term comes from the gold mining industry where concentration of gold might be predicted, but the presence of a nugget of gold is an unexpected and random occurrence. The sill is the maximum semivariance that is related to distance. The range is the maximum distance where points have some spatial relationship with each other. Some researchers use the range as a search radius, while others use 1.5 or 2 times the range as a search radius. The variogram is mysterious to many agricultural researchers, so the following exercise may help to explain some principles into its origin, structure, and usefulness (Figure 4.14).

$$\text{Semivariance} = \gamma = \frac{1}{2N_h}\sum_{i=1}^{i=n}\left[Z_{xi} - Z_{xi+h}\right]^2 \tag{4.1}$$

The first step in estimating the unknown point is graphing the variogram function over distance. Once that is done, the kriging equation uses that information to estimate the unknown.

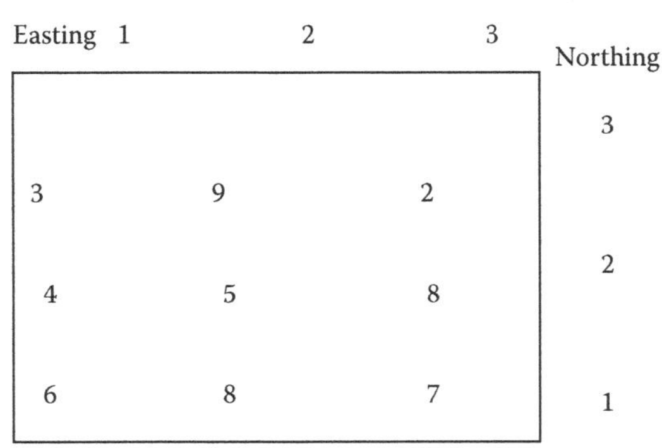

FIGURE 4.14 Illustration data set to be used for determining a simple variogram. Assume data points are 1 unit EW/NS.

We can list the difference between each of values at a certain distance away from each other. We look for pairs of values first a distance of 1 away. Other distances in our data set are those that are 1.4, 2.0, 2.2, and 2.8 away from each other. The table already shows how many (N) pairs of points are certain distances away from each other. For small data sets, such as this one, and many commercial soil testing data sets, the number of choices of distance pairs is small. A sampling strategy such as systematic unaligned grid helps to increase the number of distance pairs dramatically from a regular grid, but small data sets limit both distance and value pairs. The larger the data set, the greater confidence there will be in any estimate (Tables 4.5 and 4.6).

TABLE 4.5
To Determine the Variogram of Our Example Data Set[a]

	h = 1	h = 1.4	h = 2.0	h = 2.2	h = 2.8
	$(3-9)^2$	$(3-5)^2$	$(3-6)^2$	$(3-8)^2$	$(3-7)^2$
	$(9-2)^2$	$(9-8)^2$	$(9-8)^2$	$(4-2)^2$	$(2-6)^2$
	$(4-5)^2$	$(9-4)^2$	$(2-8)^2$	$(6-8)^2$	
	$(5-8)^2$	$(2-5)^2$	$(3-2)^2$	$(4-7)^2$	
	$(6-8)^2$	$(4-8)^2$	$(4-8)^2$	$(3-8)^2$	
	$(8-7)^2$	$(5-6)^2$	$(6-7)^2$	$(9-6)^2$	
	$(3-4)^2$	$(8-8)^2$		$(9-7)^2$	
	$(9-5)^2$	$(5-7)^2$		$(2-8)^2$	
	$(2-8)^2$				
	$(4-6)^2$				
	$(5-8)^2$				
	$(8-7)^2$				
Sums	167	60	64	113	32

[a] Each column represents a possible distance between data points. Beneath each column is the possible result of squaring the difference between a point value and a value a certain distance away from the point $[(Z_{(xi)}-Z_{(xi+h)})^2$ for each distance with sums under columns].

TABLE 4.6
Using the Data from Table 4.5, the Variogram is Calculated for Each Possible Distance from Our Example Data Set

Distance (h)	$N_{(h)}$	$1/(2N_{(h)})$	$\Sigma(Z_{(xi)}-Z_{(xi+h)})^2$	$\gamma^*_{(h)}$
1	12	1/24	167	6.96
1.4	8	1/16	60	3.75
2.0	6	1/12	64	5.33
2.8	2	1/4	32	8.0
2.2	8	1/16	113	7.1

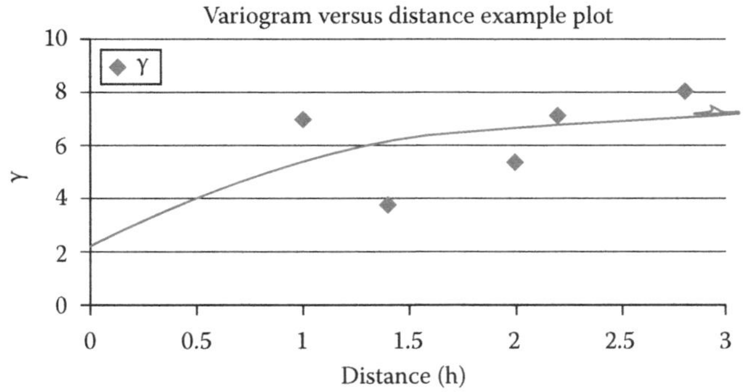

FIGURE 4.15 Example variogram versus distance plot with spherical relationship imposed.

Multiplying the $1/(2N_{(h)})$ times the $\Sigma(Z_{(xi)} - Z_{(xi+h)})^2$ for each distance gives the variogram ($\gamma^*_{(h)}$) for each distance. When the $\gamma^*_{(h)}$ are plotted with their distance, the result is the variogram versus distance relationship that can be visualized (Figure 4.15).

Wollenhaupt et al.[13] evaluated IDS, two types of kriging, and a polynomial trend surface interpolation against Delaunay triangulation as a benchmark. All of the interpolations performed well. The data sets consisted of over 200 samples each in two separate fields. Interpolations were made from the 106 ft original data set, a 212 ft grid, and a 318 ft grid.

Their decision to use Delaunay triangulation is based largely on a previously published example[18] that found triangulation provided a smaller variance than the inverse distance and was not as affected by data clustering.

Similarly, kriging was found to be a better interpolator compared to inverse distance when imposing 99 samples in hexagonal (staggered start), "inhibited," random, and clustered computer-generated patterns on a plane and two complex computer-generated surfaces.[23] The study also found that better interpolation was possible when "noise" in the data was low. Isaacs and Srivastava[18] also commented that inverse distance performs better with unclustered data.

Franzen and Peck[15] found that IDS produced similar maps for soil pH, P, and K in two Illinois fields sampled in a regular 80 ft grid and interpolated from a 220 ft and 330 ft grid for each factor. Gotway et al.[24] compared kriging with inverse distance on 12 data sets from two fields sampled in a triangular grid from 0.02 to 0.05 ac/grid depending on the site. The better method depended on the data set; however, they found that when the data coefficient of variation (CV) was less than 25%, inverse distance to a high power (4) was a better choice compared with kriging. When the data sets had higher CVs, kriging tended to have better interpolation.

Mueller et al.[25] compared the map quality produced by ordinary kriging and inverse distance on grid sample data (50 ft) from five fields in Kentucky. He concluded that for those data sets, inverse distance 1.2 (power of 1.2) was a better predictor of interpolated points from a 100 and 200 ft grid than ordinary kriging.

Geostatistics may not improve the estimates if the semivariogram does not contain spatial structure. In many situations, sampling on a 2.5 ac grid does not provide enough information to describe the spatial structure. Under these conditions,

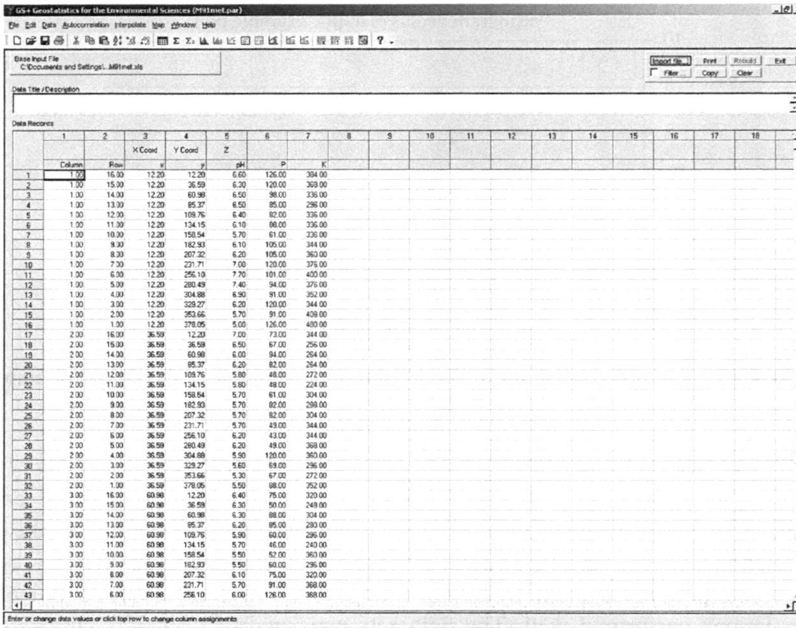

FIGURE 4.16 GS+ 7 for Windows opening data view of the Mansfield, IL, 1991 soil pH, P, and K data, with the *x*, *y*, and *z* columns chosen.

IDS may be good as interpolator of unsampled as kriging. Inverse distance and kriging are two interpolations supported by both Surfer® (Golden Software Co., Golden, CO) and ArcGIS 9.2 (ESRI®, Redland, CA). For most GIS software, semivariogram construction must be conducted on additional software. One of the most common softwares is GS+ 7® for Windows by Gamma Design Software, LLC, Plainwell, MI. It is also possible to produce one's own geostatistical variogram software.

As an example, Figure 4.16 shows a txt file, surfer x, y, z file, or excel file imported into GS+ 7 for Windows. By clicking on the appropriate northing and easting columns and clicking on to a choice of z values, in this case soil pH in column 5, the program is ready to calculate the variogram and its model.

Figure 4.17 shows the defaults of the autocorrelation analysis of the Mansfield, IL, 1991 soil pH data. The program defaults the active lag distance (the distance it notes is relevant to the search) and the lag class distance interval (the distance between interpolation estimate nodes on the chart). Both of these defaults can be changed by the user. The model information for the variograms can also be displayed and the defaults changed (Figure 4.18). The resulting nugget, sill, and range should be recorded for use in any kriging interpolation programs that might be used. GS+ 7 for Windows also has mapping capabilities, but many people prefer to use the parameter information to map in another application (Figure 4.19).

ArcGIS or ArcMap contains, or has the ability to contain, what it designates as a "tool box" containing spatial analyst tools. These tools contain several computational aids such as hydrology parameters, groundwater, and data interpolation calculation tools. In Figure 4.20, we have selected the interpolation tool and can now

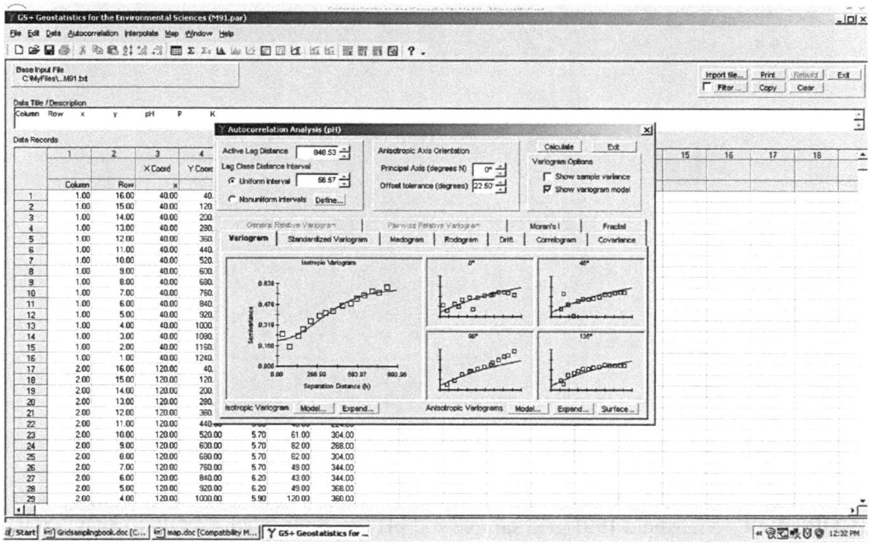

FIGURE 4.17 GS+ 7 for Windows screen showing the initial results of choosing <Autocorrelation>, then <variogram>, then <primary variable> from the top bar of the page shown behind pop-up window of autocorrelation analysis.

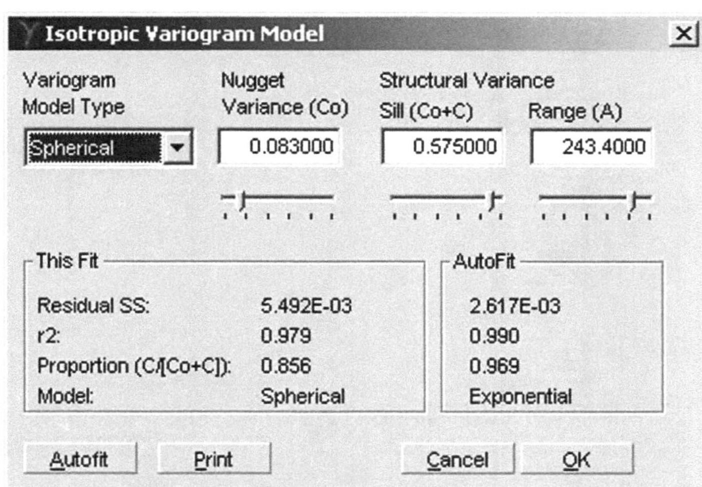

FIGURE 4.18 GS+ 7 for Windows screen of adjusted defaults of active lag distance. Also, the variogram model pop-up has been selected, the variogram type changed to spherical, and nugget, sill, and range parameters have been adjusted to maximize the model r^2.

proceed to interpolate using a choice of tools including IDW, kriging, spline, or others. From here, the options are straightforward. Select the kriging option; type in the parameters of nugget, sill, range, and spherical model obtained from GS+ 7; and select OK. The kriging equations will be computed and the contoured results of interpolation will appear similar to the Surfer maps in Figures 4.7, 4.9, and 4.11.

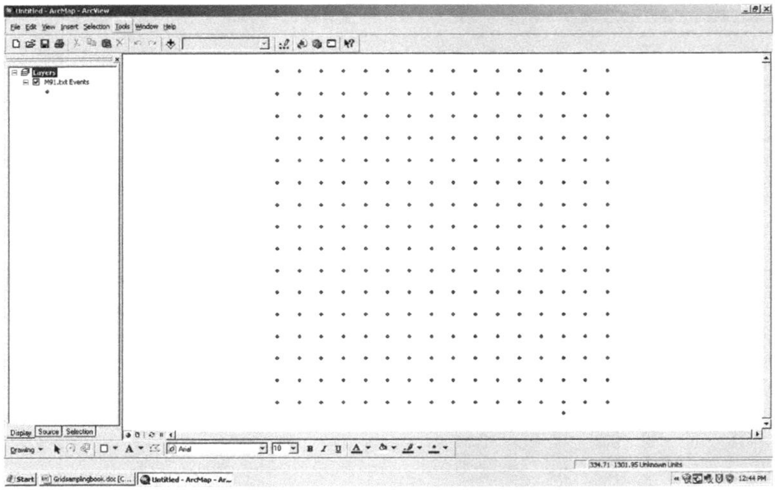

FIGURE 4.19 Mansfield 1991 grid sample soil pH locations in an ArcGIS display, showing the results of <Tools> <Add x, y, z data>, and choosing a txt version of the Mansfield 1991 soil pH.

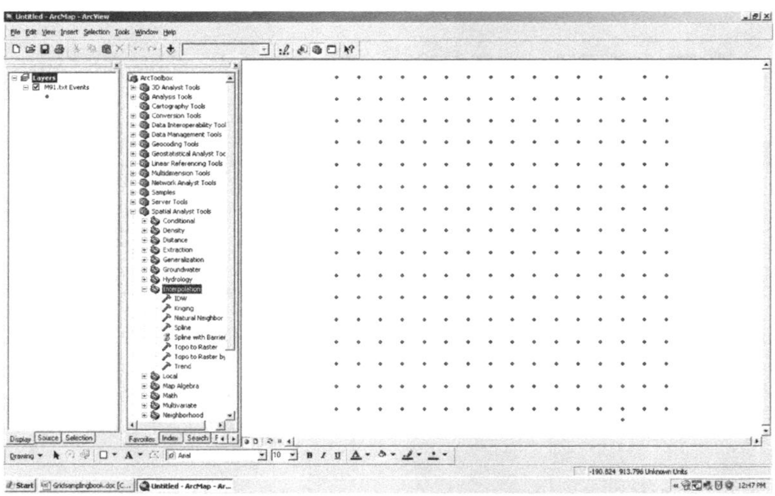

FIGURE 4.20 The results of choosing <Window> <Toolbox> <Spatial Analyst Tools> and <Interpolation> in ArcGIS.

4.3 SUMMARY

By adopting efficient sampling strategies, energy efficiency and profitability can be increased. Grid sampling is a tool that is used to reveal soil fertility patterns in farm fields. The grid sampling must be composed of a reasonable sampling strategy, including proper pattern, number of sample cores per sample, and the proper configuration of subsample collection within each grid or sampling polygon. This chapter discusses the benefits and negatives associated with several different sampling approaches. When designing a sampling protocol, consideration should be given to the interpolation process to create a recommendation map for the grower. Data and directions for calculating IDW factors and semivariances are provided. Several software packages are available that can compute interpolation, although some require parameters for kriging equations be developed outside of their software. No map should be considered "perfect," for smaller-scale variation of soil properties is usually hidden from view; however, properly gridded sample results are usually an improvement from a whole-field sampling approach.

REFERENCES

1. Peck, T.R. and P.N. Soltanpour. The principles of soil testing. In R.L. Westerman, ed. *Soil Testing and Plant Analysis*, 3rd edn. Soil Science Society of America Book Series No. 3. Soil Science Society of America, Madison, WI, 1990, pp. 1–9.
2. Mallarino, A.P. Spatial variability patterns of phosphorus and potassium in no-tilled soils for two sampling scales. *Soil Sci. Soc. Am. J.* 60, 1473–1481, 1996.
3. Franzen, D.W., A. Halvorson, and V. Hofman. Variability of soil nitrate, phosphate, chloride and sulfate-S under different landscapes—A report of work in progress. *Extension Report 35*. NDSU Extension Service, Fargo, ND, 1997.
4. Borges, R. and A.P. Mallarino. Variation of early growth and nutrient content of no-till corn and soybean in relation to soil phosphorus and potassium supplies. *Commun. Soil Sci. Plant Anal.* 29, 2589, 1998.
5. Mallarino, A.P. and D.J. Wittry. Efficacy of grid and zone soil sampling approaches for site-specific assessment of phosphorus, potassium, pH, and organic matter. *Prec. Agric.* 5, 131, 2004.
6. Webster, R. and M.A. Oliver. *Statistical Methods in Soil and Land Resource Survey*. Oxford University Press, New York, 1990.
7. Wollenhaupt, N.C., D.J. Mulla, and C.G. Gotway-Crawford. Soil sampling and interpolation techniques for mapping spatial variability of soil properties. In F.J. Pierce and E.J. Sadler, eds. *The State of Site-Specific Management for Agriculture*. ASA, CSSA, SSSA, Madison, WI, 1997, pp. 19–53.
8. Russel, D.A. and D.G. Williams. History of chemical fertilizer development. *Soil Sci. Soc. Am. J.* 41, 260, 1997.
9. Fulton, J.P., M.W. Veal, and S.A. Shearer, Performance assessment of variable-rate fertilizer technology. *Site-Specific Management Center Newsletter*, Purdue University, April 2006.
10. Franzen, D.W. Summary of forty years of grid sampling in two Illinois fields. NDSU Technical Bulletin, Fargo, ND, 2008.
11. Fulton, J.P., S.A. Shearer, T.S. Stombaugh, and S.F. Higgins. Simulation of variable-rate application of granular materials. ASAE Paper No. 021186. *ASAE Annual International Meeting*, July 28–31, Chicago, IL, 2002.
12. Fulton, J.P., S.A. Shearer, T.S. Stombaugh, and S.F. Higgins. Comparison of variable-rate granular application equipment. ASAE Paper No. 031125. *ASAE Annual International Meeting*, July 27–30, Las Vegas, NV, 2003.
13. Wollenhaupt, N.C., W.P. Wolkowski, and M.K. Clayton. Mapping soil test phosphorus and potassium for variable-rate fertilizer application. *J. Prod. Agric.* 7, 441, 1994.
14. Kastens, T., V.K. Shuyvetter, and J. Schmidt. Point vs. area grid soil sampling in the Great Plains. In Schlegel, A.J., *Proceedings of the 2004 Great Plains Soil Fertility Conference*, March 2–3, 2004, Denver, CO. Potash & Phosphate Institute, Brookings, SD, 2004, pp. 59–68.

15. Franzen, D.W. and T.R. Peck. Field soil sampling density for variable rate fertilization. *J. Prod. Agric.* 8, 568, 1995.
16. Ferguson, R. and G. Hergert. Soil sampling for precision agriculture. University of Nebraska Cooperative Extension EC 00-154, 2000.
17. Dahnke, W.C. and R.A. Olson. Soil test correlation, calibration and recommendations. In R.L. Westerman, ed. *Soil Testing and Plant Analysis*. SSSA Book Series No. 3. Soil Science Society of America, Madison, WI, 1990, pp. 45–71.
18. Isaacs, E.H. and R.M. Srivastava. *An Introduction to Applied Geostatistics*. Oxford University Press, New York, 1990.
19. Davis, J.C. *Statistics and Data Analysis in Geology*, 2nd edn. John Wiley & Sons, New York, 1986.
20. Wahba, G. *Spline Models for Observational Data*. SIAM, Philadelphia, PA, 1990.
21. Journel, A.G. and C.J. Huijbregts. *Mining Geostatistics*. Academic Press, London. 1978.
22. Cressie, N.A. *Statistics for Spatial Data*, revised edn. John Wiley & Sons, New York, 1993.
23. Zimmerman, D., C. Pavlik, A. Ruggles, and M.P. Armstrong. An experimental comparison of ordinary and universal kriging and inverse distance weighting. *Math. Geol.* 32, 375, 1999.
24. Gotway, C.A., R.B. Ferguson, G.W. Hergert, and T.A. Peterson. Comparison of kriging and inverse-distance methods for mapping soil parameters. *Soil Sci. Soc. Am. J.* 60, 1237, 1996.
25. Mueller, T.G., N.B. Pusuluri, K.K. Mathias, P.L. Cornelius, R.I. Barnhisel, and S.A. Shearer. Map quality for ordinary kriging and inverse distance weighted interpolation. *Soil Sci. Soc. Am. J.* 68, 2042, 2004.

5 Integration of USDA-NRCS Web Soil Survey and Site Collected Data

Kurtis D. Reitsma and Douglas D. Malo

CONTENTS

5.1 EXECUTIVE SUMMARY

Advancements in computer technology and global communication have increased the availability of a wide variety of digital information. Within the last several years, a number of organizations have developed digital soil data and systems that make these data available to the public. The ability to predict potentials and constraints of global food and fiber production and soil carbon measurements prompted the United Nations, Food Agriculture Organization (FAO), and International Institute for Applied Systems Analysis (IIASA) to develop the Harmonized World Soil Database.[7] The Harmonized World Soil Database provides a wide variety of data suitable for use on a large scale. Canada and Austria have developed similar soil databases and are available via the Internet at http://sis.agr.gc.ca/cansis/nsdb[1] and http://www.asris.csiro.au[2] respectively. Continental, regional, or land survey data may be available and can be found using a common Internet search engine.

Smaller local scale data sets are best accessed from local, state, provincial, or federal government data gateways.

In the United States, the U.S. Department of Agriculture, Natural Resource Conservation Service (USDA-NRCS) maintains a number of data gateways, providing soils and other tabular and spatial data products.[5,9] The most recent development is the Web Soil Survey (WSS), an interactive online interface providing user-defined reports and geographic information system (GIS) ready spatial data. This chapter guides the users through the process of integrating soil data retrieved from WSS (ver. 2.2.6, 2009)[8] with information and data collected on-site. All data are expressed in U.S. English units.

5.2 INTRODUCTION

5.2.1 Data Retrieval Systems

Methods and scale of conducting and developing soil surveys vary throughout the world, increasing the difficulty in comparing soils on a continental or global scale. Additionally, protocols for storage and delivery are often unique to a global region. For example, the goal of the Harmonized World Soil Database is to provide data at a resolution of approximately 1 km. In the United States, users can access various gateways providing generalized soil data (e.g., U.S. General Soil Map, formerly STATSGO)[9] or more detailed data (e.g., Soil Survey Geographic Database, SSURGO).[9] The U.S. General Soil Map database provides large-scale generalized information similar to the Harmonized World Database, while SSURGO provides small-scale detailed (SSURGO) spatial data with minimum size delineations of 0.6–4 ha. Spatial and tabular data provided by WSS are based on SSURGO, having scales ranging from 1:12,000 to 1:31,680. Publically available soil surveys were not developed for site-specific management purposes (http://soils.usda.gov/) but were meant to be used for planning purposes.

First released in 2005, the most current WSS version (ver. 2.2.6) includes interactive help, advanced search, linear measurement tool, explanation of interpretive values and data, and many other features. Detailed information is available on the WSS homepage (http://websoilsurvey.nrcs.usda.gov).

This chapter guides the users through the process of integrating soil information obtained from USDA-NRCS WSS (WSS, ver. 2.2.6, 2009) with site collected information. Due to the robust nature of the WSS, an Internet connection of 100 Mb/s or greater is recommended. Minimum system recommendations include IBM-PC Pentium® 4 or higher processor, with 1 Gb RAM, operating system, Microsoft Windows (Windows 7, Vista, or XP Service Pack 2), installed software, MS Excel ver. 2003 or higher, ESRI® ArcGIS™ (ver. 9.3).[3,6]

5.3 GETTING STARTED WITH WSS

A clear objective for the project should be defined before starting WSS. In this case study, the objective is to determine the land capability classification of nine 640 ac sections of a township (T 120N, R 55W, 5th Principle Meridian) located in

Day County, South Dakota. Generally, data are assembled and retrieved using the following steps:

1. Locate the Web site and navigate to the general vicinity of the area of interest (AOI).
2. Define boundaries of the study areas.
3. Define the type of soil attribute data and method of aggregation.
4. Prepare map layers and datasets of various selected soil properties/characteristics.
5. Electronically store and/or print reports of data.
6. Retrieve spatial and/or tabular data.

Spatial and complete tabular datasets are delivered by accessing an Internet data server at an address provided in an e-mail sent to the user. The WSS uses an Internet map server as an interface to aggregate tabular data and display user-defined map layers. We recommend calibrating your screen to the map scale to achieve maximum functionality of the interface. Screen calibration is performed using the following steps:

1. Open your web browser and navigating to the WSS (http://websoilsurvey.nrcs.usda.gov/) and click "Start Web Soil Survey" .
2. Click the Scale button; a dialog box opens with instructions for screen calibration.
3. Adjust the bar on the screen to 1 in. and click OK.

5.3.1 Navigating and Defining the AOI

The zoom tool can be used to locate the AOI, but users can quickly navigate to a known location using the "Quick Navigation" tools located to the left of the map display. Select a physical address, State/County, soil survey area, latitude/longitude coordinate, PLSS (Public Land Survey System—section, township, range), BLM (Bureau of Land Management) field office, DoD (Department of Defense) military installation, U.S. Forest Service (USFS) unit, national park, or hydrologic unit code (eight-digit hydrologic unit codes [HUCs]). The PLSS, BLM unit, and HUC are largely unique to the United States. The PLSS is a method for surveying and identifying land parcels for titles and deeds in most of the United States. The PLSS divides the land into 6 mi^2 townships with 36 1 mi^2 sections that may be further subdivided; more information is available using WSS interactive help ?. HUCs are a method of identifying drainage basins in the United States; more information about HUCs in the United States is available from the U.S. Geological Survey (http://water.usgs.gov/GIS/huc.html).[10] Selecting BLM field office, DoD military installation, or USFS unit zooms to lands under control of the respective agency in the United States.

Tool categories in the WSS can be expanded or collapsed by clicking on the title bar or double arrow Quick Navigation. Nearly all tool categories in the system work in this manner. Using the "quick navigation" tools is

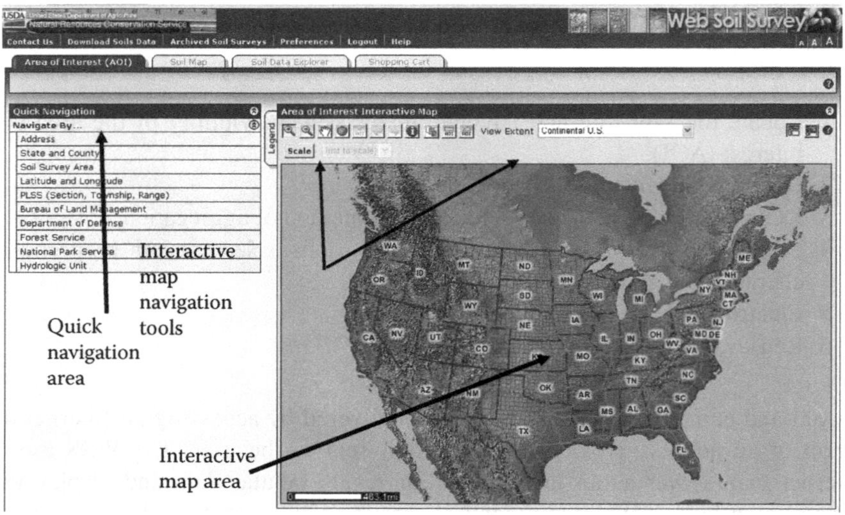

FIGURE 5.1 WSS AOI selection with quick navigation and interactive map options.

recommended as WSS will quickly zoom to the vicinity of the AOI. However, selection of the AOI is limited to 10,000 ac. Figure 5.1 pro vides a glimpse of the WSS environment.

In the following example, an area located in Day County, South Dakota, will be selected by

1. Selecting "PLSS," and entering "120" and "55" into the "Township" and "Range" input boxes respectively, select the "N" and "W" radio button appropriately. In the "Principle Meridian" combo box, select "Fifth Principle." If in doubt as to what to select as the principle meridian, click the View Meridian Map button and click in the shaded region where the AOI occurs, the combo (drop-down) box will automatically fill with the meridian selected.
2. Click on View, note that the selected township appears in the center of the map window but is slightly oversize. The simplest AOI selection method is rectangular extent, but a more precise AOI selection is possible by using the polygon definition tool. The map view should be slightly wider than the AOI.
3. Click on "Define areas of interest by Polygon" tool directly above the map view.
4. Click a point using the crosshairs on the northeast corner of Section 1, northwest corner of Section 3, southwest corner of Section 15, and southeast corner of Section 13. Note that when the crosshairs are lined up on the section boundary, it appears white. Close the polygon and select the AOI by double-clicking the last point (Section 13). A crosshatched polygon will appear around Sections 1, 2, 3, 10, 11, 12, 13, 14, and, 15 indicating the AOI has been selected (Figure 5.2).
5. Enter "120N, 49W, Day SD" in the AOI Information, "Name" input box.

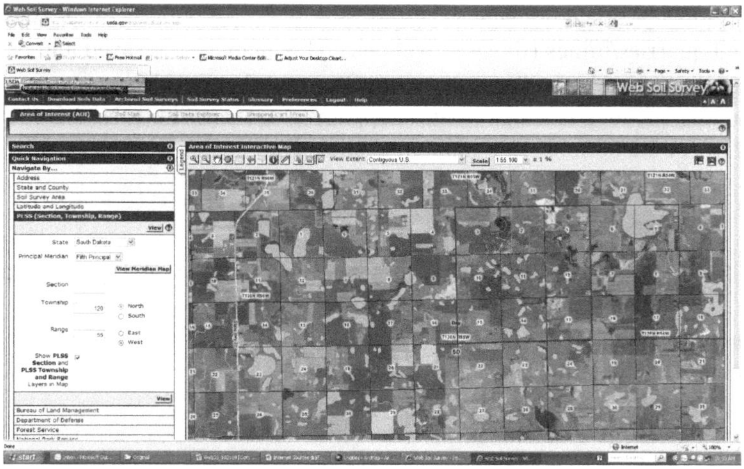

FIGURE 5.2 Selection of AOI in WSS.

5.3.2 Generating a Soil Map

Soil mapping units within the AOI are extracted or "clipped" from a much larger spatial data set. By clicking on the "Soil Map" tab in the WSS, a soil map of the AOI appears in the map window, with a table of contents showing the area and percent of each soil map unit (Figure 5.3). Reports for each soil map unit can be generated by clicking on the name of each map unit in the table of contents. Reports provide general information about the map units; note that there are several different soil components that comprise each map unit.

A variety of features are shown on the map. Clicking on the [Legend] tab in the upper left corner of the map view window provides a definition of symbols and the ability to toggle layers "on" or "off." In addition to soils, the map shows

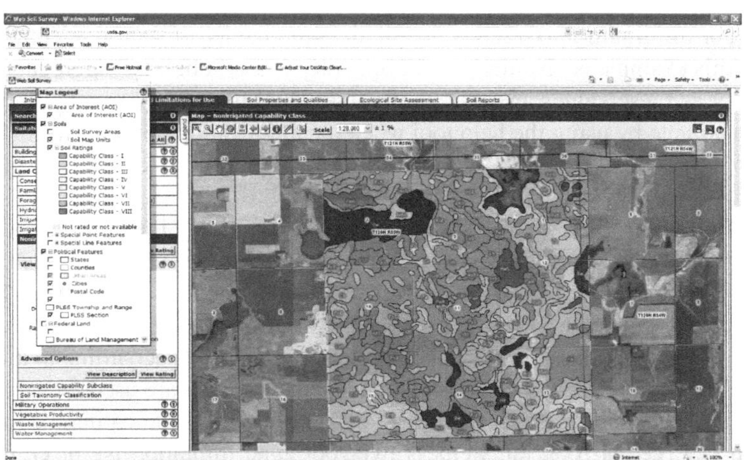

FIGURE 5.3 Soil mapping units symbolized by land Ccl in WSS.

transportation, political boundaries, water resources, and point features such as wetlands and quarries/mines (likely gravel pits). This map can be saved and printed by clicking "Printable Version" or downloaded later by clicking Add to Shopping Cart. In this example, the map is added to the shopping cart, enter "T 120N, R 55W, Day SD Soils Map" in the optional title input box, click OK.

5.3.3 Exploring Soil Data

The Soil Data Explorer is separated into the following functional categories:

1. Intro to Soils
2. Suitability and limitations for use
3. Soil properties and qualities
4. Ecological site assessment
5. Soil report

Each division can be further separated into groupings. For example, Table 5.1 shows suitability ratings. Suitability and limitation ratings have been generated through soil surveys developed over time. These are a type of "interpretive value" and give an indication of how well suited a soil would be for a particular purpose and if there are any limitations for use. A second organizational grouping that might be useful is vegetative productivity. This group may indicate the suitability of a soil for crop production or grazing. Some interpretations and values are assigned to mapping units, while others are assigned to individual components that make up a soil mapping unit. Recall that components have discrete boundaries where the definite boundaries shown on the map are soil mapping units. Therefore, component data must be aggregated in some manner to assign a representative value to individual mapping units.

Depending on the selected parameter, the WSS allows users to select a method for aggregating component data: dominate condition, dominate component, weighted average, or minimum/maximum value of all components. Since the selected parameter for this example is "Nonirrigated Capability Class/Subclass," aggregation is not necessary as these values are assigned to individual mapping units. More information is available by using the help tool when the aggregation option is available.

5.3.3.1 Intro to Soils

The "Intro to Soils" tab serves as a reference tool, providing the user with an understanding of the data provided within this division. Selecting "All Land Uses" displays the entire list of topics. In this case study, cropland suitability is selected. The Land Capability Classification System groups soils at three capability levels: class, subclass, and unit. Capability class (Ccl) is a method for grouping soils according to their limitations, risk of damage, and response to management if they are used for typical field crops. Capability subclass (CScl) groups soils within a class, indicating the limitation for cropping (i.e., erosion [e], wetness [w], root zone limitation [s], or climate [c]). Therefore, Ccl and CScl can provide insight for designing sustainable cropping systems. A search engine can be used to find parameters with each category. In this case,

TABLE 5.1
Suitability Rating Data Available from WSS (All Land Uses)

Suitability/Rating Category	Selected Examples[a]
Building site development	Corrosion of concrete; corrosion of steel; dwellings with basements; lawns, landscaping, and golf fairways; local roads/streets; and small commercial buildings
Construction materials	Gravel source, roadfill source, sand source, source of reclamation material, and topsoil source
Disaster recovery	Clay liner material source, composting facility (subsurface and surface), composting medium and final cover, and rubble and debris disposal (large-scale event)
Land classification	Conservation tree and shrub group, ecological site (forage suitability group, rangeland site), farmland classification (prime, important, unique), hydric soil rating, and land Ccl/CScl for dryland and irrigated (where appropriate) conditions
Land management	Erosion hazards for off-road, off-trail, road, and trail use; and suitability for roads
Military operations	Bivouac areas, excavation for fighting positions, helicopter landing zones, and vehicle trafficability
Recreational development	Camp areas, off-road motor cycle trails, paths and trails, picnic areas, and playgrounds
Sanitary facilities	Daily cover for landfills, sanitary landfill (area and trench), septic tank absorption fields, and sewage lagoons
Vegetative productivity	Crop productivity index, forest productivity (tree site index and production), Iowa corn suitability rating, range production (favorable, normal, and unfavorable year), irrigated crop yields by map unit or map unit component (where appropriate), and dryland crop yields by map unit or map unit component
Waste management	Manure and food processing waste
Water management	Excavated ponds (aquifer fed)

[a] All suitability or rating values may not be available for the selected AOI.

the Ccl map is created first and the subclass maps second. To create a Ccl map for the nine sections shown in Figure 5.3, use the following procedure:

1. Copy the entire "*X*:\Chapter_5" (*X*:\is the CD/DVD drive) folder to the root directory of your hard drive (C:\Chapter_5). **Important: Failure to save example data to the proper location will increase the difficulty of recreating the resultant map and data.**
2. If not already done so, click on the "Soils Data Explorer" tab.
3. Expand the 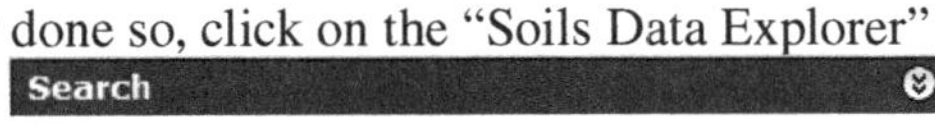

bar and enter "land capability class" in the input box, click "Search." Note that there are six matches under "Suitabilities and Limitations for Use."
4. Expand the "Suitabilities and Limitations for Use" category, expand the "Matched soil data used to generate the Rating" subcategory, and double

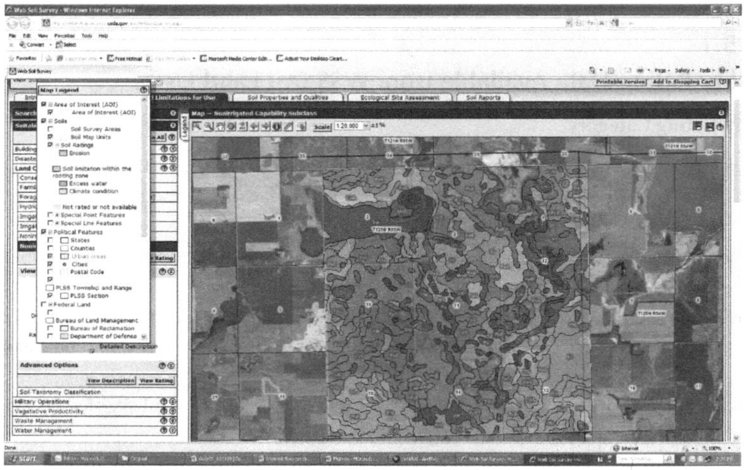

FIGURE 5.4 Soil mapping units symbolized by land CScl in WSS.

click "Land Classifications: Nonirrigated capability class" to open the data retrieval tool. This tool could be opened by clicking on the retrieval category, but the search tool quickly finds references and more discrete parameters quickly.

5. Select all of the options in the retrieval tool and click View Rating. Soil mapping unit polygons are now shaded with colors corresponding to individual Ccls (Figure 5.4).
6. Scroll down to view the amount and percent of the AOI of each soil mapping unit corresponding to Ccl.
7. Data in the summary table below the map window will need to be saved for later use in this chapter. While holding the left mouse button down, drag over the entire table to highlight all of the data. While holding the "Ctrl" key down, press "C" (keyboard command for "copy"). Open MS Excel to a new workbook, place the mouse pointer in cell A1; while holding the "Ctrl" key down, press "V" (keyboard command for "paste"), rename this worksheet "Ccl." Save this spreadsheet as "C:\Chapter_5\ArcGIS\Tab_Data\120_55_Interp.xls." A spreadsheet with this data is provided on the CD accompanying this book (120_55_Interp_Key).
8. Rename fields and arrange data on the spreadsheet to resemble that shown in Table 5.2. **Important: Failure to rename fields properly will increase the difficulty of recreating resulting map and data.**
9. Click Add to Shopping Cart and enter "T 120N, R 55W, Day SD Capability Class" in the subtitle input box and click OK.

Land capability classification generally indicates the suitability for most types of field crops. Suitability declines as ratings increase from 1 to 8 (I–VIII); soils assigned class VIII have limitations that preclude use for cropping. What may be more important is what limitations a soil has that may restrict or preclude the use of a soil for crop production. CScl provides a generalized grouping indicating limitations a soil

TABLE 5.2
Structure of MS Excel Worksheet (120_55_Interp$Ccl)

MUNAME	MUSYM	Ccl	ACRE_Ccl	MUPCT_Ccl
Buse-Barnes loams, 9%–20% slopes	BnD	6	8.3	0.001
Cubden silty clay loam	Cu	2	10.8	0.002
Divide loam	Dd	3	193.4	0.033

Not all data shown in this table; should contain 34 records (lines of data).

may have. Definitions of soil CScls are provided in the "Intro to Soils" section of the WSS. A CScl map is created using the following procedure.

1. Expand the "Nonirrigated Capability Subclass" retrieval tool by clicking the title bar. Click View Rating. The polygons are now shaded to corresponding subclasses, creating a map depicting the spatial distribution of subclasses in the AOI (Figure 5.4).
2. Scroll down to view the amount and percent of the AOI of each soil mapping unit and the CScl.
3. Data in the summary table below the map window will need to be saved for later use in this chapter. While holding the left mouse button down, drag over the entire table to highlight all of the data. While holding the "Ctrl" key down, press "C" (keyboard command for "copy"). Open MS Excel workbook created in step 6 above. Activate a new worksheet, place the mouse pointer in cell A1; while holding the "Ctrl" key down, press "V" (keyboard command for "paste"), rename this worksheet "CScl." A spreadsheet with this data is provided on the CD accompanying this book (120_55_Interp_Key).
4. Arrange the data on the spreadsheet to resemble that shown in Table 5.3. **Important: Failure to rename fields properly will increase the difficulty of recreating resulting map and data.**
5. Click Add to Shopping Cart and enter "T 120N, R 55W, Day SD Capability Subclass" in the subtitle input box and click OK.

TABLE 5.3
Structure of MS Excel Worksheet (120_55_Interp$CScl)

MUSYM	CScl	ACRE_CScl	MUPCT_CScl
BnD	e	8.3	0.001
Cu	s	10.8	0.002
Dd	s	193.4	0.033

Not all data shown in this table; should contain 34 records (lines of data).

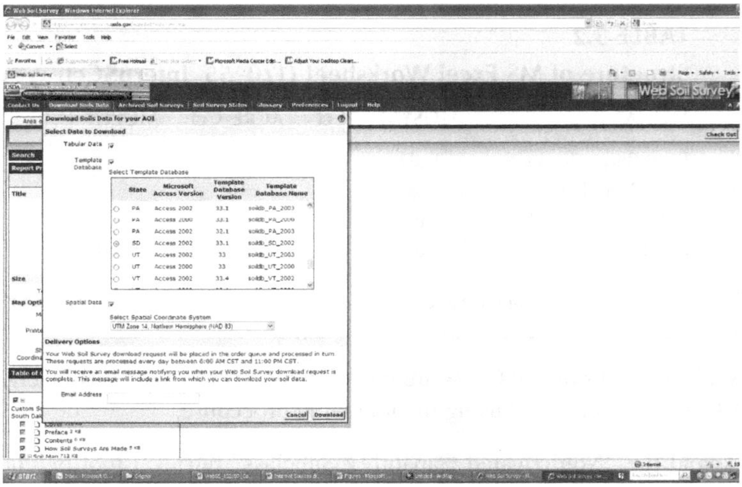

FIGURE 5.5 Preparing soil data for delivery in WSS.

6. Click on the "Shopping Cart (Free)" tab. Scrolling through the table of contents, it is evident that both of the layers created are present in the shopping cart in addition to reports and other reference material.
7. Click "Download Soils Data" in the main menu bar near the top of the web page.
8. A dialog box opens, make sure that the "Spatial Data" box is checked and the coordinate system is set to "UTM Zone 14 Northern Hemisphere (NAD 83)," enter your e-mail address where directions for downloading the compressed file can be obtained (Figure 5.5), click Download.
9. Click Check Out. After WSS has compiled the information, a dialog box will appear asking if you would like to download the report now or later. Select the radio button next to "Get Now" and click OK. A detailed report of the soils in the AOI is delivered as a pdf file that can be saved or printed. A copy of the soil report for this example is provided on the data CD accompanying this book.

Consider the soil maps created in this example. The first map depicts the spatial distribution of various soil mapping units that occur in the AOI. The distribution of soil mapping units shows a high degree of variability, indicating a complexity difficult to manage. However, considering that the properties of soil provide more meaningful information, spatial variability declines and patterns begin to emerge when mapped by property. The land Ccl map indicates that the majority of the soils have moderate limitations (class II) with inclusions of soils with more severe limitations (class III, IV). There are some prominent contiguous areas of soils that are better suited to pasture, forestland, or wildlife habitat (class V, VI) and some areas that have properties that preclude agricultural use (class VIII). The land CScl provides general limitations that in many situations, but not all, can be overcome by management. For example, soils with an erosion (e) limitation could be overcome by implementing erosion-control practices such as contour farming, no-tillage systems, or terraces. Areas with soils(s) limitations might be limited by many factors including a shallow depth, large amount of stones,

TABLE 5.4
Summary of Spatial Data Delivered by WSS

Shapefile Name	Feature Type	Description
aoi_a_aoi	Polygon	Boundary of AOI
Soilmu_a_aoi	Polygon	Soil mapping unit boundaries
soilmu_l_aoi	Line	Line feature of soil mapping units
[a]soilmu_p_aoi	Point	Location of core collection for characterization
[a]soilsf_l_aoi	Line	Special feature noted during survey
[a]soilsf_p_aoi	Point	Special feature noted during survey

[a] Data set may not be populated depending on AOI.
Metadata for all datasets provided as "soil_metadata_sd037."

and/or high salt concentrations. Soils that are shallow or droughty may benefit from perennial grass or hay crops. Areas limited by excessive water (w) may be improved with artificial drainage, water use–intensive crop rotations, or precise planting of water-intensive plant species or tree/shrub planting in recharge areas. Alternative land uses such as grazing or wildlife habitat may be considered.

The WSS is a useful planning tool but should not be used to develop crop production strategies without integrating on-site data and observations. Localized conditions, seasonal trends, land modifications, and other influencing factors may not be represented by these data. In addition, this exercise examines a narrow scope of soil parameters. Management strategies can be influenced by other soil properties or conditions induced by current land management.

5.3.4 Retrieving Spatial and Attribute Soil Data

The WSS delivers an e-mail with a hyperlink for retrieval of requested data. These data are delivered as a "zip" archive file that contains data files selected by the user. In this exercise, spatial and attribute data were selected. Spatial data is delivered as ESRI® shapefiles, attribute data is delivered as *.txt files and as an MS Access database (soildb_SD_2002). There are a number of spatial datasets delivered with the order; these datasets are summarized in Table 5.4. The MS Access database consists of soil data for the entire soil survey area (usually a county) and has automated procedures that allow the user to retrieve soil reports at a click of the mouse. Before proceeding with this exercise, ensure that spatial and tabular data have been retrieved from WSS and extracted to "C:\Chapter_5\Data."

5.4 USING SOIL SPATIAL AND ATTRIBUTE DATA IN GIS

There are many different ways that soil data can be used. Soil survey information is often most useful when it is combined with field collected data. In this example, soil survey information will be combined with point locations where information was collected. Coordinate and base data are provided on the data CD accompanying this book; these should have been copied to the root directory of your hard drive in the previous section.

Soil samples were collected in a grid pattern (200 ft) from a 155 ac field in the NE ¼, Section 13, T 120N, R 55W. A second field was sampled in a random fashion (NE ¼, Section 1, T 120N, R 55W). All soil samples were collected consistent with methods outlined in Gelderman et al.[4] A global positioning system (GPS) was used to collect point data using a geographical coordinate system (GCS), North American Dataum—1983 (NAD 83). Data was retrieved from the GPS and stored in an MS Excel file (C:\Chapter_5\ArcGIS\Tab_Data\120_55_SS.xls).

Begin by opening ArcMap™ (ESRI® ArcGIS™ ver. 9.3), select "A new empty map" when prompted. Use the following procedure to import soil sample point data to create event themes.

1. Select "Tools" in the main menu bar at the top, click "Add X Y Data." The "Add X Y Data" dialog box will appear.
2. Uncheck "Warn me if the resulting layer will have restricted functionality".
3. Click the icon next to the combo box below. "Choose a table from the map or browse for another table" and navigate to "C:\Chapter_5\ArcGIS \Tab_Data." Select the "120_55_SS.xls" MS Excel file and the "NE1$" worksheet, click "Add".
4. If "auto fill" did not detect the X Y fields, select "Longitude" for the "X Field" and "Latitude" for the "Y Field." Click "Edit..." below the coordinate system dialog and click the "Select..." button next to "Select a predefined coordinate system." Navigate to "Geographic Coordinate Systems," "North America," "North American Datum 1983.prj," click "Add", "OK", and "OK". The map should be populated with points and "NE1$ Events" should be listed under "Layers" in the table of contents (Figure 5.6).
5. Repeat this procedure to import the grid sampling locations by selecting the NE13$ worksheet.

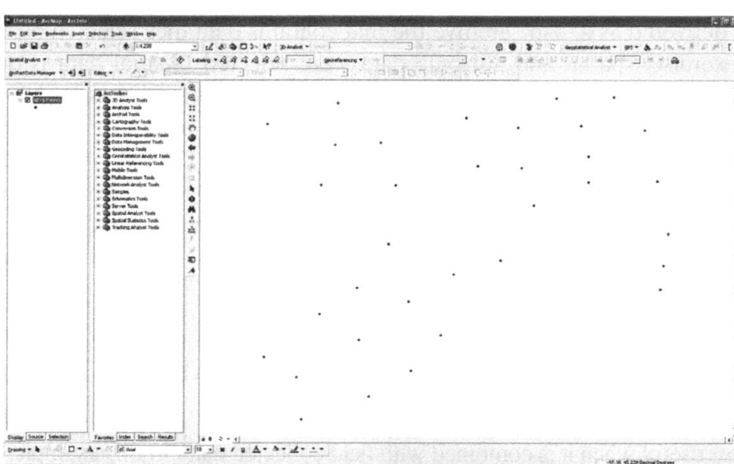

FIGURE 5.6 Soil sampling points for NE ¼, Sec. 1, T 102N, R 55W added into ArcMap™ as an event theme.

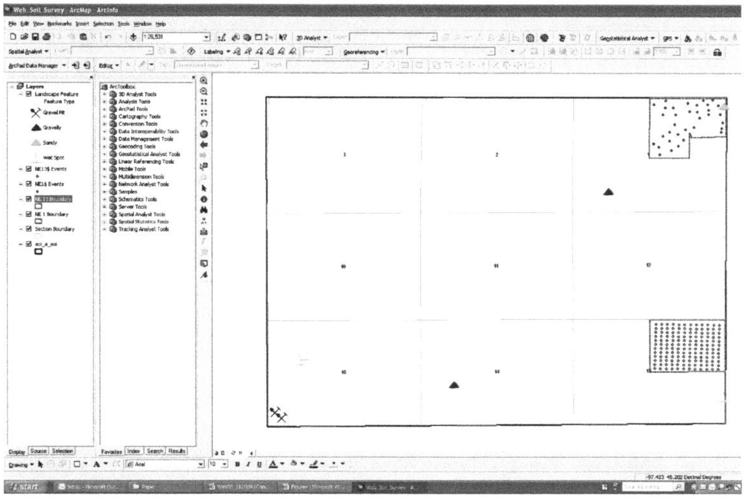

FIGURE 5.7 Soil sampling point event themes and other layers shown in ArcMap™.

After both soil sample location event themes have been created, click the globe icon to zoom to full extent; both sets of data should be visible in the map window. Tool icons can be easily identified by rolling the mouse pointer over the icon to show the icon definition. Map base layers have been created and stored in the "Layers" folder (C:\Chapter_5\ArcGIS\Layers). Layer files have predefined symbologies and layer names. The symbology and layer names can be edited or created in the properties of the layer. Add base layers to the map using the following procedure.

1. Click the ("Add Data") icon in the tool bar and navigate to C:\Chapter_5\ArcGIS\Layers.
2. Hold down the "Ctrl" key and select layers "aoi_a_aoi.lyr," "NE_1_bnd.lyr," "NE_13_bnd.lyr," "n55w13_sec.lyr," and "soilsf_p_aoi.lyr." The map should resemble Figure 5.7.

Currently, the map coordinate system is defined as geographic as the layers added to the map have a geographic coordinate system definition. Maps are more useful when a coordinate system with recognizable units is used. In this exercise, we will use a Universal Transverse Mercator (UTM) projection with "Foot_US" as the distance unit. Define the coordinate system of the data frame (map window) using the following procedure.

1. Right click in an area of white space in the data frame, select "Data Frame Properties…," click on the "Coordinate System" tab.
2. Expand the "Predefined" list. Expand the "Projected Coordinate Systems" list. Expand the "UTM" list. Expand the "NAD 1983" list. Select NAD 1983 UTM Zone 14N from the list.
3. Click Modify..., select "Foot_US" from the Linear Unit, Name combo box. Click OK and click OK.
4. The map in the data frame should appear similar to Figure 5.8.

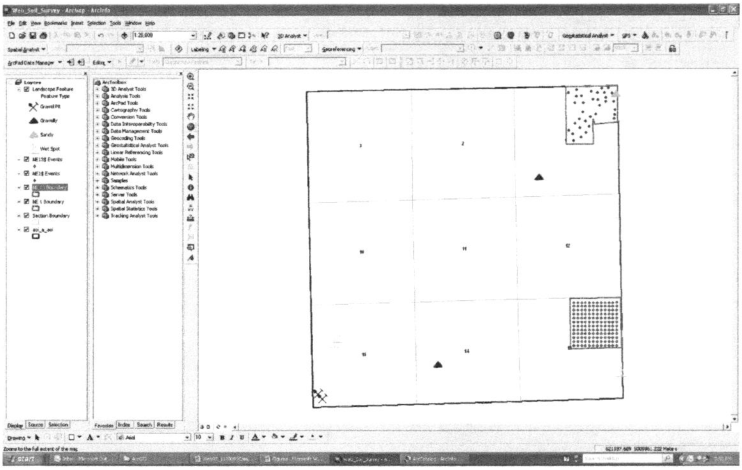

FIGURE 5.8 ArcMap™ dataframe shown projected to NAD 83, UTM Zone 14N.

5.4.1 Joining Attribute to Spatial Data

The soil mapping unit layer provided by WSS has minimal attributes. However, recall that data were copied from the WSS to an MS Excel workbook (120_55_Interp). ArcMap™ is a fully functional relational database able to relate data tables, provided that all tables within the relationship have a common field that uniquely identifies each record. In this case, MUSYM is resident in each table and will be used as the key field for the relationship. The following procedure is used to create a customized soils layer within the ArcGIS™ environment.

1. Add the "soilmu_a_aoi" layer from the geodatabase as a shapefile (C:\Chapter_5\ArcGIS\Geodatabase\Chapter_5_geodb.mdb).
2. Right click on the "soilmu_a_aoi" theme in the table of contents, select "Joins and Relate," select "Join…." The "Join Data" dialog box opens.
3. In the combo box below item 1, select "MUSYM."
4. Uncheck the box next to "Show the attribute tables of layers in this list."
5. Click  next to the combo box below item 2; navigate to "C:\Chapter_5\ArcGIS\Tab_Data\120_55_Interp.xls."
6. Select the "Ccl$" (capability class) worksheet, click Add.
7. Select "MUSYM" in the combo box below item 3, click OK.
8. Repeat this procedure to join the subclass ("CScl$") table to the map layer.
9. After joining the subclass table, right click on the "soilmu_a_aoi" map layer, select "Open Attribute Table." The attribute table should have all of the data from both worksheets (Figure 5.9).
10. Close the attribute table and click and drag the "soilmu_a_aoi" map layer to the bottom of the table of contents.

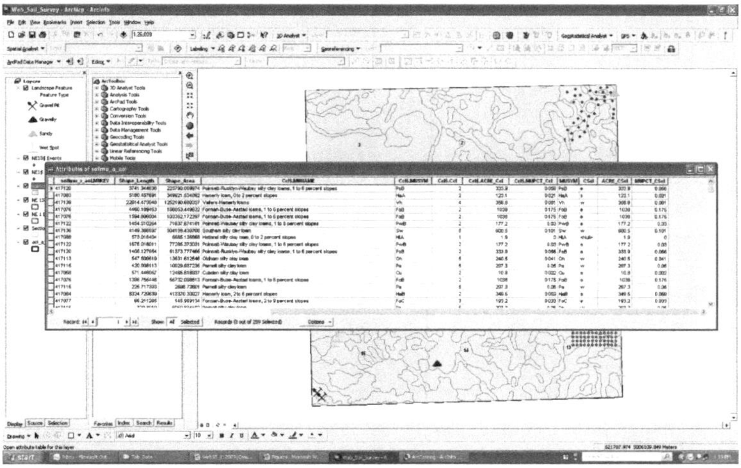

FIGURE 5.9 ArcMap™ environment showing attribute table of soilmu_a_aoi following joins created to 120_55_Interp.xls, Ccl$ and Cscl$.

5.4.2 Creating Map Layer Symbology

Managing data in a GIS ultimately produces a map that allows for visualization of data. Symbolizing data in an organizational manner helps relationships, patterns, or trends to emerge from the map. Answers to simple questions may result from visualization of data alone or may require statistical analysis for more complex questions. Regardless, symbology is a critical component when visually presenting data. Symbology for the custom soil map layer is defined using the following procedure.

1. Right click on the "soilmu_a_aoi" map layer, select "Properties." Click on the "General" tab and enter "Soil Mapping Unit" in the "Layer Name" input box. Click the "Symbology" tab, click "Categories," "Unique values, many."
2. ArcMap™ has the ability to build symbology for unique values using multiple fields. In the first combo box, select "Ccl$ Ccl" for capability class and "CScl$ CScl" in the second combo box for subclass. Click "Add All Values." Note that combinations of class and subclass appear in the symbology dialog. Colors, patterns, and labels of the symbology can be edited in the dialog box. For simplicity, a layer file has been created with predefined symbology and will be used.
3. Right click on the "soilmu_a_aoi" map layer, select "Remove." Click on the Add Data icon and navigate to "C:\Chapter_5\ArcGIS\Layers," select "soilmu_a_aoi.lyr" from the list, click "Add."
4. Click and drag the soil mapping unit layer to the bottom of the table of contents, your map should resemble Figure 5.10.

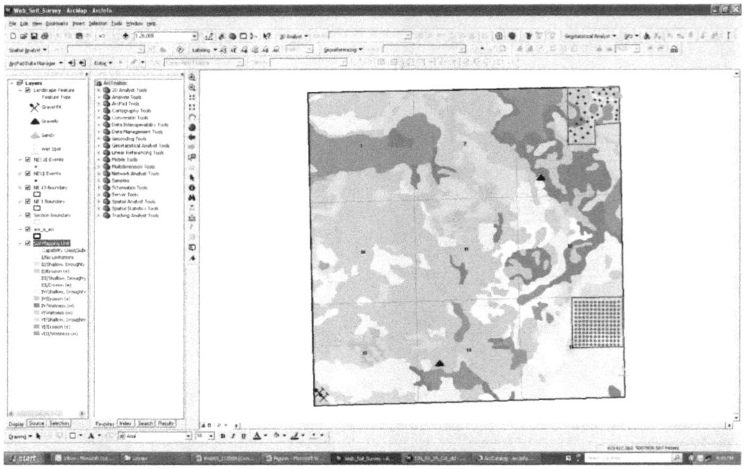

FIGURE 5.10 Soil map symbolized for land Ccl and CScl in ArcMap™.

5.4.3 Performing Spatial Joins and Exporting Attribute Data

A customized soil map has been created using data acquired from WSS and some simple techniques in ArcMap™. Although we are able to visualize where soil sampling points occur with respect to the soil properties shown on the map, it is often desirable to have these data in tabular form. The soil sample point locations currently reside in the map as event themes; these will need to be converted to ESRI® shapefiles using the following procedure.

1. Rename "NE1$ Events" and "NE13$ Events" to "NE_1_ss" and "NE_13_ss" respectively by right clicking on the theme and selecting the "General" tab. Enter the appropriate name in the "Layer Name" input box.
2. Unless you have already done so, click the to open the ArcToolbox.
3. Expand the "Conversion Tools" list, expand the "To Shapefile" tool, and double click "Feature class to Shapefile" to open the conversion dialog.
4. Select "NE_1_ss" and "NE_13_ss" respectively in the "Input features" combo box, note that they now appear in the list below.
5. Click next to the "Output folder" input box and navigate to delete fields shown in Table 5.5.

TABLE 5.5
Fields to Delete in Spatial Join Dialog Box

Object_ID	Ccl_ACRE_Ccl
SPATIALVER	Ccl_MUPCT_Ccl
Shape_Length	CScl_MUSYM
Shape_Area	CScl_ACRE_CScl
Ccl_MUSYM	CScl_MUPCT_CScl

6. "C:\Chapter_5\ArcGIS\Shapefiles," click OK. Note that these shapefiles have already been created for you and reside in the geodatabase (C:\Chapter_5\ArcGIS\Geodatabase\Chapter_5_geodb.mdb). If you are using the data that accompanies this book, click Cancel.

Assigning values to a point feature class based on spatial occurrence with respect to another feature class is possible; this procedure is called a "Spatial Join." In this instance, attributes from the "Soil Mapping Unit" feature will be joined to the soil sample location themes "NE_1_ss" and "NE_13_ss" using the following procedure.

1. Expand the "Analysis Tools" list, expand the "Overlay" list, and select "Spatial Join." The "Spatial Join" dialog opens.
2. Select "NE_1_ss" in the "Target Features" combo box.
3. Select "Soil Mapping Unit" in the "Join Features" combo box.
4. Click  next to the "Output Feature Class" input box, navigate to "C:\Chapter_5\ArcGIS\Shapefiles\" and enter "NE_1_ss_soil.shp," click Save.
5. In this case, there are duplicate and unnecessary fields in the data set, remove selected fields for inclusion in the new data set by highlighting the field(s) to be removed and click X to remove the field. Fields to remove are listed in Table 5.4.
6. Click OK to perform the spatial join and create the new shapefile.
7. Repeat to join "NE_13_ss."
8. After adding the "NE_1_ss_soil" and "NE_13_ss_soil" themes to the map, attribute tables of these themes can be exported as "*.txt" files and imported into MS Excel by opening the attribute table and selecting "Export" from the list of "Options." MS Excel files of these attribute tables have been created and stored on the accompanying CD (C:\Chapter_5\ArcGIS\Tab_Data\120_55_SS_Soils.xls).

5.5 CONCLUSION

A GIS integrates computer hardware, software, and data to capture, manage, analyze, and display data. Consider that nearly all data have location associated with it. GIS assists in determining if there are trends or relationships between datasets that can assist in answering questions. In this case, land Ccl and CScl were spatially assigned to points representing soil sampling areas in a field. These data can be used with additional data from analysis of soil samples or yield data to determine if these interpretive values explain any spatial variation within the field.

Fully developing an analysis of these fields requires additional data collection or integration of additional map layers. However, developing options for improving energy efficiency and productivity is possible considering the data presented in this exercise. Soils in the field of Section 1 are said to have limitations for row cropping due to wetness and erosion, some are extreme, precluding agricultural use (Figure 5.11). Crop productivity may be improved with management or installing artificial drainage. Wetland areas and areas with class VIII (w) may act as catchments for drainage water. Adopting a reduced tillage or no-tillage system may reduce erosion and build organic matter which may improve soils classified as III (s). These practices may improve conditions such as the sandy area in the NE corner of the field in Section 1 (Figure 5.11). Nutrient management strategies may be altered to apply nutrients during drier periods to reduce leaching, denitrification, or runoff losses.

These practices may improve productivity and efficiency of energy and resource use. However, if management or structural land modifications are not feasible, alternative land uses are an option. Redirecting land use to enterprises more suitable to indigenous soil conditions can optimize input efficiency and profitability. Spatial soil data integrated with site collected and other publically available data in a GIS can assist in designing enterprises optimally suited to the land capabilities.

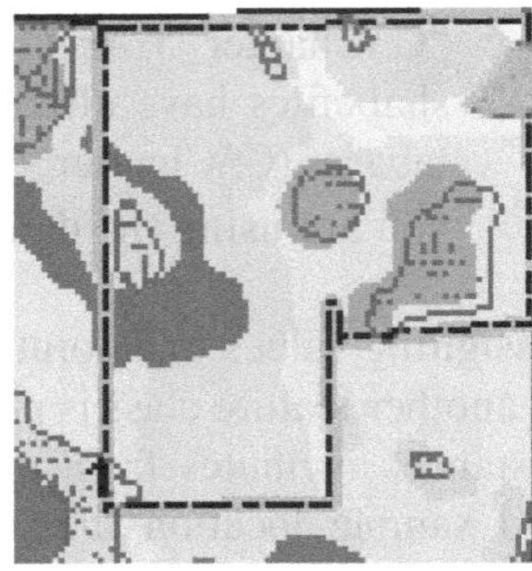

NE ¼, Sec. 1, T 120N, R 55W

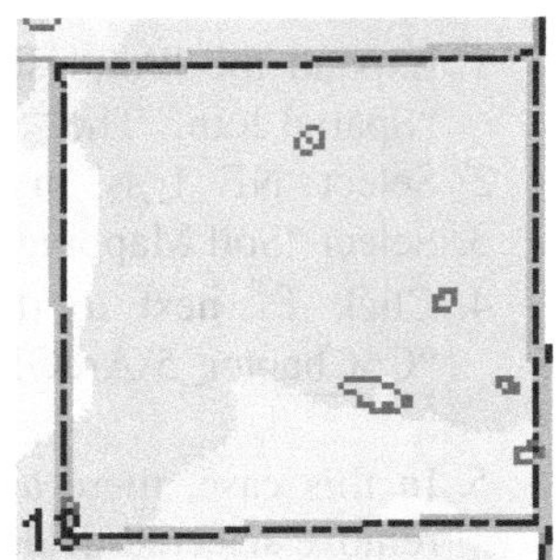

NE ¼, Sec. 13, T 120N, R 55W

FIGURE 5.11 Zoomed view of soil occurrence within AOI fields.

ACKNOWLEDGMENTS

Partial funding was provided by South Dakota Agricultural Experiment Station, South Dakota Corn Utilization Council, SunGrant Initiative—North Central Regional Center, Sustainable Agricultural Research and Education (SARE), and NRCS-CIG.

REFERENCES

1. Agriculture and Agri-Food Canada, 2008. The National Soil DataBase. Available at http://sis2.agr.gc.ca/cansis/nsdb/. Accessed November 24, 2009.
2. CSIRIO Land and Water, 2006. Australian Soil Resource Information System. Available at http://www.asris.csiro.au. Accessed November 24, 2009.
3. ESRI® ArcMap™ 9.3, ESRI® ArcGIS™ 9.3, Copyright ©1999–2008, ESRI, Inc.
4. Gelderman, R., J. Gerwing, and K. Reitsma, 2006. Recommended soil sampling methods for South Dakota. South Dakota State University. Extension publication # FS 935. Available at http://agbiopubs.sdstate.edu/pub_description.cfm?Item=FS935. Accessed November 24, 2009.
5. Malo, D.D., 2008. Online Sources of Basic Soils Information. In S. Logsdon, D.E. Clay, D. Moore, and T. Tsegaye, eds. *Soil Science Step-by-Step Field Analysis*. Soil Science Society of America, Madison, WI, pp. 35–48.
6. Microsoft® Office Excel®, Microsoft Office Enterprise 2007, Copyright ©2006, Microsoft Corporation.

7. Nachtergaele, F., H. van Velthuizen, and L. Verelst, 2009. Harmonized World Soil Database, version 1.1, Documentation. Food and Agriculture Organization of the United Nations (FAO) the IIASA, ISRIC-World Soil Information, Institute of Soil Science—Chinese Academy of Sciences (ISSCAS) or Joint Research Centre of the European Commission (JRC). Available at http://www.iiasa.ac.at/Research/LUC/External-World-soil-database. Accessed November 24, 2009.
8. Soil Survey Staff, Natural Resources Conservation Service, United States Department of Agriculture. Web Soil Survey. Available at http://websoilsurvey.nrcs.usda.gov. Accessed November 24, 2009.
9. United States Department of Agriculture, Natural Resource Conservation Service, 2009. NRCS Soils Website; Helping People Understand Soils. Available at http://soils.usda.gov/. Accessed November 24, 2009.
10. United States Department of Interior, Geological Survey, 2009. USGS Water Resources Website; Hydrologic Unit Maps. Available at http://water.usgs.gov/GIS/huc.html. Accessed December 7, 2009.

6 Space, Time, Remote Sensing, and Optimal Nitrogen Fertilization Rates: A Fuzzy Logic Approach

Nicolas Tremblay, Yacine M. Bouroubi, Bernard Panneton, Philippe Vigneault, and Serge Guillaume

CONTENTS

6.1 EXECUTIVE SUMMARY

Fuzzy logic inference systems (FISs) can help provide within-field nitrogen (N) fertilization recommendations by combining critical plant- and soil-based spatial information. This chapter describes how, based on spatially distributed information, FIS can be used to develop in-season N recommendations. A sample problem is provided. Soil and plant information considered in this analysis included apparent soil electrical conductivity (EC_a), elevation (ELE), and the remote sensing–based N sufficiency index

(NSI = $NDVI_{sample}/NDVI_{well\text{-}fertilized\ reference}$). Expert knowledge for formulating fuzzy rules was developed from corn growth data following an in-season N application. The best mid-season growth response to in-season N occurred in areas of low EC_a and high ELE. Under these favorable soil conditions, maximum mid-season growth was obtained without in-season N irrespective of the NSI values. Where soil conditions were less favorable (i.e., high EC_a and low ELE), mid-season growth benefited from high in-season N rate only when NSI was low. These relationships were modeled using a simple FIS having three inputs (EC_a, ELE, and NSI) fuzzified with only two sets (low and high), an output (optimum N rate) with three fuzzy sets (low, medium, and high) and a set of eight simple rules. The FIS appeared to be a useful and handy tool for incorporating expert knowledge into spatially variable N recommendations. An example describing a basic implementation of the FIS in ArcGIS is included.

6.2 INTRODUCTION

Nitrogen (N) management is one of the primary factors affecting crop yield and pollution from agroecosystems. A reduction in N application rates has a more beneficial impact on NO_3-N losses than an increase in drain spacing or a decrease in drain depth.[1] Nitrogen lost to denitrification, leaching, volatilization, or surface erosion is harmful to the environment. Intensive agricultural practices contribute to the degradation of surface and subsurface water quality. Row crops are particularly susceptible to N losses early in the season. An extensive survey of N fertilization practices for corn in Quebec[2] showed that the overall optimal N rate (N_{opt}), which ranged between 25 and 135 kg N ha^{-1} in 50% of the fields, was generally below the recommended rate which ranged from 120 to 170 kg N ha^{-1}.[3] To reconcile economic and environmental objectives, crop production systems must adhere to a "just enough" principle.[4] For crops where split N application is possible, a low initial dressing at sowing should be applied, leaving most of the total anticipated N dose retained for the topdressing (early growth) application. It is well known that split N applications enhance N use efficiency by crops and increase economic returns.[5,6]

Early detection of N deficiency in crops provides producers with an opportunity for more closely matching their application rates to the real-time N requirements of the crop. Remote sensing–based vegetation indices (VIs) that assess a crop's biomass/chlorophyll status can be obtained through a variety of sensors and platforms.[7–12] One of the most commonly used VIs is the normalized difference vegetative index (NDVI). NDVI is calculated with the equation, where NDVI = (NIR − RED)/(NIR + RED). Information provided by VIs can be enhanced for N assessment purposes by transforming them into an N sufficiency index (NSI), where NSI = $NDVI_{sample}/NDVI_{well\text{-}fertilized\ control}$,[8,11,13,14] or into a response index (RI), where RI = $VI_{sample}/VI_{unfertilized\ control}$.[15] The uniform application of fertilizer N can result in areas where fertilizer is overapplied and other areas where it is underapplied. Optimal N fertilization is bound to the spatial variability of field parameters such as topography[16] or EC_a.[17,18] Low EC_a tends to be associated with soils more favorable to plant growth.[18] Topography also influences soil water content, temperature, nutrient use efficiency, water infiltration, and crop growth and development.[17,19–22] In addition, interactions among climatic conditions, EC_a, and topography are likely to further influence growth and yield.[23] The spatially variable application of N fertilizers is the only strategy capable of optimizing (both economically and environmentally) overall N use in this context.[12,19,24–27] The economically optimum N rate (EONR) is the N rate that maximizes producer profits. Despite the abundance of studies showing the influence of plant N deficiency level and soil conditions on EONR, very few actually translate this knowledge into recommendations systems for actual N rates. One example of an exception is a polynomial model for estimating EONR based on parameters such as crop biomass, EC_a, and topographical features.[28] However, the rigidity associated with the polynomial equation did not have the flexibility to predict N requirements outside the boundary conditions of the original work.

Artificial intelligence systems offer an option for predicting N recommendations. One study showed the potential and the suitability of the fuzzy logic approach for the classification and mapping of loosely defined soil phenomena and parameters.[29] Other studies found that fuzzy logic was the most adapted artificial intelligence strategy for describing a system characterized by imprecision, particularly where expert knowledge can be valued.[30,31] Some studies applied fuzzy logic to the cartography of the risks of overfertilization to the environment (nitrate leaching, soil acidification, and ammonia gas losses).[30] Others developed a fuzzy logic-based decision-support system integrating remote sensing data and plant growth models for management of within-field spatial variability.[32] The purpose of this chapter is to demonstrate how an FIS can be used to modulate N rates based on soil (EC_a and ELE) and crop (NSI) information.

6.3 MATERIALS AND METHODS

6.3.1 Extracting the Field Parameters

The example used herein is extracted from a study called GAPS #333 conducted between 2004 and 2007 by Agriculture and Agri-Food Canada. The experiment was conducted on corn (*Zea mays* L.) in St.-Valentin QC, Canada (UTM coordinates: 629589.38 E 4994395.04 N). The 2005 field size was 8.66 ha.

In this example, the N rate treatments were established lengthwise (Figure 6.1) in the field. One 4.5 m strip without any N fertilization was established in the center of the fields. All remaining areas in the field received 30 kg N ha^{-1} at sowing (May 12) and from 0 to 157.5 kg N ha^{-1} during in-season N application. Two 4.5 m wide N-saturated strips were established with 220 kg N ha^{-1} broadcasted at sowing for a total of 250 kg N ha^{-1}. Nonlimiting fertility conditions in other nutrients were

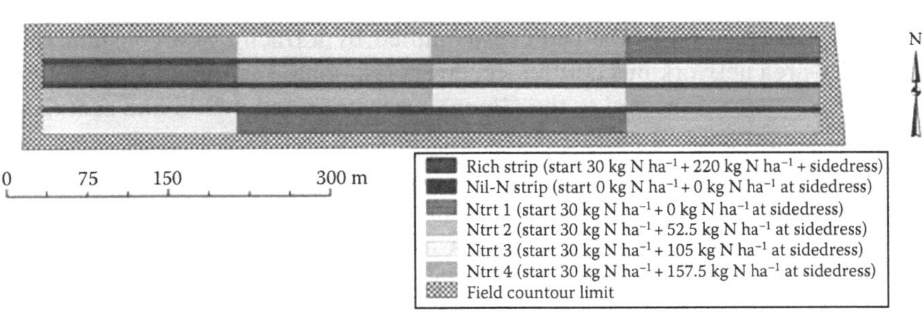

FIGURE 6.1 N treatment map of the 2005 corn crop.

FIGURE 6.2 Veris sensor cart for EC_a measurement.

achieved following provincial standards.[3] The trial was planted in conventional tillage (ploughing in autumn and standard soil preparation before sowing in spring) with 0.75 m row spacing.

A Veris™ model 3100 sensor cart system (Veris Technologies, Salina, KS) was used to measure EC_a (Figure 6.2). The measurement electrodes were configured to provide both shallow (0–30 cm in depth) and deep (0–90 cm in depth) EC_a readings; the shallow readings are the ones reported in this chapter. EC_a was collected on bare soil in spring from perpendicular transects, approximately 10 m apart.

ELE were measured with a differential global positioning system (DGPS; Pathfinder™ Pro XRB, Trimble Navigation Ltd., Sunnyvale, CA; accuracy ±15 cm) mounted on an all-terrain vehicle. The distance between the ELE transects was approximately 9 m. The sampling density along the transect was between 6 and 12 m, and depended on the speed of the all-terrain vehicle (5–8 km h^{-1}).

EC_a data (Figure 6.12A) and DGPS ELE data (Figure 6.12B) were processed by kriging. The software programs GS^{+}™ (Gamma Design Software, LLC, Plainwell, MI) and ArcGIS™ Geostatistical Analyst (ESRI, Redlands, CA) were used for generating semivariograms, extracting geostatistical parameters, and selecting interpolation models.

6.3.2 NDVI and Calculating the NSI

In this example, the canopy reflectance measurements and output NDVI were obtained from five GreenSeeker™ sensors (NTech Industries, Inc., Ukiah, CA) fixed side by side on a boom, spaced 0.75 m apart and mounted on an aluminum trailer (Figure 6.3). The trailer was pulled by a tractor at a constant speed. Controller-area network bus emitter–receivers (National Instruments Corporation, Austin, TX) were connected to the GreenSeeker units and a GPS and sent data to a PCMCIA card hooked up to a portable computer. The GPS link allowed NDVI readings to be mapped in real time. The raw NDVI data were corrected for positional inaccuracies owing to the lack of precision (±1.5 m) of the WAAS (Wide Area Augmentation System) GPS system and were interpolated by kriging. The map is shown in Figure 6.12C.

As mentioned in the introduction, raw NDVI values were converted to NSI values by dividing each number by the corresponding NDVI value of well-fertilized zone.

FIGURE 6.3 GreenSeeker™ sensors mounted on a trailer and pulled by a tractor designed for NDVI measurement early in the season. The setup was designed to operate over five corn rows simultaneously.

For this calculation, the NDVI value in the denominator was represented by the 90th percentile of the 20 nearest neighbors, within the well-fertilized strips, of the point of interest. In all cases, nearest neighbors were within 43.5 m from any point. A soil similarity condition was added: NSI is retained only if EC_a and ELE for the numerator and the denominator are close (less than one standard deviation of the whole values of EC_a or ELE, respectively). This condition was valid at 98% for EC_a and at 88% for ELE. It follows that NSI calculation was made using saturated strip data with similar soil EC_a and topography as the ones at the point of interest. The map is shown in Figure 6.12D.

6.3.3 Background Knowledge about Soil and Plant Status Needed to Determine N Needs

To establish proper N requirements, a prior knowledge of soil textural and landscape characteristics as well as of plant N status is required. For example, Kravchenko et al.[20] reported that EC_a was negatively correlated with plant growth. Another study showed that favorable growing conditions for corn were found in areas with low EC_a and high ELE, and that N requirements were higher in areas with high EC_a or low ELE.[2] High EC_a and low ELE are characteristic of clay areas, while low EC_a and high ELE are characteristic of sandy soils. The NSI can be used to identify N requirements in these zones. The NSI is less variable than absolute NDVI with regard to soil condition, type of sensor, and growing stage.[3,14] The same authors found that NSI was related to N response but only in soil conditions naturally unfavorable to plant growth (high EC_a and low ELE).[14] Under low and medium EC_a conditions, N rate did not have a significant influence on growth regardless of NSI status at stages V5–V7. For high EC_a levels, a high N rate stimulated growth, in proportion with N limitation intensity, as measured by the NSI. This relationship with NSI held for ELE as well. For medium and high ELE, the response to N rate was not significant, regardless of deficiency (NSI) level. For low ELE, a high N rate increased growth, particularly when NSI was small (strong N deficiency). When NSI was high, a moderate N rate was adequate.

6.3.4 FIS for Estimating Spatial N Needs

For completeness, a basic introduction to FISs is presented. Fuzzy inference systems are used to convert input data to output data using fuzzy logic. Fuzzy inference systems can be defined from expert knowledge[33] or from knowledge extracted from a training data set.[34] These two approaches can also be combined.[35] Here, the approach based on expert knowledge was implemented.

As a first step, the domain of each input variable is partitioned into a collection of fuzzy sets. Depending on its value, an input is assigned a membership degree to each of the set using a membership function (MF). This process is called fuzzification of the inputs, and an example is provided in Figure 6.4. In this example, the domain of the input variable has been partitioned into three fuzzy sets. For each fuzzy set, simple trapezoidal MFs were defined. Using these functions, the membership to each set is calculated. In the example, a variable with a value of 155 has membership values of 0, 0.28, and 0.72 in fuzzy sets 1, 2, and 3, respectively. After fuzzification, an input is a member of all the sets with varying degrees of membership. In principle, more than two fuzzy sets can overlap and the sum of memberships for one variable can take any value. However, limiting overlaps to two neighboring sets and forcing the sum of memberships to 1 result in inference systems where rules are easier to interpret.[33]

Once all inputs are fuzzified, fuzzy logic is used to combine these inputs to arrive at a conclusion using fuzzy operators. Inputs combined to arrive at a single conclusion form a rule. For example, assume our system has two input variables (X_1 and X_2) and a single output (Y), partitioned as in Figure 6.5. Then, one of the rules can be (RULE 1 in Figure 6.5):

$$\text{IF} \underbrace{(X_1 \text{ is Low}) \text{ AND } (X_2 \text{ is Low})}_{\text{Antecedent}} \text{ THEN } \underbrace{Y \text{ is Low}}_{\text{Consequent}} \tag{6.1}$$

The antecedent is a combination of the membership of the inputs and the consequent is the outcome from the rule. Various operators can be used, but here the minimum

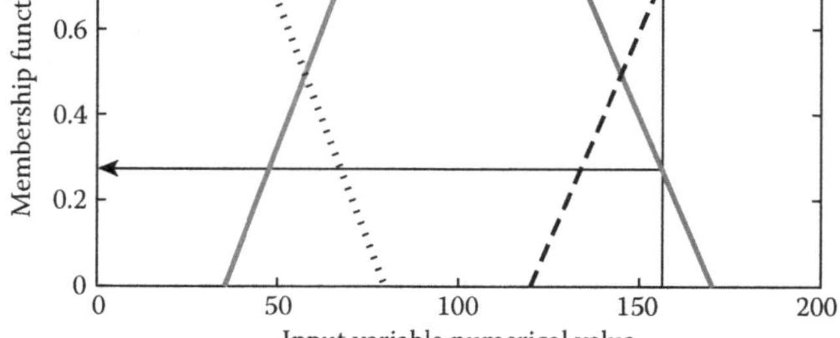

FIGURE 6.4 Example of fuzzy partitions and associated MFs.

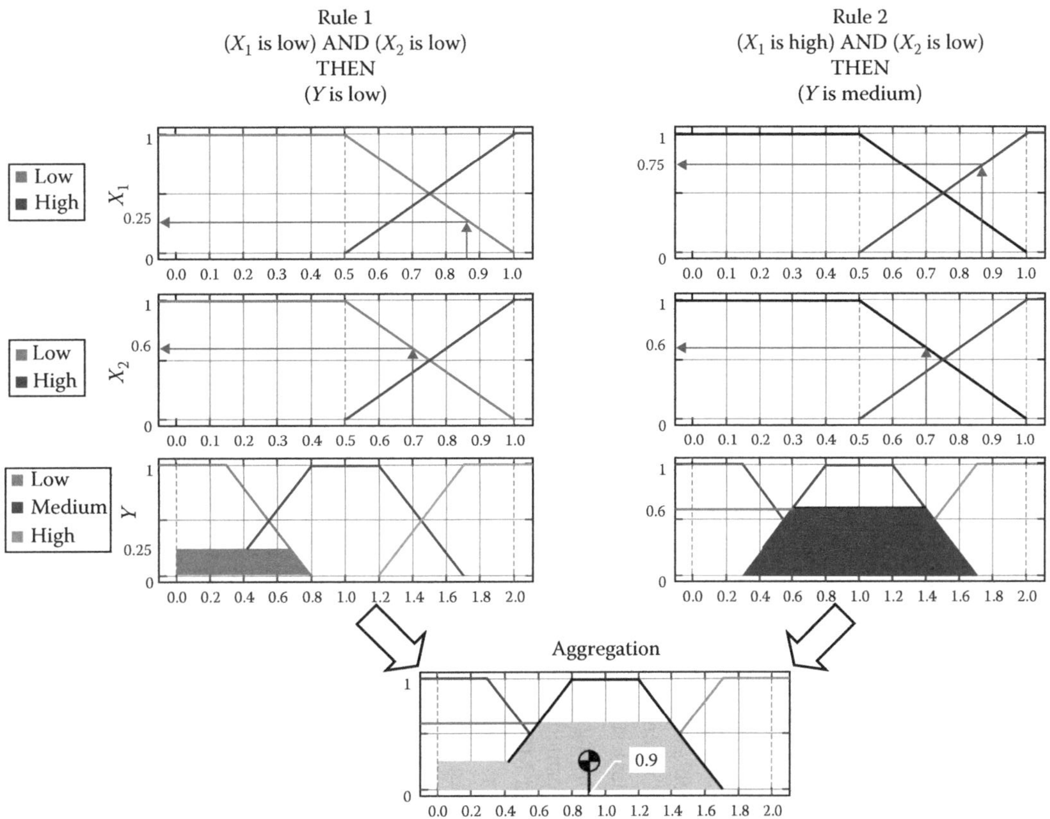

FIGURE 6.5 A two-input–one-output basic FIS with two rules. Example for $X_1 = 0.875$ and $X_2 = 0.7$ yielding $Y = 0.9$ as output.

value operator was used for the fuzzy AND. More than a single rule can yield the same consequent. In this case, these rules can be combined with a fuzzy OR operator as in

$$\text{IF} \quad \text{Antecedent 1} \quad \text{OR} \quad \text{Antecedent 2} \quad \text{THEN} \quad \text{Consequent 1} \tag{6.2}$$

Equation 6.2 is equivalent to using two separate rules:

$$\begin{aligned} &\text{IF} \quad \text{Antecedent 1} \quad \text{THEN} \quad \text{Consequent 1} \\ &\text{IF} \quad \text{Antecedent 2} \quad \text{THEN} \quad \text{Consequent 1} \end{aligned} \tag{6.3}$$

Here, fuzzy AND was the minimum of the memberships and fuzzy OR was the maximum of the memberships.

In our case, the output is a continuous variable (nitrogen dose). Fuzzy sets and associated MFs have to be defined for the output variable as well. The value of the antecedent is the degree of support for the rule. Returning to the example in Figure 6.5, for $X_1 = 0.875$ and $X_2 = 0.7$, the degree of support for RULE 1 would be 0.25 (min(0.25, 0.60)). For this rule, the MF of the consequent would be truncated to 0.28 yielding the red area in Figure 6.5. In the end, for one data point (i.e., all input variables each at a fixed value), a set of truncated MFs will be created from the consequents of all activated rules (grey and dark grey areas in Figure 6.5). A rule is activated when its degree of support is greater than 0.

The next step is to combine or aggregate all consequents from activated rules corresponding to a data point. Several methods can be used. Here, the aggregation was performed using the MAX rule. Graphically, this is equivalent to overlapping all truncated MFs to define a region enclosed within the resulting outer boundary. An example is shown in Figure 6.5 as a grayed area.

The last step is to retrieve a numerical value out of the result of the aggregation. This was achieved by computing the centroid of the area defined by the aggregation and taking the corresponding value of the output variable (0.9 in the example of Figure 6.5).

6.3.4.1 Design of the FIS

The first step in the implementation of a fuzzy inference system is to define the partitions for the input and output variables. In our case, the inputs were EC_a, ELE, and NSI and the output was N rate. Considering that the knowledge base is not extensive, partitioning of the input variables was kept simple. Only two fuzzy sets were used for each input and three sets were used for the output. (Figure 6.6: examples in this section were generated using FisPro 3.1.2.)[36]

The next step is to define a set of rules. Based on expert knowledge developed from experimental trials, eight rules were formulated (Table 6.1). In summary, the rules reflect the following knowledge:

- NSI was the more important factor and in areas of low NSI, crop growth benefits from higher N rate.
- In both, areas of high EC_a and areas of low ELE, crop growth benefits from higher N rate.

When low NSI combined with either low ELE or high EC_a, it was decided that N rate should be high (rules 1 and 2). When a high NSI was combined to either a high ELE

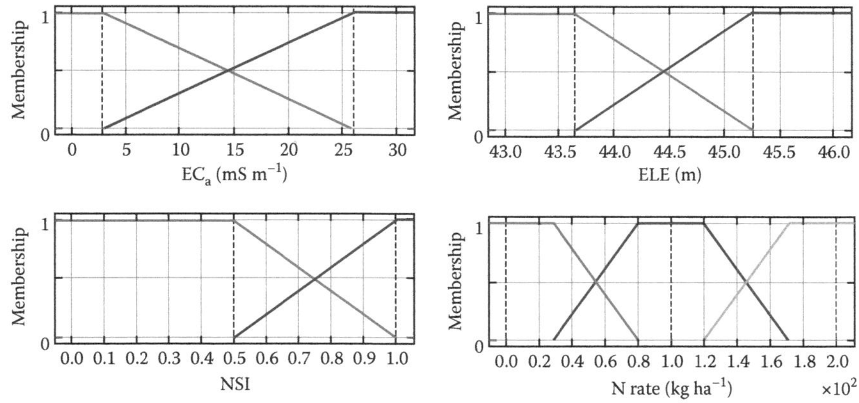

FIGURE 6.6 MFs for the three inputs and the output.

TABLE 6.1
Rule Set Used to Develop Fuzzy Inference System

Rule	EC_a	ELE	NSI	N rate
1		Low	Low	High
2	High		Low	High
3	Low		High	Low
4		High	High	Low
5	Low		Low	Medium
6		High	Low	Medium
7	High		High	Medium
8		Low	High	Medium

or low EC_a, the N rate was low. In all other cases where one variable was calling for a low N rate and the other for a high N rate, it was decided to set N rate to the mid level (Med in Table 6.1).

The inner workings of the FIS for two cases are illustrated in Figure 6.7. In case 1, both EC_a and NSI were set at the lowest level and ELE was set at mid level. In this case, rules 1, 5, and 6 were activated. Rule aggregation is shown in Figure 6.7 in the upper-right corner. The centroid was 139.8 kg ha^{-1}, a fairly high dose of N. In case 2, EC_a was kept at the low end of the range and both, ELE and NSI, were set at the top of the range. In this case, rules 3, 4, and 8 were activated. The estimated N rate was 15.1 kg ha^{-1}. Surfaces illustrating the response of the inference system at high and intermediate values of NSI are shown in Figure 6.8.

6.3.5 Step-by-Step Exercises Using ArcGIS 9.2

The objective of these exercises is to generate a map of N rates recommendation for a corn crop, using previously known terrain characteristics and an assessment of crop N deficit derived from a remotely sensed parameter. Fuzzy logic concepts

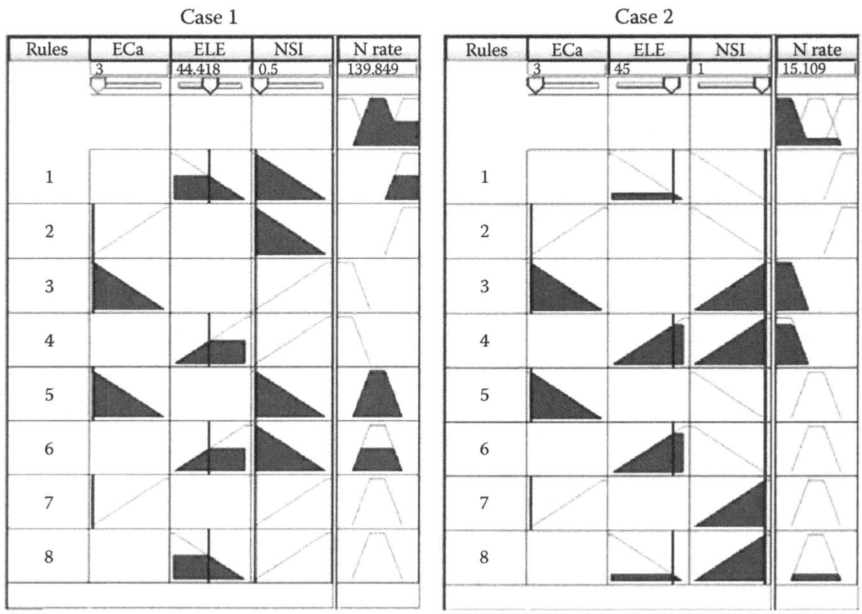

FIGURE 6.7 Two illustrative cases for the FIS.

have been described above, and the purpose of the following section is to demonstrate how the calculations are conducted using the Raster Calculator application in ArcGIS. It is strongly recommended that all steps be performed carefully and in the given order. The exercises require ArcGIS 9.2 with the Spatial Analyst extension activated. To activate the Spatial Analyst extension, click **Tools > Extensions**. Select **Spatial Analyst**. To see the extension on the toolbar, click **View > Toolbars > Spatial Analyst**.

Exercise 6.1: MF for EC_a, ELE, and NSI

Explanations

For the sake of demonstration with ArcGIS, two simple MFs are used for input parameters (EC_a, ELE, and NSI). Triangular (or trapezoidal) MFs evaluate the trueness that an input variable X is in "Low" or "High" value sets (Figure 6.6) and are expressed as

$$\text{MF } (X \text{ in Low}) = 1 - \frac{X - X_{\min}}{X_{\max} - X_{\min}} \tag{6.4}$$

$$\text{And since } \sum \text{MFs} = 1, \quad \text{MF } (X \text{ in High}) = \frac{X - X_{\min}}{X_{\max} - X_{\min}} \tag{6.5}$$

X represents the input variable EC_a, ELE, and NSI. And always, if MF < 0, take MF = 0, and if MF > 1, take MF = 1.

NSI = 1.0

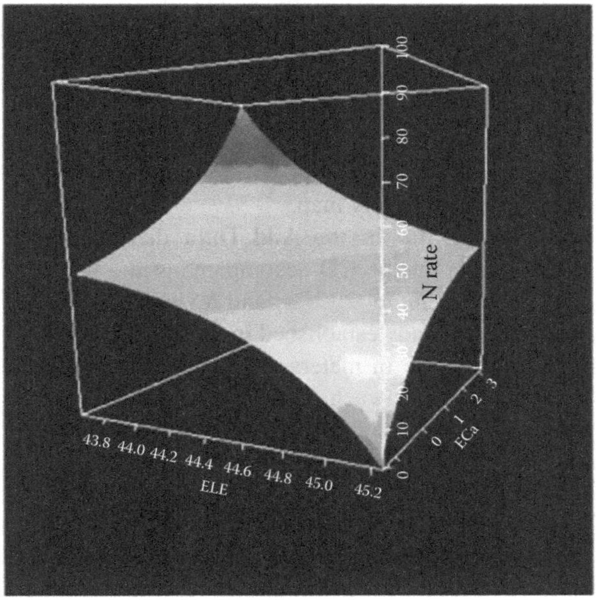

NSI = 0.75

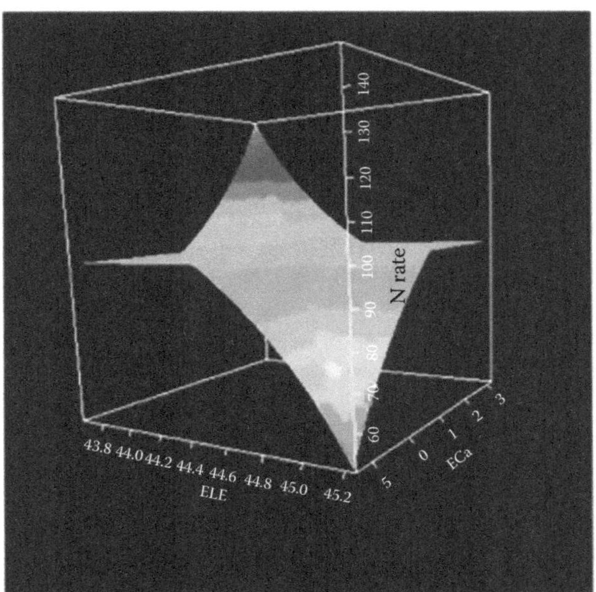

FIGURE 6.8 Response surfaces at two values of NSI. Note that at NSI = 1, the N rate axis ranges from 0 to 100 and that at NSI = 0.75, the range is from 50 to 150.

In the case of EC_a and ELE, X_{min} and X_{max} are equal to the minimum and the maximum of measured values. For NSI and based on literature and our experience, minimum and maximum values were set to 0.5 and 1.0, respectively.

For N rate, it is not necessary to give the formulation of the three trapezoidal MFs proposed (Figure 6.6) because we will use discrete values of N rate and their MFs for the aggregation–defuzification step.

Step-by-Step in ArcGIS

1. Open ArcMap to a new, empty map.
2. Click **File > Add Data**. In the **Add Data** dialogue box, navigate to ChapterX\ExerciseX\ on the CD accompanying the book and open files *ECaMap_2005.tif*, *ELEmap_2005.tif* and *NSImap_2005.tif*. Note that the projection has been already established in the metadata of each TIFF files. The projection should be in meters (UTM NAD 83 zone 18). The three maps should open on top of each other.
3. Click **Spatial Analyst** on the Spatial Analyst toolbar and select **Raster Calculator**.
4. In the **Raster Calculator** dialogue box, copy paste the formula 1 – *([ECaMap_2005.tif] – 3)/23* in the text box under the **Layers** label. The **Raster Calculator** dialogue box should look like the example shown in Figure 6.9. Once it does, click the **Evaluate** button.
5. A new layer will appear in the workspace. This temporary layer will be named *Calculation*. To continue the exercise, you have to export the

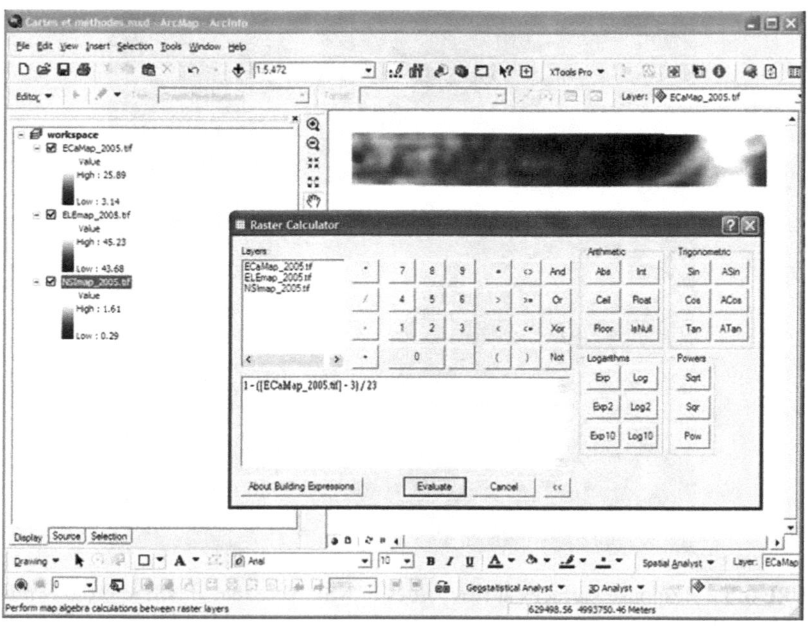

FIGURE 6.9 Raster Calculator settings for the MF that generates a raster map.

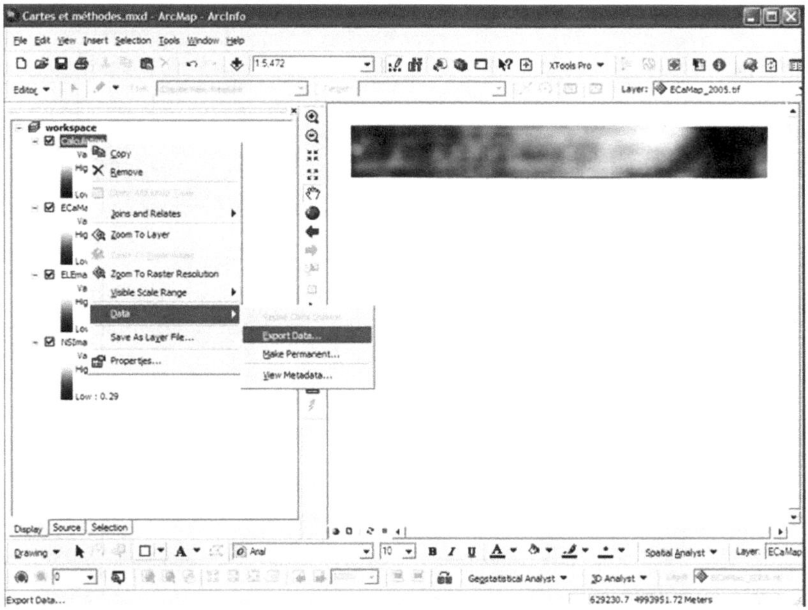

FIGURE 6.10 Exporting a temporary raster file into another file format in ArcMap.

map in TIFF format and rename it, following the suggestion in Table 6.1. Right-click with the mouse cursor on the *Calculation* layer. On the conceptual menu, click **Data > Export Data** (Figure 6.10). In the **Export Raster Data—Calculation** dialogue box (Figure 6.11), change the default filename in the **Name** label to *mfECa_Low* and change the default file format in the **Formats** label to *TIFF*. Click the **Save** button. The *mfECa_Low.tif* map file will appear as a new layer in the workspace. Keep the file open.

6. The *Calculation* layer is no longer necessary. Right-click with the mouse cursor on the *Calculation* layer. On the conceptual menu, click **Remove**.
7. Repeat steps 1–6 with all the formulas shown in Table 6.2 using copy paste. Rename the resulting maps with the filenames indicated in the "Resulting files (suggested filenames)" column in Table 6.2. *It is important that each expression be given the corresponding filename shown in the table.* At the end, you should have eight TIFF files (Table 6.2). These files will be used for the next steps.

Exercise 6.2: Using the rules

Explanations

The implementation of the inference rules must follow a mathematical logic that ArcGIS can understand. The set of rules in Table 6.1 was reformulated as four rules using the fuzzy OR operator (MAX) as First rule: If (EC_a is High OR ELE is Low)

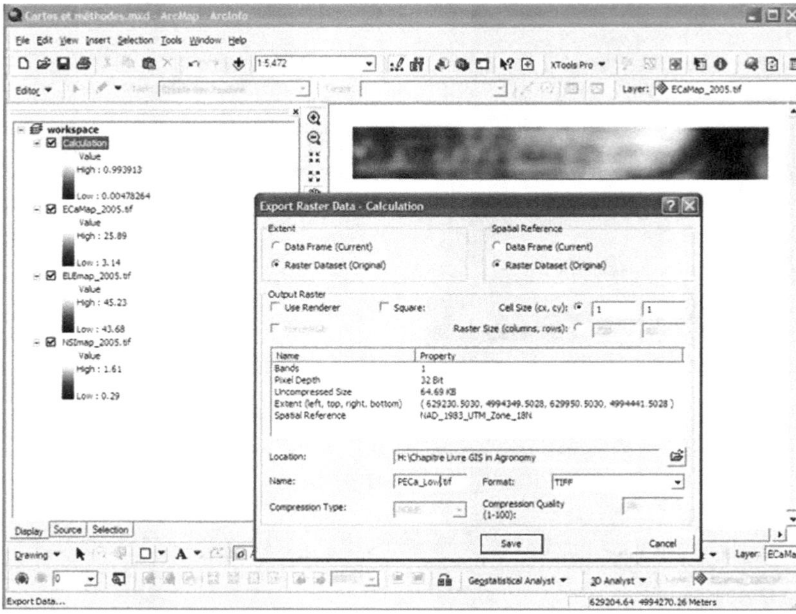

FIGURE 6.11 Writing a TIFF raster file in ArcMap.

TABLE 6.2
Mathematical Expression of the Defuzzification Formula Used in Exercise 6.1

Expressions	Resulting Files (Suggested Filenames)
1 – ([ECaMap_2005.tif] – 3)/23	mfECa_Low.tif
([ECaMap_2005.tif] – 3)/23	mfECa_High.tif
1 – ([ELEmap_2005.tif] – 43.65)/1.6	mfELE_Low.tif
([ELEmap_2005.tif] – 43.65)/1.6	mfELE_High.tif
1 – ([NSImap_2005.tif] – 0.5)/0.5	mfNSI_Low_linear.tif
([NSImap_2005.tif] – 0.5)/0.5	mfNSI_High_linear.tif
MAX (MIN ([mfNSI_Low_linear.tif], 1), 0)	mfNSI_Low.tif
MAX (MIN ([mfNSI_High_linear.tif], 1), 0)	mfNSI_High.tif

AND NSI is Low THEN N rate is High. Second rule: If (EC_a is Low OR ELE is High) AND NSI is High THEN N rate is Low. Third rule: If (EC_a is Low OR ELE is High) AND NSI is Low THEN N rate is Med. Fourth rule: If (EC_a is High OR ELE is Low) AND NSI is High THEN N rate is Med.

AND/OR operators are materialized by MIN and MAX operations, respectively. Table 6.2 presents all formulas resulting from the eight maps obtained in Exercise 6.1.

TABLE 6.3
Mathematical Expression of the Defuzzification Formula Used in the Second Exercise

Expressions	Resulting Files (Suggested Filenames)
MIN (MAX([mfECa_High.tif], [mfELE_Low.tif]), [PNSI_Low.tif])	mfNopt_High.tif
MIN (MAX([mfECa_Low.tif], [mfELE_High.tif]), [PNSI_High.tif])	mfNopt_Low.tif
MIN (MAX([mfECa_Low.tif], [mfELE_High.tif]), [mfNSI_Low.tif])	mfNopt_Med1.tif
MIN (MAX([mfECa_High.tif], [mfELE_Low.tif]), [mfNSI_High.tif])	mfNopt_Med2.tif

Step-by-Step in ArcGIS

8. Click **Spatial Analyst** again on the Spatial Analyst toolbar, and select **Raster Calculator**. In the **Layers** label, you should have eight TIFF files in addition to the three original ones. These files come from Exercise 6.1.
9. In the **Raster Calculator** dialogue box, copy paste the formula *MIN (MAX([mfECa_High.tif], [mfELE_Low.tif]), [mfNSI_Low.tif])* in the text box under the **Layers** label. Click the **Evaluate** button.
10. Again, a new layer will appear in the workspace. This temporary layer will be named *Calculation*. Export the temporary layer in TIFF format and rename the new map, following the suggestion in Table 6.3. Right-click with the mouse cursor on the *Calculation* layer. On the conceptual menu, click **Data > Export Data**. In the **Export Raster Data—Calculation** dialogue box, change the default filename in the **Name** label to *mfNopt_High* and change the default file format in the **Formats** label to *TIFF*. Click the **Save** button. The map *mfNopt_High.tif* map file will appear as a new layer in the workspace. Keep the file open.
11. The *Calculation* layer is no longer necessary. Right-click with the mouse cursor on the *Calculation* layer. On the conceptual menu, click **Remove**.
12. Repeat steps 8–11 with all the formulas shown in Table 6.3 using copy paste. Rename the resulting maps with the filenames indicated in the "Resulting files (suggested filenames)" column in Table 6.3. *It is important that each expression be given the corresponding filename shown in the table.* At the end, you should have four TIFF files (Table 6.3). These files will be used in Exercise 6.3.

Exercise 6.3: Aggregation

Explanations
The output value is the result of all rules implemented in the FIS. One of the possible strategies for aggregation is to use the envelope (MAX) bordering the MFs

(Figure 6.5). The defuzification is then the calculation of the N rate output value from the integral:

$$N \text{ rate} = \frac{\int_{N_{min}}^{N_{max}} N \cdot \text{MF}(N)\, dN}{\int_{N_{min}}^{N_{max}} \text{MF}(N)\, dN} \quad (6.6)$$

This integral is substituted by the sum of discrete values of N rate (N_{discr}). In our case, they span from 20 to 200 kg N ha^{-1} with 20 kg N ha^{-1} increments

$$N \text{ rate} = \frac{\sum_{N_{discr}=20}^{200} N \text{ rate} \times \text{Trueness}(N \text{ rate} = N_{discr})}{\sum_{N_{discr}=20}^{200} \text{Trueness}(N \text{ rate} = N_{discr})} \quad (6.7)$$

Thus, the trueness degree that a value N rate of a pixel in the map is equal to a value N_{discr} is

$$\begin{aligned} \text{Trueness}(N \text{ rate} = N_{discr}) = \text{MAX}\{&\text{MIN}[\text{MF}(N \text{ rate in Low}), \text{MF}(N_{discr} \text{ in Low})], \\ &\text{MIN}[\text{MF}(N \text{ rate in Med})_{\text{rule } 3}, \text{MF}(N_{discr} \text{ in Med})], \\ &\text{MIN}[\text{MF}(N \text{ rate in Med})_{\text{rule } 4}, \text{MF}(N_{discr} \text{ in Med})], \\ &\text{MIN}[\text{MF}(N \text{ rate in High}), \text{MF}(N_{discr} \text{ in High})]\} \end{aligned}$$

Step by Step in ArcGIS

13. Click **Spatial Analyst** on the Spatial Analyst toolbar and select **Raster Calculator**. In the **Layers** label, you should have four more TIFF files in addition to the three original ones and the eight calculated earlier. These four maps come from Exercise 6.2.
14. In the **Raster Calculator** dialogue box, copy paste the formula *MAX (MIN ([mfNopt_Low.tif], 1), MIN ([mfNopt_Med1.tif], 0), MIN ([mfNopt_Med2.tif], 0), MIN ([mfNopt_High.tif], 0))* in the text box under the **Layers** label. Click the **Evaluate** button.
15. Again, a new layer will appear in the workspace. This temporary layer will be named *Calculation*. Export the temporary layer in TIFF format and rename the new map, following the suggestion in Table 6.4. Right-click with the mouse cursor on the *Calculation* layer. On the conceptual menu, click **Data > Export Data**. In the **Export Raster Data—Calculation** dialogue box, change the default filename in the **Name** label to *trNequal_20* and change the default file format in the **Formats** label to *TIFF*. Click the **Save** button. The *trNequal_20.tif* map file will appear as a new layer in the workspace. Keep the file open.
16. The *Calculation* layer is no longer necessary. Right-click with the mouse cursor on the *Calculation* layer. On the conceptual menu, click **Remove**.
17. Repeat steps 13–16 with all the formulas shown in Table 6.4 using copy paste. Rename the resulting maps with the filenames indicated in the "Resulting files (suggested filenames)" column in Table 6.4. *It is important*

TABLE 6.4
Mathematical Expression of the Defuzzification Formula Used in Exercise 6.3

Expressions	Resulting Files (Suggested Filenames)
MAX (MIN ([mfNopt_Low.tif], 1), MIN ([mfNopt_Med1.tif], 0), MIN ([mfNopt_Med2.tif], 0), MIN ([mfNopt_High.tif], 0))	trNequal_20.tif
MAX (MIN ([mfNopt_Low.tif], 0.8), MIN ([mfNopt_Med1.tif], 0.2), MIN ([mfNopt_Med2.tif], 0.2), MIN ([mfNopt_High.tif], 0))	trNequal_40.tif
MAX (MIN ([mfNopt_Low.tif], 0.4), MIN ([mfNopt_Med1.tif], 0.6), MIN ([mfNopt_Med2.tif], 0.6), MIN ([mfNopt_High.tif], 0))	trNequal_60.tif
MAX (MIN ([mfNopt_Low.tif], 0), MIN ([mfNopt_Med1.tif], 1), MIN ([mfNopt_Med2.tif], 1), MIN ([mfNopt_High.tif], 0))	trNequal_80.tif
MAX (MIN ([mfNopt_Low.tif], 0), MIN ([mfNopt_Med1.tif], 1), MIN ([mfNopt_Med2.tif], 1), MIN ([mfNopt_High.tif], 0))	trNequal_100.tif
MAX (MIN ([mfNopt_Low.tif], 0), MIN ([mfNopt_Med1.tif], 1), MIN ([mfNopt_Med2.tif], 1), MIN ([mfNopt_High.tif], 0))	trNequal_120.tif
MAX (MIN ([mfNopt_Low.tif], 0), MIN ([mfNopt_Med1.tif], 0.6), MIN ([mfNopt_Med2.tif], 0.6), MIN ([mfNopt_High.tif], 0.4))	trNequal_140.tif
MAX (MIN ([mfNopt_Low.tif], 0), MIN ([mfNopt_Med1.tif], 0.2), MIN ([mfNopt_Med2.tif], 0.3), MIN ([mfNopt_High.tif], 0.8))	trNequal_160.tif
MAX (MIN ([mfNopt_Low.tif], 0), MIN ([mfNopt_Med1.tif], 0), MIN ([mfNopt_Med2.tif], 0), MIN ([mfNopt_High.tif], 1))	trNequal_180.tif
MAX (MIN ([mfNopt_Low.tif], 0), MIN ([mfNopt_Med1.tif], 0), MIN ([mfNopt_Med2.tif], 0), MIN ([mfNopt_High.tif], 1))	trNequal_200.tif

that each expression be given the corresponding filename shown in the table. At the end, you should have 10 TIFF files (Table 6.4). These files will be used for the final exercise, Exercise 6.4.

Exercise 6.4: Defuzzification

The final defuzification step consists in the calculation of the *N* rate by the weighted summation as explained in the previous exercises.

Step-by-Step in ArcGIS

18. Click **Spatial Analyst** again on the Spatial Analyst toolbar, and select **Raster Calculator**. In the **Layers** label, you should have 10 more TIFF files in addition to the three original ones, the eight from Exercise 6.1, and the four calculated in Exercise 6.2. These 10 maps come from Exercise 6.3.
19. In the **Raster Calculator** dialogue box, copy paste the formula *(20 * [*tr*Nequal_20.tif] + 40 * [*tr*Nequal_40.tif] + 60 * [*tr*Nequal_60.tif] +*

80 * [trNequal_80.tif] + 100 * [trNequal_100.tif] + 120 * [trNequal_100.tif] + 140 * [trNequal_140.tif] + 160 * [trNequal_160.tif] + 180 * [trNequal_180.tif] + 200 * [trNequal_200.tif])/([trNequal_20.tif] + [trNequal_40.tif] + [trNequal_60.tif] + [trNequal_80.tif] + [trNequal_100.tif] + [trNequal_100.tif] + [trNequal_140.tif] + [trNequal_160.tif] + [trNequal_180.tif] + [trNequal_200.tif]) in the text box under the **Layers** label. Click the **Evaluate** button.

20. A new layer will appear in the workspace. This temporary layer will be named *Calculation*. Export the temporary layer in TIFF format and rename the new map, following the suggestion in Table 6.4. Right-click with the mouse cursor on the *Calculation* layer. On the conceptual menu, click **Data > Export Data**. In the **Export Raster Data—Calculation** dialogue box, change the default filename in the **Name** label to *Nopt_Fuzzy* and change the default file format in the **Formats** label to *TIFF*. Click the **Save** button. The *Nopt_Fuzzy.tif* map file will appear as a new layer in the workspace. Keep the file open.
21. The *Calculation* layer is no longer necessary. Right-click with the mouse cursor on the *Calculation* layer. On the conceptual menu, click **Remove**.
22. End of the exercise.

6.4 RESULTS

As shown in Figure 6.12E, the calculated N requirements vary between 36.6 and 159.9 kg N ha^{-1}, according to the field attributes and crop status, as measured by

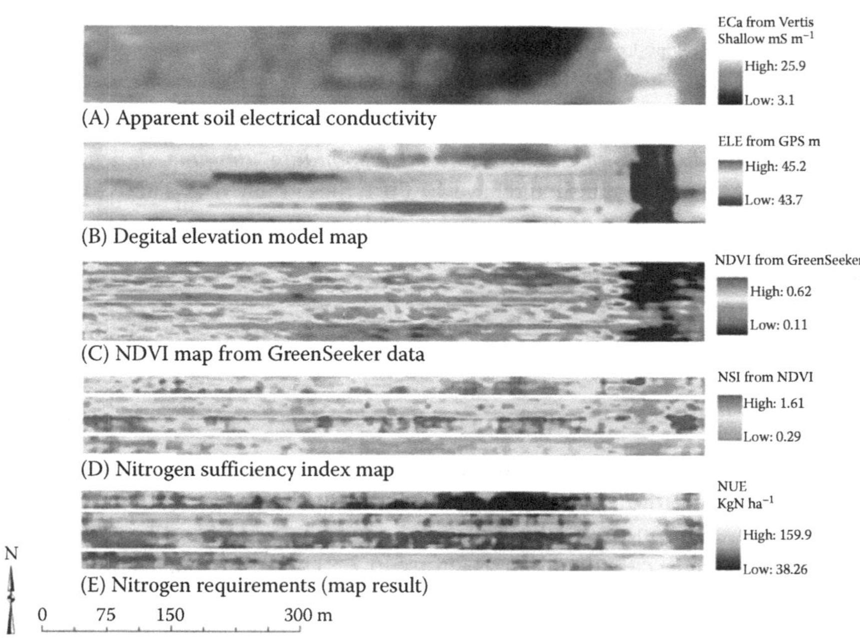

FIGURE 6.12 Based maps (A, B, C, and D) used to calculate N requirements (E) in a corn field.

canopy reflectance (NDVI) and translated into NSI. This map shows the high spatial variability of N requirements. The variable *N* rate proposed by the output map should help enhance nitrogen use efficiency and reduce the environmental impact of overfertilization.

6.5 CONCLUSIONS

In 2005, the implementation of this strategy would have resulted in the application of 557 kg N to this 8.66 ha corn field. This represents a 40% reduction compared to the 1386 kg that the grower would have normally applied (160 kg N ha^{-1}) with no anticipated reduction in crop productivity. As a result of the development of GPS, systematic soil sampling, geographic information systems, and precision farming, it has been determined that fields are highly spatially variable. The effect of this variability on fertilizer management is that the current uniform application practice results in areas where N is applied in excess of needs. Technologies are now available to account for this spatial variability and better match N supplies with crop requirements and help benefit the environment. Remote sensing technologies using airborne, satellite, or on-board platforms have been developed for that purpose. Such technologies can provide the information on the degree of N limitation endured by the crop at the moment of N fertilizer application. This time-critical information on crop status can be then added on top of the other layers of information that were previously obtained and are known to influence crop response of to N fertilizer applications. As a result, the spatial and timing aspects of a sound N fertilizer management can be considered fully. The purpose of the example used in this chapter was to provide a simple, hands-on overview of the way in which an FIS can be developed and adapted within a geographic information system as a decision-making tool for agricultural input management on the field scale. In order for the use of FISs in ArcGIS to become more flexible, access to or development of a fuzzy toolbox (using ArcObjects™) would be welcome.

ACKNOWLEDGMENT

The authors would like to acknowledge the support of the GAPS program of Agriculture and Agri-Food Canada.

REFERENCES

1. Davis, D.M., Gowda, P.H., Mulla, D.J., and Randall, G.W. Modeling nitrate nitrogen leaching in response to nitrogen fertilizer rate and tile drain depth or spacing for southern Minnesota, USA. *Journal of Environmental Quality* 29, 1568, 2000.
2. Tremblay, N., Fallon, E., Bélec, C., Tremblay, G., and Thibault, É. Growing season and soil factors related to predicting corn nitrogen fertilization in Quebec. In Bruulsema, T.W., ed. *Managing Crop Nitrogen for Weather*, Chapter 8. *Proceedings of the Symposium "Integrating Weather Variability into Nitrogen Recommendations"*. Sponsored by the Soil Science Society of America. Published by the International Plant Nutrition Institute, Norcross, GA, 2007, 12 pp.

3. CRAA: Centre de Références en Agriculture et Agroalimentaire du Québec (CRAAQ) 2003. Fertilisation reference guide (in French). ISBN 2-7649-0034-1.
4. Schroder, J.J., Neeteson, J.J., Oenema, O., and Struik, P.C. Does the crop or the soil indicate how to save nitrogen in maize production? Reviewing the state of the art. *Field Crops Research* 66, 151, 2000.
5. Randall, G.W., Vetsch, J.A., and Huffman, J.R. Nitrate losses in subsurface drainage from a corn-soybean rotation as affected by time of nitrogen application and use of nitrapyrin. *Journal of Environmental Quality* 32, 764, 2003.
6. Tran, T.S., Giroux, M., and Cescas, M.P. Utilisation de l'engrais azoté marqué au 15N par le maïs selon les modes d'application et les doses d'azote. *Canadian Journal of Soil Science* 77, 9, 1997.
7. Doerge, T.A. Management zone concepts. *Site-specific Management Guidelines*, SSMG-2:4. Potash & Phosphate Institute, Norcross, GA, 2000.
8. Doerge, T.A. Nitrogen measurement for variable-rate N management in maize. *Communications in Soil Science and Plant Analysis* 36, 23, 2005.
9. Hawkins, J.A., Sawyer, J.E., Barker, D.W., and Lundvall, J.P. Using relative chlorophyll meter values to determine nitrogen application rates for corn. *Agronomy Journal* 99, 1034, 2007.
10. Hong, N., White, J.G., Weisz, R., Crozier, C.R., Gumpertz, M.L., and Cassel, D.K. Remote sensing-informed variable-rate nitrogen management of wheat and corn: Agronomic and groundwater outcomes. *Agronomy Journal* 98, 327, 2006.
11. Raun, W.R., Solie, J.B., Martin, K.L., Freeman, K.W., Stone, M.L., Johnson, G.V., and Mullen, R.W. Growth stage, development, and spatial variability in corn evaluated using optical sensor readings. *Journal of Plant Nutrition* 28, 173, 2005.
12. Welsh, J.P., Wood, G.A., Godwin, R.J., Taylor, J.C., Earl, R., Blackmore S., and Knight, S.M. Developing strategies for spatially variable nitrogen application in cereals, part II: Wheat. *Biosystems Engineering* 84, 495, 2003.
13. Raun, W.R., Solie, J.B., Taylor, R.K., Arnall, D.B., Mack, C.J., and Edmonds, D.E. Ramp calibration strip technology for determining midseason nitrogen rates in corn and wheat. *Agronomy Journal* 100, 1088, 2008.
14. Tremblay, N. and Belec, C. Adapting nitrogen fertilization to unpredictable seasonal conditions with the least impact on-the environment. *Horttechnology* 16, 408, 2006.
15. Mullen, R.W., Freeman, K.W., Raun, W.R., Johnson, G.V., Stone, M.L., and Solie, J.B. Identifying an in-season response index and the potential to increase wheat yield with nitrogen. *Agronomy Journal* 95, 347, 2003.
16. Pennock, D.J., Anderson, D.W., and de Jong, E. Landscape-scale changes in indicators of soil quality due to cultivation in Saskatchewan, Canada. *Geoderma* 64, 1, 1994.
17. Heiniger, R.W., McBride, R.G., and Clay, D.E. Using soil electrical conductivity to improve nutrient management. *Agronomy Journal* 95, 508, 2003.
18. Kitchen, N.R., Drummond, S.T., Lund, E.D., Sudduth, K.A., and Buchleiter, G.W. Soil electrical conductivity and topography related to yield for three contrasting soil-crop systems. *Agronomy Journal* 95, 483, 2003.
19. Dharmakeerthi, R.S., Kay, B.D., and Beauchamp, E.G. Spatial variability of in-season nitrogen uptake by corn across a variable landscape as affected by management. *Agronomy Journal* 98, 255, 2006.
20. Kravchenko, A.N., Thelen, K.D., Bullock, D.G., and Miller, N.R. Relationship among crop grain yield, topography, and soil electrical conductivity studied with cross-correlograms. *Agronomy Journal* 95, 1132, 2003.

21. Kravchenko, A.N., Robertson, G.P., Thelen, K.D., and Harwood, R.R. Management, topographical, and weather effects on spatial variability of crop grain yields. *Agronomy Journal* 97, 514, 2005.
22. Marques da Silva, J.R. and Alexandre, C. Spatial variability of irrigated corn yield in relation to field topography and soil chemical characteristics. *Precision Agriculture* 6, 453, 2005.
23. Derby, N.E., Steele, D.D., Terpstra, J., Knighton, R.E., and Casey, F.X.M. Interactions of nitrogen, weather, soil, and irrigation on corn yield. *Agronomy Journal* 97, 1342, 2005.
24. Isik, M. and Khanna, M. Variable-rate nitrogen application under uncertainty: Implications for profitability and nitrogen use. *Journal of Agricultural and Resource Economics* 27, 61, 2002.
25. Kyveryga, P.M., Blackmer, A.M., and Morris, T.F. Disaggregating model bias and variability when calculating economic optimum rates of nitrogen fertilization for corn. *Agronomy Journal* 99, 1048, 2007.
26. Mamo, M., Malzer, G.L., Mulla, D.J., Huggins, D.R., and Strock, J. Spatial and temporal variation in economically optimum nitrogen rate for corn. *Agronomy Journal* 95, 958, 2003.
27. Scharf, P.C., Kitchen, N.R., Sudduth K.A., and Davis, J.G. Spatially variable corn yield is a weak predictor of optimal nitrogen rate. *Soil Science Society of America Journal* 70, 2154, 2006.
28. Berntsen, J., Thomsen, A., Schelde, K., Hansen, O.M., Knudsen, L., Broge, N., Hougaard, H. et al. Algorithms for sensor-based redistribution of nitrogen fertilizer in winter wheat. *Precision Agriculture* 7, 65, 2006.
29. McBratney, A.B. and Odeh, I.O.A. Application of fuzzy sets in soil science: Fuzzy logic, fuzzy measurements and fuzzy decisions. *Geoderma* 77, 85, 1997.
30. Assimakopoulos, J.H., Kaliva, D.P., and Kollias, V.J. A GIS-based fuzzy classification for mapping the agricultural soils for N-fertilizers use. *Science of the Total Environment* 309, 19, 2003.
31. Mertens, M. and Huwe, B. FuN-Balance: A fuzzy balance approach for the calculation of nitrate leaching with incorporation of data imprecision. *Geoderma* 109, 269, 2002.
32. Jones, D. and Barnes, E.M. Fuzzy composite programming to combine remote sensing and crop models for decision support in precision crop management. *Agricultural Systems* 65, 137, 2000.
33. Mandani, E.H. and Assilian, S. An experiment in linguistic synthesis with a fuzzy logic controller. *International Journal of Man-Machine Studies* 7, 1, 1975.
34. Guillaume, S. Designing fuzzy inference systems from data: An interpretability-oriented review. *IEEE Transactions on Fuzzy Systems* 9, 426, 2001.
35. Alonso, J.M., Magdalena, L., and Guillaume, S. Hilk: A new methodology for designing highly interpretable linguistic knowledge bases using the fuzzy logic formalism. *International Journal of Intelligent Systems* 23, 761, 2008.
36. Guillaume, S., Charnomordic, B., and Lablée, J.L. FisPro: Logiciel open source pour les systèmes d'inférence floue, 2008, available at http://www.inra.fr/internet/Departements/MIA/M/fispro/

7 Digital Northern Great Plains and Zone Mapping Application for Precision Agriculture

Xiaodong Zhang

CONTENTS

7.1 EXECUTIVE SUMMARY

To lower risk, enhance productivity, increase energy efficiency, and improve profitability, agricultural producers are increasingly turning to the use of information technology (IT) to aid their decision-making processes. To overcome the traditional bottleneck impeding prompt and easy access by agricultural producers to information like satellite imagery, we have developed two web-based data and decision support systems, digital northern great plains (DNGP) and zone mapping applications for precision farming (ZoneMAP). The two systems are connected with each other, driven by a common database system for remote sensing imagery (RSI) from both aircrafts and satellites. DNGP is primarily for data delivery augmented with common GIS capabilities and ZoneMAP focuses on decision support with automatic determination of the number of management zones and their delineation. The focus of this chapter is the applications of the two systems on the Northern Great Plains region of the United States and how the techniques and methodology can easily be used for other areas of the country and world.

7.2 INTRODUCTION

While recent increased demand on global food supply has led to improved economic returns for the agricultural industry, there has also been an increase in risk associated with food production systems because input costs have also increased significantly.[1] To lower risk, increase energy efficiency, enhance productivity, and improve profitability, agricultural producers are increasingly turning to the use of IT to aid their decision-making processes.[2] In addition to land, labor, and capital, which have long been agriculture's traditional assets, information management has become the fourth asset,[3] of increasing importance and has come to be known as precision agriculture. According to the National Research Council, "precision agriculture is a management strategy that uses information technologies to bring data from multiple sources to bear on decisions associated with crop production."

Information of both spatial and temporal dimensions is required for precision agriculture. RSI acquired from satellites and aircrafts offers tremendous potential in providing information that is needed for precision crop management. Of course, information is valuable to producers only if it is timely, accurate, and can be (a) easily accessed, (b) straightforwardly integrated with multiple sources, (c) analyzed with software and hardware a typical information-seeker possesses, and (d) used with a minimum of training. These challenges are magnified in the case of precision agriculture by the digital size of satellite scenes and the limited bandwidths available to producers in many rural areas.

Typically, remote sensing data is delivered by scene, i.e., one IMAGE of the surface of the Earth. A single Landsat scene covering ~740,000 ac and contains about 500 MB of data. For a 56 k dialup connection to the Internet, still used by some rural residents, downloading one scene would take 20 h. A typical farm field of 3000 ac, though, only occupies 1/250th of a scene. Obviously, if an image can be partitioned, even a slow connection can provide enough bandwidth for access. Adoption of remote sensing technology is also affected by the depths of scientific and technical knowledge required to analyze and interpret the data. One way to lower the barrier is to develop value-added products that can be easily interpreted. To a producer, the value of remote sensing images lies in how to convert them into a management decision that enhances productivity. To do so, images or their products have to be integrated with other applications. Compatibility of the data and product with others is important to ensure that we deliver not just data but productivity.

In response to these challenges, we designed and developed two web-based systems, DNGP and ZoneMAP. Both systems follow a client–server design of architecture, offering a simple web interface that is easy to use and a server technology that handle data access, processing, and presentation. Also, the two systems are linked together by the same database system of RSI.

7.3 METHODS

7.3.1 Data and Database System

The Upper Midwest Aerospace Consortium (UMAC, http://www.umac.org) has collected a rich archive of RSI over the northern Great Plains, including North and South Dakota, Minnesota, Montana, Wyoming, and Idaho spanning more than 30 years. Data include medium-resolution (20–250m) multispectral images from satellite sensors landsat multispectral scanner (MSS), thematic mapper (TM) and enhanced thematic mapper plus (EMT+), advanced space-borne thermal emission and reflection radiometer (ASTER), moderate-resolution imaging spectroradiometer (MODIS), surface relief from shuttle radar topography mission (SRTM), and high-resolution (1–2 m) images from AeroCam,[4] a multispectral airborne camera that was developed and operated by UMAC. To ensure consistency in temporal and spatial comparisons, all images have been atmospherically corrected. The final product is reflectance at the surface.

All images are managed though an Oracle database system with a spatial operation extension. While Oracle database system offers the capability of storing binary data such as images, we chose to store images as external GeoTIFF files and to link the images to database through their attributes in both regular and spatial dimensions. An important spatial attribute is the border lines of an image, which are used for spatial operation such as intersection or containment. Often the polygon of the border lines of a geo-referenced image does not match the geo-extension of the image file, due to the inclination of the orbit. For these images, the border lines of the image are preferred. We have developed an algorithm to automatically detect the image extensions from Landsat-type data file. The database system also saves the information about users, among which a very convenient piece of information is the area of interest (AOI) that a user may use frequently.

7.3.2 Digital Northern Great Plains

In designing the DNGP system, Zhang[5] adopted "thin-client" architecture to ensure the minimum footprint on a client computer; with all computing and analysis carried out by the host server and results presented through a web interface accessible via a web browser. There are three tiers in the design: client presentation, data processing, and data storage. The client tier is the graphical user interface, through which users interact with DNGP and its database. The web interface features a modularized design with three independent panels for map, image list, and layers. Since the contents of the interface are constantly changing due to, e.g., a change of image, we use AJAX to ensure dynamic and targeted update of content. Figure 7.1 shows the interface.

The data processing tier is the main part of the DNGP implementation where the server side process takes place, interacting with the client and database. This tier was implemented using the open source applications. Raster and vector data are processed using Geospatial Data Abstraction Library and OpenGIS Simple Features Reference Implementation library, respectively. Various raster and vector layers are combined into a final image using another open-source package, MapServer.

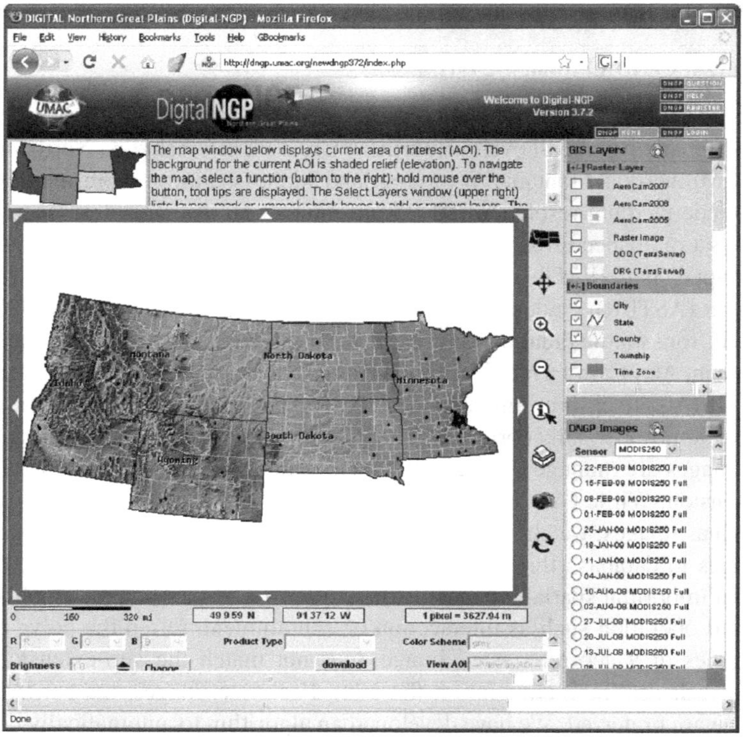

FIGURE 7.1 The DNGP web user interface.

We also used PROJ.4 Cartographic Projections Library for coordinate conversion between different projections and datum. The data storage is managed by an Oracle database management system with spatial extension.

The DNGP system has several features that enable users with low-bandwidth Internet connections to search and download remote sensing data (RSD) and products easily and quickly. These features include

- Intuitive user interface, which allows users to conduct searches via spatial coordinates. This is critical because essentially a farm is a spatial object. With thousands of images archived, this allows a user to focus quickly on his or her particular AOI.
- Remote sensing images can be subset either spatially or spectrally. A significant amount of bandwidth can be saved, and hence a timely delivery achieved, by eliminating data of no interest to a user.
- Products (e.g., sugar beet yield) are generated on the fly, which not only simplifies the database design but provides dynamic update capability as well. Whenever algorithms used in developing products are improved, users have access to them.
- Images or products can be downloaded in a variety of formats to ensure compatibility with other application software.

- A multitude of vector layers, including township, highways, lakes, rivers, and other geographic features has been incorporated into the system to help users identify their targets.
- "One interface" design ensures a smooth user experience and simplifies the learning curve.

7.3.3 Zone Mapping: Precision Farming

With DNGP system providing data access that is tailored to the needs of agriculture producers, ZoneMAP was designed to offer a decision support tool to further enhance productivity.[6,7] By connecting to the same database system with a rich archive of RSD spanning the past 30 years, ZoneMAP allows a user to define variability within a field using spatial information that is multispectral and multitemporal.

We chose fuzzy c-means as the clustering algorithm for ZoneMAP. It is basically the same as that used by Fridgen et al.[8] except for the method of estimating the measure of similarity between observations and centers of the clusters that are to be determined. Typically, measure of similarity can be estimated using Euclidean distance, diagonal distance, or Mahalanobis distance. Since the Euclidean distance algorithm requires variables to be of equal variances and noncorrelated, which are rarely true in reality, we only implemented the latter two algorithms. The diagonal distance algorithm compensates for the difference in variances of different variables, but it is still sensitive to correlated variables. The Mahalanobis distance algorithm is designed for data sets that are both correlated and have different variance by adjusting the distance with variance–covariance matrix within the pool of data sets. We found that both the diagonal and Mahalanobis distance algorithms gave very similar results in our testing. We used the diagonal distance algorithm in classification because it is much faster and therefore highly suitable for a web-based application, such as ZoneMAP.

Determining the most appropriate number of zones is difficult in the interpretation of unsupervised classification. Fridgen et al.[8] used the convergence of fuzziness performance index (FPI) and normalized classification entropy (NCE) to determine the optimal number of management zones. Theoretically, the best classification occurs when membership sharing (FPI) and the amount of class disorganization (NCE) is at a minimum with the least number of classes used. However, sometimes NCE and FPI do not converge and the optimal number of zones suggested by one parameter is significantly different from the one suggested by the other.[9]

Another method to evaluate classification success is to estimate how much within-cluster variability is reduced for a number (n) of clusters as compared with $n-1$ clusters. We have found that generally the percentage of total within-cluster variability with respect to the total initial variability decreases as the number of clusters increases. A similar trend for the variance reduction was found by Brock et al.[9] We also found that typically the total within-cluster variance decreases rapidly initially and then approaches an asymptotical value slowly as the number of clusters continues to increase. The optimal number of zones is therefore decided as the number of clusters that reduces the variance significantly as compared to the initial variability, yet changes little when the number of zones is further increased. By trial and error, we came to two criteria that can capture this turning point in a relatively consistent manner: (1) overall reduction of variance is >50% and (2) consecutive reduction of variance is <20% or the trend is broken, i.e., within-cluster variability increases instead of decreases.[7]

7.4 EXERCISE

The purpose of this chapter is to introduce the systems that we have developed and to demonstrate how the tools can best be used for practical applications. The second objective will be addressed in this section.

7.4.1 DNGP

To access DNGP, go to http://dngp.umac.org. Figure 7.1 shows the web interface of DNGP, which consists of three panels, the image panel, the GIS layer panel, and the DNGP list of images panel. The image panel is the primary interface showing an image that a user is interested in and the associated navigation and image operations. The list of images is organized by the sensors and only those images that cover at least a portion of the current AOI are listed. Therefore, the content of the list will be dynamically updated as a user changes the geographic extension, which can be achieved either through navigation (zoom in/out or pan) and by selecting an AOI that has been saved. The GIS layer offers a variety of layers with information ranging from city boundaries to soil type. Once an image is selected for viewing, the control panel immediately below the image window becomes active, and some simple image processing operations (e.g., band combination, intensity) are available. For example, for multispectral images from Landsat TM sensor, the default combination of bands for the colors of red, green, and blue is band 4, band 3, and band 2, respectively. With this combination, healthy vegetation would apprear reddish. This combination of bands can be changed and potentially different information regarding a surface property can be highlighted in different colors.

Also, based on a selected image, several popular vegetation indexes such as normalized differential vegetation index (NDVI) or green NDVI (GNDVI) can be calculated. NDVI is defined as (NIR band − red band)/(NIR band + red band) and GNDVI is defined as (NIR band − green band)/(NIR band + green band). The design of DNGP is such that these calculations are carried out on-the-fly and, therefore, the existing algorithms can be easily updated and new algorithms can be easily introduced.

For an image or its product that a user would like to download, there are two options: as an ASCII file or a GeoTIFF image. For the ASCII format, the first two columns are longitude and latitude, and the rest columns are the values. For example, the third column for an NDVI product will be the values of NDVI; while for a typical Landsat image, the third column will be its band 1 data, fourth column its band 2 data, and so on.

While using DNGP does not require registration, users who register do have an advantage of having their AOIs saved for later retrieval. For a registered user, the image window can automatically focus on a saved AOI once it is selected from

"View AOI" drop-down menu. This is very convenient for regular users who need to revisit the same location frequently. To date, DNGP has had over 800 registered users. Further help on how to use DNGP can be found by clicking "help" button located in the upper right corner.

7.4.2 ZoneMAP

While DNGP focuses on RSD delivery, ZoneMAP intends to be a decision support tool. Both of them are internally linked and supported by the same spatial data processing architecture. To access ZoneMAP, go to http://zonemap.umac.org.

Figure 7.2 shows the login screen of ZoneMAP. Because the use of ZoneMAP starts with an AOI, it is required that a user have an account with DNGP, to which an AOI can be defined and saved. There are three clickable buttons at the top of the login interface, HELP, FEATURES, and TUTORIALS. For the first-time users, we highly recommend going through each of these. The FEATURES highlights the latest improvement and new features that have been developed. The TUTORIALS is still an ongoing project and so far we have finished five tutorials on "how to login," "change your AOI," "preview a satellite image," "upload an image," and "upload a text file." All the tutorials are written in Flash animation, and are straightforward to follow. The HELP summarizes the ZoneMAP project. Users may find that throughout the ZoneMAP interfaces, there are red question marks associated with each panel. These are context-sensitive help links that provide information on that particular panel or operation. For example, there is a red question mark in the upper right corner of the LOGIN panel (Figure 7.2). Clicking it leads to additional information regarding login and its associated operations.

Once logged in, the user needs to select an AOI to start with (Figure 7.2B). If no AOIs have been defined, the user will be asked to create an AOI. AOI is defined by rectangle bounding box with latitudes and longitudes. Because the algorithm described above to delineate a field into zones is computational intensive, it is highly

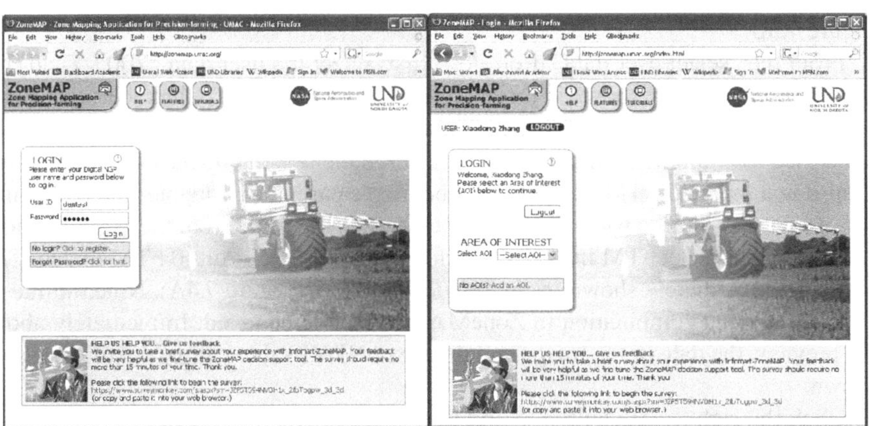

(A) (B)

FIGURE 7.2 ZoneMAP web interface: (A) before login and (B) after login.

recommended to limit the sizes of AOIs within a section (1 mi × 1 mi). The zones created are also more representative because unrelated information such as roads that divide sections is not included. To name the AOIs, use letters, numbers or their combinations only. While special characters can be used, they are not recommended. Also, do not include spaces in the name.

The next step will be to select or upload data that will be used for creating zones. There are two major panels: Manage User Data (Figure 7.3) and Create Zone Map (Figure 7.4). By definition, the User Data include the data that a user provides, such as soil electro-conductivity, pH values, yield data, etc., and the zone maps that a user would have created. The data that a user provides can be either in a text or imagery format. But data in the text format will be transformed into an image within the boundaries defined by the AOI. The image data from the user does not have to fit exactly with the AOI; ZoneMAP will transform the image. In the "view/delete user data" panel (Figure 7.3A), a user data "2003 Yield 02" is highlighted. The data was originally in a text format and was converted to a GeoTIFF by ZoneMAP. In the "data description," a text message of "The corn yield data in 2003" was typed and will be saved along with the yield data.

Figure 7.3B shows a zone map, highlighted as "example" and created using this yield data and the NDVI map derived from a Landsat TM image on August 25, 2004. From this panel, a user can also download user data. Data can be downloaded in three formats, as an image, an ASCII text file, or an ESRI Shapefile, and all of them are geo-referenced. For text or Shapefile download, the application rates for each zone can be prescribed and will be downloaded together with the zone map (see lower right corner of Figure 7.3B). In the exercise, arbitrary values between 10 and 60 are assigned to the different zones as defined by different colors.

As its name implies, zones are created in the Create Zone Map panel, from which the images to be used for classification are selected. Images can be selected from DNGP database system or from user data. For each image to be selected, its content can be viewed in the preview window. In the following, we will repeat the procedures that have been used to create the zone map "example" as shown in Figure 7.3B.

From the "select user data" drop-down list, select the user data "2003 Yield 02" and its content will be shown in the "preview user data/image." Click the right arrow button in the "building list of sources" panel to add this data into the list of images to be used for classification. From the "select remote sensing images/select a sensor" drop-down list, select TM. Note the number in the parenthesis denotes the total number of images acquired by the TM sensor that cover the AOI. From the "select remote sensing images/select TM image" drop-down list, select 25-Aug-04 TM image. The image is immediately shown in the preview window (Figure 7.4A). Note, unlike in DNGP, the band combination in ZoneMAP cannot be changed. Immediately above the preview, the date of the image and the mean NDVI value within the AOI are also shown.

Click the right arrow button to add this image to the list and because the image has multiple bands, a user needs to decide what bands are to be used. NDVI and GNDVI can also be used for classification. For this exercise, select NDVI (lower right corner of Figure 7.4A).

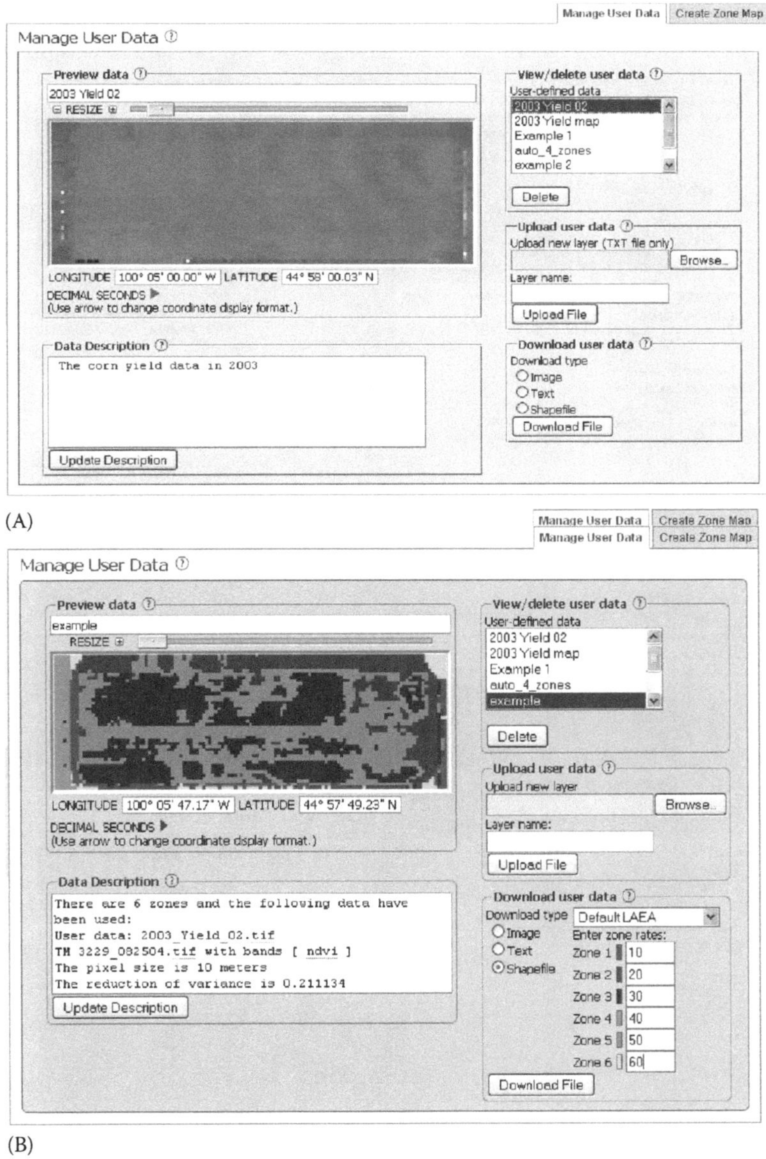

FIGURE 7.3 Manage User Data panel: (A) a 2003 yield data uploaded by the user and (B) the zone map created by using the yield data and NDVI derived from a Landsat TM image on August 25, 2004, and its download with an option of inputting application rate for each zone.

The other parameters that are required are listed in "create zone map preview" panel. Here, a user needs to define the number of zones and the pixel size in meters for the final zone map. The default pixel size is 10m and the default number of zones is 3. We recommend a user to test different values and evaluate which ones give the best results. For the exercise, choose 6 for the number of zones and use 10m as the pixel size

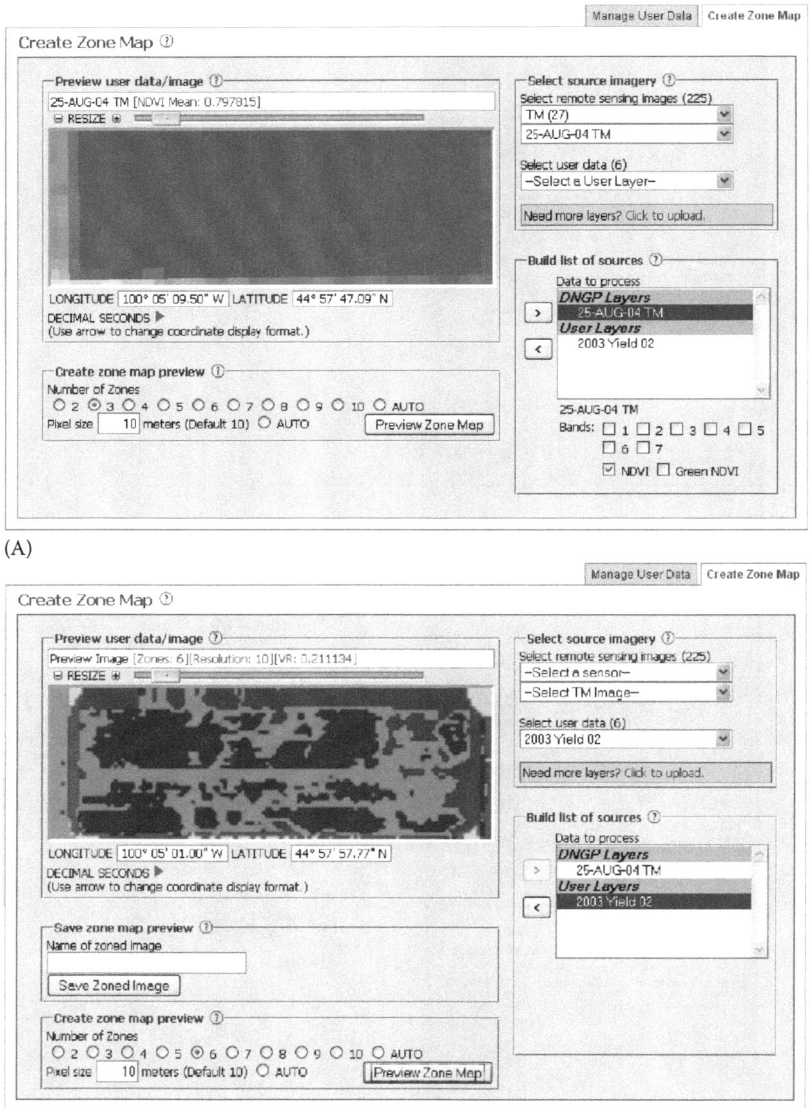

(B)

FIGURE 7.4 Create Zone Map panel: (A) a TM image on August 25, 2004 was selected and (B) a zone map with six zones and a 10 m pixel size was created using the NDVI estimated from the selected TM image and a user data of yield for corn in 2003.

(Figure 7.4B). Click "Preview Zone Map" button. After the calculation is done, the zone map will be displayed in the preview window. Once the user thinks the zone map makes sense and wants to save it, simply type a name ("example" without quotes) in the "name of zoned image" text box (note no spaces), and click "saved zoned image" button. The zone map will be saved and is linked to the current AOI. Also saved are the metadata associated with this zone map and this information can be viewed and modified in "data description" under the "manage user data" panel. The zone map, example, that we just created, the metadata reads "There are six zones and the following data have been used: User data: 2003_Yield_02.tif, TM 3229_082504.tif with bands [NDVI]. The pixel size is 10m. The variance reduction of, 0.211134, suggests that management could be improved by separating the field into management zones.

7.5 CONCLUSIONS

Both systems of NDGP and ZoneMAP were designed to bring space technology to ordinary users who are considering the use of precision agriculture technologies. The guideline of the design has been intentionally made simple, effective, and easy to use. Since the inception of DNGP in 2003 and ZoneMAP in 2007, we have seen a steady increase in their usages. For example, the average number for daily unique visits of DNGP site has increased from under 10 when it first started to over 70 now. However, all these tools are intended for decision support and not decision making. It is important for users to evaluate the results, especially the zones created by ZoneMAP, and to make sure that the delineation conforms to their knowledge of the field.

ACKNOWLEDGMENTS

The funding for the projects was provided by Universities Space Research Association (USRA) through grants NCC8-259 and 03450-08 and by NASA through grant NNX06AE16G. We would like to thank Dan Forgey, Gary Wagner, and many other end users for sharing their data as well as knowledge. Without their help, this research would not be possible.

REFERENCES

1. Tilman D., Cassman K.G., Matson P.A., Naylor R., and Polasky S. Agricultural sustainability and intensive production practices. *Nature* 418, 671, 2002.
2. Hendrickson J., Sassenrath G.F., Archer D., Hanson J., and Halloran J. Interactions in integrated US agricultural systems: The past, present and future. *Renewable Agriculture and Food Systems* 23, 314, 2008.
3. Hanson J.D., Hendrickson J., and Archer D. Challenges for maintaining sustainable agricultural systems in the United States. *Renewable Agriculture and Food Systems* 23, 325, 2008.
4. AeroCam. http://www.umac.org/sensors/aerocam/index.html, 2009, accessed April 3, 2009.
5. Zhang X. http://dngp.umac.org, 2004, accessed May 20, 2009.
6. Zhang X., Shi L., Seielstad G., and Helgason C. http://zonemap.umac.org, 2007, accessed May 20, 2009.
7. Zhang X., Shi L., Jia X., Seielstad G., and Helgason C. Zone mapping application for precision-farming: A decision support tool for variable rate application. *Precision Agriculture*. DOI: 10.1007/s11119-009-9130-4.
8. Fridgen J.J., Kitchen N.R., Sudduth K.A., Drummond S.T., Wiebold W.J., and Fraisse C.W. Management zone analyst (MZA): Software for subfield management zone delineation. *Agronomy Journal* 96, 100, 2004.
9. Brock A., Brouder S.M., Blumhoff G., and Hofmann B.S. Defining yield-based management zones for corn-soybean rotations. *Agronomy Journal* 97, 1115, 2005.

8 Spatial Variability of Field Machinery Use and Efficiency*

Viacheslav I. Adamchuk, Robert D. Grisso, and Michael F. Kocher

CONTENTS

8.1 EXECUTIVE SUMMARY

In site-specific crop management, it is a common practice to log the geographic coordinates of agricultural machinery measured using a global satellite navigation system (GNSS) such as the global positioning system (GPS). Yield, fertilizer application, and seed placement maps provide useful data for making agronomic decisions. However, the travel path itself reveals valuable information about machinery performance. Often, during field operations, odd field shapes, obstacles, or contour farming will require operators to increase the complexity of the machinery maneuvering. This usually reduces field efficiency. This chapter presents a methodology to parameterize the spatially variable characteristics of traffic patterns, and to define field areas where field efficiency is significantly reduced. Geographic positions recorded during the harvesting of a field with a complex shape are provided to illustrate the method developed. The information obtained can be used to optimize traffic patterns, or to reevaluate the potential profitability of field areas that require different degrees of complexity in machinery maneuvering and therefore require varying energy use.

* The original work was presented at the 2004 ASAE Annual International Meeting as Paper No. 041149.

8.2 INTRODUCTION

Implementing precision agriculture practices in modern crop production generates a large volume of records containing the coordinates of the agricultural machinery's locations during various field operations. Historically, these coordinates were used to determine the location of the physical value associated with the corresponding field operation (i.e., crop yield, application or seeding rate, soil characterization, tractor performance, implement draft, etc.). Developing and processing numerous layers of spatial data has proven popular for making use of geographical coordinates. The log of the times and geographic coordinates of agricultural machinery within a field provides valuable information on machinery performance that can (and probably should) be used to determine the spatially variable cost of field operation. According to MAX® (Farming for MAXimum Efficiency, Conservation Technology Information Center, West Lafayette, IN), field operation costs can be as high as 25% of the total cost of crop production. Since field geometry frequently causes farmers to invest greater effort and time in operating within non-rectangular areas of the field (waterways, terraces, etc.), it can be misleading when developing profitability maps to assume a uniform distribution of machinery operation costs across the entire field area.

Field capacity (FC) (effective and theoretical) and field efficiency (FE) (ASABE, 2008a) are the primary parameters used to evaluate machinery performance. While FC represents the area of land processed per unit time for a particular field operation, FE is defined as the ratio between effective and theoretical field capacities and relates the estimated and actual time required to complete a field operation (with no reference to the area). In the past, field capacity or efficiency were evaluated only on a field basis, either by using machinery operation parameters or by simply using a reference table (ASABE, 2008b). For example, Renoll (1981) used a conventional recording method (a stop watch and a clipboard) to determine field machinery performance. Alternatively, Grisso et al. (2002) as well as Taylor et al. (2002) used records of machinery location determined with a GPS receiver. They proved geospatial field records to be an effective resource for evaluating overall machinery performance. In a study conducted by Grisso et al. (2004), the positions of agricultural machinery logged during harvesting and planting operations were used not only to evaluate FE but also to define parameters representing the complexity of traffic patterns. Steering angle, steering angle per distance traveled, steering rate, and radius of curvature were the primary indices introduced. Their field averages indicated some correlation with the overall FE when fields with various types of traffic patterns were analyzed.

The primary objective of this chapter is to present a methodology for using records of agricultural machinery positions and times to evaluate the spatial variability of machinery performance. Specifically, analytical tools are presented for the construction of spatial maps representing the variability of FE.

8.3 METHODS

Several approaches are available for processing machinery position log files, including filters and geometrical transformations. However, in every case, the efficiency of farm machinery operation can be affected by three factors: (1) the travel speed,

(2) the effective swath width, and (3) the field traffic pattern. The position logging interval is assumed to be constant, but can affect the results reported as well. In this work, area coverage was used as the primary parameter in combining the three factors.

8.3.1 Algorithm Development

To develop an algorithm for traffic pattern processing, the following assumptions were made: first, that the entire log was created using a fixed time interval between successive machinery location coordinates (a logging option not always available); and second, that the resulting map should show the area coverage of the machine for every field location. It was also assumed that the coverage of farm machinery can be simplified using a sequence of rectangular segments defined by the recorded geographic positions. This is not the case when turning and the logging interval is relatively large. Figure 8.1 illustrates a route represented by four points: A, B, C, and D. Each segment of this route can be represented either as a sequence of rectangles with constant width (w) and variable distance (d), or as rectangles with constant width and fixed distance (d_f). The variable distance can be used to determine the actual coverage provided by the machine, and the fixed distance can be used to assess deviation from the theoretical coverage if a constant travel speed was maintained while operating along the same route. More complex segments incorporating travel pass curvature could also be used to better represent the actual coverage. In such a case, a minimum of three consecutive points would be used to estimate the radius of curvature and then determine the area of a sector of an annulus better representing the true ground coverage.

To assure that every field location has a defined coverage, an equally spaced grid with minimum and maximum coordinates corresponding to the endpoints of the field was used to construct the output. Figure 8.2 shows such a grid representing a section of a field with points corresponding to the machinery route A–B–C–D. Every linear segment (i.e., AB, BC, and CD) of the route was represented by the rectangular area coverage and was related to the grid cells overlapped by this rectangle. To illustrate the calculation procedure, a linear segment between points B

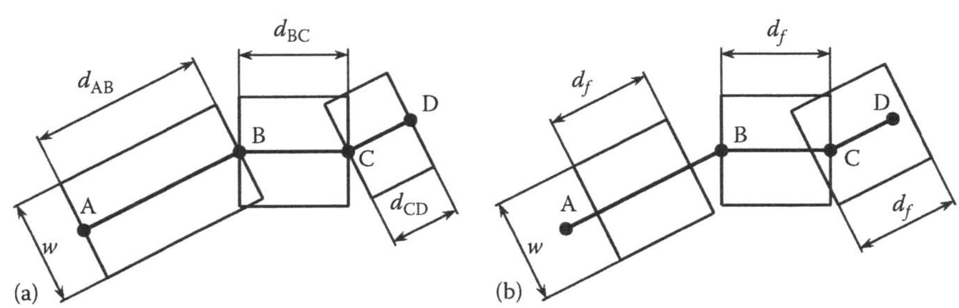

FIGURE 8.1 Simplified segments of equipment route ABCD using (a) variable (actual) travel speed and (b) constant (theoretical) speed of operation.

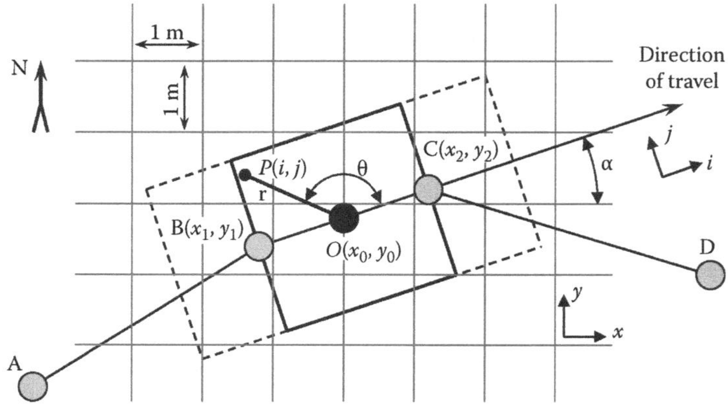

FIGURE 8.2 Area coverage computation diagram.

(with coordinates x_1, y_1) and C (with coordinates x_2, y_2) was considered. Values x and y corresponded to easting and northing coordinates, expressed in linear units (m). The simplified coverage segment was represented by a rectangle with width (w) corresponding to the physical width of the implement and the travel length (d) calculated as the distance between B and C:

$$d = \sqrt{(x_2 - x_1)^2 + (y_2 - y_1)^2} \tag{8.1}$$

The center of this rectangle $O(x_0, y_0)$ had coordinates

$$x_0 = \frac{x_1 + x_2}{2}, \quad y_0 = \frac{y_1 + y_2}{2} \tag{8.2}$$

By contrast, the same rectangular coverage area was viewed as an array of infinite points P with local coordinates i and j (Figure 8.2). It was assumed that the direction of i corresponded to the direction of travel and that for the center of rectangle O both i and j were equal to zero. Therefore, the i coordinate for point P ranged from $-d/2$ to $d/2$, and the j coordinate ranged from $-w/2$ to $w/2$. If the increments of the i and j coordinates are set to a finite number, a defined array of points P (i,j) can be obtained. In this example, the increment for both coordinates was set 10 times smaller (0.1 m) than the side of a square grid cell (1 m). This allowed the total of $100 * d * w$ points P arranged in a $10d \times 10w$ array to represent the entire rectangle. Through a method illustrated in Figure 8.2, each point was assigned to the respective grid cells using x and y coordinates. This resulted in each of the $100 * d * w$ points P for a given rectangle being assigned to the respective grid cells covered by that rectangle. The number of points from the given rectangle that were within the boundaries of each grid cell, was added to the point total for that grid cell.

The (x, y) and (i, j) coordinate systems were related using the angle α between the travel direction and the positive x axis (true east):

$$\alpha = \begin{cases} \dfrac{\pi}{2} & \text{for } x_1 = x_2 \text{ and } y_1 < y_2 \\ \dfrac{3\pi}{2} & \text{for } x_1 = x_2 \text{ and } y_1 > y_2 \\ \tan^{-1}\left(\dfrac{y_2 - y_1}{x_2 - x_1}\right) & \text{for } x_1 < x_2 \\ \pi + \tan^{-1}\left(\dfrac{y_2 - y_1}{x_2 - x_1}\right) & \text{for } x_1 > x_2 \end{cases} \tag{8.3}$$

For cases in which both x and y coordinates remained the same between consecutive data records (e.g., stops), α values determined for the preceding records were used. In addition to the local rectangular coordinates i and j, every point P was defined using local polar coordinates r and θ with respect to the center of the rectangle O and the positive direction of i. This provided a relatively simple way to account for changes in travel direction for each new rectangle. Points with coordinates $i = 0$ or $j = 0$ were avoided to reduce the number of logical operators. Therefore

$$r = \sqrt{i^2 + j^2} \quad \text{and} \quad \theta = \begin{cases} \tan^{-1}\left(\dfrac{j}{i}\right) & \text{for } i > 0 \\ \pi + \tan^{-1}\left(\dfrac{j}{i}\right) & \text{for } i < 0 \end{cases} \tag{8.4}$$

To determine the grid cell coordinates (x, y) associated with a point $P(r, \theta)$, the following equations were used with values rounded to the nearest integer:

$$\begin{aligned} x &= \text{round}(x_0 + r\cos(\theta + \alpha)) \\ y &= \text{round}(y_0 + r\sin(\theta + \alpha)) \end{aligned} \tag{8.5}$$

After running the algorithm (Figure 8.3), a two-dimensional array Coverage 1 was obtained, with values corresponding to the percent coverage for each square meter of the field. This array represents the physical coverage of each grid cell, with greater than 100% coverage signifying more than $100 * d * w$ covered points P per grid cell (potential for overlaps) and less than 100% coverage indicating fewer than $100 * d * w$ covered points P per grid cell (skips). In contrast, the FE can be related to the coverage that would have been achieved if the travel speed remained constant across

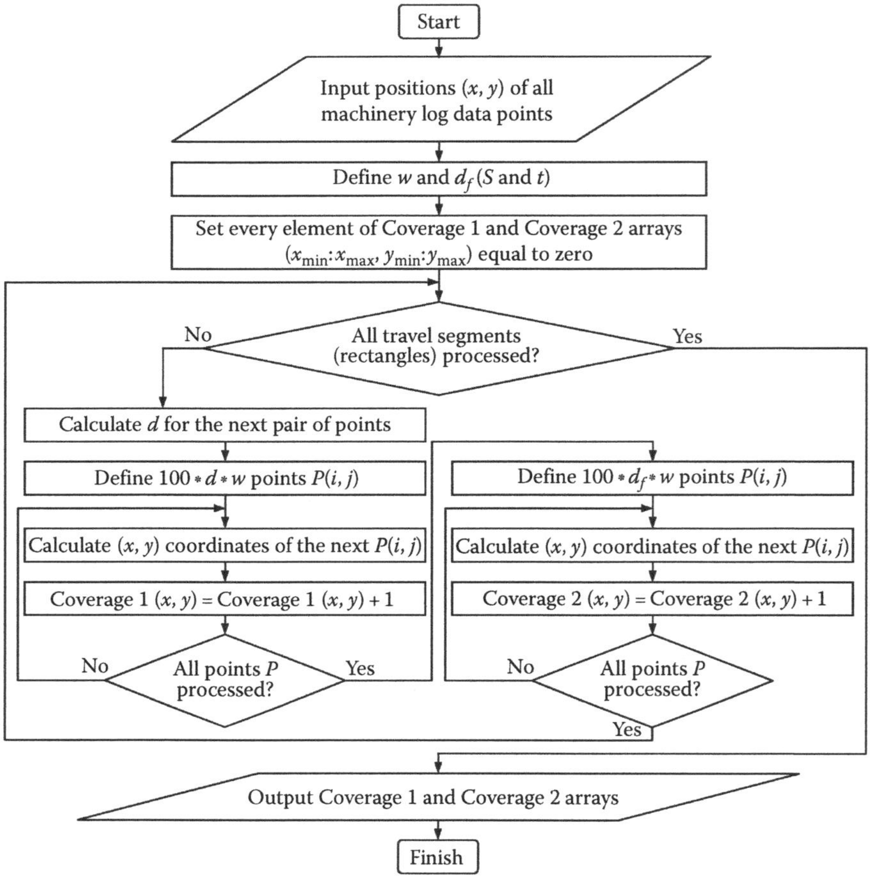

FIGURE 8.3 Algorithm for calculating Coverage 1 and 2 data layers.

the entire field (Coverage 2). Therefore, a fixed (theoretical) distance between two consecutive records (d_f), as shown in Figure 8.2, was defined as the product of the average operation travel speed (S), and the position logging interval (t):

$$d_f = S \cdot t \tag{8.6}$$

The described algorithm was executed using MATLAB® 2007a (The MathWorks, Natick, MA). The input delimited text file contained three columns (easting—x, northing—y, and time interval between two consecutive records). The output file had four columns containing the coordinates x and y for the center of each grid cell as well as the corresponding values for Coverage 1 and Coverage 2. Coverage 1 was calculated using the variable (actual) distance traveled d. The fixed (theoretical) distance (d_f) traveled was used in calculating the values corresponding to Coverage 2.

8.3.2 Supplemental Code

As long as the form of the delimited text data input remains unchanged, the supplemental *Efficiency count.m* script can be used to process and summarize an actual machinery travel log data set. The log file *Example.txt* contains three columns representing field easting and northing coordinates in meters (obtained through local projection of WGS-84 geographic coordinates with 10^{-6} degree resolution) as well as the log time interval in seconds. A zero value for the easting coordinate corresponds to the record for the western-most point of the field and a zero value for the northing coordinate corresponds to the record for the southern-most point in the field. Conversion of the GNSS coordinates to these local field coordinates was completed using Adamchuk (2001).

The output file *Example_count.txt* consists of five columns: (1) easting coordinate in meters for each 1 m^2 area of the field that has at least one coverage event, (2) the corresponding northing coordinate in meters, (3) the percent coverage according to the Coverage 1 layer, (4) the percent coverage according to the Coverage 2 layer, and (5) the percent coverage corresponding to complete stops only. Both the input and the output text files can be displayed within any geographic information system environment using an orthogonal projection, or after converting the local projection coordinates back to longitude and latitude, or using standard projections. If conventional projection method is applied to geographic coordinates in the input travel log file, geographic coordinates in the same projection will be used to generate the output file.

8.3.3 Example Field Data

To illustrate the algorithm output, an agricultural field with a complex shape (Field R1, Rogers Memorial Farm, Eagle, NE) with a total area of 4.24 ha was selected. A soybean harvesting operation was used in this example. The combine header was 4.6 m (15 ft) wide (w = 4.6 m). A total of 11.4 Mg of soybean (average yield of 2.69 Mg/ha) was harvested and removed from the field. The combine stopped and unloaded grain three times. Two data files were simultaneously generated. A PF3000™ (Ag Leader Technology, Inc., Ames, IA) yield monitor with an AgLeader Add-On GPS 3100 receiver (beacon differential correction) was used to collect yield data while harvesting. The position of the center of the combine was recorded in 4 s (0.25 Hz) intervals with the header down (during harvesting only). A standard begin/end of row delay filter was applied. In addition, a GPS 16 (Garmin International, Inc., Olathe, KS) receiver with Wide Area Augmentation System (WAAS) differential correction was placed 1.1 m to the right of the Ag Leader antenna. The receiver output was recorded independently in 1 s (1 Hz) intervals from the beginning to the end of field harvesting (including stops, maneuvering, and unloads).

Initial data processing included conversion of the geographic longitude and latitude into the local rectangular coordinates according to Adamchuk (2001), and correction of the position offset of the Garmin GPS 16 receiver. Figure 8.4 illustrates the nonstop position log (Garmin receiver at 1 Hz sampling rate) and positions recorded by the yield monitor (Ag Leader GPS receiver at 0.25 Hz sampling rate).

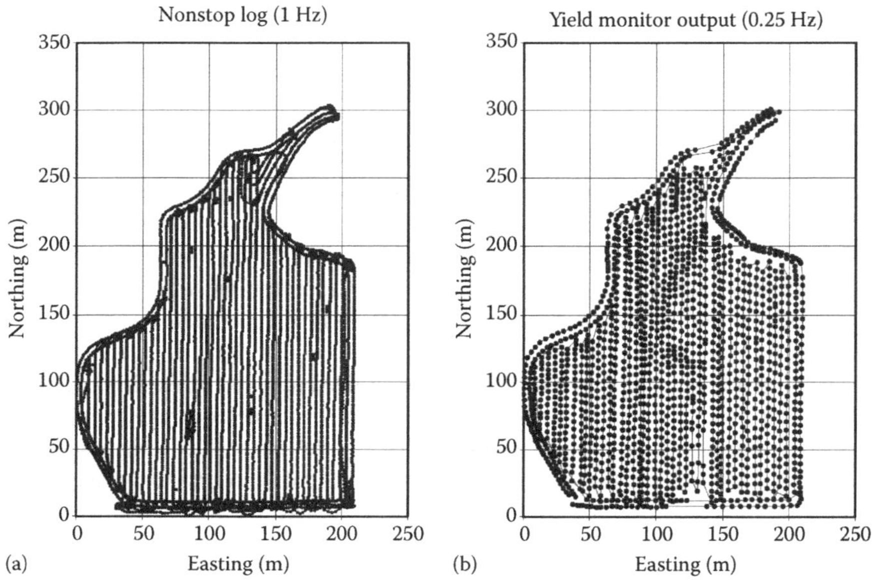

FIGURE 8.4 Combine positions logged (a) nonstop and (b) only while harvesting with the start and end of pass setting.

The continuous log contained records from the beginning to the end of harvesting; the yield monitor output included harvesting locations only.

The algorithm developed can be applied to any logged travel route. However, the meaning of the Coverage 1 and 2 values will change depending on which positions are excluded from the input file. It is critical to identify whether or not turns, stops, and unexpected field maneuvering are included in the log file. In this study, the nonstop (1 Hz) log was used to analyze the spatial variability of the harvest efficiency. The yield monitor recordings were used primarily to define the average harvesting speed (S). Time gaps in the yield monitor recordings (times when the header was up so the combine was not engaged in harvesting the crop) were used to mark non-harvesting records in the nonstop (1 Hz) log. The non-harvesting records were ignored and the remaining speed data were averaged to determine the average harvesting speed, S. The average harvesting speed was 1.4 m/s, and, therefore, $d_f = 1.4$ m (for the continuous log in which $t = 1$ s).

The Coverage 1 and 2 maps reveal the spatial variability of the machinery performance. However, for decision making, this information should be converted into the conventional terms of effective capacity, field efficiency, and cost. From the overall evaluation of the recorded data, the field with an actual area of 4.24 ha was harvested in 2.78 h. This resulted in an effective field capacity (EFC) of 4.24 m^2/s (1.53 ha/h). On the other hand, the theoretical field capacity of the harvest operation (header width times average harvesting speed) was 6.44 m^2/s (2.32 ha/h). The ratio of the EFC to the theoretical field capacity (TFC) for the average speed of operation was 0.66. According to Jose and Brown (2002), \$49.42/ha (\$20/ac) is the most common farm custom rate for soybean harvesting. Therefore, the total cost of this operation was \$210.

Analysis of the algorithm outputs revealed that the sums of all nonzero grid cell point values from the Coverage 1 and 2 arrays were 5.80 and 6.42 ha, respectively. This means that the 37% increase in the Coverage 1 area compared to the actual area resulted from overlaps during maneuvering and reduced width of cut during harvest. In like manner, the 51% increase in the Coverage 2 area compared to the actual area was caused by both overlaps and overestimation of travel speed (including stops). The average travel speed (computed based on the entire nonstop data set) was 1.26 m/s (90% of the average harvesting speed). Therefore, the overall field efficiency of 0.66 (approximately equal to the ratio of actual field area to the Coverage 2 area) can be considered the product of the efficiency resulting from the combine route ($e_{overlap}$ = actual field area/Coverage 1 area = 4.24/5.80 ha = 0.73) and the efficiency resulting from the inconsistent travel speed (e_{speed} = Coverage 1 area/Coverage 2 area = 5.80/6.42 ha = 0.90).

8.4 RESULTS

Figure 8.5 shows spatial maps of Coverage 1 and 2 produced using the continuous log (Figure 8.4a). According to the color scheme, <75% coverage corresponds to the areas with incomplete coverage due to issues such as potential skips (Coverage 1 and 2) and operation at higher-than-average speed (Coverage 2). Similarly, >125% coverage indicates the potential for multiple coverage's due to overlaps, several passes, and stops (Coverage 1 and 2), and operation at lower-than-average speed (Coverage 2). The rest of the field shows areas with normal coverage (Coverage 1 and 2)

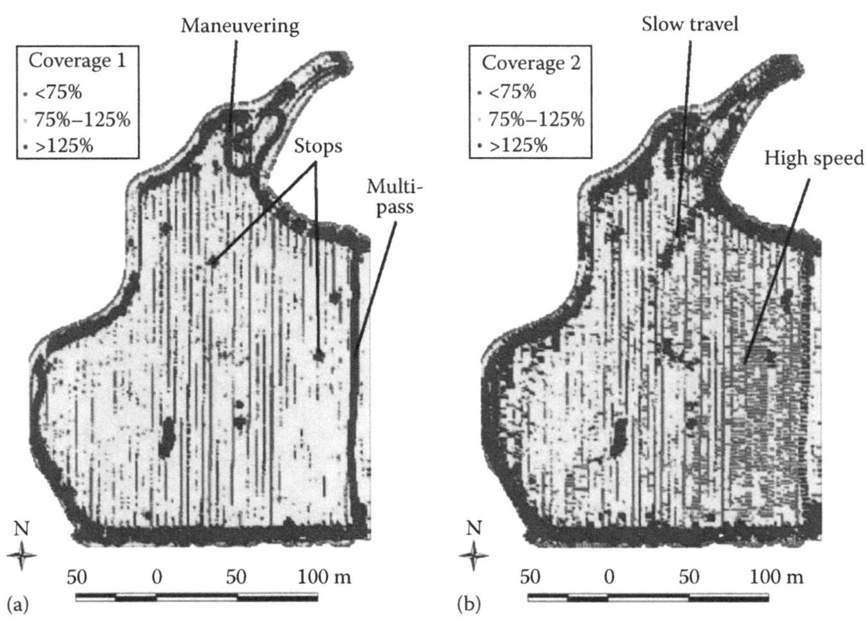

FIGURE 8.5 Grid of field coverage calculated using (a) variable (Coverage 1) and (b) fixed distances between consecutive points (Coverage 2).

and average harvesting speed (Coverage 2). The Coverage 1 map indicates the physical presence of the harvester. The Coverage 2 map also indicates areas where travel speed deviated from the average harvest speed. Since the size of a grid cell was 1 m^2 (less than the GPS receiver accuracy), some indications of potential skips and overlaps could result from the imprecise measurement of the harvester position. The rectangular representation of route segments between two consecutive points during high-speed turns might present additional noise. Map smoothing using conventional interpolation techniques (not presented) could improve the visual appearance of these maps.

Figure 8.6 shows categorical maps produced based on Coverage 1 and 2. A field coverage efficiency map (Figure 8.6a) was derived from Coverage 1 as the inverse of all grid cell point totals with higher than 100% coverage. It indicates coverage efficiency, which was categorized into three intervals: <0.5—excessive coverage for maneuvering; 0.5–0.9—overlaps; and 0.9–1.0—normal coverage. This map can be used to improve traffic patterns through optimization of the harvester route during non-harvest portions of the operation (unloads, turns, etc.).

Since the Coverage 2 map was developed based on the assumption of a constant speed, dividing its coverage values for each field location by the TFC resulted in the theoretical time spent in each field location. If time is used as the major indicator of investment (both labor and energy), the total cost of harvesting can be redistributed according to the time ($210 distributed over 2.78 h is equal to $0.021/s).

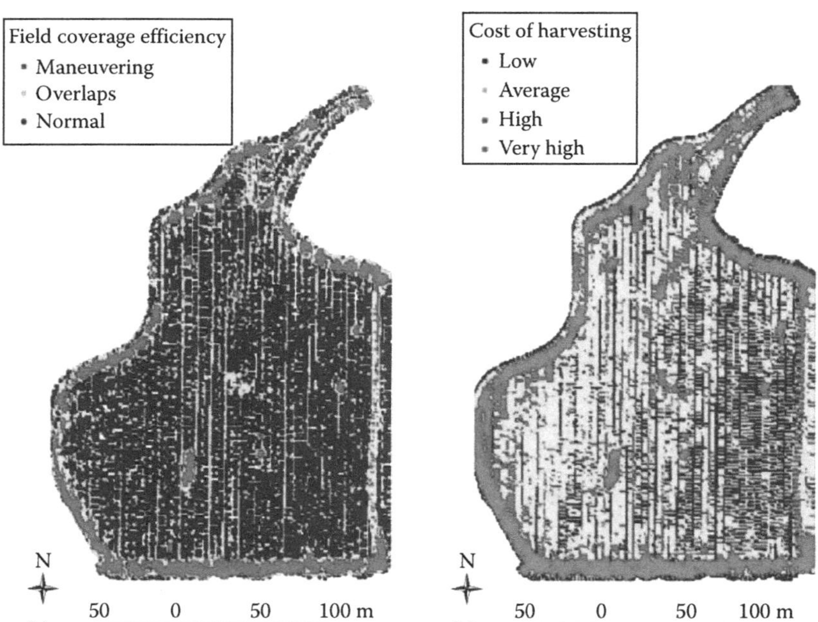

FIGURE 8.6 Categorized maps of (a) field coverage efficiency and (b) cost of harvesting. In (a), normal maneuvering is observed at 0.9–1.0, overlaps can be found at 0.5–0.9, and maneuvering corresponds to values below 0.5. In (b), low cost of harvesting is <$40/ha, medium is $40–60/ha, high is $60–10/ha, and very high is >$100/ha.

Figure 8.6b illustrates a cost of harvesting map categorized with <\$40/ha as the low cost, \$40–60/ha as average, \$60–100/ha as high, and >\$/100/ha as very costly harvesting. Conceptually, a yield map layer for each field operation could be used in conjunction with this kind of map to calculate an overall profitability map. If profit map values show losses or small gains in particular field areas (such as the northern portion of the illustrated field, Figure 8.6), alternative traffic patterns and/or land usage should be considered.

EFC and FE can be calculated to evaluate the overall performance of the operation as well as to investigate operational variation effects. Thus, Coverage 1 divided by the theoretical time (the same as the time used to calculate the cost of harvesting) corresponds to the physical coverage of each square grid area in a unit of time, or locally defined EFC. After dividing by the field average of these locally defined field capacity values ($5.82\,m^2/s$ or 2.10 ha/h), a map can be constructed showing relative machinery efficiency related to the variable speed effect (Figure 8.7a). Areas with a high speed of operation are represented with relative machinery efficiency values greater than 1 (more efficient than an average field location), while grid cells with relative machinery efficiency less than 1 showed the locations where the actual coverage area was smaller than what theoretically should have been covered using average TFC. This map removes most of the effects of maneuvering, and can be used to determine the locations of the actual slowdowns.

Overall machinery efficiency—the ratio between effective (area of a grid cell divided by time) and theoretical ($6.44\,m^2/s$ or 2.32 ha/h) field capacities—indicates the overall machinery performance efficiency (Figure 8.7b). This map relates to the

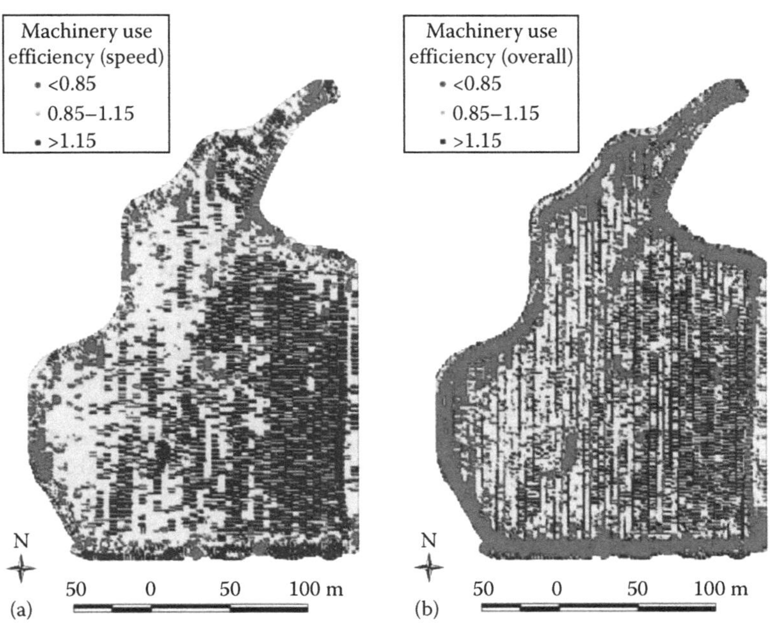

FIGURE 8.7 Machinery use efficiency maps representing (a) speed effect and (b) overall performance.

cost map and can be used to make judgments about operating in low-performing areas of the field (low efficiency and high cost). Although there are many ways to utilize these data layers, two major strategies can be pursued. The areas with relatively low machinery efficiency due to a systematic nonproductive machinery operation (involving extra turns, travel around obstacles, point rows, etc.) can be evaluated to determine more effective traffic patterns. And machinery efficiency expressed in energy use and/or monetary terms can be used to evaluate the spatial profitability while accounting for the inconsistent cost of field operations.

8.5 CONCLUSIONS

The maps presented in this work are examples of various types of information contained within the records of geographic positions logged during various field operations. The algorithm developed allows the producer to transform this information into two coverage maps (Coverage 1 and 2). The Coverage 1 map indicates field areas affected by repeated passes and variable true swath width. The Coverage 2 map shows the effect of variable travel speed as well. These generated maps can be converted into a set of data layers associated with conventional categories evaluating machinery performance (cost of operation, capacity, efficiency, etc.). These processed data layers will be complementary for decision-making strategies to improve site-specific crop management.

REFERENCES

Adamchuk, V.I. Untangling the GPS data string. *Precision Agriculture Extension Circular EC 01-157*. Lincoln, NE: University of Nebraska Cooperative Extension, 2001.

ASABE Standards, 49th edn. S495.1 NOV2005. *Uniform Terminology for Agricultural Machinery Management*. St. Joseph, MI: ASABE, 2008a.

ASABE Standards, 49th edn. D497.5 FEB2006. *Agricultural Machinery Management Data*. St. Joseph, MI: ASABE, 2008b.

Grisso, R.D., Jasa, P.J., and Rolofson, D. Field efficiency determination from spatial data. *Applied Engineering in Agriculture* 18, 171, 2002.

Grisso, R.D., Kocher, M.F., Adamchuk, V.I., Jasa, P.J., and Schroeder, M.A. Field efficiency determination using traffic pattern indices. *Applied Engineering in Agriculture* 20, 563, 2004.

Jose, H.D. and Brown, L.J. Nebraska farm custom rates: Part II. *Extension Circular EC02-826-A*. Lincoln, NE: University of Nebraska Cooperative Extension, 2002.

Renoll, E.S. Predicting machine field capacity for specific field and operating conditions. *Transactions of the ASAE* 24, 45, 1981.

Taylor, R.K., Schrock, M.D., and Staggenborg, S.A. *Extracting Machinery Management Information from GPS Data*. Paper No. 02-10008. St. Joseph, MI: ASAE, 2002.

9 Precision Manure Application Requirements

John Nowatzki

CONTENTS

9.1 EXECUTIVE SUMMARY

Precision manure application uses the global positioning system (GPS) and electronic control systems to monitor manure application rate, avoid misses or overlaps, vary application rates, and record information about the application process. Computer-generated maps installed in the in-cab controller prescribe the desired manure application rate at each location across fields. Precision liquid manure application is more effective than precision application of solid manure. Adopting precision manure management techniques can help improve the energy efficiency of the overall system. This chapter will help readers understand the methods, required equipment, and procedures to use geospatial technologies when applying manure.

9.2 INTRODUCTION

9.2.1 WHY PRECISION APPLICATION OF MANURE?

Precision manure application using geographic information system (GIS) and GPS can reduce energy requirements by allowing producers to avoid overlappping or missing applications areas. Other advantages for precision manure application includes avoiding application in environmentally sensitive areas, turning the applicator off when traveling outside field boundary areas and varying the application rate based on projected crop nutrient needs at different locations across fields. Precision manure applications require management practices that are similar to commercial fertilizer. Manure has significant value as a field crop input and that value is easily diminished by improper applications. Missing areas of fields during manure application can result in crop nutrient deficiencies causing lower crop yields. Overlapping manure application can also reduce crop yields because of too much vegetative growth resulting in increased crop diseases or lodging. Geospatial technologies can enhance implementation of efficient manure management practices including determining the optimum amount of manure to apply at specific locations in fields for specific crops and yield goals, applying prescribed rates, and recording where and when manure was applied.

9.3 METHODS

9.3.1 GPS APPLICATIONS IN PRECISION MANURE APPLICATION

Precision manure application could include manually monitoring manure applications, using a vehicle GPS guidance system to avoid overlapping or missing areas, using cluster analysis to develop management zone maps, and using variable-rate controllers to control applications.

Variable-rate manure application involves using both GIS computer programs and GPS.[1,2] GIS computer programs manage and present data that are linked to locations and are used in personal computers to develop field zone maps and precision manure application maps. Zone maps divide fields into areas that require unique management practices. Precision manure application maps are geo-referenced, digital maps used in variable-rate controllers to prescribe the correct amount of manure at each field management zone. In-cab computer controllers used with GPS on manure application equipment send signals to flow control valves on manure applicators to change the application rate as the equipment moves from one management zone to the next. The computer controller also creates a digital "as-applied" map that records the time and rate of manure application at each point in the field.

9.3.2 MANAGEMENT ZONES

Management zones used with variable-rate fertilization (VRF) assign a unique yield goal to each zone in a field. The variable-rate equipment applies the appropriate treatment to that zone. This often requires the capability to change the rate during field operation. VRF uses a GPS integrated with a computer in the tractor cab to signal the fertilizer application equipment to change rates as it moves from one zone to another.

Management zone maps can be developed based on the producer's perspective or by a computer program that processes spatial information. One commonly used approach relies on cluster analysis. In this analysis, areas with similar characteristics are grouped together. Factors that can be included in a cluster analysis include topography, soil physical and chemical properties, cropping history, remotely sensed images of previous years' vegetative growth, historical crop yield data, soil survey information, and grower knowledge of the field.

Producers have several options when developing management zone and fertilizer application maps (Figure 9.1). A management zone is a subdivision of a field that is regarded as homogeneous.

From the perspective of precision agriculture, management zones are the smallest subdivisions between which the seeding, application of chemicals, and other management parameters are to be varied. For example, a producer could purchase software, hire crop consultants, or use freeware. Currently, software can be purchased from farm equipment and GIS companies. A partial listing of software can be viewed at http://www.spatialhydrology.com/software_gisrsgps.html. Many of these programs have the ability to combine various data layers and conduct spatial analysis. There are also online tools that can assist in spatial analysis and identify management zones. ZoneMap and Management Zone Analysis are available at

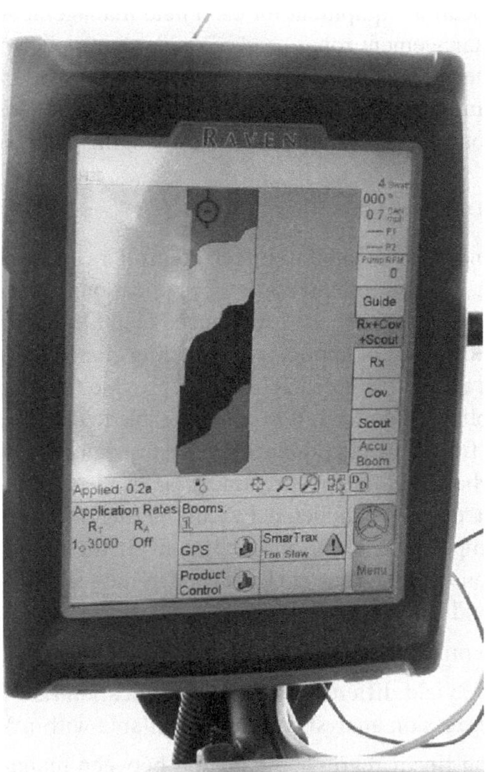

FIGURE 9.1 Management zones as shown on the monitor.

http://zonemap.umac.org/ and http://www.ars.usda.gov/services/software/download.htm?softwareid=24&modecode=36-20-15-00. The ZoneMap program is available for use without charge, and is capable of developing both management zone maps and fertilizer application maps. ZoneMap allows users to build zone maps from historical vegetative growth patterns and also allows users to upload and incorporate their own geo-referenced field information to help delineate field variability to develop management zones. Both zone and application maps are digital computer files and therefore have unique digital formats. Users need to be cognizant of the file format required by the in-cab controller and export and save them in the appropriate format from the GIS program. Management zone maps can be used for a variety of purposes including fertilizer, manure, weed, variety selection, and pest management techniques. It is important to refine the zone boundaries for different problems and incorporate additional field information as it is collected.

9.3.3 Application Maps

Manure application maps indicate the desired manure application rate for each field management zone. Application maps are generally created in a GIS program on an office computer and exported in the digital format usable in in-cab controller. The controller uses GPS to provide vehicle real-time location in order to signal rate changes to the application equipment for each field management zone.

The size of the management zones needs consider the applicator width and the time needed to change the application rate. For practical equipment operation, the smallest area requiring a manure application rate change should be at least twice the width of the applicator and long enough to allow the rate changing technology to function.

9.3.4 As-Applied Maps

Digital as-applied maps are created in in-cab controllers as the applicator equipment operates in the field. The controller records the application rate and time of the manure application and saves the data on a geo-referenced map file. As-applied maps can also serve as an historical record of application including showing field setback distances and environmentally sensitive areas. The maps provide the operator assurance that the manure was applied as planned. It may be possible to test the as-applied map with yield data collected from the field (Table 9.1). This test can be improved by conducting a difference map, where pre- and posttreatment yields are subtracted from each other.

This assessment can be conducted by subtracting the pre- and posttreatment yields and then comparing them to the treatment. If the treatment is directing the yields, then the amount applied should be related to the yield increase. The following steps are used to conduct this analysis.

Step 1. Subtract the pre- and post-yields.

Step 2. Compare the yield difference with the application rate. This can be accomplished using the regression analysis program available within Microsoft Excel.

Step 3. The resulting linear regression equation between manure treatment and the yield increase is $y = -11.8 + 7.8$ (manure rate), $r^2 = 0.50$. This analysis indicates that the manure treatments and resulting yield increases were correlated to each other.

TABLE 9.1
Hypothetical Case Study Where Manure Rates Are Compared with Yield Increases

Management Zone	Manure Rate (ton/ac)	Pretreatment Yield (bu/ac)	Posttreatment Yield
1	7	120	145
2	6	110	155
3	4	130	148
4	2.5	125	157
5	3	140	168
6	1	160	144
7	1.5	170	150

9.4 REQUIRED EQUIPMENT AND PROCEDURES

9.4.1 GPS Antenna and Receiver

VRF relies on GPS to maintain constant equipment positions allowing the in-cab computer to signal rate changes as the equipment moves from one field management zone to the next. When selecting GPS units, costs and accuracy are directly related to each other. As a rule of thumb, GPS units suitable for vehicle guidance can also be used for variable-rate treatments.

For most agricultural purposes, a differential GPS is required. Differential correction uses a radio signal broadcast from known locations on Earth. These Earth-based stations receive radio signals from the GPS satellites and determine the error from their known positions. The error is calculated and transmitted to individual receivers. Correction signals can be obtained from a variety of sources. In the United States, the U.S. Coast Guard provides a free differential correction beacon signal. The Coast Guard signal is an AM radio signal that is broadcast from several locations and travels as a "ground wave" over the Earth's terrain. Each station has a radial coverage of approximately 300 mi. As the distance increases, the accuracy of the signal decreases. Another method for obtaining real-time differential correction data in the field is by using geostationary satellites. Before purchasing a GPS receiver, it is best to identify the source of base station data.

Mounting of the GPS receiver antenna is important. Where possible, it should be mounted in the center of the equipment and the applicator. If mounting on the applicator is not possible, then the separation distance between the GPS antenna and applicator should be programmed into the application software.

9.4.2 In-Cab Computer/Controller

The central component of precision application and variable-rate application (VRA) equipment is the computer/controller normally mounted in the vehicle cab. The controller performs several functions including (1) operating a software program to display the application map, (2) recording the GPS signal to recognize the vehicle real-time position on the map, (3) communicating signals with the rate control device to alter the application rate, (4) monitoring vehicle speed, and (5) recording the manure application map.

Unfortunately, many older controllers are not capable of performing VRA requiring producers interested in VRA to purchase new controllers. Several functions are important when choosing a controller including compatibility with other brands, compatible digital data formats, ease of interaction, ability to record and save application maps, and capabilities to perform other functions besides variable rating. Other controller functions include operating spinner fertilizer spreaders, variable rate and boom section control on sprayers, and vehicle guidance systems.

9.4.3 Equipment for Precision Application of Solid Manure

It is more difficult to accomplish precision application of solid manure than liquid manure and is generally less accurate. However, there are management practices that can be employed to increase the precision and accuracy of solid manure. Using GPS guidance along with solid manure spreader innovations such as horizontally mounted rear throwers, along with drive sprockets on augers and expellers that can be changed to match the consistency of the manure being spread allow for more even distribution of manure. Other new solid manure spreaders incorporate vertically mounted rear manure throwers that spread manure more evenly than traditional spreaders.

Rate changes with solid manure application can be accomplished by either altering the operation rate of the spreader or altering the rate of travel of the spreader. Load cells can provide a continuous data stream of weight during the manure applicator discharge (Figure 9.2). Measured weight differences during unloading can be used to determine applicator discharge rate. In-cab controllers can calculate the application rate by knowing the discharge rate, application width, and travel speed (Figure 9.3).

9.4.4 Equipment for Precision Application of Liquid Manure

Manure in liquid form is normally used for precision and VRA. The components needed to accomplish precision and variable-rate liquid manure application are an in-cab controller, a section control switch, a liquid flow meter, and a flow control

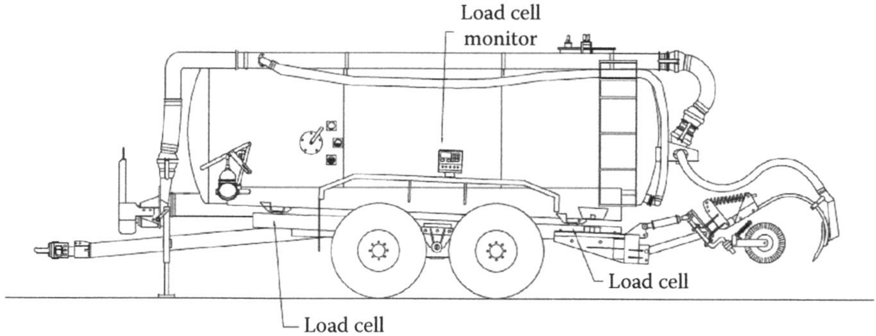

FIGURE 9.2 A diagram of a liquid manure applicator equipped with load cells. (From Ess, E.R. *Implementing Site-Specific Management: Liquid Manure Application*, SSM-1-W, Purdue University, West Lafayette, IN, 2001.)

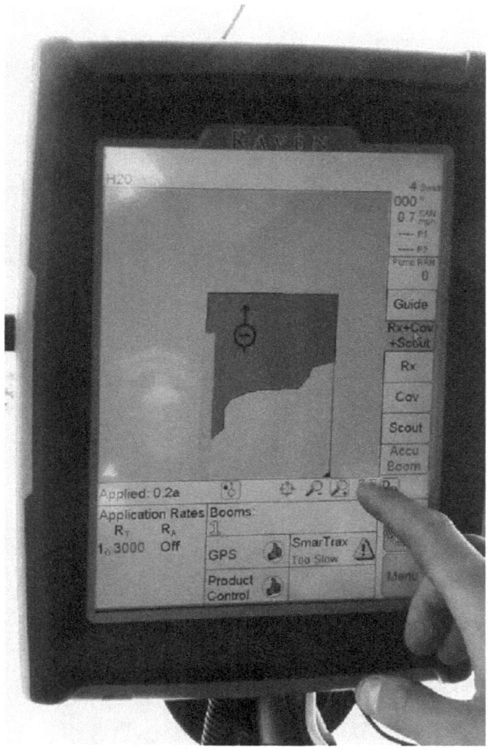

FIGURE 9.3 In-cab computer/controller.

valve. The controller can be used to maintain a constant application rate as travel speed varies or vary the rate based on previously defined application maps. VRA is accomplished by varying applicator discharge rates with flow control valves.

A flow control valve and an electromagnetic flow meter can be plumbed in line after the pump on liquid or slurry manure applicators to measure and change the manure flow. Flow meters and controllers automatically adjust a valve or change the pump speed to keep a constant application rate as tractor speed or topography changes. Both the flow valve and flow control valve are connected to the in-cab controller that functions to continually monitor the flow rate and signal the flow control valve to change the application rate to correspond to the rate prescribed in the manure application map (Figure 9.4). Electromagnetic flow meters function better than turbine-type meters with liquid manure because they are not affected by solids in the stream flow. Doppler flow meters can also be used to monitor the liquid flow.[3]

9.4.5 Commercial Equipment Options

Commercial equipment is available to accomplish different levels of precision manure application (Figure 9.5). A flow meter plumbed inline after the pump can be used with an in-cab monitor to gauge how much manure is flowing through the applicator. Using a monitor capable of monitoring the speed of the tractor, width of the spread

FIGURE 9.4 Flow meter and control valve.

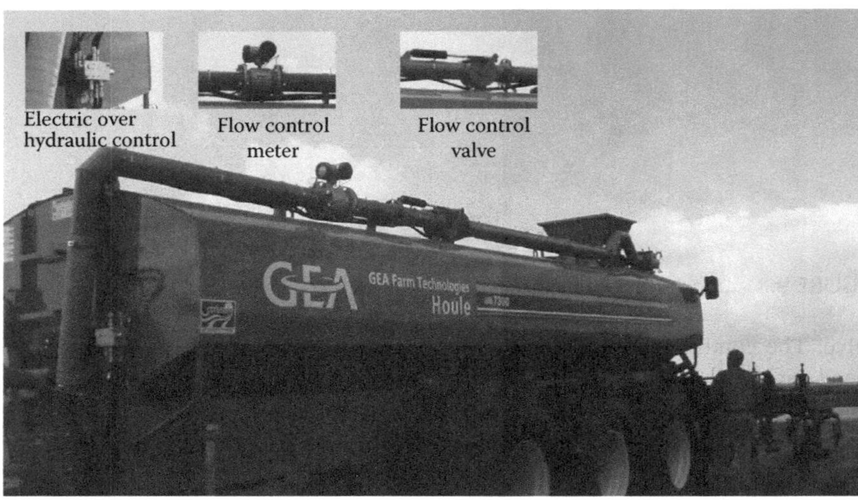

FIGURE 9.5 Commercial applicator with variable-rate controls.

pattern, and flow rate allows the operator to know how much manure is being applied per acre. Connecting a GPS to the system allows the operator to produce a map of where manure was applied on the land. Finally, by adding a flow control valve inline after the flow meter, and an electronic switch, the system can vary the application rate at different field zones. Manure application equipment manufacturers to date are only marketing machines equipped with flow meters, in-cab monitors, and GPS units allowing operators to monitor application rate and prepare "as-applied" manure application maps. However, some operators are modifying their equipment by incorporating the components necessary to accomplish variable-rate manure application.[4]

A hydraulic cylinder is used to change the flow rate of the flow control valve. An electric over hydraulic control switch is incorporated with the system to prevent the cylinder from operating when the manure injectors are not engaged in the ground.

9.5 SUMMARY

Variable-rate equipment is continuously being improved.[4] Critical components of variable-rate equipment include GPS, computer controller, and the system for changing the application rate. In the simplest systems, it may be possible to change the rate by modifying the speed, while in more complex system, a computer controller will change the rate. Different systems are suitable for different uses and system requirements. When selecting a GPS system, the source of the differential correction should be identified before purchasing the GPS system.

REFERENCES

1. Ess, E.R. *Implementing Site-Specific Management: Liquid Manure Application*. SSM-1-W. Purdue University, West Lafayette, IN, 2001.
2. Laatsch, T.P. *Mapping Manure Application*. Environmental Services, Carlyle, IL, 2008. Available at http://www.mnpork.com/producers/2008/ManureMapping2008.pdf
3. Krohne Messtechnik GmbH & Co. KG, Duisburg. http://www.krohne.com/
4. Eldon, C. Stutsman, Inc. Web site: http://www.stutsmans.com/index.cfm?show=10&mid=18

10 Case Study for Improving Nutrient Management Efficiency by Optimizing the Plant Population

Gregg Carlson, David E. Clay, and Joseph Schefers

CONTENTS

10.1 EXECUTIVE SUMMARY

A goal of precision farming is to maximize profitability, increase energy gains, and minimize the impacts of agriculture on the environment. To achieve this goal, appropriate recommendations must be followed. One precision farming adoption barrier is the "poor" accuracy of many site-specific application rate models. Numerous field experiments have been conducted across the world with the goal of defining the relationships between inputs and outputs over landscapes. To integrate this information into site-specific recommendations, these data must be analyzed. This chapter provides a case study that demonstrates how data from numerous site years can be used to develop a regional site-specific application recommendation model for corn (*Zea mays*) plant populations. A similar approach can be used for developing locally derived site-specific recommendations for fertilizers and pesticides. Data collected from multiple on-farm studies are provided with this case study.

10.2 INTRODUCTION

Our ability to improve agricultural energy efficiencies using geographic information system techniques is hindered by the ability to process on-farm data and the ability of the prediction models to deliver improved yields and profits.[1–3] The development of improved recommendation models is complicated by (1) many factors interacting to influence yields, (2) the wide-scale use of single-factor models that do not consider synergistic relationships between yield-limiting factors, and (3) the difficulty of conducting, processing, and analyzing data collected from multiple sites. One approach to maximize nutrient efficiency is to select the appropriate plant population. Techniques to overcome multiple questions are needed.

It makes intuitive sense that site-specific recommendations are a function of the yield potential. Areas with higher yield potentials often require higher inputs, while areas with lower yield potentials require less. Many fields contain significant yield spatial variability, which is the direct result of landscape position. Across these landscapes, water is often the factor that most controls yield. For example, yields in summit/shoulder areas are often reduced by too little water, while yields in footslope/toeslope areas are reduced by too much water. In many fields, nutrient and pest stresses can further reduce yields. To maximize yields and the efficient use of nutrients, plant population levels must be matched to the unique conditions existing at each location.

Complex relationships between yields and the numerous limiting factors often result in relatively poor correlations between predicted and measured yields.[1–3] Developing site-specific recommendations that consider the complexity of the problem is complicated by the difficulty of conducting multifactor experiments in complex terrains. It may be possible to solve this problem by using data sets collected from multiple experiments conducted across geographic regions. For example, if water stress is the primary limiting factor, then it may be possible to define the response function using experiments conducted in different rainfall environments.[4,5] The advantage of experiments conducted across a geographic region is that the boundary conditions might be wider and more data might be available for analysis. Over the past 20 years, thousands of corn population studies have been conducted. This chapter provides a case study that demonstrates how data from numerous site years can be used to develop a regional plant population site-specific application model. A similar approach can be used for other problems.

A goal of this case study is to provide an approach for analyzing data from field experiments conducted at numerous locations. The proposed approach is based on defining a plant population yield response curve for each individual field experiment and using an iterative approach to develop a site-specific model that maximizes profitability.

10.2.1 Corn Productivity and Plant Populations

The development of a site-specific plant population is one step in developing a nutrient management strategy that maximizes energy efficiency. Planting the economic optimum plant population will maximize energy efficiency and minimize carbon footprints. The optimum plant population is influenced by many factors including optimum yield, genetics, available resources, early season growth, and maturity rating. In a discussion of plant populations, it is important to understand that the optimum plant population is impacted by the degree of competition between adjacent plants. Corn has been bred to grow at high populations, with yields per field gradually increasing as population increases to an optimum value.[6] Increasing the population beyond that value can result in yield reductions. Water stress can complicate the problem. Associated with the increasing yield per area is a reduction in the yield per plant. Smaller per plant yields that are not etiolated might be the result of adaptive evolution where plants with a low shade avoidance mechanism are selected for propagation.[7–9] Corn's response to competition is not typically of many wild plants, which often become etiolated (i.e., becoming taller).[10–13]

The complexity of the relationship among seeding rate, maturity rating, hybrid, and yield potential has resulted in region-specific recommendations. For example, Pioneer hybrid specific population density calculator is available at http://www.pioneer.com/web/site/portal/menuitem.0cbb3257343891dc86738673d10093a0/, whereas in Iowa the recommendation is between 28,000 and 32,000 plants ac (69,000–79,000 plants/ha). For silage production, planting densities are 2000–4000 plants/ac higher (4900–9900 plants/ha).[14]

Variable seeding rate experiments have had mixed findings. Bullock et al.[15] reported that variable rate seeding will remain economically unfeasible until the costs associated with obtaining and processing information necessary for the recommendation model decrease. Abendroth and Elmore[16] reported that, in Iowa, the optimum plant population should increase 485 plants/ac (1200 plants/ha) for every 16 bu (1000 kg/ha) increase in yield. Coulter[17] reported that the maturity rating influences the optimum density. Corn with a maturity rating of 92–96 days had a higher optimum density than corn with a maturity rating of 102 days. Paszkiewicz and Butzen[18] reported that in areas with a yield potential of 130 bu/ac (8140 kg/ha) or greater, the corn response to increasing density was similar regardless of yield potential. Shanahan et al.[19] reported that the economic optimum plant population was approximately 2000 plant/ac (4920 plants/ha) less in low- than in high-yielding areas. Analysis of these studies suggests that benefits can be derived by implementing variable rate planting. However, the benefit from adopting variable rate planting will be dependent on the accuracy of the prediction model and zone classification approach. This chapter presents an approach for increasing the accuracy of site-specific prediction models. Related chapters in this series address approaches for identifying yield zones.

10.2.2 Plant Population Case Study: Model Derivation

For many cultivated crops, there is a direct relationship between population and yield (Figure 10.1). Based on this relationship, recommendations for the most profitable population level can be determined.

One approach for determining the optimum level for a single experiment is summarized in the following text. In these calculations, gy is the grain yield, pp in the seeding rate, \$gy is the selling price of a bushel of corn, and \$pp is the purchase price of a corn seed. A common approach to define the optimum value is to first determine the relationship between population level and yield. A polynomial equation

Plant pop	Yield
18	128
24	150
30	162
36	167

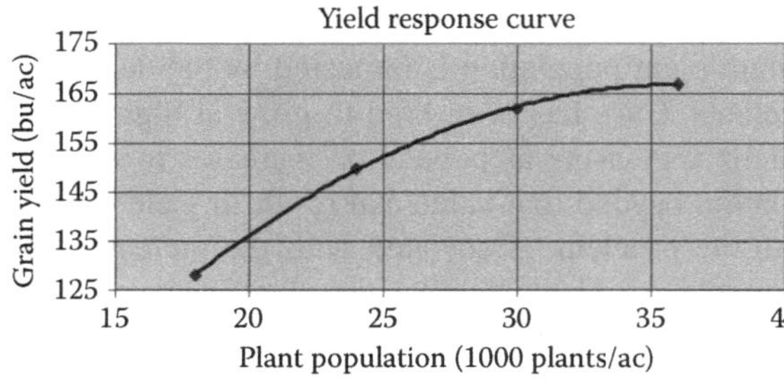

FIGURE 10.1 A hypothetical relationship between corn plant population and measured yields.

such as, $\text{gy} = a + b \times \text{pp} + c \times \text{pp}^2$, can be used for this purpose. Using a second-order polynomial equation, the ratio between the rate of yield ($d$gy) and population change is calculated using the first-order derivative, $d\text{gy}/d\text{pp} = b + 2 \times c \times \text{pp}$.

Now, because the change in the value of the corn/ac [$d(\$\text{gy} \cdot \text{gy})$] and total cost of the seed/ac [$d(\$\text{pp} \cdot \text{pp})$] are defined by the equations

$$d(\$\text{gy} \times \text{gy}) = d\text{gy} \times \$\text{gy} \quad \text{and} \quad d(\$\text{pp} \times \text{pp}) = d\text{pp} \times \$\text{pp}$$

respectively, the relationship

$$\frac{d(\$\text{gy} \times \text{gy})}{d(\$\text{pp} \times \text{pp})}$$

can be defined by the equation

$$\frac{d\text{gy} \times \$\text{gy}}{d\text{pp} \times \$\text{pp}}$$

The equation

$$\frac{d(\$\text{gy} \times \text{gy})}{d(\$\text{pp} \times \text{pp})} = \frac{d\text{gy} \times \$\text{gy}}{d\text{pp} \times \$\text{pp}}$$

can be rearranged to

$$\frac{d(\$\text{gy} \times \text{gy})}{d(\$\text{pp} \times \text{pp})} \times \frac{\$\text{pp}}{\$\text{gy}} = \frac{d\text{gy}}{d\text{pp}}$$

Now, because $d(\$\text{gy} \times \text{gy})/d(\$\text{pp} \times \text{pp}) = 1$ at the optimum population, $d\text{gy}/d\text{pp}$ must equal \$pp/\$gy. The resulting equation is $d\text{gy}/d\text{pp}$ = \$pp/\$gy. After substituting $b + 2 \times c \times \text{pp}$ for $d\text{gy}/d\text{pp}$, the equation

$$b + 2 \times c \times \text{pp} = \frac{\$\text{pp}}{\$\text{gy}}$$

was derived. This equation can be rearranged to

$$\text{pp} = \frac{(\$\text{pp}/\$\text{gy}) - b}{2 \times c}$$

By solving this equation, the optimum plant population for a single field location and price structure can be determined. This solution is based on the use of the polynomial equation. The same solution may be obtained using other yield response equations and Excel's nonlinear optimization procedure, Solver, which is used to maximize the $\text{Partial}_{\text{profit}}$. Note that solving the differential equation $d\$\text{gy}/d\$\text{pp} = 1$ will maximize the $\text{Partial}_{\text{profit}}$. When one uses Solver to maximize the $\text{Partial}_{\text{profit}}$, the analyst will be using a more generalized solution since one may want to use an equation describing gy = f(pp) that may not be easily differentiable. In Microsoft Office 2003, solver is installed by selecting tools, add-ins, checking solver add-in, and clicking on OK. In Microsoft Excel 2007, solver is installed using the following directions:

1. Click the Add-ins category.
2. In the Manage Box, click Excel Add-ins, and then click Go.
3. To load an Excel Add-in, do the following:
   a. In the Add-ins available box, select the check box next to the add-in that you want to load and then click OK.
   b. If the add-in you want to use is not listed, click browse, and then locate the add-in. Add-in not currently available can be downloaded from Downloads on Office Online. Click yes to install the download.

After you have successfully installed solver, you will find it on the far right under the Data tab (Figure 10.2). In Microsoft Excel 2003, solver is installed by double-clicking on tools and selecting add-ins.

Figure 10.3 shows a spreadsheet that utilizes the mathematics developed above. In cells A8 through C8 are the a, b, and c constants for the polynomial yield response equation (gy $= a + b \times \text{pp} + c \times \text{pp}^2$). On the left side of Figure 10.3 are the spreadsheet results of this analysis and on the right are the spreadsheet formulas used to determine the most profitable level of input, in this case the most profitable planting population, pp. Note that in cell A11 is the optimum population, pp. We will discuss a second method to obtain the same solution as found in cell A11. The Partial profit ($\text{Partial}_{\text{profit}}$ = corn value − seeding costs) equation is entered in cell C11. After the equation has been entered, run solver to obtain the maximized $\text{Partial}_{\text{profit}}$, the solution to the problem, Figure 10.4. Before running solver, an initial guess must have

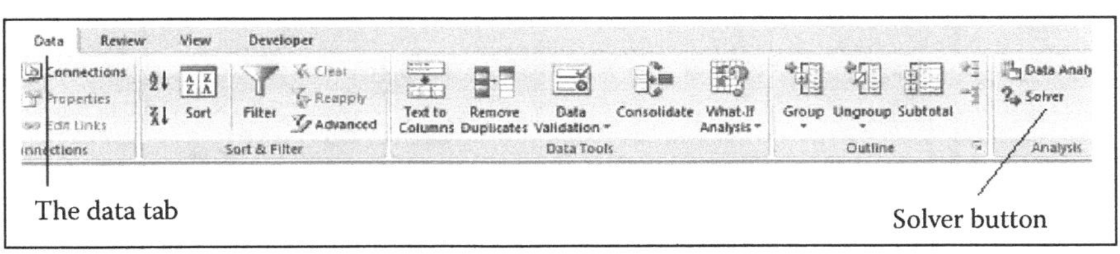

FIGURE 10.2 The location of the Solver button on the Data tab in Microsoft Excel.

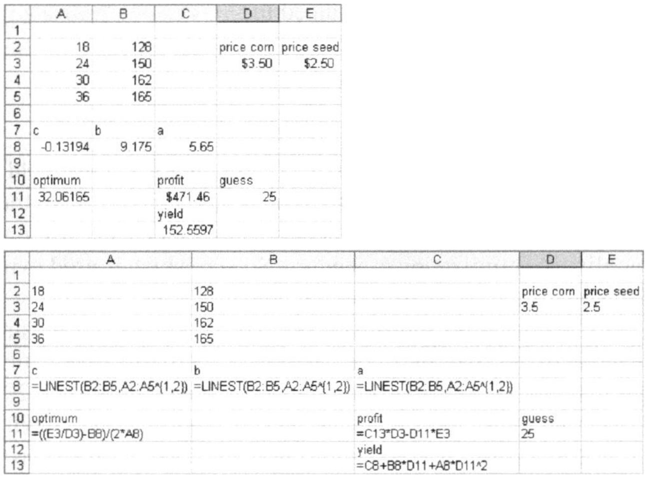

	A	B	C	D	E
1					
2	18	128		price corn	price seed
3	24	150		$3.50	$2.50
4	30	162			
5	36	165			
6					
7	c	b	a		
8	-0.13194	9.175	5.65		
9					
10	optimum		profit	guess	
11	32.06165		$471.46	25	
12			yield		
13			152.5597		

	A	B	C	D	E
1					
2	18	128		price corn	price seed
3	24	150		3.5	2.5
4	30	162			
5	36	165			
6					
7	c	b	a		
8	=LINEST(B2:B5,A2:A5^{1,2})	=LINEST(B2:B5,A2:A5^{1,2})	=LINEST(B2:B5,A2:A5^{1,2})		
9					
10	optimum		profit	guess	
11	=((E3/D3)-B8)/(2*A8)		=C13*D3-D11*E3	25	
12			yield		
13			=C8+B8*D11+A8*D11^2		

FIGURE 10.3 Screen for developing parameters and solution. This figure shows the data set and equations used in the following example. Data is available on the data disk under GIS Chapter to population data.XLS.

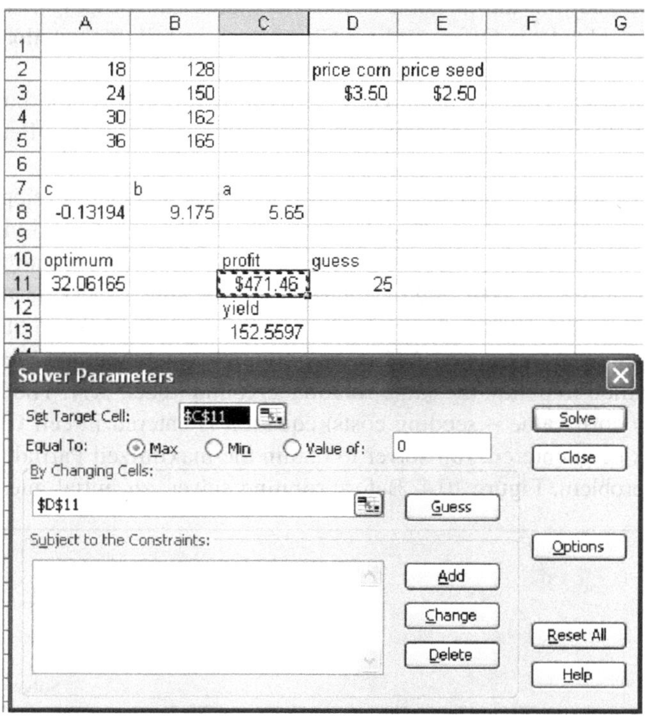

	A	B	C	D	E	F	G
1							
2	18	128		price corn	price seed		
3	24	150		$3.50	$2.50		
4	30	162					
5	36	165					
6							
7	c	b	a				
8	-0.13194	9.175	5.65				
9							
10	optimum		profit	guess			
11	32.06165		$471.46	25			
12			yield				
13			152.5597				

FIGURE 10.4 Using solver to determine the optimum population for a single experiment.

been entered into cell D11. In this example, 25 (25,000 plants/ac) was the guess and was typed into cell D11. After starting solver, you must first, select C11 as "Set Target Cell", Figure 10.4. This cell contains the equation $\text{Partial}_{\text{profit}} = \text{gy} \times \$\text{gy} - \text{pp} \times \$\text{pp}$, where grain yield (gy) was calculated in cell C13. Recall that that value for grain yield is calculated with the equation $\text{gy} = \text{a} + \text{b} \times \text{pp} + \text{c} \times \text{pp}^2$. Also recall that a, b, and c values are available in cells A8 through C8. As mentioned earlier, cost of seed and the value of corn are entered into cells (D3 and E3). Second, in Figure 10.4, in the line starting "Equal To" select "Max" because we wish to maximize the $\text{Partial}_{\text{profit}}$. Third, under the text "By Changing Cells" type D11. Fourth, press the Solve button. The program will ask if you want to save the solution, click on OK. A new number, the solution, appears in cell D11 and note that it is the same solution that was calculated by the calculus derived method, the solution of which is found in cell A11 (Figure 10.5).

These calculations were used to determine the optimum plant population for one yield potential experiment. Similar calculations can be used for a second or third experiment. If the yield potentials in the different environments are controlled by a common limiting factor, i.e., water stress, then it may be possible to determine the relationship between water stress or yield potential and plant population by combining the data from the different experiments. The following section demonstrates one approach for conducting an analysis of numerous experiments. In these calculations, optimum plant populations will be determined using two equations, exponential and polynomial equations. The exponential solution is on sheet 1, while the polynomial solution is on sheet 2. The data set and programs are available in Chapter 10 of the accompanying disk.

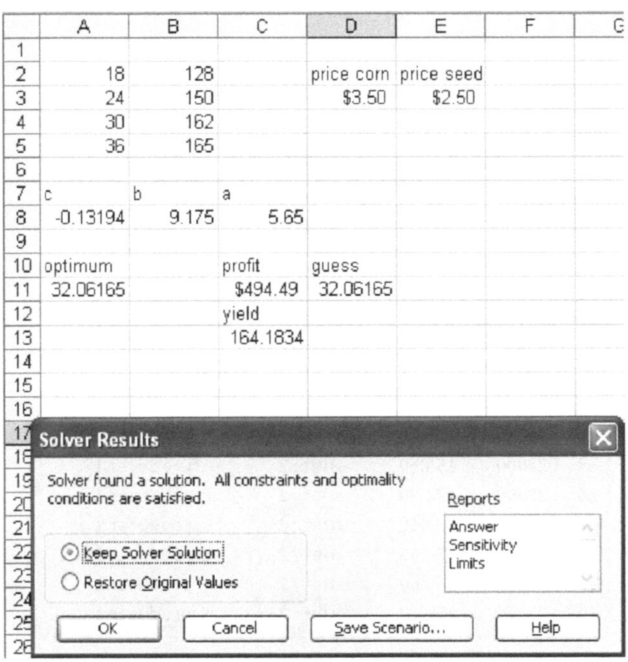

FIGURE 10.5 The Excel spreadsheet after solver has been run.

10.2.3 Step-by-Step Guide for Developing a Site-Specific Population Equation

This same approach can be used to determine the equation that defines the seeding rate as a function of yield potential. A step-by-step guide for this process is given in the following text. In these calculations, yield is defined as bushels/acre. To calculate the seed rate per hectare, seeding rate, yields, and corn selling price must be converted to SI units. Bushels/acre are converted to Mg/ha by multiplying the yields by 0.06259, seeds/ac are converted to seeds/ha by multiplying the rate by 2.47, and $/bu are converted to $/Mg by multiplying the selling price by 44.

Step 1. Load your raw data from the disk (labeled Chapter 10). Note that there is a VBA (visual basic) program under Commandbutton1. To load this macro, you must go to Tools, Options, Security, and Macro Security and then set the level from Very High to High. Figure 10.6 shows typical input information. For this program to work, seeding population data must be in column A, yield data must be in column B, and delimiters must be in columns C and D. If one or both of the two delimiters change, the program assumes a break from one experiment to the next. For this application, the format must be followed as shown with data starting in row 6. This data set contains two sheets. Sheet 1 is the solution using the exponential model, while sheet 2 contains the solution using the polynomial equation. Do not change the sheet names.

	A	B	C	D
1				
2	CommandButton1			
3				
4				
5	X population	Y yield	Hybrid	Cooperator
6	20000	152.60	variety 1	producer 1
7	24000	162.10	variety 1	producer 1
8	28000	167.70	variety 1	producer 1
9	32000	173.30	variety 1	producer 1
10	36000	174.60	variety 1	producer 1
11	40000	175.90	variety 1	producer 1
12	44000	163.10	variety 1	producer 1
13	48000	161.50	variety 1	producer 1
14	20000	156.30	variety 2	producer 1
15	24000	156.60	variety 2	producer 1
16	28000	172.90	variety 2	producer 1
17	32000	173.10	variety 2	producer 1
18	36000	177.60	variety 2	producer 1
19	44000	168.00	variety 2	producer 1
20	48000	155.70	variety 2	producer 1

FIGURE 10.6 Raw data loaded from disk.

Step 2. Run the VBA by pushing the button, Commandbutton1. The program does calculations that will result in the populated spreadsheet as seen in Figures 10.7 and 10.8. The program does all work discussed in steps 3 through 10 below.

Step 3. In this application, the relationship between yield and plant population is defined by the equation. Note that in the case of yield = f(seed population) that 0 seed results in 0 yield. Also note that 0 added fertilizer will not usually result in 0 yield so this equation may not be the appropriate equation for the yield response of fertilizer. Obviously, different problems should use different equations. This equation is transformed by dividing both sides of the equation by planting population and then by taking the natural logarithm of both sides of the equation. The transformed equation is linear in Ln(A) and B. For each experiment, column E below row 6 contains the terms, Ln(yield/population).

Step 4. In cells F6:G6 (and for each separate experiment below line 6) Linest, Excel's linear regression procedure, is used to determine the constant Ln(A) and the coefficient B.

Step 5. Cell H6 (and for each separate experiment below) contains the maximum yield value in the experiment. Our recommendation is to use the long-term median yield at the site of the experiment for input into cell H6. We would calculate this from 5 or more years of yield monitor data files at the specific site of each experiment within the field of concern. Five years of yield monitor data was not available for this data set so an alternative, maximum yield, was used.

Step 6. Ln(A) is transformed to A in cell I6 (and for each separate experiment below).

Step 7. The users best guess for the relationship between seeding rate and yield [seeding rate = constant + yield potential (factor)] is placed in I3 and J3. The constant is placed in I3 and the factor (slope) is placed in J3. In this example, 25,000 is placed in I3 and 0 is placed in J3.

Step 8. In cell J6 (and for each separate experiment below), Equation 4, the planting population will be calculated based on the initial guessed inputs from cells I3 and J3. In cell K6, the estimated yield (and for each separate experiment below) will be calculated using the planting population, cell J6, and the coefficients for A and B from cells I6 and F6.

Step 9. We use the estimated seeding rate and yield for the first experiment and then for all experiments, which were calculated in cells J6 and K6, respectively and then below. The partial profit ($\text{Profit}_{\text{partial}}$ = returns − inputs) for the first experiment is calculated in cell L6 and then all experiments below.

Step 10. We then use the partial profit's from each experiment, cells L6 and below, to calculate the average $\text{Profit}_{\text{partial}}$ from all of the separate experiments.

Step 11. If your yield response curves (the coefficients of which are in columns F and I below row 5) have adequately described the field, you are now ready to do the real analysis. Leaving cell J3 at 0, you may change cell I3 to any population you wish and the average partial profit for all experiments combined, $\text{Profit}_{\text{partial}}$, cell E2, will be calculated. Note that if you did this twice for two different populations guesses, the differences between the two guesses will be the real differences in total profit,

	A	B	C	D	E	F	G	H	I	J	K	L	M	N	O	P
1					Prof $/A	CommandButt			op pop=f(yld pot)		cst seed	val corn		yield		
2					450.043				a1	b1	$/seed	$/bu	80	120	160	200
3									25000	0	0.0025	3.5	25000	25000	25000	25000
4																
5	X population	Y yield	Hybrid	Cooperator	Ln(yld/pl B		Ln(A)	maximum A		Est po	Est yld	Profit?				
6	20000	152.60	variety 1	producer 1	-4.87567	-2.9073E-05	-4.2936478	175.90	0.01365502	25000	165.0351	515.1228				
7	24000	162.10	variety 1	producer 1	-4.9976											
8	28000	167.70	variety 1	producer 1	-5.11778											
9	32000	173.30	variety 1	producer 1	-5.21847											
10	36000	174.60	variety 1	producer 1	-5.32878											
11	40000	175.90	variety 1	producer 1	-5.42672											
12	44000	163.10	variety 1	producer 1	-5.59758											
13	48000	161.50	variety 1	producer 1	-5.69445											
14	20000	156.30	variety 2	producer 1	-4.85171	-2.9918E-05	-4.2657836	177.60	0.01404086	25000	166.1515	519.0302				
15	24000	156.60	variety 2	producer 1	-5.03211											
16	28000	172.90	variety 2	producer 1	-5.08725											
17	32000	173.10	variety 2	producer 1	-5.21962											
18	36000	177.60	variety 2	producer 1	-5.31174											
19	44000	168.00	variety 2	producer 1	-5.56798											
20	48000	155.70	variety 2	producer 1	-5.73103											

FIGURE 10.7 Spreadsheet showing data after VBA program is run. If the data are changed, push the command button to recalculate values.

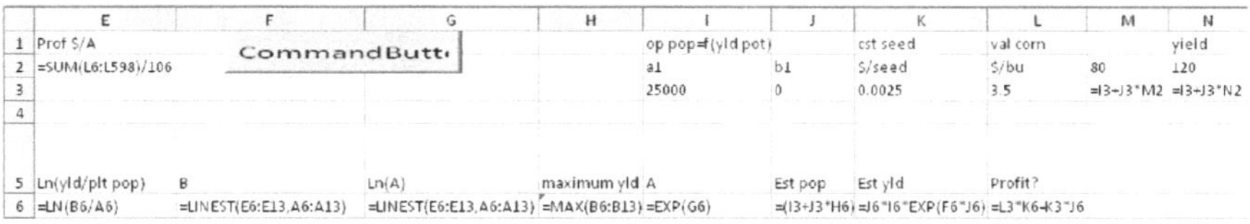

	E	F	G	H	I	J	K	L	M	N
1	Prof $/A	CommandButt			op pop=f(yld pot)		cst seed	val corn		yield
2	=SUM(L6:L598)/106				a1	b1	$/seed	$/bu	80	120
3					25000	0	0.0025	3.5	=I3+J3*M2	=I3+J3*N2
4										
5	Ln(yld/plt pop)	B	Ln(A)	maximum yld	A	Est pop	Est yld	Profit?		
6	=LN(B6/A6)	=LINEST(E6:E13,A6:A13)	=LINEST(E6:E13,A6:A13)	=MAX(B6:B13)	=EXP(G6)	=(I3+J3*H6)	=J6*I6*EXP(F6*J6)	=L3*K6-K3*J6		

FIGURE 10.8 The Microsoft Excel spreadsheet showing appropriate formulas used to calculate the values in Figure 5.7.

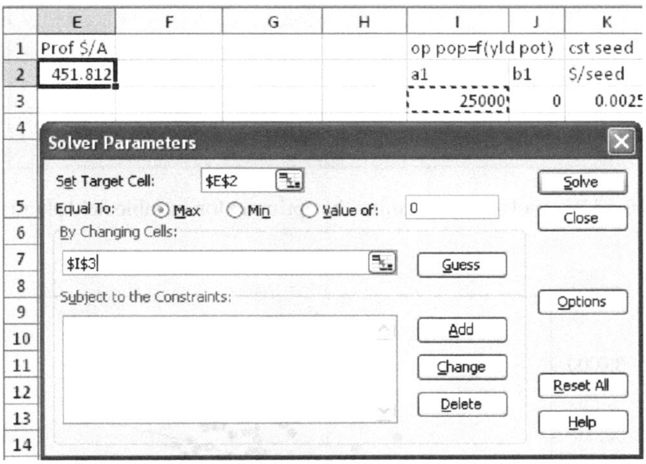

FIGURE 10.9 Using solver to determine the optimum for a single-field planting population.

not partial profit. The difference is what is important. To determine the equation for determining the optimum population, select solver (in Excel 2003 under tools). Run solver by selecting E2 ($\text{Profit}_{\text{partial}}$) as the variable "Set Target Cell" and I3 as "By Changing Cells". When the population level is defined as a single value, the optimum yield potential is 27,455 plants/ac (67,927 plants/ha) (Figures 10.9 and 10.10).

Step 12. Figure 10.9 is identical to Figure 10.10 with the exception that "By Changing Cells" is now I3:J3 which will maximize giving a site-specific variable rate rather than a single population rate for the entire field as discussed earlier. The resulting solution for seed cost of \$0.0025/seed and a corn selling price of \$3.50/bu (\$154/Mg) is seeding rate (seeds/ac) = 12,927 + 92.97 · yield potential (bu/ac) or seeding rate (seeds/ha) = 31,929 + 3,669 · yield potential (Mg/ha). The estimated partial profit for this equation is \$453.62/ac (\$1120/ha).

Step 13. Assess the impact of seed costs on the recommendation. For this calculation, change K3 from \$0.0025/seed to \$0.005/seed. In this calculation, the seeding prediction equation was seeding rate (seeds/ac) = 9569 + 92.3 · yield goal (bu/ac) or seed rate (seeds/ha) = 23,639 + 3.642 · yield goal (Mg/ha). Increasing the corn value has the opposite impact on the seeding rate. For example, doubling the corn value from \$3.5/bu to \$7.0/bu produced the original results because the ratio of \$seed/\$corn was the same (0.0025/3.5 = 0.005/7).

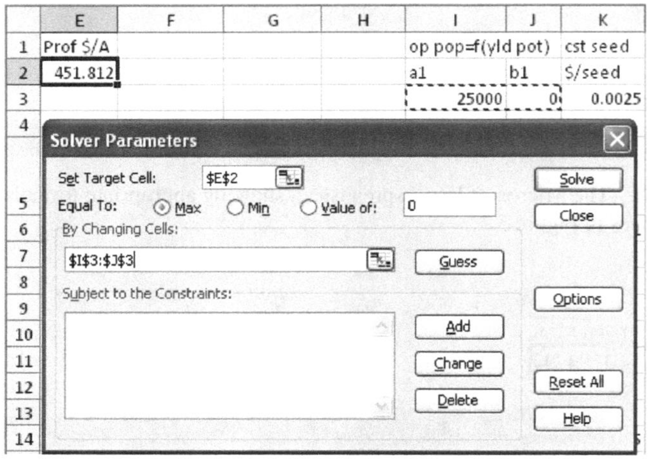

FIGURE 10.10 Using solver to determine the optimum for variable field planting population.

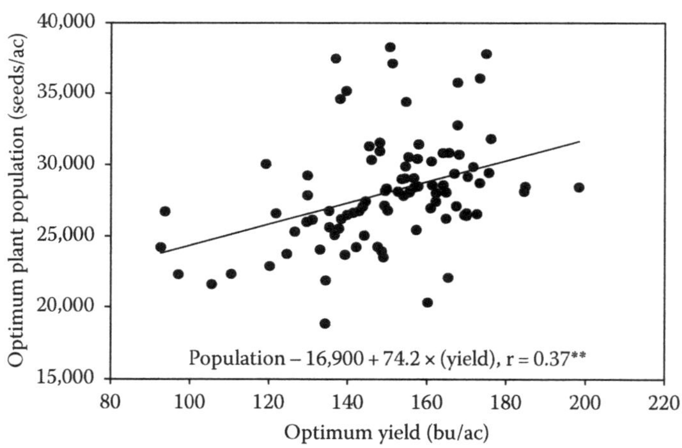

FIGURE 10.11 Relationship between optimum plant population and yield in experiments conducted in the northern Great Plains between 1998 and 2002.

This section showed an approach for determining site-specific plant population recommendations using an exponential model. There was a strong relationship between the predicted and observed responses (Figure 10.11). This analysis would suggest that in a deep silt loam soil that produces a uniform yield of 180 bu/ac (11.300 Mg/ha), the seeding rate should be 29,662 seeds/ac (seeding rate = 180 · 92.97 + 12,927 seeds/ac) or 74,100 seeds/ha. In a coarse textured soil that produced a uniform yield of 120 bu/ac (7510 kg/ha), the predicted seeding rate was 24,000 seeds/ac (24,000 seeds/ac = 120 · 92.9 + 12,900 seeds/ac) or 59,300 seeds/ha. Many fields contain both high- and low-yielding areas. In fields with variable yields, a manager may choose to plant 24,000 seeds/ac (59,300 seeds/ha) in the low-yielding soil and 30,000 seeds/ac (74,100 seeds/ha) in the high-yielding soil.

The solution depends on the model selected to fit the data. Sheet 2 provides a similar solution using a polynomial equation. When the exponential model was used to develop the site-specific plant population equation, a slightly different equation was produced. The polynomial equation is, optimum seeding rate = 9943 + 111 · yield goal (bu/ac) or seed rate (seeds/ha) = 24,559 + 4,380 · yield goal (Mg/ha). The estimated partial profit for this equation over the data set was $458.24/ac or $1132/ha. This equation produced a slightly higher estimated population of 29,923 seeds/ac or 73.910 seeds/ha for a field with a yield goal of 180 bu/ac (11.3 Mg/ha) when seed was purchased for $0.0025/seed and corn was sold at the rate of $3.5/bu ($138/Mg).

10.3 IMPLEMENTING ON-FARM RESEARCH

The success of implementing site-specific management can be assessed using on-farm research. Figure 10.12 shows an example of a two-block on-farm study. To implement this proposed procedure, plots are planted and documented relative to a line of GPS points at planting. Each field trial should be placed perpendicular to the major landscape formations (yield zones) within the field, and it is desirable to have three or more randomized replications.

To increase the confidence of the populations, each plant population needs to be verified by a minimum of three stand counts (for ease of counting, in an early vegetative stage) to document the actual emerged plant population. GPS-referenced yield monitor data need to be collected at harvest, and the data will be partitioned by treatment and environment. Stip-trial research studies should be harvested with combines equipped with a yield monitor/GPS/data-recording system. Within each block, there are areas that can be harvested from different landscape positions. Each of these landscape positions likely has a different yield potential. Analysis of on-farm experiments is beyond the scope of this chapter and is available in Knighton[20] and Wittig and Wick.[21]

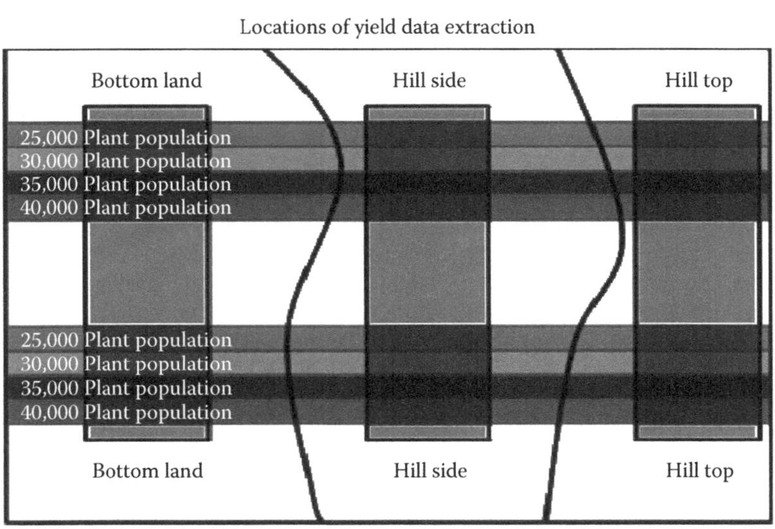

FIGURE 10.12 Potential experimental design for an on-farm study.

10.4 CONCLUSION

In this approach, a large number of data sets could be analyzed simultaneously, and the primary factor for selecting a model was profitability. The resulting site-specific equation is dependent on the equation used to define the yield response function. Different response functions were obtained for the two equations. This analysis is based on data collected from field experiments conducted between 1998 and 2002. These experiments included a much wider range of plant populations and yield levels than what is typically found in most plant population trials. This analysis procedure is applicable for many other input parameters. In this analysis, the impact of other limiting factors (N, pests, weeds, and P) must be minimized.

ACKNOWLEDGMENTS

Support for this chapter was provided in part by South Dakota Soybean Research and Promotion Council, South Dakota Corn Utilization Council, NASA through the Upper Midwest Aerospace Consortium, USDA-NCSARE, the South Dakota Agricultural Experiment Station, USDA-CSREES, and SD 2010 initiative.

REFERENCES

1. Derby, N.E., Steele, D.D., Terpstra, J., Knighton, R.E., and Casey, F.X.M. Interactions of nitrogen, weather, soil, and irrigation on corn yield. *Agronomy Journal* 97, 1342, 2005.
2. Lory, J.A. and Scharf, P.C. Yield goal versus delta yield for predicting fertilizer nitrogen need in corn. *Agronomy Journal*, 95, 994, 2003.
3. Sawyer, J., Nafziger, E., Randall, G., Bundy, L., Rehm, G., and Joern, B. *Concepts and Rationale for Regional Nitrogen Rate Guidelines for Corn (PM 2015)*. Iowa State University Extension, Ames, Iowa, 2006, Available at http://www.extension.iastate.edu/Publications/2015.pdf (accessed October 2007).
4. Clay, D.E., Clay, S.A., Lyon, D.J., and Blumenthal, J.M. ^{13}C discrimination in corn grain can be used to separate and quantify yield losses due to water and nitrogen stress. *Weed Science* 53, 23, 2005.
5. Sarlangue, T., Andrade, F.H., Calvino, P.A., and Purcell, L.C. Why do maize hybrids respond differently to variations in plant densities? *Agronomy Journal* 99, 984, 2007.
6. Nafziger, E.D. Inter- and intraplant competition in corn. Crop management network. doi: 1094/cm-2006-0227-05-RV, 2006. http://www.plantmanagementnetwork.org/pub/cm/review/2006/complete/
7. Pierik, R., Voesenek, A.C.J., DeKroom, H., and Visser, E.J.W. Density-induced plant size reduction and size inequality in ethylene-sensing and ethylene insensitive tobacco plant. *Biology* 6, 201–205, 2004.
8. Clay, S.A., Clay, D.E., Horvath, D., Pullis, J., Carlson, C.G., Hansen, S., and Reicks, G. Corn (*Zea mays*) responses to competition: Growth alteration vs limiting factor. *Agronomy Journal* 101, 1522, 2009.
9. Schmitt, J. Is photomorphogenic shade avoidance adaptive? Perspectives from population biology. *Plant Cell Environment* 20, 826, 1997.
10. Smith, H. and Whitelam, G.C. The shade avoidance syndrome: Multiple responses mediated by multiple phytochromes. *Plant Cell Environment* 20, 840, 1997.
11. Balleré, C.L. and Casal, J.J. Light signals perceived by crop and weed plants. *Field Crops Research* 67, 149, 2000.

12. Rajcan, I., Chandler, K.J., and Swanton, C.J. Red-far red ratio of reflected light: A hypothesis of why early-season weed control is important in corn. *Weed Science* 52, 774, 2004.
13. Horvath, D.P., Llewellyn, D., and Clay, S.A. Heterologous hybridization of cotton (*Gossypium hirsutum*) microarrays with velvetleaf (*Abutilon theophrasti*) reveals physiological responses due to corn competition. *Weed Science* 55, 546, 2007.
14. Farnham, D. *Corn Planting Guide*. Iowa State Extension, 2001. Available at http://www.extension.iastate.edu/Publications/PM1885.pdf
15. Bullock, D.G., Bullock, D.S., Nafziger, E.D., Doerge, T.A., Paszkiewicz, R., Carter, P.R., and Peterson, T.A. Does variable rate seeding in corn pay? *Agronomy Journal* 90, 830, 1998.
16. Abendroth, L. and Elmore, T. Corn seeding rates and variable rate seeding. Integrated Crop Management IC-498 (5). Available at http://www.ipm.iastate.edu/ipm/icm/2007/4-9/seedingrate.html
17. Coulter, J. *Optimum Plant Population for Corn in Minnesota*. University of Minnesota Extension, M1244, 2009. Available at http://www.extension.umn.edu/distribution/cropsystems/M1244.html
18. Paszkiewicz, S. and Butzen, S. Corn hybrid response to plant population. *Crop Insights* 17, 16, 2009.
19. Shanahan, J.F., Doerge, T.A., Johnson, J.L., and Vigil, M.F. Feasibility of site-specific management of corn hybrid and plant densities in the Great Plains. *Precision Agriculture* 5, 1575, 2004.
20. Knighton, R.E. Setting up on-farm experiments. In Clay, D.E. et al. (eds.). *Site-Specific Management Guidelines*, Chap. 17. Available at http://www.ipni.net/ppiweb/ppibase.nsf/$webindex/06E070534CA50AB2052569DB0033F992, 1999.
21. Wittig, T.A. and Wick, III Z.W. Simple on-farm comparisons. In Clay, D.E. et al. (eds.). *Site-Specific Management Guidelines*, Chap. 18. Available at http://www.ipni.net/ppiweb/ppibase.nsf/$webindex/06E070534CA50AB2052569DB0033F992, 1999.

11 Soil Water Status Maps for Variable Rate Irrigation

C.B. Hedley and I.J. Yule

CONTENTS

11.1 EXECUTIVE SUMMARY

Energy requirements and security of food production will be improved by increasing the water use efficiency of existing irrigation systems. Agriculture uses 70%–80% of allocated global freshwaters for irrigation, and in recent years increased dependence on irrigation to sustain food production for the global community is depleting freshwater resources in many parts of the world below sustainable limits. These pressures can be reduced by optimizing irrigation scheduling. Technological advances in irrigation, mainly in automated and semiautomated sprinkler irrigation systems, allow application of water with millimeter precision at a fixed position. In addition, variable rate sprinkler systems can apply different depths of water to different positions. This chapter presents and demonstrates a geographical information systems (GIS) decision support tool for a soil-based automated variable rate irrigation (VRI) sprinkler system.

The case study presented illustrates how (1) soil total available water-holding capacity (TAWC) maps can be derived from soil apparent electrical conductivity (EC_a) maps and (2) a daily time step can be added to the TAWC map to produce a soil water status map, updated daily, available for uploading to a VRI system. Soil water status is defined as the millimeters of plant available water stored in the soil on any one day. This chapter also discusses some advantages of a VRI system, including benefits for nutrient management, water savings, and energy savings.

The accuracy of these soil water status maps is dependent on how well each EC_a-defined zone can be hydrologically characterized. VRI is desirable where highly variable soils exist under one irrigation system and has further applications for irrigating different crops under one irrigator, shutting off or reducing irrigation over exclusion areas such as raceways, gateways, and low-lying and/or compacted poorly draining areas, as well as for chemigation and fertigation.

11.2 INTRODUCTION

The case study for this chapter is a 40 ha irrigated dairy pasture near Christchurch, on the Canterbury Plains of South Island, New Zealand. The Canterbury Plains are the largest alluvial plains in New Zealand covering 750,000 ha. Here highly variable young alluvial soils have developed in different depths of fine materials (sands and silts) over outwash gravels. Recent alluvial soils (<3000 years old) occur on low terraces that can be highly variable in depth and stoniness. A recent trend has been the conversion of these stony terrace soils, traditionally used for dry-land extensive grazing, to dairy pastures. This is accomplished by augmentation of natural annual rainfall (500–600 mm) with irrigation using center pivot and lateral sprinkler irrigation schemes (Figure 11.1). This significant land-use change has occurred since 1990 when sheep farming predominated (Figure 11.2) so that in 2006 Canterbury had become a national leader in dairy production.

One irrigation system typically irrigates an area of more than 300 ha so that where the depth of fine material to gravels is highly variable, so will be the ability of the soil to hold and retain available water for plant use. A frequent consequence is overwatering of very stony soils, and in this case VRI becomes desirable. The 2008 Environment Report of the Canterbury Regional Council[1] recognized the significant pressure being applied to the Canterbury's lakes and rivers due to intensification of land use, allocation of freshwaters, changing climatic factors, and changes arising from nutrient inputs and sedimentation. Surface water takes have doubled from about 100–200 $m^3\ s^{-1}$ and groundwater takes have trebled from about 50–150 $m^3\ s^{-1}$ since 1990. In New Zealand, irrigation demands 77% of all allocated freshwaters, and 66% of this irrigation water is used in the Canterbury region. The amount of consented irrigated land in New Zealand increased by 52% over the period 1999–2006, an annual rate of increase of 7%.[2] This national increase is also largely explained by demand for irrigation by growth in the dairy industry.

In Canterbury, river flows have dropped from the long-term mean at nearly all monitored sites, and in some cases flows are 10%–25% below the long-term mean. One of the largest recorded drops is on the Selwyn River just south of Christchurch (37% decrease), while the Rakaia River south of the Selwyn has drops of 10%. The case

FIGURE 11.1 The introduction of center pivot irrigation systems into Canterbury, New Zealand, provides opportunities for VRI technology to be used. (Photo courtesy of Google Earth, 2009.)

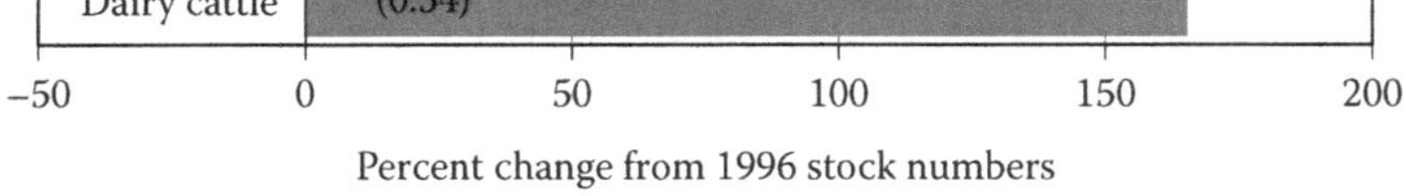

FIGURE 11.2 Relative change to stock numbers in Canterbury, New Zealand, between 1996 and 2006, with estimated 2002 population totals (millions) in brackets. (LIC, New Zealand Dairy Statistics 2007–2011. Prepared by Livestock Improvement Corporation (LIC), Hamilton, New Zealand. Document available from DairyNZ www.dairynz.co.nz/dairystatistic, 2009, 51 p.; Canterbury Regional Council. *Canterbury Regional Environment Report 2011*. ECan, Christchurch, New Zealand. Document available at www.ecan.govt.nz, 2011, 246 p.)

study for this chapter is situated between the Rakaia and Selwyn Rivers, where a center pivot sprinkler system is used to irrigate dairy pastures, on Waimakariri soils.

The Waimakariri soils are widespread in this region and are relatively young alluvial soils on low terraces. The single most important factor determining their properties, like many other soils in the Canterbury Plains, is the depth of fine material over the outwash gravels. The variable nature of these soils and the need to increase irrigation and energy efficiency makes this area an excellent site for a precision irrigation management case study.

In this study, an on-the-go electromagnetic (EM) mapping system was used to collect high-resolution (<10m) global positioning system (GPS)-located EC_a data, which was then kriged into EC_a maps using the Geostatistical Analyst extension tool in ArcGIS. Soil EM surveys are very useful for investigating soil salinity issues;[3,4] however, EC_a information can also be used to characterize soil texture and moisture variability.[5–7] The approach used to determine the available water-holding properties of soil management zones from the EC_a information is discussed and demonstrated.

VRI scheduling requires knowledge of plant demand as well as soil supply of available water at any position on any day.[8] Two methods for monitoring the drying and wetting patterns are (1) a daily water balance model and (2) *in situ* soil moisture monitoring. TAWC can then be adjusted on a daily basis to produce soil water status maps for spatial irrigation scheduling. The map is available to upload to an automated PC-controlled variable rate sprinkler system. Sprinkler systems with VRI control of individual sprinklers are being developed with varying degrees of automation,[9,10] to address the need for improved water use efficiency.

Precision irrigation decisions can be crop-based or soil-based. The emphasis of this chapter is the development of a soil-based decision support tool, so that precision irrigation can be applied on the basis of soil differences. This is particularly important in regions where highly variable, young, alluvial soils exist, such as in many parts of New Zealand.

11.3 METHODS

11.3.1 Site Selection

The dairy farm lies 50km WSW of Christchurch. Here a 600m center pivot sprinkler irrigation system was installed to irrigate dairy pastures for the seasonally dry part of the year, typically November–March. A 40ha portion of the irrigated zone was selected for this study. Regional statistics suggest that irrigation can increase pasture dry matter (DM) production from 6.4T DM ha^{-1} yr^{-1} to at least 11.1T DM ha^{-1} yr^{-1}. Stocking rates are 3.3 cows ha^{-1} with irrigation. To minimize feed deficits, 200kg N ha^{-1} yr^{-1} is typically applied.

The soils at this site are mapped as Waimakariri soils (weathered fluvial recent soils), characterized by varying depths of fine sandy material with wide-ranging stoniness over outwash gravels which occur at varying depths.

11.3.2 Apparent Electrical Conductivity Mapping

A Geonics electromagnetic EM38 sensor (Geonics Ltd., Mississauga, Ontario, Canada) and real-time kinematic-differential GPS with Trimble Ag170 field computer (Trimble Navigation Ltd., Sunnyvale, California) on-board an all-terrain vehicle (ATV) were used for on-the-go soil EC_a mapping. The ATV was driven at 12km h^{-1} at swath widths of 10m. Soil EC_a, measured by the EM38 sensor, was logged simultaneously with high-resolution positional data every second. The EC_a and positional data were collected on a data memory card in the Ag170 field computer.

11.3.3 GIS Manipulation of EM Data

Data were extracted from the data memory card and manipulated into a suitable file format in an Excel spreadsheet for interpolation and map production in ArcGIS. The process for importing data into ArcGIS, transforming the WGS84 coordinates to local coordinates (in this case NZMG'49), setting a field boundary, and raster interpolation for EC_a map development, followed the method detailed by Cassel[4] in Chapter 8. In the first book of this series, "GIS Applications in Agriculture,"[11] Cassel[4] provides a step-by-step account of a procedure similar to the one which is summarized as follows:

- Positional and sensor data were imported into ArcMap as a csv file. This data was then displayed and converted into a shapefile.
- A new polygon shapefile was created in Arc Catalogue and moved into the map document. This file was used to define the outline shape of the sensor data using Editor.
- Ordinary kriging was used in Geostatistical Analyst to develop the EC_a prediction map. Variogram models of the sensor data were compared to assess which one best characterized the data, and in this case a spherical variogram was used. The kriged surface was exported as a raster file.
- The raster surface was clipped to the shape of the polygon outline file using Spatial Analyst/Extraction/Extract by Mask.
- The clipped raster was then classified into three zones, by right-clicking on the file and selecting Symbology in Properties. Color coding was also set in Symbology.

The three soil EC_a classification zones were targeted for soil sampling for TAWC estimation.

11.3.4 Estimation of Soil TAWC in EC_a-Defined Management Zones

The soil TAWC of each EC_a-defined management zone (high EC_a zone, intermediate EC_a zone, low EC_a zone) was estimated by soil sampling at three replicate sites within each zone. TAWC is defined as the difference between the equivalent depths of water that the effective root zone (or a sampling depth within that root zone) contains at field capacity (FC) and at permanent wilting point (WP). It is preferable to determine FC in situ, i.e., in the field, to account for interactions between horizons in the soil profile. It is estimated here as the equivalent water depth held after 2 days of free drainage following a heavy rain (or irrigation) event. WP is also best estimated in the field using the crop of interest, although this was not possible at this irrigated site.

The soil was sampled to 60 cm depth to include the majority of roots, which extract the soil water for plant use. Soil samples (0–15, 15–30, 30–45, 45–60 cm) were collected when the soils were at FC in August 2008, for estimation of field moist soil water content (i.e., FC), bulk density, percent stones, and a laboratory estimation of WP. Small intact soil cores (55 mm × 25 mm) were collected, where possible, to assess bulk density. In cases where the soils were too stony to enable intact soil cores to be collected for bulk density estimation, the bulk density of a neighboring sample was extrapolated to this sample depth. Percent stones were determined by collecting bagged samples to a defined depth.

Larger stones were sieved out and weighed in the field, and the remaining samples were taken back to the laboratory for calculation of the remaining percent stones. WP was assessed from the gravimetric soil water content of repacked 2 mm sieved soil samples equilibrated at a tension of 1500 kPa, on a 15 bar pressure plate extractor (Soilmoisture Equipment Corp. 1500, Santa Barbara, California).

11.3.5 Soil Water Balance

A soil water balance was used to estimate the soil water deficit on any one day. Model inputs are (1) soil TAWC value; (2) daily rainfall; (3) daily regional calculation for potential evapotranspiration (PET); and (4) irrigation events. Model outputs are (1) soil moisture deficit (SMD) on any one day and (2) expected runoff and deep percolation (estimated as the excess water when soils reach FC) (Figure 11.3).

The soil water balance was initiated on August 3, 2007, a day when the soils were expected to be at FC, after a cool, wet winter period and large winter rainfall event (34 mm). Capillary rise was assumed to be zero in these soils, as the depth of water table was sufficient to maintain the capillary fringe below the lower level of the root zone.

The Excel spreadsheet (Table 11.1; Chapter 11/GIS Chapter 11 data) was used to calculate the daily SMD for each EC_a-defined zone. A simplified version of the soil water balance spreadsheet used in this study is provided on the CD which

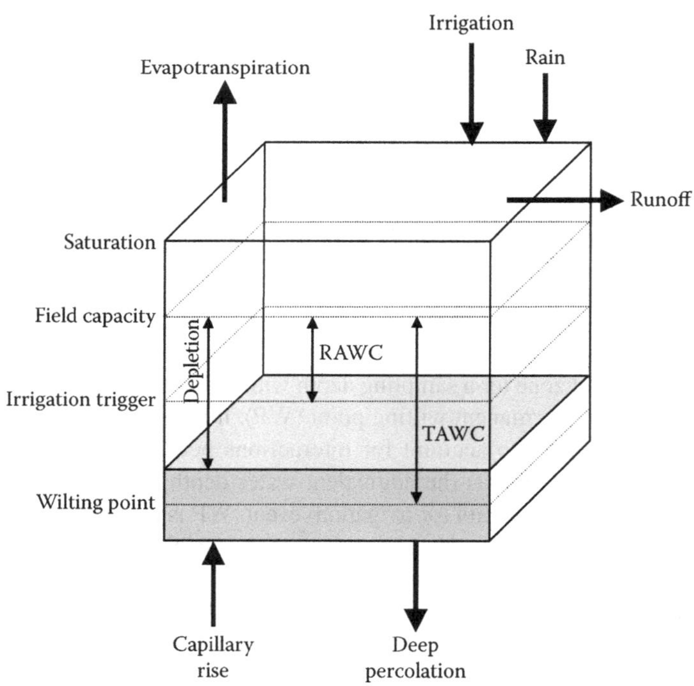

FIGURE 11.3 Diagram to represent the soil water balance of a root zone. (Adapted from Allen, R.G. et al. Crop evapotranspiration. Guidelines for computing crop water requirements, in *FAO Irrigation and Drainage Paper 56*, p. 169, FAO, Rome, 1998, 300 p. With permission.) RAWC, readily available water-holding capacity; TAWC, total available water-holding capacity.

TABLE 11.1
Part of the Excel Spreadsheet Used to Predict Daily Soil Moisture Deficit and Irrigation Events for Each Soil Zone (see Appendix)

				With Irrigation					No Irrigation			
Year	Imput Rain (mm)	Irrigation (mm)	Imput Er (mm)	E (mm)	D (mm)	S (mm)	SMD (mm)	Irrigation Event (IE) (mm)	E (mm)	D (mm)	S (mm)	SMD (mm)
3/8/2007	0	0	1	1	0	101	−3	0	1	0	101	−3
4/8/2007	34	0	1	1	33	101	0	0	1	33	101	0
5/8/2007	0	0	1	1	0	100	−1	0	1	0	100	−1
6/8/2007	0	0	0	1	0	99	−2	0	1	0	99	−2
7/8/2007	0	0	1	0	0	99	−2	0	0	0	99	−2
8/8/2007	0	0	1	1	0	98	−3	0	1	0	98	−3
9/8/2007	0	0	2	1	0	97	−4	0	1	0	97	−4
10/8/2007	0	0	2	2	0	95	−6	0	2	0	95	−6
11/8/2007	0	0	2	2	0	93	−8	0	2	0	93	−8
12/8/2007	6	0	1	2	0	97	−4	0	2	0	97	−4

Er, reference ET; E, actual ET; D, drainage; S, available water stored in the soil; SMD, soil moisture deficit.

accompanies this book, and step-by-step details for its use are given in the Appendix to this chapter. It uses climatic data obtained from a nearby meteorological station, which is freely accessed from a web site (http://cliflo.niwa.co.nz/). The first worksheet in this Excel file provides the data specific to this case study. The second worksheet provides a soil water balance template for the user to enter other specific data of interest. Required inputs are site daily rainfall and PET data, soil TAWC (≤250 mm), and trigger point for irrigation. The worksheet automatically calculates the optimal irrigation schedule. The depth of irrigation can be changed (≤50 mm).

As a rule of thumb, irrigation should commence when approximately half the plant-available water (TAWC) has been used up. Specific depletion factors and methods for determining this "irrigation trigger," "critical deficit," or "stress point" (SP) are given in Allen et al.[12] The trigger point marks the point where water in the soil is no longer readily available to the plant, and evapotranspiration (ET) from that plant falls below its potential rate. The transpiration (T) component of ET is directly related to plant yield. If the plant transpires at its potential rate, it will usually achieve potential yield, assuming no other limiting factors. If the plant transpires below its potential rate, due to inadequate supply of readily available water, then yield will usually be reduced proportionally.[12]

11.4 RESULTS AND DISCUSSION

11.4.1 EC_a Map and Soil Water-Holding Properties

The EC_a map separates the field into three areas with different soil characteristics (Figure 11.4). Soils in the high EC_a zone are deep sandy soils with no stones, characterized by 20 cm deep topsoil over 10–30 cm sandy loam subsoil. Below this, a transition BC horizon extended to a depth varying between 45 cm and >1 m over outwash gravels. In contrast, the low EC_a zone is characterized by stony–very stony topsoil and subsoil, each typically about 15 cm in depth, overlying a transition BC horizon, with outwash gravel parent materials occurring at 45 cm depth (Table 11.2). The relationship shown in Figure 11.5 was used to predict soil TAWC from the high-density EC_a data set, and the resulting TAWC map is shown in Figure 11.6.

The soil TAWC varies considerably at this site, ranging from 36 to 144 mm. Assuming a mean ET rate of 4 mm for the hottest months of November to February, the most stony soils with lowest TAWC will reach the irrigation trigger point (0.5 TAWC) after just 4–5 days of drying from FC, while the stone-free sandy soils are able to store enough readily available water for about 18 days.

11.4.2 Soil Water Balance and Its Application for Irrigation Scheduling

A soil water balance was used to track the wetting and drying patterns of soils in the three management zones. Daily rainfall and PET data were obtained from the closest meteorological station, and soil TAWC was adjusted for each soil zone. The soil water balance was updated daily to provide a real-time irrigation scheduling tool.

The water balance predicts the day on which the soil trigger point is reached, and irrigation should commence. Hypothetical VRI scheduling applies 10 mm of irrigation to Zone 3 (TAWC 44 mm) on September 1, 2007. The first irrigation event scheduled to Zone 2 (TAWC 73 mm) for the 2007/2008 summer season is on 30 September, and

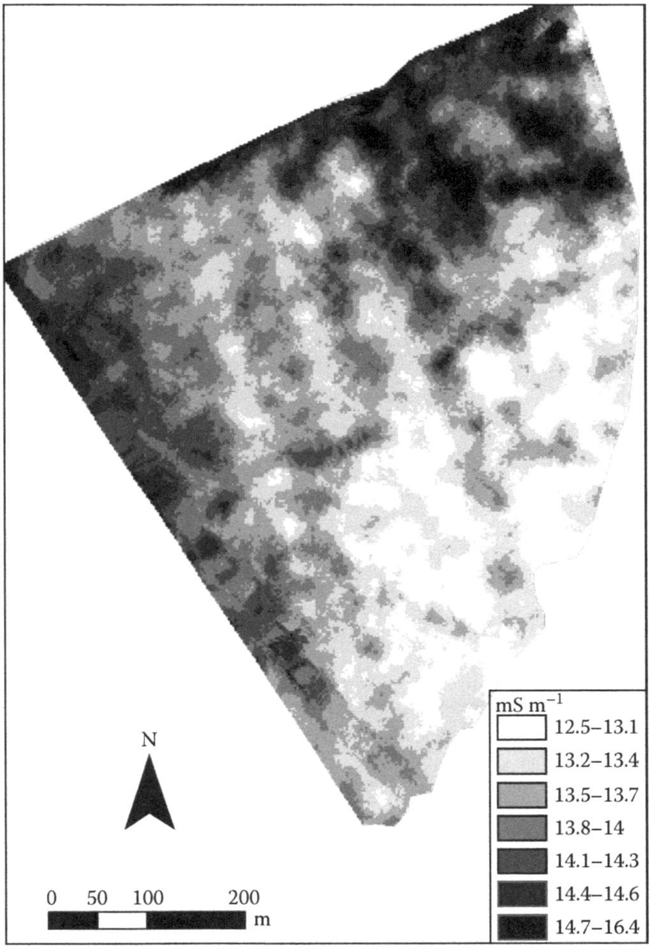

FIGURE 11.4 EC_a map of the 40ha dairy pasture, irrigated by a center pivot sprinkler irrigation system.

Zone 1 (TAWC 101mm) soil water storage is sufficient to readily supply water to pasture until its first scheduled irrigation event on October 25, 2007 (Figure 11.7). The water balance was also used to compare total irrigation water demands under a VRI system compared with a URI (uniform rate irrigation) system. URI assumes that to obtain potential yield, the whole site is irrigated to the trigger point of the zone with the smallest TAWC, maintaining readily available water in these stony soils and all other zones. In contrast, VRI irrigates each soil zone to its specific trigger point. We have calculated hypothetical 10mm irrigation events for VRI and URI, and the results for the 2007/2008 season are presented in Figure 11.10. The water savings using VRI are quantified in Table 11.3 for August 2007 to December 2008. The relationship between soil water storage and EC_a on any day can be determined and used to predict the soil water status map. EC_a values were extracted from the EC_a map, for each sampling site position, as described in the following text.

TABLE 11.2
Soil TAWC, FC, WP, and Percent Stones for Each EC_a-Defined Zone

Soil	ECa Range (mS m^{-1})	Sample Depth (cm)	TAWC (mm)	Field Capacity (mm)	Wilting Point (mm)	Percent Stones (%)
Waimakariri sandy loam	14.7–16.7	0–15	46	58	12	0
		15–30	21	28	7	0
		30–45	17	20	4	0
		45–60	17	20	3	0
Site mean		0–60	101	126	26	0
Waimakariri stony sandy loam	13.7–14.6	0–15	32	39	7	43
		15–30	20	23	3	68
		30–45	11	13	2	23
		45–60	11	13	2	23
Site mean		0–60	74	88	14	39
Waimakariri very stony sandy loam	12.5–13.6	0–15	12	20	7	42
		15–30	18	21	3	75
		30–45	7	9	1	81
		45–60	7	9	1	81
Site mean		0–60	44	59	12	70

Notes: TAWC, FC, and WP are expressed as a depth of water (mm) in each sampling depth, and in the site mean depth (60 cm). Data from Table 11.2 were used to develop a relationship between soil EC_a and TAWC (Figure 11.5).

- A shapefile of the soil sampling GPS positions was imported into the EC_a map document in ArcMap.
- EC_a values were extracted using Spatial Analyst in Arc Toolbox, and selecting the option Extraction/Extract Values to Points. The input point feature file is the GPS position shapefile, the input raster is the EC_a raster, and a new output point feature file is created with EC_a values extracted for each GPS coordinate.
- The extracted EC_a data point values can be viewed by right-clicking on the new file and viewing the new "raster value" column in the attribute table. The attribute table can be exported as a dbf file and opened in Excel.

Figure 11.8 shows the relationship between soil EC_a and predicted soil water storage in the root zone (mm) for August 3, 2007 (FC) and September 1, 2007, September 29, 2007, and October 24, 2007 when Zones 1, 2, and 3 received their first spring-time irrigation, respectively.

The linear regression model is used to interpret the high-resolution EC_a data set in terms of soil water status on any one day, and the maps produced are available for uploading to a VRI system to control irrigation scheduling to specified zones on specified days (Figure 11.9).

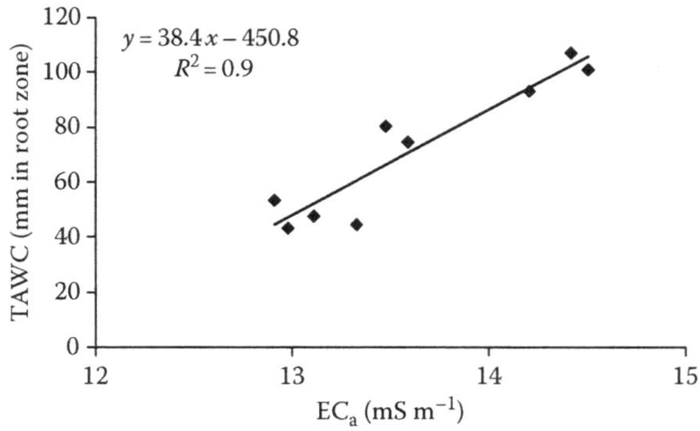

FIGURE 11.5 Relation of soil EC_a to TAWC for Waimakariri soils, at the study site.

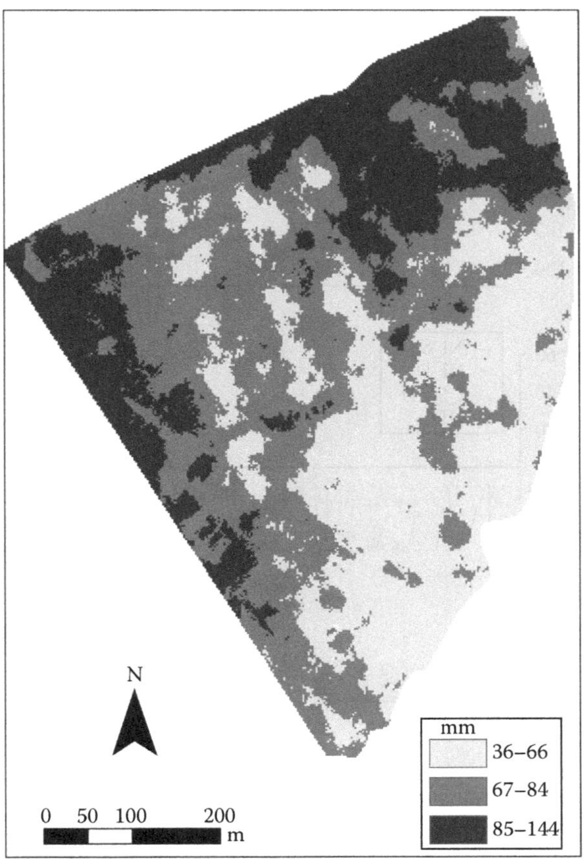

FIGURE 11.6 TAWC map (millimeter available water in the root zone) of the 40 ha dairy pasture, irrigated by a center pivot sprinkler irrigation system.

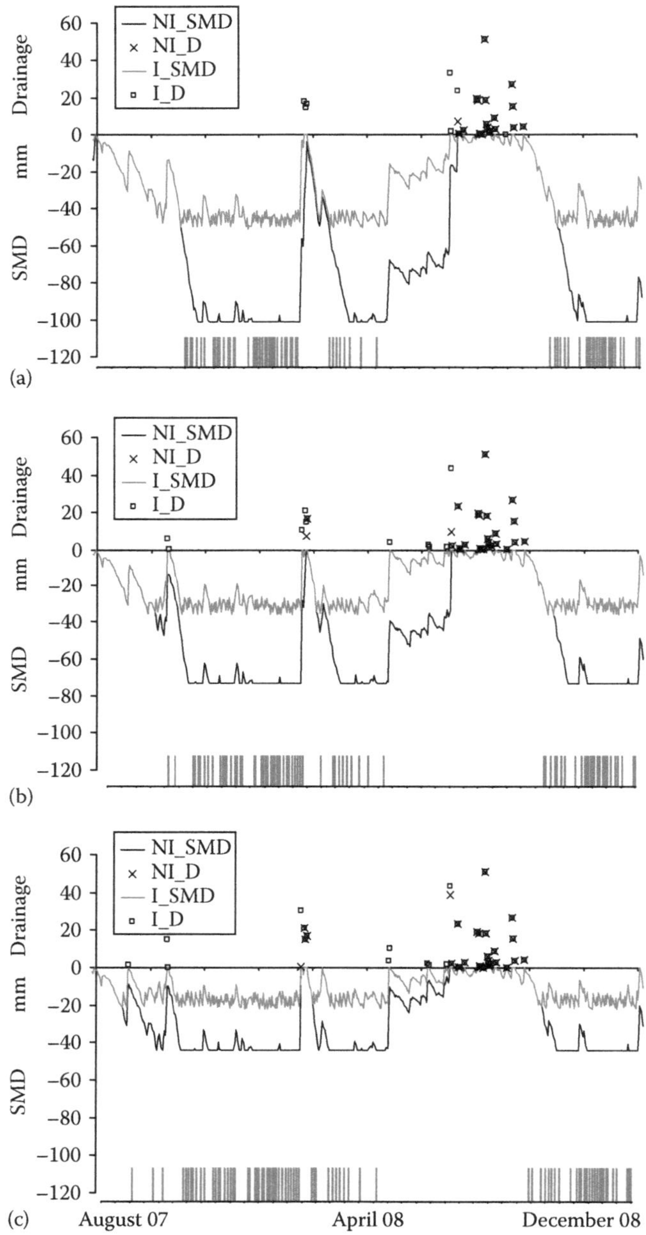

FIGURE 11.7 Soil moisture deficit (SMD, mm), drainage (D, mm), and 10 mm irrigation events (vertical bars at the bottom of each graph) for the three soil zones (a = Zone 1—TAWC 101 mm; b = Zone 2—TAWC 73 mm, c = Zone 3—TAWC 44 mm) from August 2007, when the soils were at FC, to December 2008. NI, no irrigation; I, irrigation; D, drainage/runoff.

TABLE 11.3
Potential Water Savings of VRI Compared with URI for Waimakariri Soils at the Study Site

Site	Zone	Area (ha)	Irrigation mm	Irrigation $m^3\ ha^{-1}$	No. Irrigation Days (Days)	Total Volume Water Applied (mL)	Mean Depth Applied (mm)
URI							
Total		40	520	5200	52	208	520
VRI							
Waimakariri sandy loam	1	9	440	4400	44	40	
Waimakariri stony sandy loam	2	20	470	4700	47	94	
Waimakariri very stony sandy loam	3	11	520	5200	52	57	
Total		40				191	477
VRI–URI					%	ML	$mm\ ha^{-1}$
water saving					8	17	43

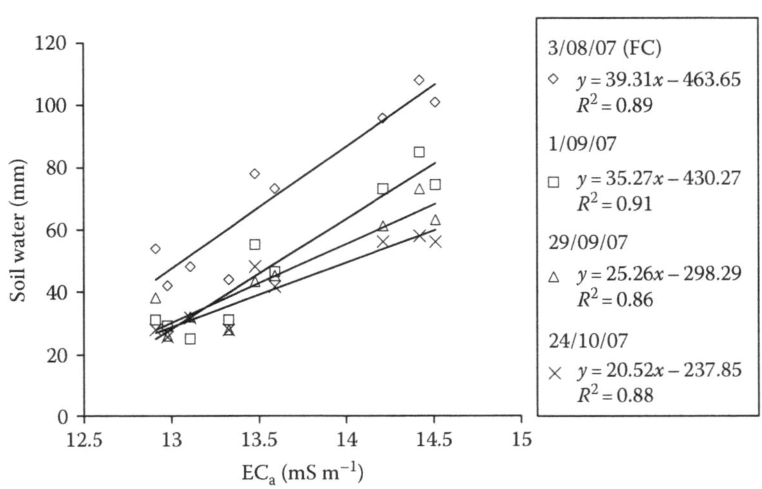

FIGURE 11.8 Linear regression models of soil water balance-predicted soil available water status (millimeter available water in the root zone) against soil EC_a on August 3, 2007, September 1, 2007, September 29, 2007, and October 24, 2007 for each soil profile analyzed.

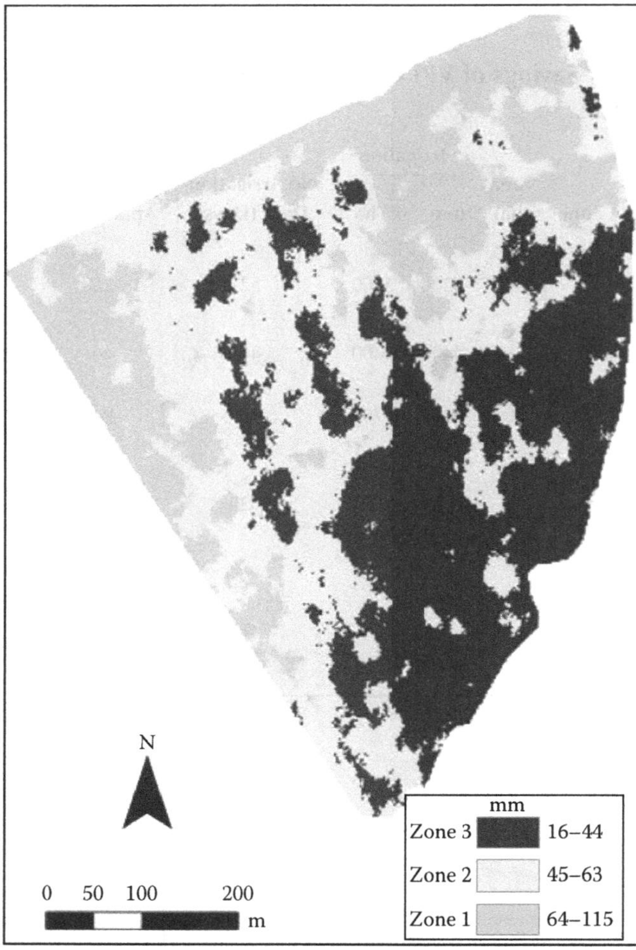

FIGURE 11.9 Soil water status map for September 1, 2007, delineating the management zone (Zone 3) which requires irrigation.

11.4.3 Comparison of VRI and URI Key Performance Indicators

A key performance indicator (KPI) of any irrigation system is its water use efficiency, defined here as the millimeters of irrigation water applied per ton of DM production. URI and VRI aim for maximum potential yield by maintaining soil water status above the plant SP so that water is readily available to the plant. Other irrigation strategies, such as deficit irrigation, aim for efficient conversion of each millimeter of irrigation to DM production, with strategic timing, but withhold water at certain times beyond the SP, so that potential yield may not be achieved. Deficit irrigation strategies for pasture may be used when allocated freshwaters are limited, precluding URI or VRI. Section 11.4.1 discussed the potential water savings of VRI compared with URI, and this section explores further potential benefits of a VRI system compared with a URI system, using KPIs.

TABLE 11.4
KPIs for VRI and URI of the Case Study Irrigated Dairy Pasture in North Canterbury, New Zealand

KPI	Units	Description
Water use	mm $season^{-1}$	Total amount of irrigation water applied in one season (July 1, 2007 to June 30, 2008)
Drainage/runoff	mm $season^{-1}$	Drainage and runoff during one season (July 1, 2007 to June 30, 2008), calculated as excess above FC by the soil water balance model; implications for nutrient leaching
Energy usage	kg CO_2-eq	Energy usage (kWh m^{-3}) for operation of irrigation system (largely cost of pumping water) converted to equivalent CO_2 emissions
Irrigation water use efficiency	mm T^{-1} DM	Reports millimeter of irrigation water applied per ton DM produced

KPIs considered are

1. Water use
2. Irrigation water lost as drainage and runoff
3. Energy usage
4. Irrigation water use efficiency (IWUE)

Water use and drainage/runoff KPIs are calculated from the soil water balance (see Table 11.4 for details). Energy use is calculated as kWh m^{-3} water applied, using a factor of 0.42 kWh m^{-3} irrigation water applied, based on a recent survey of New Zealand irrigation systems[13] and then converted to kg CO_2-eq using an implied emission factor of 0.18 kg CO_2-eq per kWh electricity consumed.[14] The IWUE index uses a reported dairy pasture production of 17.6 T DM ha^{-1} yr^{-1} for a similar soil in this region[15] to calculate millimeters of irrigation water applied per ton DM produced (Table 11.4; Figure 11.10).

An 8% reduction in water use by VRI compared with URI is accompanied by a 43% reduction in excess water lost as drainage and runoff over the 2007–2008 period. This implies that any potential nutrient leaching is reduced under VRI. Increased dissolved N concentrations in lowland rivers have been observed in this region over the past decade, associated with the introduction of high-productivity dairy farming systems, and VRI is one method of mitigating against increasing levels of N leached into waterways, resulting in better utilization of nutrients applied by minimizing drainage and runoff.

Energy used by VRI scheduling was 181 kWh ha^{-1} (650 MJ ha^{-1}) less than URI, due primarily to the reduced requirement for pumping, directly related to water savings, with an estimated financial return of about NZ\$86 ha^{-1}. This mitigates greenhouse gas emissions, with a saving of 33 kg CO_2-eq ha^{-1} yr^{-1}. The IWUE KPI indicates that VRI gave more DM production per millimeter of irrigation water applied. These KPIs illustrate the overall benefits of VRI scheduling based on soil TAWC differences.

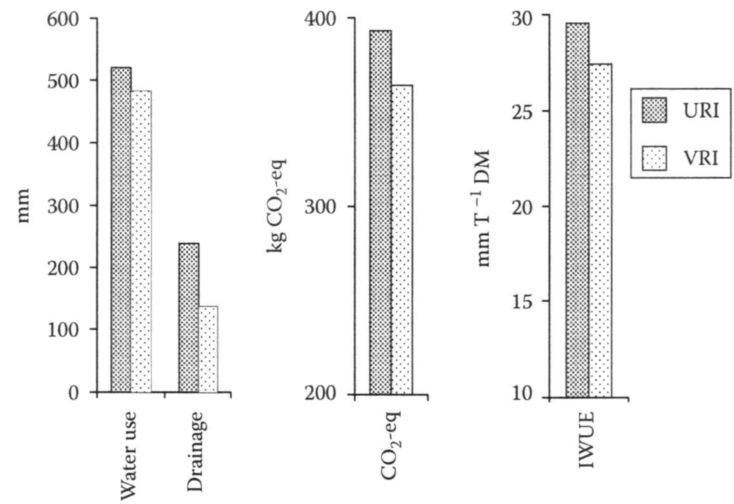

FIGURE 11.10 A comparison of KPIs of VRI and URI for a 40 ha irrigated dairy pasture in North Canterbury, New Zealand.

11.5 CONCLUSION

Soil EC_a increased with decreasing percent stones in the soil profile at this irrigated dairy pasture site on weathered fluvial recent, Waimakariri soils, in North Canterbury. In addition, soil EC_a was strongly related to soil TAWC ($R^2 = 0.9$) so that a regression model could be used to predict TAWC from soil EC_a for the production of a TAWC map. A daily time step was added to the TAWC map using a soil water balance model to produce daily soil water status maps. These maps are available to upload to an automated VRI system for spatial irrigation scheduling. At this site, VRI applies the first irrigation in late Spring to the very stony soil (TAWC = 44 mm) on September 1, 2007 and delays irrigation to the intermediate TAWC zone by 29 days and to the highest TAWC soil zone by 54 days, optimizing the use of stored soil water. Overall irrigation water savings using VRI scheduling was 8% for the season July 1, 2007 to June 30, 2011. This was accompanied by a 43% reduction in drainage and runoff, because VRI uses stored available water above the SP before applying irrigation, reducing the likelihood and frequency of soil moistures greater than FC. This implies that VRI will reduce the risk of nutrient leaching through the soil profile, with increased nutrient use efficiency compared with URI. This is an important mitigation strategy for N leaching in this region which has experiencing increases in groundwater N due to increased N leaching from dairy pastures over the last decade. In addition, in comparison with URI, VRI gave improved water use efficiency and reduced overall energy use, mitigating CO_2 emissions. VRI optimizes the amount, timing, and positioning of irrigation scheduling under one irrigation system, and this optimization is accompanied by a number of environmental and resource use benefits.

ACKNOWLEDGMENTS

The authors acknowledge Trevor Webb, Todd White, and John Dando for assistance with this research, and Dave Scotter for assistance with soil water balance modeling. Financial support was supplied by MAF SFF, Landcare Research, and AGMARDT.

APPENDIX

1. The SWBsimple.xls file can be found in Chapter 8 folder on the CD accompanying this book. Worksheet 1 provides the soil water balance (SWB) (with and without irrigation) for this case study. Worksheet 2 provides an SWB template for the user to enter their own site-specific data.

Worksheet 1

2. Worksheet 1 "Soil WB" contains the rainfall and PET climatic data for the Selwyn case study and calculates and plots the daily soil moisture deficit (SMD, mm) for the period August 3, 2007 to the end of 2008 (with irrigation and without irrigation).
3. The SWB example in "Soil WB" worksheet starts in mid-winter, after a heavy rainfall (34 mm), and assumes that the SMD is brought to 0 (i.e., FC). At this time, the soil storage of plant available water (S) = 101 mm (i.e., TAWC).
4. Hypothetical irrigation events are input to the spreadsheet to maintain the SMD above the irrigation trigger (S = 50.5 mm). The irrigation trigger in this example is set at a depletion factor of 0.5 TAWC.
5. Different values for TAWC and "irrigation trigger" are input to this worksheet to assess the total irrigation requirement for one season of each soil management zone (with a unique TAWC value) for the Selwyn case study.

Worksheet 2

6. Worksheet 2 "SoilWB_template" is available for calculating the SWB for any soil and climatic region. Input site name to cell C1, a value for soil TAWC to cell C2 (≤250 mm), a value for the irrigation trigger to cell C3, and a depth of irrigation into cell I1 (≤50 mm).
7. Input daily rainfall data into column B and a regional or site-specific reference value for PET into column D. Reference PET is calculated for a uniform grass sward cover, and this can be adjusted for any crop type and stage (see Ref. [12]). Dates can be adjusted, but the SWB should start at a point where the soils are considered to be at FC (SMD = 0).
8. Worksheet 2 will automatically calculate an irrigation schedule for this period of time. It also provides the total irrigation requirement for the period of interest in cell I4.

REFERENCES

1. Canterbury Regional Council. *Canterbury Regional Environment Report 2011*. ECan, Christchurch, New Zealand. Document available at www.ecan.govt.nz, 2011, 246 p.
2. New Zealand Ministry for the Environment, Environment New Zealand. A Report published by MfE in December 2007, Wellington, New Zealand. Document available from www.mfe.govt.nz, 2007, 456 p.
3. Rhoades, J.D. Electrical conductivity methods for measuring and mapping soil salinity. *Advances in Agronomy* 49, 201, 1993.
4. Cassel, F.S. Soil salinity mapping using ArcGIS. In Pierce, F.J., and Clay, D.E., eds., *GIS Applications in Agriculture*. CRC Press, Boca Raton, FL, 2007, pp. 141–162.
5. Kitchen, N.R., Sudduth, K.A., and Drummond, S.T. Mapping of sand deposition from 1993 midwest floods with electromagnetic induction measurements. *Journal of Soil and Water Conservation* 51, 336, 1996.
6. Hedley, C.B., Yule, I.Y., Eastwood, C.R., Shepherd, T.G., and Arnold, G. Rapid identification of soil textural and management zones using electromagnetic induction sensing of soils. *Australian Journal of Soil Research* 42, 389, 2004.
7. Sudduth, K.A., Kitchen, N.R., Wiebold, W.J., Batchelor, W.D., Bollero, G.A., Bullock, D.G., Clay, D.E. et al. Relating apparent electrical conductivity to soil properties across the north-central USA. *Computers and Electronics in Agriculture* 46, 263, 2005.
8. Hedley, C.B. and Yule, I.J. Soil water status mapping and two variable-rate irrigation scenarios. *Precision Agriculture*, 57, 10: 342–355, 2009.
9. Kim, Y., Evans, R.G., and Iversen, W.M. Remote sensing and control of an irrigation system using a distributed wireless sensor network. *IEEE Transactions on Instrumentation and Measurement* 57, 1379, 2008.
10. Bradbury, S. Variable rate irrigation for centre pivot and linear move irrigators. Document available at www.precisionirrigation.co.nz/index.php, 2009.
11. Pierce, F. and Clay, D.E. *GIS Applications in Agriculture*. CRC Press, Taylor & Francis, New York, 2007.
12. Allen, R.G., Pereira, L.S., Raes, D., and Smith, M. Crop evapotranspiration. Guidelines for computing crop water requirements. *FAO Irrigation and Drainage Paper 56*, p. 169, FAO, Rome, 1998, 300 p.
13. FAR, Cost of irrigation. *A Survey Report for Growers*, February 2008. Foundation for Arable Research, New Zealand, 2011, 2 p.
14. New Zealand Ministry of Economic Development. *New Zealand Greenhouse Gas Emissions 1990–2007*. Prepared by Ministry of Economic Development (MED), Wellington, New Zealand, 2011, 31 p. ISSN 1177-9764.
15. LIC, New Zealand Dairy Statistics 2007–2011. Prepared by Livestock Improvement Corporation (LIC), Hamilton, New Zealand. Document available from DairyNZ www.dairynz.co.nz/dairystatistic, 2009, 51 p.

12 Maximizing Nutrient Efficiency through the Adoption of Management Practices That Maintain Soil Organic Carbon: Calculating Carbon Turnover Kinetics

David E. Clay, Gregg Carlson, and Sharon A. Clay

CONTENTS

12.1 EXECUTIVE SUMMARY

A critical component of maximizing nutrient management efficiency is the adoption of management practices that maintain the soil organic carbon (SOC). Management practices that degrade soil resources may have low energy efficiency, while systems that improve soil quality may have the opposite impact. One practice that could reduce soil sustainability is the harvesting of crop residues for animal feed, animal bedding, and ethanol production. Residue harvesting can reduce SOC, which in turn can lead to reduced sustainability and productivity. Yield monitor data sets when combined with spatial and temporal SOC measurements can provide the information needed to develop site-specific carbon management plans. This chapter (1) develops and demonstrates approaches for calculating SOC maintenance requirements and mineralization rate constants at different landscape positions and (2) discusses techniques for conducting site-specific experiments and determining site-specific C sequestration potentials.

12.2 INTRODUCTION

12.2.1 Calculating Carbon Turnover in Nonisotopic Experiments

Because SOC is continuously being degraded by soil microorganisms, crop residues must be added to maintain current levels. The SOC maintenance requirement is the amount of nonharvested carbon (NHC) that must be added to maintain the SOC content at the current level (NHC_m). Management practice that decreased SOC contents can reduce soil health and ultimately productivity.

The carbon cycle is driven by photosynthesis that produces organic biomass that is respired by microorganisms. Only a portion of the NHC that is returned to soil ends up in SOC and only a portion of the SOC ends up as CO_2 (Figure 12.1). The rates that nonharvested biomass is converted from fresh biomass to SOC and SOC is converted to CO_2 are functions of many factors including management, climate, and biomass composition. The first-order rate mineralization constants for NHC (k_{NHC}) and SOC (k_{SOC}) can be used to calculate mineralization rates.[1] An approach for calculating the mineralization rates in historic studies is presented in the following.

Based on the carbon flow diagram shown in Figure 12.1, two equations can be defined. The first equation is

$$\frac{d\text{SOC}}{dt} = k_{\text{NHC}}\left[\text{NHC}_{\text{a}} - \text{NHC}_{\text{m}}\right]$$

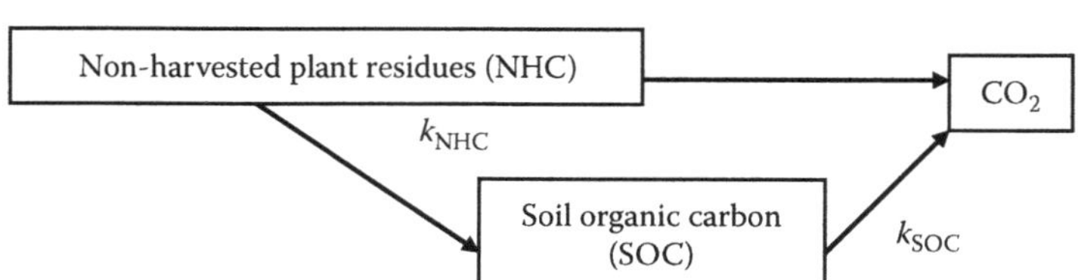

FIGURE 12.1 A relational diagram showing the relationship between three carbon pools and the associated rate constants.

This equation states that the temporal change in SOC (dSOC/dt) is equal to the NHC mineralization rate constant (k_{NHC}) times the difference between the amounts of carbon added to the soil (NHC_a) and the maintenance requirement (NHC_m). The second equation is

$$k_{SOC} \cdot SOC_e = k_{NHC} \cdot NHC_m$$

This equation states that at the SOC equilibrium point (SOC_e), the rate that NHC is converted into SOC ($k_{NHC} \cdot NHC_m$) is equal to the rate that SOC is mineralized into CO_2 ($k_{SOC} \cdot SOC_e$). These two equations were combined to produce the equation

$$\frac{NHC_a}{SOC_e} = \frac{k_{SOC}}{k_{NHC}} + \frac{dSOC}{dt}\left[\frac{1}{k_{NHC} \cdot SOC_e}\right]$$

This equation was solved by defining either SOC_i (initial SOC) or SOC_{final} as SOC_e, NHC_a/SOC_e as y, and dSOC/dt as x. After these substitutions, the equation

$$\frac{NHC_a}{SOC_{time0}} = \frac{k_{SOC}}{k_{NHC}} + \left[\frac{dSOC}{dt} \cdot \frac{1}{k_{NHC} \cdot SOC_e}\right]$$

was derived. When fit to a zero-order equation, the y-intercept and slopes are k_{SOC}/k_{NHC} and $1/(k_{NHC} \cdot SOC_e)$, respectively (Figure 12.2). After determining the slope (m) and y-intercept (b), the SOC maintenance (NHC_m) and rate constants can be calculated with the equations:

$$NHC_m = b \cdot SOC_e$$

$$k_{NHC} = \frac{1}{(m \cdot SOC_e)}$$

$$k_{SOC} = \frac{b}{(m \cdot SOC_e)}$$

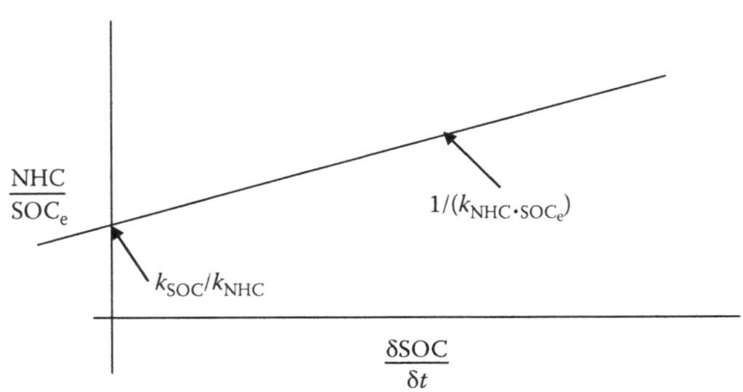

FIGURE 12.2 A graphical representation of the maintenance calculations. (Data from Clay, D.E. et al., *Agron. J.*, 102, 443, 2006.)

This approach assumed that above- and below-ground biomass make equal contributions to SOC, that the amount of below-ground biomass is known, and that NHC is known and that initial (SOC_e) and final (SOC_{final}) SOC values are near the equilibrium point. Advantages with this approach are that k_{SOC} and k_{NHC} are calculated directly from the data, and the assumptions needed for these calculations can be tested. A disadvantage with this solution is that surface and subsurface NHC are combined, and therefore $k_{NHC\ Below}$ and $k_{NHC\ Surface}$ cannot be calculated directly. Combining, $k_{NHC\ Below}$ and $k_{NHC\ Surface}$ into a single rate constant (k_{NHC}) may contain errors,[2,3] resulting from surface and below-ground biomass having different mineralization rate constants. If an experiment contains a no-plant control area, it may be possible to separate these values.[4]

The Clay et al.[1] approach requires an accurate measure of above- and below-ground C inputs.[5,6] Obtaining dependable values for above-ground biomass is relatively easy and typically accomplished by weighing above-ground biomass or estimating the value using the harvest index. However, obtaining accurate measurements of below-ground biomass is very difficult.[7–9] In the past, nearly all attempts have underestimated this value.[10] For mass balance purposes, below-ground values should include all plant parts not included in above-ground measurements. Root-to-shoot ratios and/or simulation models are the most common approaches for estimating below-ground biomass.[7,11,12]

12.3 CASE STUDY 1: DETERMINING MINERALIZATION RATE CONSTANTS ON DATA REPORTED BY LARSON ET AL.

The Larson et al.[13] study was conducted in Clarinda, Iowa. Soil at the site was a fine, silty, mixed, mesic, Typic Hapludoll. The soil was plowed, disked, and cultivated several times. Yield data for this experiment are available in Morachan et al.[14] When using root-to-shoot ratios to calculate below-ground biomass, grain and stover values must be added. For this calculation, grain weights must be converted to dry weight. Data shown in Table 12.1 can be used to convert yield data from one moisture percentage to another.

TABLE 12.1
Corn, Soybean, and Wheat Weights Are Various Moisture Contents

	Grain Moisture (%)					
Commodity	**20%**	**18%**	**15.50%**	**13%**	**10%**	**0%**
	Weight (lb bu^{-1})					
Corn	59.15	57.71	56	54.39	52.58	47.32
Soybeans	65.25	63.65	61.78	59.65	58.0	52.2
Wheat	64.88	63.29	61.42	60	57.67	51.9

Note: Corn is typically reported at 15.5% moisture, while soybean and wheat are reported at 13% and 13.5% moisture, respectively. The wheat weight (lbs/bu) at 13.5% moisture is 60 lbs.

Step 1. Calculate the average grain dry yields from 1953 to 1964. This can be done in Microsoft Excel by using = average(C4:M4). In this example, the average yields from 1955 to 1964 range from 5385 to 5517 kg grain ha^{-1}. Dry grain weights can be converted to bu ac^{-1} by converting to lb ac^{-1} and then dividing by 47.32 (column O, Table 12.2).

(a) Grain yields are often reported on a wet-weight basis. The equation for percent moisture is 100 × (wet grain – dry grain)/(wet grain). Weights for grains at different moisture percentages are shown in Table 12.1. The amount of dry grain contained in a bushel of corn at 15.5% moisture is determined by substituting 0.155 for moisture percentage and 56 lb for wet grain weight and then solving for dry grain. The resulting equation is dry grain = 0.845 × 56 lb bu^{-1}.

Step 2. Estimate above-ground NHC; this can be estimated by assuming there is a 50% harvest index (grain/grain + stover).

Step 3. Estimate below-ground biomass (roots + exudates) (column C, Table 12.3). In this example, it was estimated that the root + exudates-to-shoot ratio (k_{rec} = [roots + exudates]/[grain + stover]) was 0.55.[11] In this case study, roots consist of all above- and below-ground plant parts that are not included in the measured above-ground portions. This may include the small amount of stalk above the soil surface. Johnson et al.[11] used k_{rec} values of 0.82, 0.55, and 0.62 for wheat, corn, and soybean, respectively. Total root biomass (column C, Table 12.3) was converted to root-C (column D, Table 12.3) by assuming that roots contained 43% carbon.

Step 4. Calculate $d\text{SOC}/dt$ (surface 15 cm) and NHC/SOC (Table 12.4). For these calculations, subtract SOC final from SOC initial and divide the difference by 11 years. NHC/SOC is determined by dividing the NHC values (column D, Table 12.4) by SOC (column B, Table 12.4). Note, the SOC and NHC units must be identical.

Step 5. Determine the regression equation between NHC/SOC (y) and $d\text{SOC}/dt$ (Table 12.5).

(a) In Microsoft Excel, select tool, data analysis, and regression.
(b) Insert the cell ranges for x ($d\text{SOC}/dt$) and y (NHC/SOC) and select OK. The results of this analysis are shown in Table 12.5. The resulting equation is NHC/SOC = 0.131 + 0.000284 ($d\text{SOC}/dt$). The values for this equation are in B17 and B18. The adjusted r^2 for this equation is 0.984 (cell B6).

Step 6. Calculate SOC maintenance requirement and the mineralization rate constants.

(a) The maintenance requirement is the y-intercept times SOC, SOC_{main} = $0.131 \cdot 26{,}750 = 3{,}504$ kg C (ha year)$^{-1}$.
(b) In this analysis, approximately 1000 kg C ha^{-1} are provided by the roots, resulting in an above-ground requirement of 2500 kg surface residue carbon/year. This carbon can be provided by returning crop residues or applying manure.
(c) Calculate k_{SOC} and k_{NHC} values

TABLE 12.2
Yield Data Associated with the Larson et al. and Morachan et al. Experiment

A	B	C	D	E	F	G	H	I	J	K	L	M	N	O	P
Residue Type	NHC Added (Mg ha^{-1})	54	55	56	57	58	Year 59 (Mg ha^{-1})	60	61	62	63	64	Ave. Grain (15.5%)	Dry Grain (0%)	Est. Stover
CK	None	0	5.09	5.09	5.09	5.09	5.09	5.09	6.60	6.82	7.52	6.35	5.26	4.44	4.44
Alfalfa	2	0	5.28	5.28	5.28	5.28	5.28	5.28	7.06	7.52	7.89	6.31	5.50	4.67	4.67
Alfalfa	4	0	5.22	5.22	5.22	5.22	5.22	5.22	7.56	7.49	7.52	6.78	5.52	4.66	4.66
Alfalfa	8	0	5.36	5.32	5.32	5.32	5.32	5.32	7.03	7.53	7.60	6.25	5.48	4.63	4.63
Alfalfa	16	0	5.22	5.22	5.22	5.22	5.22	5.22	7.55	7.19	7.51	6.03	5.41	4.58	4.58
Corn	2	0	5.37	5.37	5.37	5.37	5.37	5.37	6.92	7.00	7.62	6.22	5.45	4.61	4.61
Corn	4	0	5.28	5.28	5.28	5.28	5.28	5.28	6.97	7.09	7.48	5.79	5.36	4.53	4.53
Corn	8	0	5.32	5.32	5.32	5.32	5.32	5.32	7.05	7.38	7.40	5.71	5.41	4.57	4.57
Corn	16	0	5.42	5.42	5.42	5.42	5.42	5.42	7.45	6.71	7.12	5.44	5.38	4.55	4.55

Source: Data from Larson, W.E. et al., *Agron. J.*, 64, 204, 1972; Morachan, Y.B. et al., *Agron. J.*, 64, 199, 1972.

TABLE 12.3
Amounts of Calculated Root and Residue Carbon Added to Each Plot

A	B	C	D	E	F	G	H
Residue Type	**Amount Residue** (Mg ac^{-1})	**Total Root + Ex Biomass**	**Total Root + ex-C**	**50% Roots 0–15 cm** (kg Root-C ha^{-1})	**Biomass Total** (kg BS ha^{-1})	**Biomass Carbon** (kg BS-C ha^{-1})	**Total NHC** (kg C ha^{-1})
CK	None	4,886	2,101	1,050	0	0	1,050
Alfalfa	2	5,110	2,197	1,098	2,000	860	1,958
Alfalfa	4	5,128	2,205	1,103	4,000	1,720	2,823
Alfalfa	8	5,096	2,191	1,096	8,000	3,440	4,536
Alfalfa	16	5,033	2,164	1,082	16,000	6,880	7,962
Corn	2	5,070	2,180	1,090	2,000	860	1,950
Corn	4	4,988	2,145	1,073	4,000	1,720	2,793
Corn	8	5,024	2,160	1,080	8,000	3,440	4,520
Corn	16	5,006	2,153	1,077	16,000	6,880	7,957
Equations in the columns			=C × 0.43	=D/2		=F × 0.43	=G + E
			Assumes 43% carbon in the samples	Assumes roots are 50% of above ground biomass	Treatments added to the soils	Assumes 43% carbon in the samples	Total carbon added = roots + treatment

Source: Data from Larson, W.E. et al., *Agron. J.*, 64, 204, 1972.
Note: In these calculations the root + exudate/shoot ratio was 0.55 and carbon content was 43%.

TABLE 12.4
NHC and NHC/SOC Values Study

A	B	C	D	E
SOC Final (kg ha^{-1})	**kg C ha^{-1}(i)**	***dSOC/dt* kg (ha Year)$^{-1}$**	**NHC kg ha^{-1}**	**NHC/SOC**
23,700	26,750	−277.4	1050	0.0392
24,820	26,750	−175.4	1958	0.0732
25,580	26,750	−106.6	2823	0.1055
28,080	26,750	112.6	4536	0.1696
33,350	26,750	599.5	7962	0.2976
24,080	26,750	−242.5	1950	0.0729
26,020	26,750	−66.5	2793	0.1032
27,340	26,750	53.7	4520	0.1690
33,400	26,750	598.9	7957	0.2975
Equation in columns		=(col A − col B)/11		=col D/col B

Source: Data from Larson, W.E. et al., *Agron. J.*, 64, 204, 1972.

$$k_{NHC} = \frac{1}{\text{slope} \times \text{SOC}_e} = \frac{1}{0.000284 \times 26{,}750} = 0.132\,\text{g NHC (g} \times \text{NHC} \times \text{year)}^{-1}$$

$$k_{SOC} = \frac{\text{intercept}}{\text{slope} \times \text{SOC}_e} = \frac{0.131}{0.000284 \times 26{,}750} = 0.0173\,\text{g SOC (g} \times \text{SOC} \times \text{year)}^{-1}$$

(d) The rate constants can be used for many different purposes. For example, based on the equation, $k_{SOC} \cdot SOC_e = k_{NHC} \cdot NHC_m$, changes in SOC based on NHC values can be estimated.

Step 1. If NHC is 4000 what will the new SOC value be?

Step 2. SOC = (k_{NHC} · NHC)/k_{SOC} = (0.132 · 4000)/0.0173 = 30,520 kg C ha^{-1}. Based on these rate constants, returning 4000 kg C ha^{-1} would increase SOC from 26,750 to 30,520 kg C ha^{-1}.

12.4 CASE STUDY 2: LANDSCAPE POSITIONS IMPACT ON CARBON TURNOVER AT STRATTON, COLORADO

History: Findings used in this case study were obtained from Sherrod et al.[15] This study conducted near Stratton was initiated in 1985 and continued for 12 years. This site had an average open pan evaporation (PET) value of 1270 mm. Treatments at the site were wheat-fallow (WF), wheat-corn-fallow (WCF), wheat-corn-millet-sunflower (WCMF), and continuous corn (CC). No-tillage was used at the site. Each treatment was replicated twice. Soil samples were collected from 0 to 5 cm at the beginning and end of the study. Soil samples were analyzed for SOC by wet oxidation, and the efficiency of the oxidation process was not stated. Samples were dried and ground during which visible plant parts, >2 mm, were removed from the sample. These samples were

TABLE 12.5
Regression Analysis of NHC/SOC(y) and dSOC/dt(x)

A	B	C	D	E	F	G	H	I
				Summary				
Regression Statistics								
Multiple *R*	0.993007							
R square	0.986064							
Adjusted *R* square	0.984073							
Standard error	0.012016							
Observations	9							
				ANOVA				
	df	**SS**	**MS**	***F***	**Significance *F***			
Regression	1	0.071511	0.071511	495.2906	9.35E–08			
Residual	7	0.001011	0.000144					
Total	8	0.072522						
	Coefficients	**Standard Error**	***t* Stat**	***P*-value**	**Lower 95%**	**Upper 95%**	**Lower 95.0%**	**Upper 95.0%**
Intercept	0.131635	0.004068	32.35534	6.97E–09	0.122015	0.141255	0.122015	0.141255
X variable 1	0.000284	1.27E–05	22.25513	9.35E–08	0.000253	0.000314	0.000253	0.000314

TABLE 12.6
Initial Soil Organic C at Stratton

A	B	C	D	E
Landscape Position	**Sampling Depth (cm)**	**Carbon (g kg^{-1})**	**Initial C Bulk Density (g cm^{-3})**	**Organic C (kg ha^{-1})**
Summit	5	10.58	1.37	7,247
Side-slope	5	10.44	1.3	6,786
Toe-slope	5	19.21	1.15	11,046
Math in column				= col B × col C × col D × 100

converted to kg ha^{-1} using measured bulk densities. The study reported annualized stover production which were used to estimate roots using the equation,

$$\text{Roots} = \text{annualized stover} \times 1.1.$$

The root calculation value was based on a 50% harvest index and a 0.55 root-to-shoot ratio. The calculations assumed that 30% of the roots were contained in the surface 5 cm of soil.

Step 1. Calculate initial soil carbon levels for the Stratton site. The gravimetric values are converted to kg ha^{-1} by multiplying the concentration by the bulk density (column D, Table 12.6) and sampling depth (multiply columns B, C, and D by each other). The resulting answer is then converted to kg ha^{-1} by using appropriate constants and measured values.

Step 2. Calculate NHC and *d*SOC/*dt*. Based on reported values,[15] NHC was estimated using the following assumptions. First, roots were estimated by multiplying the annualized above-ground biomass returned by 1.1 (column F, Table 12.7). Second, roots contained in the surface 5 cm are estimated by multiplying below-ground biomass by 0.3. Third, root carbon was estimated by multiplying roots in the segment by 0.4. Based on these assumptions, NHC (column H) is equal to

$$\text{Column F} \times 0.4 + \text{Column G} \times 0.3 \times 0.4$$

It is important to consider that for a mass balance, the roots contain all biomass not typically measured in the above-ground portions. In corn, this generally includes all biomass below the top of the brace roots. *d*SOC/*dt* is calculated by subtracting column D from column C and then dividing this difference by 12 years.

Step 3. Calculate NHC/SOC (column I, Table 12.7). This is determined by dividing column H by column C.

Step 4. Calculate the mineralization rate constants for the different landscape positions. The rate constants are determined by determining the relationship between NHC/SOC and *d*SOC/*dt*. The relationship is quantified by determining the linear equation between *d*SOC/*dt* and NHC/SOC. This can be conducted in Microsoft Excel. In this case study, equations were calculated separately for each

TABLE 12.7
Calculated Change in Soil Organic C over the 12 Years of the Experiment and the Ratio between Non-Harvested C (NHC) and Soil Organic C (SOC) at Stratton

A	B	C	D	E	F	G	H	I
	Rotation	SOC_i (kg ha^{-1})	SOC_f (kg ha^{-1})	$dSOC/dt$ (kg Year)$^{-1}$	Annualize above Ground Biomass Returned (kg C ha^{-1})	Estimated Root Biomass Returned (kg ha^{-1})	NHC C Returned (kg ha^{-1})	NHC/SOC
Summit	WF	7,247	6,785	−38.5	2,060	2,266	1,095.92	0.151224
	WCF	7,247	7,000	−12.5833	2,290	2,519	1,218.28	0.168108
	WCMF	7,247	7,835	49	2,285	2,513.5	1,215.62	0.167741
	CC	7,247	8,400	96.08333	3,905	4,295.5	2,077.46	0.286665
Side-slope	WF	6,786	5,545	−103.417	1,825	2,007.5	970.9	0.143074
	WCF	6,786	6,685	−8.41667	2,120	2,332	1,127.84	0.166201
	WCMF	6,786	6,900	9.5	2,150	2,365	1,143.8	0.168553
	CC	6,786	7,410	52	3,470	3,817	1,846.04	0.272037
Toe-slope	WF	11,045	8,985	−171.667	2,820	3,102	1,500.24	0.13583
	WCF	11,045	9,770	−106.25	3,465	3,811.5	1,843.38	0.166897
	WCMF	11,045	10,170	−72.9167	3,480	3,828	1,851.36	0.16762
	CC	11,045	10,555	−40.8333	4,735	5,208.5	2,519.02	0.228069
Math in column						=col F × 1.1 Assumes 50% harvest index	=0.4 × (F + G × 0.3) Assumes 40% carbon and 30% of the roots are in the segment	=col H/col C

TABLE 12.8
NHC Maintenance Requirement (NHCm) and Calculated Mineralization Rate Constants of Non-Harvested C (K_{NHC}) and Soil Organic C (K_{SOC}) for the Three Landscape Positions at Stratton

A	B	C	D	E	F	G	H	I	J	K
	Rotation	$dSOC/dt$ (kg Year)$^{-1}$	NHC/SOC	y-Intercept	Slope		SOC_i (kg ha^{-1})	k_{soc} g/(g Year)	knhc g (g Year)$^{-1}$	NHCm kg C (ha Year)
Summit	WF	−38.525	0.151218	0.189561	0.00101		7,247	0.025902	0.13664	1,373.748
	WCF	−12.6083	0.168101							
	WCMF	48.975	0.167734							
	CC	96.05833	0.286653							
Sideslope	WF	−103.417	0.143074	0.222709	0.000862		6,786	0.038073	0.170954	1,511.301
	WCF	−8.41667	0.166201							
	WCMF	9.5	0.266136							
	CC	52	0.272037							
Toe-slope	WF	−171.729	0.135821	0.235107	0.000618		11,046	0.034461	0.146578	2,596.989
	WCF	−106.313	0.166886							
	WCMF	−72.9792	0.167608							
	CC	−40.8958	0.228053							
		Not-included in analysis								

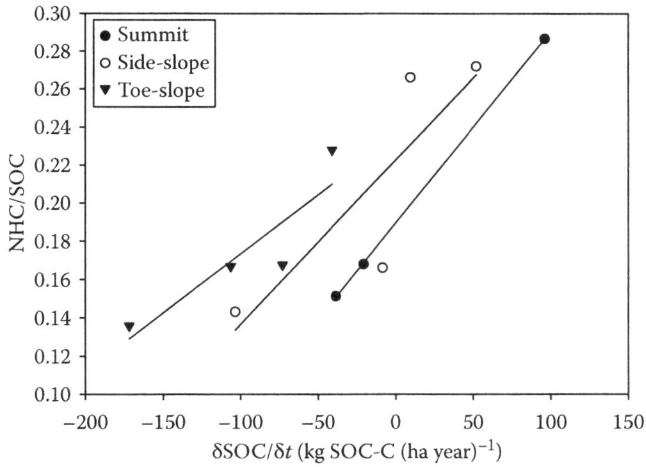

FIGURE 12.3 Relationship between *dSOC/dt* and NHC/SOC at Stratton. Equations and data for the relationships are shown in Table 12.8. (From Sherrod, L.A. et al., *Soil Sci. Soc. Am. J.* 67, 1533, 2003.)

landscape position, and the resulting slope and y-intercept values are located in columns E and F in Table 12.8. Based on the y-intercept and slope values, the k_{SOC} [(y-intercept)/(SOC · slope) = (Col E)/(Col I · Col H)] and k_{NHC}[1/(SOC · slope) = 1/(Col I · Col H)] values were calculated. These values are in columns I and K (Table 12.8) (Figure 12.3).

Step 5. Interpret the mineralization rate constants. The SOC mineralization rate constants were larger for side- and toe-slope than the summit area. Lower k_{SOC} values in the summit indicate that this carbon was more stable than side- and toe-slope carbon. Clay et al.[16] had similar results.

Step 6. Calculate the half-lives for SOC in the different landscape positions. These values are calculated as follows:

$$\text{Summit: } 1/2 \text{ life} = \frac{\ln 2}{k} = \frac{0.6931}{0.0259} = 26.8\,\text{years}$$

$$\text{Side-slope: } 1/2 \text{ life} = \frac{0.6931}{0.03807} = 18.2\,\text{years}$$

$$\text{Toe-slope: } 1/2 \text{ life} = \frac{0.6931}{0.03446} = 20.1\,\text{years}$$

12.5 CASE STUDY 3: LANDSCAPE IMPACTS ON C TURNOVER AT STERLING

Data used in this case study were obtained from Sherrod et al.[15] These data complement the data reported in case study 2. The Sterling site has a 100-year average precipitation of 420 mm and a PET value of 1015 mm. The Sterling Colorado site has a lower PET value than that of the Stratton Colorado site.

TABLE 12.9
Influence of Landscape Position on Soil Organic Carbon and the NHC/SOC Ratios

A	B	C	D	E	F
Landscape Position	**NHC (kg C ha^{-1})**	**SOC$_i$ (kg ha^{-1})**	**SOC$_f$ (kg ha^{-1})**	***d*SOC/*dt* 12 Years**	**NHC/SOC**
Summit	835.24	7,105	7,830	60.41667	0.117557
	1,146.46	7,105	6,810	−24.5833	0.16136
	1,141.14	7,105	7,290	15.41667	0.160611
	1,606.64	7,105	8,175	89.16667	0.226128
Side-slope	864.5	6,492	6,925	36.08333	0.133164
	1,095.92	6,492	6,690	16.5	0.168811
	1,231.58	6,492	6,965	39.41667	0.189707
	1,532.16	6,492	8,745	187.75	0.236007
Toe-slope	1,037.4	9,264	7,160	−175.333	0.111982
	1,439.06	9,264	8,800	−38.6667	0.155339
	1,449.7	9,264	9,715	37.58333	0.156487
	2,016.28	9,264	11,825	213.4167	0.217647
[illegible]	Not included				

Step 1. Set up the data set. The data set should be constructed using the approach described above. The initial and final amounts of SOC contained in the surface 5 cm after 12 years are in columns C and D (Table 12.9). The *d*SOC/*dt* values determined by subtracting SOC$_i$ from SOC$_f$, and dividing this difference by 12 years (column E, Table 12.9). NHC values were calculated using the same assumptions discussed in case study 2.

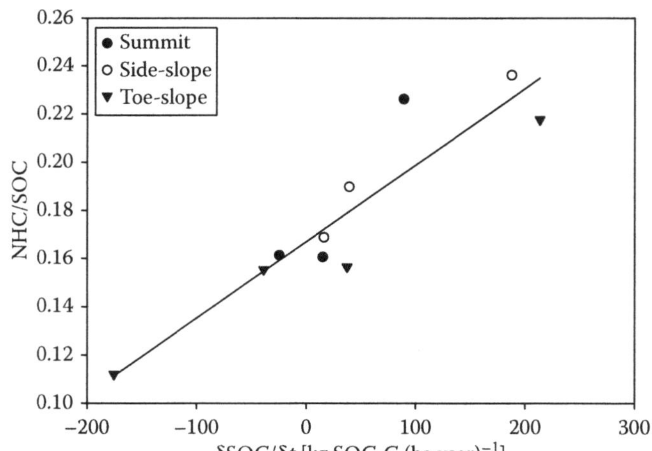

FIGURE 12.4 The relationship between *d*SOC/*dt* and NHC/SOC for Sterling. The resulting equation was NHC/SOC = 0.1669 + 0.000317 *d*SOC/*dt* (r = 0.84**).

TABLE 12.10
NHC Maintenance Requirement and Rate Constants for Sterling

Landscape Position	y-Intercept	Slope	k_{soc} g SOC (SOC Year)$^{-1}$	k_{NHC} Gnhc-C (g NHC Year)$^{-1}$	NHCm kg NHC-C ha^{-1}
Summit	0.1669	0.000317	0.0741	0.4440	1185.8245
Sideslope	0.1669	0.000317	0.0811	0.4859	1083.5148
Toeslope	0.1669	0.000317	0.0568	0.3405	1546.1616

Step 2. Determine the relationship between *d*SOC/*dt* and NHC/SOC. In this analysis, there were several points that did not fit the resulting curve and were not included in the analysis. In this analysis, data from all landscape positions were combined (Figure 12.4). These results are different than Stratton where equations were determined for each landscape position.

Step 3. Determine the maintenance requirement and rate constants for Sterling. Use the relationship between *d*SOC/*dt* and NHC/SOC to calculate these values (Table 12.10). These values show that the NHC and SOC turnover rates are landscape specific. Based on these calculations, more carbon is needed in the toe-slope than summit and side-slope areas to maintain SOC. A comparison between Stratton and Sterling suggests that SOC is turning over much more rapidly at Sterling than Stratton. The half-lives for this Sterling ranged from 8.55 years in the side-slope to 12.2 years in the toe-slope area.

12.6 CALCULATING SITE-SPECIFIC CARBON SEQUESTRATION POTENTIALS

The calculation approach described above relies on two types of measurements, temporal changes in SOC and the amount of NHC returned to soil. In whole-field experiments, this information can be obtained from strip experiments containing two treatments, crop residue removed and crop residue returned. In these experiments, the amount of crop residues returned (NHC) can be measured or estimated from the yield monitor data (using a harvest index and residue removal rate). By combining temporal changes in SOC with annual NHC additions, the rate constants can be calculated using the approaches described above. These calculations assume that the current SOC level is near the equilibrium value (SOC_e). Once the rate constants are calculated, the impact of different residue management approaches on SOC levels can be determined.

12.6.1 Rate Constants Are Known

The long-term average for the amount of above-ground biomass returned to soil located in at two sites located in a field are 9000 and 6000 kg C (ha year)$^{-1}$. What is the potential to sequester additional C is the soils contain 60,000 and 50,000 kg C ha^{-1}? If it is assumed that 50% of the roots are contained in the zone, the harvest index is 50%, the root-to-shoot ratio is 0.55, and biomass contains 40% carbon, then NHC for site 1 is 7560 kg C ha^{-1} (NHC = 9 · 0.4 + 18 · 0.55 · 0.4) and NHC for site 2 is 5040 kg C ha^{-1} (NHC = 6 · 0.4 + 12 · 0.55 · 0.4). If the rate constants are known (k_{SOC} = 0.02 and k_{NHC} = 0.20) then the equation

$$k_{SOC} \times SOC_e = k_{NHC} \times NHC_m$$

can be used to solve for the SOC content at the equilibrium point. In this example, SOC_e for site 1 is 75,600 kg C ha^{-1} [(7560 · (0.2/0.02)] and SOC_e for site 2 is 50,400 kg C ha^{-1} [5400(0.2/0.020] kg C ha^{-1}. If these sites contain 60,000 and 50,000 kg C ha^{-1}, then the potential to sequester additional C is 15,600 kg C ha^{-1} (75,600–60,000) at site 1 and 400 kg C ha^{-1} (50,400–50,000) at site 2.

12.6.2 Rate Constants Are Unknown

In this case study, the rate constants must be measured. The approach described above requires at least two treatments where different amounts of NHC are returned to the soil. In whole-field experiments, three NHC rates can be implemented (residue removed, residue returned, and residue returned plus residue from the harvested site) by removing the NHC from one strip and adding this carbon to a second strip. Two rates can be implemented by removing the carbon from one strip. At differentially corrected global positioning system-identified sampling points, NHC, SOC, and bulk density must be measured at adjacent points in the different treatments. For NHC, it must be measured or estimated from the annual yield monitor data, while for SOC it must be measured at the beginning and end of the experiment. Soil depths often measured are the 0–15 and 0–30 cm soil zones. Protocols for collecting soil samples are to composite at least 10 individual cores from each depth interval. Care must be used when collecting and processing these samples. Removal of biomass or roots from the sample can bias results. In our laboratory, samples are ground and sieved. The portion of the sample that does not pass the sieve is ground with a mortar and pestle. Soil samples should be analyzed for organic carbon using the appropriate technology. Wet oxidation or combustion in an oven may not provide accurate values. At many sites, 4–5 years may be required to produce measurable changes in soil carbon.

NHC can either be estimated from the yield monitor data or measured. Once the $dSOC/dt$ and NHC values are known, the rate constants can be calculated (see above).

12.7 SUMMARY

This chapter demonstrates an approach that can be used to calculate mineralization rate constants for SOC at different landscape positions. The approach is only as accurate as the information used to derive the models. For accurate SOC turnover prediction, accurate measures of SOC and NHC are required. Once the rate constants are known, they can be used within a GIS program to calculate the impact of different management scenarios (within a tillage system).

When evaluating carbon turnover data, it is important to consider that the findings are only as good as the data. Good soil sampling procedures must be followed. In carbon experiments, accurate information from the beginning and end of the experiment is needed. Details about the efficiency of carbon analysis approach are required. Wet oxidation procedures are often not 100% effective. Sieving, grinding, and the removal of plant parts from the ground soil samples may bias results.

ACKNOWLEDGMENTS

Support for this chapter was provided by NC SARE, USDA-CSREES, USDA-NRCS, SD Corn Utilization Council, SD Wheat Commission, and SD Soybean Research and Promotion Council.

REFERENCES

1. Clay, D.E., Carlson, C.G., Clay, S.A., Reese, C., Liu, Z., and Ellsbury, M.M. Theoretical Derivation of new stable and non-isotopic approaches for assessing soil organic C turnover. *Agron. J.* 102, 443, 2006.
2. Barber, S.A. and Martin, J.K. The release of organic substance by cereal roots in the soil. *New Phytol.* 76, 69, 1976.
3. Gale, W.J. and Cambardella, C.A. Carbon dynamics of surface residue- and root-derived organic matter under simulated no-till. *Soil Sci. Soc. Am. J.* 64, 190–195, 2000.
4. Barber, S.A. Corn residue management and soil organic matter. *Agron. J.* 71, 625, 1979.
5. Ortega, R., Peterson, G.A., and Westfall, D.J. Residue accumulation and changes in soil organic matter as affected by cropping intensity in no-till dryland agroecosystems. *Agron. J.* 94, 944, 2002.
6. Prakash, V., Kundu, S., Ghosh, B.N., Singh, R.D., and Gupta, H.S. Annual carbon input to soil through rainfed soybean (*Glycine max*)–wheat (*Triticum aestivum*). *Indian J. Agric. Sci.* 72, 14, 2002.
7. Kuzyakov, Y. and Domanski, G. Carbon inputs by plants into the soil. Review. *J. Plant Nutr. Soil Sci.* 163, 421, 2000.
8. Kuzyakov, Y.V. Tracer studies of carbon translocation by plants from the atmosphere into the soil. *Eurasian Soil Sci.* 34, 28, 2001.
9. Amos, B. and Walters, D.T. Maize root biomass and net rhizodeposited carbon: An analysis of the literature. *Soil Sci. Soc. Am. J.* 70:1489, 2006.
10. Torbert, H.A., Prior, S.A., Rogers, H.H., and Wood, C.W. Review of elevated atmospheric CO_2 effects on agro-ecosystem: Residue decomposition processes and soil C storage. *Plant Soil* 224, 59, 2000.
11. Johnson, J.M.F., Allmaras, R.R., and Reicosky, D.C. Estimating source carbon from crop residues, roots, and rhizodeposits using the national grain-yield data-base. *Agron. J.* 98, 622, 2006.
12. Bolinder, M.A., Janzen, H.H., Gregorich, E.G., Angers, D.A., and Vanden Bygaart, A.J. An approach for estimating net primary productivity and annual carbon inputs to soil for common agricultural crops in Canada. *Agric. Ecosystem* 118, 29, 2007.
13. Larson, W.E., Clapp, C.E., Pierre, W.H., and Morachan, Y.B. Effect of increasing amounts of organic residues on continuous corn: Organic carbon, nitrogen, phosphorous, and sulfur. *Agron. J.* 64, 204, 1972.

14. Morachan, Y.B., Moldenhauer, W.C., and Larson, W.E. Effects of increasing amount of organic residues on continuous corn: 1. Yields and soil physical properties. *Agron. J.* 64, 199, 1972.
15. Sherrod, L.A., Peterson, G.A., Westfall, D.G., and Ahaja, L.R. Cropping intensity enhances soil organic carbon and nitrogen in no-tillage agroecosystems. *Soil Sci. Soc. Am. J.* 67, 1533, 2003.
16. Clay, D.E., Carlson, C.G., Clay, S.A., Chang, J., and Malo, D.D. Soil organic C maintenance in a corn (*Zea mays* L.) and soybean (*Glycine max* L.) as influenced by elevation zone. *J. Soil Water Conserv.* 60, 342, 2005.

13 Predictive Mapping of Soil Organic Carbon: A Case Study Using Geographically Weighted Regression Approach

Umakant Mishra and Rattan Lal

CONTENTS

13.1 EXECUTIVE SUMMARY

The development of mathematical or statistical models of relationship between environmental variables and soil organic C (SOC) at observation sites and their application to environmental data sets of entire study areas to generate predictive SOC maps is known as predictive mapping of SOC. This process has progressed dramatically due to computational advances made over the past few decades. For instance, advances in geographic information systems (GIS), remote sensing, and digital terrain modelings have created tremendous improvement over the way SOC maps have been produced. Prediction approaches have varied from statistical approaches of linear regression to more complex machine learning techniques (MLT) such as random forest and artificial neural networks. The research in SOC prediction in terrestrial ecosystems is driven by the important role these models will play in better understanding and managing the global C cycle. This chapter demonstrates a geographically weighted regression (GWR) approach to map the SOC pool down to 0.5 m depth for the state of Ohio in the United States as a case study. In this approach, a varying relationship is considered between the environmental variables and the SOC pool over the study area. Topographic attributes, land use map, climate data, normalized difference vegetation index (NDVI), and bedrock geology map were used to predict the SOC pool and predicted SOC was validated using independent samples.

13.2 INTRODUCTION

The pedologic C pool comprises SOC and soil inorganic C (SIC) components. Of the two components, the SOC pool is highly reactive and is a strong determinant of numerous ecosystems services. World soils are an important C pool (2500 Pg of SOC and SIC pools in 1 m depth), with strong impacts on the global C cycle.[1,2] Estimates of SOC pool and their spatial variability in terrestrial ecosystems are essential to better understand the global C cycle, to estimate the soil C sink capacity, and to identify effective soil C sequestration strategies. Similarly, measurement, monitoring, and verification (MMV) of the amount of C sequestered in soil is a critical factor to the development of C trading efforts where farmers opting sequestration management practices can be economically rewarded. But the amount of C stored in the soil per unit area is highly variable[3] as the magnitude of SOC pool at a location depends on a range of factors such as soil type, land use, annual input of biomass into the soil, topographic features, and climatic conditions. These factors all vary at different locations. Therefore, several approaches are needed to develop a reliable estimate of SOC stock at different spatial scales.[4–6]

Soil sampling to adequately characterize spatial variability of the SOC pool on a landscape, watershed, or a regional scale is a major challenge, requiring intensive sampling schemes and costly analyses. However, modern computational advances have enhanced our capacity in applying geospatial techniques for the study of SOC pool and its dynamics. Digital soil mapping and other innovations can be used to predict SOC pools for diverse land uses, and different scales. These techniques can also be used in assessing the SOC sequestration rates and total potentials for different time intervals, and to produce continuous maps of SOC pool as representation of its spatial patterns and associated uncertainty. Likewise, the relationship between greenhouse gases (GHGs) emission and SOC content can be developed by using spatial analysis. The overall objective of this chapter is to critically review advances in the field of SOC prediction and mapping, and demonstrate a GWR approach to map the SOC pool for the state of Ohio in the United States as a case study.

Specifically, this chapter describes the importance of estimating SOC pool at regional scales (Section 13.3), discusses applicability of different techniques used in SOC pool prediction and mapping (Section 13.4), outlines the range of environmental variables used in prediction of SOC (Section 13.5), identifies research priorities in assessing SOC pool studies (Section 13.6), and demonstrates a GWR approach to predict the SOC pool for the state of Ohio (Section 13.7).

13.3 IMPORTANCE OF ESTIMATING SOC POOL AT A REGIONAL SCALE

The modern civilization has been called a "carbon civilization,"[7] because the production of modern amenities is derived from cheap and easy access to the energy derived from C-based fossil fuels. Therefore, ecological problems of major concern to soil science are climate change, water pollution and scarcity, loss of biodiversity, food insecurity, and soil degradation and desertification.[8] Therefore, soil resources must be managed to offset anthropogenic emissions of GHGs, produce lignocellulosic feedstock and oil seeds for biofuels as alternative to fossil fuel, improve quality and quantity of renewable fresh water resources, dispose industrial/nuclear and urban wastes, enhance biodiversity, and improve ecosystem services.

The SOC pool is an important indicator of soil quality, and also an important determinant of several ecosystem services. For example, increase in SOC pool in degraded soils improves soil structure and tilth, reduces soil erosion, increases plant available water capacity, stores plant nutrients, provides energy for soil fauna, purifies water, denatures pollutants, increases soil biodiversity, improves crop/biomass yields, enhances use efficiency of inputs, and moderates climate. It makes soil a living ecosystem. Therefore, credible estimation of SOC pool at regional scale has important environmental and economic implications.

13.3.1 OFFSETTING EMISSION

Global earth surface temperatures have increased by 0.88°C since the late nineteenth century, and 11 out of the 12 warmest years on record have occurred since 1995.[9] Earth's mean temperature is projected to increase by 1.5°C–5.8°C during the twenty-first century.[20] In conjunction with increase in global temperature, sea level has risen by 15–23 cm during the twentieth century,[9] shifts in ecosystems have occurred,[10] and frequency and intensity of occurrence of wild fires and extreme events have increased.[11,12] These changes, indicative of the global climate change, are being attributed to enhanced emission of GHGs through a range of anthropogenic activities including land-use conversion, deforestation, biomass burning, draining of wetlands, soil cultivation, and fossil fuel combustion. Among the GHGs, the concentration of carbon dioxide (CO_2) has increased by 31% from 280 ppmv in 1850 to 380 ppmv in 2005, and is presently increasing at 1.7 ppmv $year^{-1}$ or 0.46% $year^{-1}$.[9,13] Concentrations of methane (CH_4) and nitrous oxide (N_2O) have also increased steadily over the same period. Among the GHGs, CO_2 is the most important infrared absorbing, anthropogenic gas in the atmosphere and is responsible for 62% of total radiative forcing of earth. Therefore, there is a strong interest in stabilizing the atmospheric abundance of CO_2 and other GHGs to mitigate the risks of global warming.[14,15]

Removing atmospheric CO_2 through plant photosynthesis and storing it in soils is a recognized option for offsetting anthropogenic emissions.[16,17] In contrast to the engineering techniques of CO_2 capture and storage, soil C sequestration is a win–win strategy. It is a natural process in which biomass C is converted to humus and incorporated into SOC pool. Progressive increase in SOC pool improves soil quality, sets in motion the land restorative processes, and advances food security.[18] Therefore, soils can be an important sink of atmospheric CO_2.[19] The C sink capacity of world soils is estimated about 1 Pg C $year^{-1}$, which can annually offset 0.47 ppm of CO_2 increase in the atmosphere.[20,21]

13.3.2 Trading Carbon Credits

Sequestration of atmospheric CO_2 as SOC pool is also called "farming C" or "growing C," especially if it can be traded as any other farm commodity (e.g., corn, soybean, milk, or meat). Gains in soil C pool with reference to a baseline period can be traded through World Bank, Chicago climate exchange (CCX), European climate exchange, national and local industry, and the clean development mechanism of the Kyoto Treaty. At present, C is being traded using a cap and trade system. In this system, the buyer of C credits are industries that want to offset fossil fuel emissions, whereas the sellers are farmers who have accrued C credits from the adoption of recommended management practices that can be sold (Figure 13.1).

FIGURE 13.1 A corn field under no-till agriculture in Midwest United States.

Buyers purchase the credits if it is cheaper than it would cost them to alter their operation. Currently, C trading is being done on compliance and voluntary basis. The countries that have signed Kyoto Protocol such as European Union have compliance market and those who have not signed Kyoto Protocol (e.g., United States) have a voluntary system of market. The lower price of C in CCX (\$1.95 Mg^{-1} of CO_2 versus \$16.75 Mg^{-1} of CO_2 in Europe), is expected to increase if a cap on industrial emissions is enforced. As the C trading market is just developing, there is a vast scope for growing soil C as a tradable commodity.[22]

There are several factors responsible for a slow progress in trading credits of C sequestered in soils. The process of making SOC a commodity has been hindered primarily because of the lack of a credible methodology for assessing C credits in soils and terrestrial ecosystems. Other factors include lack of MMV methodology for up scaling the point SOC observations to landscape, watershed, and regional scales with reasonable prediction accuracies, evaluating changes in SOC pool in terms of Mg ha^{-1} $year^{-1}$ with reference to a baseline for specific land unit over a short period of 2–5 years, and verifying that the C sequestered is permanent and not emitted back into the atmosphere because of changes in land use and management practices.

13.3.3 Upscaling the Soil Processes at Landscape and Watershed Scales

Usually, the environmental processes leading to global warming, pollution, biodiversity loss, and elemental cycling occur over large spatial scales. However, the soil data characterizing ecological research are typically collected over a point or plot scales. Thus, reconciling this mismatch in scales is the most formidable challenge to environmental scientists.[23] Therefore, upscaling or extrapolation of information from smaller to larger scales is essential.[24,25]

Techniques of spatial extrapolation used in SOC mapping can be divided into two general categories. The "measure and multiply approach" (MMA) is the most prominent method of producing coarse predictions of SOC storage at regional to global scales.[26] In MMA, the study area is stratified into different strata using expert knowledge of researcher and the point measurements of SOC within each stratum are multiplied with the aerial extent of that stratum. The MMA has been used to estimate SOC pool at a global,[27,28] national,[29,30] state,[31,32] and regional[33,34] scales. Estimates of SOC pool obtained from the MMA do not account for the spatial variability caused by soil heterogeneity within each stratum. Furthermore, errors associated with assigning the mean SOC concentration from a small number of samples to a mapping area can be an important source of discrepancy.[35]

An alternative to MMA is the "soil landscape modeling" (SLM) techniques. In the SLM techniques, the variability of soils is analyzed with respect to changes in environmental variables known to influence variation in soil property (e.g., topography, land use, or climatic factors). The data of multiple environmental factors are used as input variables to calibrate the models. These models are then used to make predictions over the entire study area.[36,37] A limited number of observations of the dependent variables are used to develop a model (equation), which is then applied to predict the dependent variable for the entire area. The estimates of the SOC pool obtained from the SLM techniques have lower estimation errors than those from the MMA as these techniques take into account changes in environmental variables.[38] In the SLM, however, the relationship between the environmental variables and SOC pool is assumed to be constant over the study area. As a result, the SLM technique generally shows the global trends of SOC pool in the study area. Therefore, the estimation errors are expected to increase in large-scale studies (watersheds, regional) because these models do not account for the varying relationships between the environmental variables and the SOC pool over the space.

13.4 APPROACHES OF PREDICTING SOC POOL

Efforts are being made to map the SOC pool using a variety of techniques. Table 13.1 summarizes a range of predictive mapping efforts for SOC pool in a chronological order, and it describes prediction techniques, lists independent variables used, and explains the scale at which the SOC pool was estimated. Considerable progress has been made in obtaining more quantitative estimates of SOC pool using techniques ranging from linear regression to MLT. However, the studies using more quantitative techniques (e.g., SLM and geostatistics) have mostly been used on small areas in comparison to MMA. Such selectivity is primarily due to sparse availability of data of SOC concentration and lack of information about environmental variables. Thus, the following discussion focuses on the methods used in predictive SOC mapping.

13.4.1 Statistical Methods

In statistical methods, correlation between SOC concentration and environmental variables is used for prediction purpose.[39] The environmental variables that are expected to affect soil forming process are used as model predictors in generating the SLMs. The statistical model is mathematically represented as

$$\text{SOC} = f(E_i) \tag{13.1}$$

where

SOC is soil organic carbon pool
f is some empirical function linking observed SOC with environmental variables
E_i are the environmental variables ranging from 1 to i

The model can be used to extend the SOC predictions from the existing point data set to the entire region, if these environmental variables are exhaustively available throughout the study area. The success of this approach depends on the strength of the relationship between the SOC and environmental variables for which the data are available. The model will also be strong and the associated prediction errors less where these relationships are strong. Moreover, the benefit lies on these environmental variables being less costly to obtain than commissioning the collection of soil profile data and using some surface interpolation or smoothing procedure.[40]

TABLE 13.1
Summary of Previous Studies Showing Predictive SOC Stock Mapping, Describing Prediction Techniques, Predictive Factors Used, and the Scale of Study

Prediction Techniques	Predictive Factors	Study Area (km^2)	Location	Studies
Clustering and regression	Terrain attributes	500	USA	101
Regression and kriging	Terrain attributes, geographical coordinates	0.26	USA	102
Linear regression	Terrain attributes	0.054	USA	64
Linear regression	Soil property, terrain attributes, climate		France	103
Bayesian methods	Terrain attributes	0.054	USA	104
Bayesian rule	Terrain attributes	260	Australia	105
Regression tree	Terrain attributes, climate, vegetation	500	Australia	39
Linear regression	Soil property, vegetation, terrain attributes		USA	67
Linear regression	Reflectance	1.15	USA	70
Linear regression	Terrain attributes	0.02	France	106
Landform segmentation	Terrain attributes		Canada	107
Linear regression	Terrain attributes	Various	Canada	108
Linear regression	Climate, vegetation, terrain attributes	56,000	Croatia	109
Linear regression	Terrain attributes		Costa Rica	110
Kriging, co-kriging, kriging with external drift	Terrain attributes	0.13	USA	50
Regression kriging	Terrain attributes	56,000	Croatia	48
Regression tree	Terrain attributes, soil, vegetation		USA	111
Regression tree	Terrain attributes, climate, reflectance		Australia	40
Regression/kriging/fuzzy logic	Reflectance	0.52/0.31	USA	76
Linear regression	Terrain attributes	15	USA	38
Linear regression, ordinary kriging, regression kriging	Terrain attributes	0.40	Spain	51
Depth distribution function Artificial neural network	Reflectance, terrain attributes	1,500	Australia	3
Kriging, co-kriging, regression kriging	Soil series, reflectance, terrain attributes	0.65	USA	49
Linear regression	Terrain attributes	Various	USA	112
Kriging		1,039	China	113

(*continued*)

TABLE 13.1 (continued)
Summary of Previous Studies Showing Predictive SOC Stock Mapping, Describing Prediction Techniques, Predictive Factors Used, and the Scale of Study

Prediction Techniques	Predictive Factors	Study Area (km^2)	Location	Studies
Principal component regression	Reflectance, terrain attributes	0.5	USA	114
Kriging, stratification	Reflectance	10.25	Italy	115
Kriging	Terrain attributes	933	China	116
Random forest	Terrain attributes, geology, soil, landuse history	15	Panama	57
PLS regression	Reflectance		Australia	42
Regression kriging	Terrain attributes, parent material, climate, land cover	13,500	Kenya	117
Depth distribution function, ordinary kriging		94,319	USA	118

Multiple linear regression (MLR) is the most widely used among various statistical techniques in SOC prediction literatures (Table 13.1). The MLR model is written as

$$C = X\beta + e \tag{13.2}$$

where
- C is vector of response (predicted SOC)
- X is matrix of environmental variables
- β is vector of parameters
- e is associated error of prediction (deviations of the model to the observed value)

The parameter is usually solved using ordinary least squares with assumptions that the error is independently distributed, having zero mean and finite variance, and is normally distributed. The discrete variables such as land use or bedrock geology can be included into MLR by coding the K factors into $K-1$ variables.

The preference for multiple regression techniques stems from their simplicity, ease of use, computational efficiency, and straightforward interpretation.[41] Gomez et al.[42] used partial least square (PLS) regression to predict SOC from reflectance. In PLS, a new set of components as regressor variables are developed, which are linear combinations of the original variables. Unlike principal component regression, which only used the combination of the predictors, the components of PLS are determined by both the response variables and the predictor variables. Apart from these statistical techniques, clustering, landform segmentation, and Bayesian statistical techniques are also employed in SOC prediction. However, these techniques have been used in smaller landscapes[36,43] in comparison to the MMA. A major drawback of these statistical approaches is that these methods do not account for the spatial position where the information of SOC was obtained. Furthermore, these methods do not account for the spatial dependence between the observations, which is rather a rule than exception for the environmental data sets.

13.4.2 Geostatistical Methods

Geostatistical methods are used in SOC mapping in presence of spatial autocorrelation in the data obtained. The presence of spatial autocorrelation can be assessed by using Moran's index[44] or by modeling a variogram. These methods show the spatial variability in SOC, which is important for various environmental applications as shown in the following equation:

$$\mathrm{SOC}(x, y) = f(\mathrm{SOC}(x+u, y+v)) \tag{13.3}$$

where

$\mathrm{SOC}(x, y)$ is the unmeasured SOC at some location (x, y)

$\mathrm{SOC}(x+u, y+v)$ is the observed SOC at some locations apart by u and v magnitude $(x+u, y+v)$

Geostatistics have been used in soil mapping research to spatially interpolate the SOC values at unmeasured locations from field collected data. Burgess and Webster[45,46] introduced ordinary kriging to the soil science community. Along with the maps of SOC, geostatistical techniques are also used to produce maps of prediction errors, which show the variation in accuracy of predictions over the study area. This property gives geostatistical methods a unique importance in SOC mapping, as the map of errors is not produced when other techniques are used.

Several hybrid methods have been developed from the combination of geostatistics and environmental correlation, where the observations or the residuals of the regression are interpolated using co-kriging[47] or regression kriging.[48] Simbahan et al.[49] mapped SOC pool at a farm scale (49–65 ha) based on multivariate secondary data using a variety of kriging algorithms. They reported that regression kriging outperformed other techniques and stressed that environmental variables should be used to predict SOC pool at larger spatial scales. Mueller and Pierce[50] compared ordinary kriging, kriging with trend model, co-kriging, and kriging with external drift to map SOC pool at 12.5 ha field. They reported that those techniques that used secondary terrain information produced maps of higher quality and suggested the use of terrain attributes in SOC mapping. Lopez-Granados et al.[51] reported from a 40 ha field study that even when a moderately correlated secondary attribute is available, methods that incorporate it to map the SOC performed better than the approaches that only use the spatial component of SOC. These studies support the conclusion that within geostatistics, soil forming factors in terms of ancillary environmental predictors should be used to estimate the spatial distribution of SOC pool using hybrid prediction techniques (e.g., regression kriging).

13.4.3 Machine Learning Methods

Data mining and MLT are ideally suitable for the analysis of complex ecological data, which requires flexible and robust analytical methods to deal with nonlinear relationships, high-order interactions, and missing values.[52] Various data mining and MLT such as artificial neural networks[3,53] and classification and regression trees[39,40,54] are being used in SOC mapping. These techniques are suitable for the analysis of complex ecological data, which requires flexible and robust analytical methods to deal with nonlinear relationships, high-order interactions, and missing values.[52] Other potential MLT such as bagging[55] and boosting[56] can be applied for SOC predictions in order to enhance prediction accuracy. Grimm et al.[57] reported that random forest, a new method of data mining, has several advantages over other modeling techniques. This technique can model high-dimensional nonlinear relationships, handle categorical and continuous predictors, resistant to over fitting, has relative robustness with respect to noise features, implements unbiased measure of error rate, implements measures of variable importance, and has only few user-defined parameters.

13.4.4 Summary of Approaches

Among different techniques, linear regression approaches between SOC and environmental variables have been most extensively used in SOC prediction (about 50% of cited studies in this chapter). This is mainly due to its simplicity in application and ease of interpretation. After regression techniques, geostatistical and hybrid geostatistical techniques incorporating environmental variables into kriging system are being used. Apart from regression and geostatistical techniques, clustering, landform segmentation, and Bayesian modeling have also been used. Recently, MLT are being used that cover larger spatial scales and use a range of environmental variables that have not been included in prior SOC studies.

13.5 ENVIRONMENTAL VARIABLES USED FOR PREDICTION

The soil factor equation developed by Jenny[58] is the basis for quantitative prediction of soil properties using environmental variables, and has wide acceptance amongst the soil science community. Jenny described soil as a function of climate, organisms, topographic relief, parent material, and time. Hudson[59] stated that Jenny's soil factor equation is a general statement implying that soils are natural bodies that are distributed in a predictable way in response to a systematic interaction of environmental factors. The SOC prediction studies have used Jenny-like approach using "scorpan" factors[60] to predict the SOC pool at various spatial scales. These factors include soil properties, and spatial coordinates of a location in addition to Jenny's soil formation factors. These environmental data sets form the basis for the extrapolation of observed SOC concentration at sampling site to entire study area.

13.5.1 Topographic Attributes

Topography is a dominant control on the earth surface processes. It directly moderates the flow of water over and through the earth's surface, moderating soil wetness and soil erosion potential and thereby influencing soil properties.[61] Topographic attributes are the important factors controlling soil formation and most extensively used environmental variable in the predictive mapping of SOC concentration and pool (Table 13.1). Hall and Olson[62] reported that landscapes have a strong nonrandom variability component, which makes landscapes predictable, and, since soils are strongly related to landscapes, they should also be predictable. Therefore, the spatial distributions of terrain attributes are useful in capturing the variability of soil properties. Terrain analysis is more useful in environments where topographic shape is strongly related to the processes driving soil formation,[63] for instance, areas where there is considerable relief. Several studies are based on using terrain attributes derived from digital elevation data as predictive factors in SLM techniques.[38,64–67] McKenzie et al.[63] focused on the role of terrain analysis in soil mapping. The studies in Table 13.1 show that terrain attributes were used as predictors of SOC in almost all of the predictive SOC mapping efforts. The terrain attributes are generally derived from the digital elevation models (DEMs). Various primary (slope, aspect, curvatures, upslope area, etc.) and secondary terrain attributes (compound topographic index, sediment transport index, terrain characterization index, etc.) are used in SOC mapping. McBratney et al.[60] suggested upslope contributing area to be an important factor for regional-scale studies as it shows the area above a certain length of contour that contributes flow across the contour.

13.5.2 Remote Sensing Data

Remote sensing data (RSD) are a potentially important component of predictive mapping of SOC, as they provide quantitative measure of surface reflectance in a spatial context, which is related to soil properties.[68] Both physical (e.g., particle size and surface roughness) and chemical factors (e.g., surface mineralogy, soil organic matter or SOM content and moisture) control soil spectral reflectance.[69] In remote sensing, spectral signatures of materials are defined by their reflectance or absorbance as a function of wavelength in the electromagnetic spectrum. The relationship between the SOM and remotely sensed measurement has been an area of considerable research.[70]

Some studies have shown a negative correlation between surface total C and reflectance in the visible and near infrared range.[71,72] This is because increasing C has darkening effect, consequently reducing the amount of energy reflected. Furthermore, increasing soil moisture content is often associated with increasing total C and that tends to depress surface reflectance. It has been reported that the relationship between SOC and reflectance is poor if soil samples are collected from large geographic areas of different landscapes such as soil samples from an entire state.[72,73] Most research to date has been focused on the use of image intensity to derive soil properties. Chen et al.[70] utilized image intensity in three bands (red, green, and blue) to develop a logarithmic linear relationship for SOC. Strongest correlation between SOC and reflectance is often found in visible spectral region. Recent research indicates that red and near infrared reflectance or image intensity can be used in deriving soil properties from remotely sensed images and guiding soil sampling.[74] More research is needed to quantify the changes in SOC from changes in land management activities.[75] Geostatistical analyses that integrate high-resolution RSD with measured surface properties may improve our ability to resolve finer differences among surface soil properties.[76]

There are many obvious factors that can affect the spectral signal. For instance, some of the soil properties are subject to variation both in time and in space such as the degree of soil crusting can be affected as a result of raindrop impact, soil texture, soil moisture regime, roughness, and vegetation canopy or crop residue cover. These factors induce changes in soil reflectance that approach or exceed the spectral response of SOM.[77] In addition, the soil properties estimation can also be subject to degradations due to radiometric and atmospheric effects, spectral and spatial resolutions.[78] Because of the confounding effect of the previously mentioned factors, it has been difficult to establish a strong relationship between SOC concentrations and data from airborne-hyperspectral sensors.[79–81]

13.5.3 Land Cover/Land Use Data

Soil can be source or sink of atmospheric C depending on the land use.[82,83] Therefore, land cover/land use data also provide information regarding the anthropogenic influence on the SOC pool. Several studies have indicated the importance of land use on the SOC dynamics. Forest land has higher SOC pool than agricultural land.[84,85] Similarly, in a study for the conterminous United States, Guo et al.[86] observed that the forest land cover contained the highest SOC pool, followed by agricultural land use, wetlands, grass and pasture land uses. Tan et al.[32] reported that the SOC pools are confounded by preferential selection of land for different land uses and concluded that human influence on the land use also plays an important role in the SOC pool and its dynamics. However, the SOC prediction/mapping literatures shows limited use of land cover/use data mainly because of the poor availability of land cover information at desired scales.

13.5.4 Climatic Data

Climatic data sets are important predictors of SOC pool. Mean annual rainfall and mean annual temperature data are extensively used in SOC predictions. Apart from these, radiation, mean moisture index, and isothermality have also been used. Mishra et al.[87] used remote-sensing-based evapotranspiration map in SOC prediction in a plot-scale study. Apart from these, measures of solar radiation and soil moisture content can also be used in SOC pool predictions.

13.5.5 Summary of Data Sources and Types

The environmental data sources for SOC prediction in the literature mainly involve using Jenny's state factors and McBratney's "scorpan" factors. There is a scarcity of availability of these data sets in many parts of the world. However, SOC predictions using a variety of above-mentioned modeling techniques can be done for locations for which these data are available.

13.6 RESEARCH PRIORITIES

Soil sampling for regional studies is time and cost-intensive. Extensive data sets of environmental parameters are increasingly becoming available due to technological improvements in data collection techniques.[3] Such data are likely to support environmental studies where sampling to record the variable of interest is constrained by the time and cost factors. Therefore, future efforts must focus on modeling the relationship between SOC pool and readily available environmental variables and utilize the relationship for regional-scale SOC pool predictions. Several studies have indicated the need of further research on measurement and projections of soil C dynamics to advance the science of C sequestration and its application on large spatial scales.[88] Likewise, developed and improved methodologies are needed to quantify the changes in SOC pool at multiple scales and to make regional-, national-, and global-scale projections.[89] In this scenario, remote sensing and GIS permit more accurate descriptions of spatial patterns and suggest direction for future research.[90] These efforts can also facilitate the trading of C credits.

Most of the environmental data sets including SOC concentration show the property of nonlinearity and nonstationarity, which prohibit the use of normal spatial statistical techniques. Heuvelink and Webster[91] suggested that the future modeling efforts must address these problems. Similarly, SOC pool is dynamic in nature, and exhibits both spatial and temporal variability. Goovaerts[47] suggested further efforts should be devoted toward modeling of the space/time variability. McBratney et al.[60] also identified several knowledge gaps for further research in digital soil mapping. They suggested additional research to identify other environmental covariates for mapping the SOC pool in different environmental settings, to design sampling methods for creating dynamic digital soil maps for SOC pools, and to assess the uncertainty and cost involved in preparing maps.

Thorough understanding of the sources and sinks of C in terrestrial ecosystems is also required. Bell et al.[67] emphasized the need for better understanding of mechanisms responsible for C sequestration at landscape scales. Likewise, agronomic and societal values of soil C in relation to ecosystem services must be assessed so that land managers practicing soil C sequestration could be adequately compensated. Similarly, C sink capacity of soils to specific depth for different ecosystems needs to be predicted for designing efficient C sequestration and trading programs.

13.7 CASE STUDY: PREDICTING SOC AT A REGIONAL SCALE

In the case study described in this chapter, we used a GWR approach to predict and map the SOC pool at a scale of a state (Ohio). The GWR approach considers varying relationships between the SOC pool (dependent variable) and its predictors (environmental variables) within the study area; as a result, it gives different weight to different environmental variables at different locations. Thus, the SOC pool estimates obtained are supposed to have lower global prediction errors in comparison to other approaches (e.g., MMA and global regression methods) in large-scale studies. The spatial distribution of SOC pool is mapped at 30 m spatial resolution using the environmental variables such as topographic attributes (elevation and slope), climatic factors (temperature and precipitation), land use, bedrock geology layers, and NDVI.

13.7.1 Study Area and Soil Database

This study area comprised the entire state of Ohio in the United States, representing a land area of 106,055 km^2. The distribution of SOC stocks was assessed down to 0.5 m depth. Regional physiography, geological characteristics, and parent materials of Ohio have been described by Calhoun et al.[92] The soils are under udic or an aquic moisture regime and a mesic temperature regime.

A total of 370 georeferenced soil profile data was extracted from the United States Department of Agriculture (USDA), Natural Resources Conservation Service database,[93] and Ohio soil survey characterization database.[94] All pedons are classified as alfisols, entisols, inceptisols, mollisols, ultisols, and histosols. The whole data set (n=370) was then divided into calibration (n=314) and validation data sets (n=58). Figure 13.2 shows the spatial distribution of calibration and validation sites of SOC profiles over the study area.

13.7.2 Environmental Variables

A DEM of the study area, with 30 m spatial resolution, was obtained from the U.S. geologic survey database.[95] Primary terrain attributes elevation (m) and slope gradient (degree) were calculated using spatial analyst function of ArcGIS 9.2 (ESRI, Inc.,

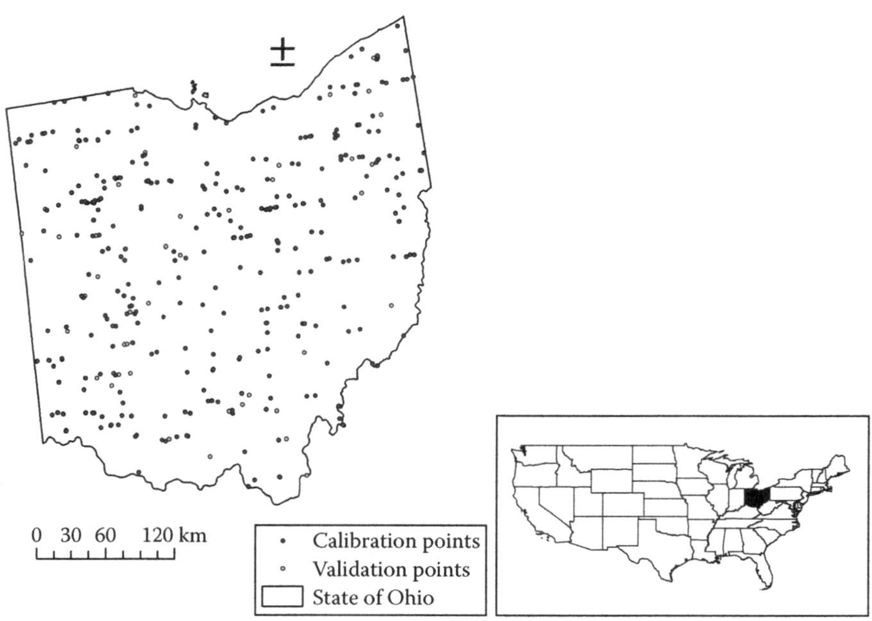

FIGURE 13.2 Distribution of calibration (n = 314) and validation (n = 56) sample points across the state of Ohio and the location of study site within the United States.

Redlands, CA). The elevation ranged from 133 to 470 m in the study area, with low-elevation areas residing in the northwestern corner (133 m) and higher-elevation areas in the eastern part of Ohio (470 m). The average slope gradient that covered a majority of study area is <5°. The most sloping areas are located in the southeast part of the state (up to 59°).

Land cover data of similar spatial resolution (30 m) were extracted for the study area from the USGS database. The land cover map showed 15 different land covers across the study area, which for the purposes of this study were reclassified into 6 major land covers. This resulted in the deciduous, evergreen, and mixed forests being classified as a forest category, pasture and crop land were recategorized as cultivated land, shrub and grasslands were kept into one category, and woody and herbaceous wetlands were categorized into one wetland category. The largest land area was classified under cultivated (50.3), followed by forest (31.3%), developed (14.1%), shrub and grasslands (2%), wetland (1%), and barren lands (0.12%), respectively.

The climate data, such as the long-term (1971–2000) mean annual air temperature (MAAT) and mean annual precipitation (MAP), were obtained from the database of spatial climate analysis service at Oregon State University.[96] MAAT and MAP were produced by the parameter-elevation regression on independent slopes model with the data. Lower temperatures (8.3°C) are generally observed in northern part of study area and higher temperatures (12.7°C) in the southern region. In general, the northern portion of study area receives lower precipitation (787 mm) than the south and eastern parts (1193 mm).

Bedrock geology data of 30 m spatial resolution covering the study area was provided by the Ohio Department of Natural Resources, Division of Geological Survey. The geology data showed 13 types of bed rocks in the study area. The largest area is under Mississippian series (19.5%), followed by Upper Silurian (15.3%), and Middle Silurian (11.1%), respectively. Remaining 10 types of bed rocks covered 54% of land area.

The NDVI (06-26-2001 to 07-11-2001) data of 250 m spatial resolution, derived from moderate-resolution imaging spectroradiometer bands 1 (red) and 2 (near infrared), was collected from the global land cover facility.[97] In this product, the NDVI values have been scaled using the formula (NDVI $*$ 200) + 50, which yields values from 0 to 250. The lower values suggest low vegetation and higher values indicate higher vegetation. The climate data, bedrock geology, and NDVI values were resampled at 30 m spatial resolution and used in this study.

13.7.3 Data Modeling and SOC Pool Estimation

Soil bulk density (ρ_b) values were available only for 26% of the pedons in the data set. Values of ρ_b for the remaining pedons were predicted from soil texture, SOC concentration, and depth of the horizons by using the pedotransfer function developed by Calhoun et al.[92] ($R^2 = 0.56$). This model gave unrealistic predictions for the horizons with high SOC concentrations. Therefore, the pedotransfer function developed by Adams[98] was used for horizons with a SOC concentration >6%.

The SOC pool was estimated in each profile by summing the C stock of each horizon from the surface to the depth of 0.5 m using the following equation:

$$C_s = \sum_{j=1}^{n} (C_m \rho_b) D \tag{13.4}$$

where
- C_s is the carbon stock (kg m^{-2})
- j is the soil horizon number (e.g., 1, 2, 3, …, n)
- C_m is the carbon concentration on mass basis (kg kg^{-1})
- ρ_b is the soil bulk density corrected for rock fragments (kg m^{-3})
- D is the thickness of each horizon (m)

13.7.4 GWR Approach

GWR was used to interpolate the SOC pool values from point observations to entire study area. It provides a method for the local analysis of relationships into the regression framework in multivariate environment. In this approach, a search region is described around a regression point and the data points within the region are used to calibrate a model. Here, each data point is inversely weighted by its distance from the regression point; as a result, the points closer to regression point get more weight than the points farther away. Therefore, GWR is sensitive to the bandwidth of the particular weighting function chosen. In regional studies, availability of data sets is not uniform. To reduce the smoothness problem incase of sparse data sets, the bandwidths can be made to adapt themselves in size to variations in density of data. Equation 13.5 describes the GWR procedure used in our data set:

$$\hat{C}_i = \beta_0(u_i, v_i) + \beta_1(u_i, v_i)X_{i1} + \beta_2(u_i, v_i)X_{i2} + \cdots + \beta_k(u_i, v_i)X_{ik} + \varepsilon_i \tag{13.5}$$

where
- $\hat{C}_i$ is predicted SOC stock at location i
- $\beta_0(u_i, v_i)$ is intercept at location I
- (u_i, v_i) is coordinates of the ith point in space
- β_1 to β_k are regression coefficients
- X_{i1} to X_{ik} are environmental variables at location I
- ε_i is associated error with the estimation

The details of GWR procedure are provided by Fotheringham et al.[99]

13.7.5 Results and Discussion

The data in Table 13.2 show the descriptive statistics of SOC pool of both the calibration and validation sites. The SOC pool shows a higher variability at both sites (50%), due to heterogeneity of soil types and environmental variables of the study

TABLE 13.2
Summary Statistics of SOC Pool Data Calibration and Validation Sites

Parameters	Calibration Sites (n = 314)	Validation Sites (n = 56)
Minimum (kg m^{-2})	2.86	2.04
Maximum (kg m^{-2})	48.25	32.08
Average (kg m^{-2})	9.75	9.5
Median (kg m^{-2})	8.57	8.38
Standard deviation (kg m^{-2})	4.88	4.79
Coefficient of variability (%)	50	50.4

area. The SOC pool ranges from 2.86 to 88.25 kg m^{-2} with an average content of 9.75 kg m^{-2} in the calibration sites. The range and mean values of SOC pool at validation sites also show a similar trend. The skewness values for SOC pool at calibration sites were more than 1 (positive skewness); therefore, it was transformed to common logarithms before model fitting.

Figure 13.3 shows the spatial distribution of SOC pool predicted using GWR approach. Higher SOC pools are located toward the north or northwest part of Ohio.

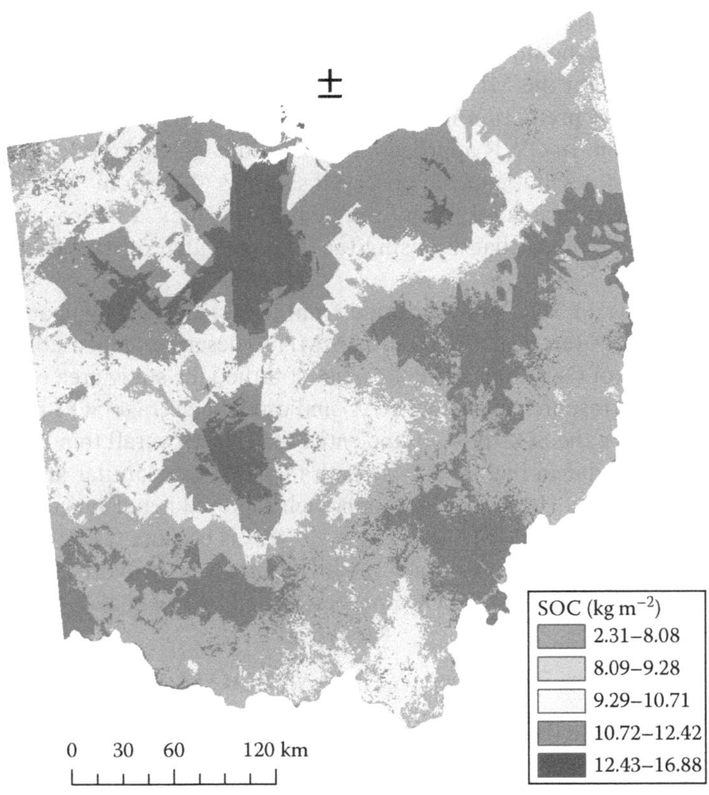

FIGURE 13.3 Predicted SOC pool (0–0.5 m) using GWR approach for the state of Ohio.

TABLE 13.3
Validation Indices (n = 56) of SOC Pool (0–50 cm) Predicted Using GWR Approach

Indices	Values
R	0.44
MEE (kg m^{-2})	−0.51
RMSE (kg m^{-2})	3.36

These areas are associated with landscapes of lower slope (<5° slope gradient) and soils formed under cooler temperatures (8°C). Meanwhile, the eastern and southern parts of state have low SOC pools, which are associated with more sloping landscapes (>12° slope gradient) and warmer temperatures (10°C–12°C).

No general relationship was observed between land use and SOC pool, which may be due to the preferential selection of soils for arable land use by farmers, as reported by Tan et al.[32] In general, cropland soils had higher SOC pool than forests soils, which does not necessarily mean that croplands have a higher SOC sequestration potential than forests. This difference is most likely due to the selection of land with high SOC pool as arable land, leaving the marginal lands for forests. That the forested sites possessed slopes >10°, further suggested these sites were on more marginal soils. These sites are characterized by well-drained soils, which would favor lower SOC pool.

The global predictive quality of the SOC estimates was evaluated by using an independent validation data set. For this purpose, different validation indices, such as r between the observed and predicted SOC values, the mean estimation error (MEE), and the root mean square error (RMSE) were calculated. The MEE and RMSE values should approach zero for an optimal prediction. The data in Table 13.3 show that the MEE is close to zero (−0.51 kg m^{-2}), which shows the lack of bias of the prediction method. Similarly, RMSE represents the average error of prediction. In GWR approach, RMSE value is 3.36 kg m^{-2}, which is less than the standard deviation of the observed samples (4.88 kg m^{-2}). This suggests that by using the information of environmental variables, a better estimation of SOC pools can be made than by using the average value of the observations as a prediction.

The variability of SOC pool obtained from this study was compared with the results obtained by Tan et al.,[32] who used MMA to describe the variability of SOC pool for the state of Ohio. Tan and colleagues stratified the study area into eight different major land resource areas (MLRAs) and used the average SOC pools of each MLRA to describe the SOC pool for the entire state. The overall trend of SOC pool distribution is similar in both results as both maps showed similar areas of higher and lower SOC pools. However, GWR approach captured better range of SOC pool (2.3–16.8 kg m^{-2}) than the MMA (8–12 kg m^{-2}). This is because GWR approach uses the variation in environmental variables, which is not used by MMA. No validation measures were provided for the SOC pools estimated by Tan et al.[32].

13.7.6 Step-by-Step Procedure to Generate SOC Map

GWR3 software for GWR[100] was used in this study to generate the model coefficient values that vary over the space. These model coefficients were then used in raster calculator of ArcGIS (ESRI, Inc., Redlands, CA) to generate final SOC pool map. The data file for GWR is an ASCII file that uses the file type of.dat or.csv.

The first row of the data file is a comma separated list of the names of the variables. The variable names should not contain spaces and no more than eight characters in length.

1. As the program starts, a window will pop up asking to select a task to perform. Select **Create a new model** and click **Go**.
2. Specify the **data file** from the location where data is located and select **Open** to proceed (GIS Chapter 13 data).
3. Determine the type of GWR model that needs to be fitted. We selected a **Gaussian** model and then click **Go**.
4. Name the file into which the results will be written, **Parameter diagnostic file**. Appropriate file type should be specified for this file (.csv) and then click **Continue**.
5. The **Calibration location file** is kept blank as the model was fitted at the data points.
6. In **Model editor** window, the GWR model parameters such as dependent and independent variables, location variables (two variables representing the coordinates of the data points), kernel type, and kernel bandwidth (adaptive or fixed) are specified.
7. The model has to be saved before it is run by clicking on **Save model**. Type the name of file on **File name** box and then click **Save**.
8. Click on the **Run** button in the **Model Editor Window** and this brings the **Run window** in which a name must be specified for the **Model listing file** (.txt).
9. Once the program is run, the user is asked whether the **output listing file** (.txt) is to be viewed, and appears in a separate window, which can be saved.
10. The model coefficients of each variable were converted into shape file (point) in **ArcGIS**.
11. Separate maps of the model coefficients were prepared using ordinary kriging function of **spatial analyst** in ArcGIS.
12. The maps of environmental variables and model coefficients were used in a **raster calculator** to generate the predicted SOC map using GWR.

13.8 CONCLUSIONS

Quantifying SOC pool at a particular location and time is essential to improve our understanding of environmental processes and related issues. Most of the predictive SOC mapping research outlined in this chapter has been conducted over small areas and for shallow depths (<0.5 m), primarily to assess the spatial variability of SOC within individual fields or across soil toposequences. The primary driving force behind this type of research has been the need to provide accurate soil information for agricultural and ecological applications. These mapping efforts are successful at the filed scale as many of the soil forming factors (climate, parent material, time, and organism) are considered constant. However, additional research must focus on predicting SOC pool at watershed and regional scales to better understand the SOC sequestration dynamics. Such predictions of SOC pool can also facilitate trading of C credits.

Among various prediction approaches, statistical methods use environmental correlation between predictors and SOC pool. Geostatistical methods use spatial autocorrelation between data points and also incorporate the environmental variables in prediction of SOC pool. Considering the technological advancements in data collection methods and computation efficiency, machine learning techniques seem to be more promising in SOC pool prediction at larger spatial scales. However, SOC pools vary spatially and temporally and their distribution is scale dependent. Therefore, different approaches of predictions and predictors usually perform better at different scales and it is unlikely that a single model can be developed and applicable at all soil and landscape units.

Previous SOC studies have shown that the terrain attributes are powerful predictors of SOC pool. A variety of terrain attributes can be derived from the DEMs that are available at a range of scales and resolutions. Thus, several options exist depending on the specific study objectives. However, these attributes are often not powerful predictors in flat landscapes where soils do not develop in well-defined toposequences. For instance, surface reflectance may be better predictor of SOC in desert landscapes where the soil is bare. Likewise, surface geology information may perform better where the soil depth is shallow.

This chapter demonstrated the use of GWR approach to map the SOC pool down to 0.5 m depth for the state of Ohio in the United States. Spatial variation of SOC pool was predicted at 30 m spatial resolution using topographic attributes, climate data, landuse map, bedrock geology, and NDVI map. In this approach, varying relationship was considered between the environmental variables and the SOC pool over the study area; as a result, better range of SOC pool variability was captured in comparison to MMA.

ACKNOWLEDGMENT

This study was conducted as a partial requirement of the PhD research. The lead author would like to acknowledge the U.S. Department of Energy for providing the necessary funding to complete this study.

REFERENCES

1. Batjes, N.H. Total carbon and nitrogen in the soils of the world. *Eur. J. Soil Sci.* 47, 151, 1996.
2. Grace, J. Understanding and managing the global carbon cycle. *J. Ecol.* 92, 189, 2004.
3. Minasny, B.M. et al. Prediction and digital mapping of soil carbon storage in the lower Namoi valley. *Aust. J. Soil Res.* 44, 223, 2006.
4. Van Meirvenne, M. et al. Regional characterization of the long term change in soil organic carbon under intensive agriculture. *Soil Use Manage.* 12, 86, 1996.
5. Post, W.M. et al. Monitoring and verifying changes of organic carbon in soil. *Climatic Change* 51, 73, 2001.
6. Lal, R. Soil carbon sequestration to mitigate climatic change. *Geoderma* 123, 1, 2004.
7. Lal, R. Soil science and carbon civilization. *Soil Sci. Soc. Am. J.* 71, 1425, 2007.
8. Lal, R. Soils and sustainable agriculture a review. *Agron. Sustain. Dev.* 28, 1, 57, 2008.

9. IPCC. *Climate Change 2007, Climate Change Impacts, Adaptation and Vulnerability.* Working group II, Geneva, Switzerland, IPCC, 2007.
10. Greene, C.H. and A.J. Pershing. Climate drives sea change. *Science* 315, 1084, 2007.
11. Running, S.M. Is global warming causing more large wildfires? *Science* 13, 927, 2006.
12. Westerling, A.L. et al. Warming and earlier spring increase western U.S. forest wildfire activity. *Science* 313, 940, 2006.
13. WMO. *Greenhouse Gas Bulletin: The State of Greenhouse Gases in the Atmosphere Using Global Observations up to December 2004.* World Meteorological Organization, Geneva, Switzerland, 2006.
14. Kerr, R.A. Scientists tell policy makers we're all warming the world. *Science* 315, 754, 2007.
15. Kintisch, E. New congress may be warming up to plans for capping emissions. *Science* 315, 444, 2007.
16. Watson, R.T. et al. *Land Use, Land-Use Change, and Forestry* (*A Special Report of the IPCC*). Cambridge University Press, Cambridge, 2000, 377 pp.
17. Lal, R. Carbon sequestration. *Philos. Trans. Roy. Soc.* 363, 815, 2008.
18. Lal, R. Enhancing crop yields in developing countries through restoration of soil organic carbon pool in agricultural lands. *Land Degrad. Dev.* 17, 197, 2006.
19. Smith, P. Soils as carbon sinks: The global context. *Soil Use Manage.* 20, 212, 2004.
20. IPCC. *Land Use, Land Use Change and Forestry.* Intergovernment Panel on Climate Change. Cambridge University, Press, Cambridge, 1999.
21. Lal, R. For forest soils and carbon sequestration. *Ecol. Manage.* 220, 242, 2005.
22. Bohannon, J. From greener production to carbon trading: Sustainable energy careers. *Science.* 315, 868, 2007.
23. Peterson, D.L. and V.T. Parker. *Ecological Scale: Theory and Applications.* Columbia University Press, New York, 1998.
24. Schneider, D.C. The rise of the concept of scale in ecology. *Bioscience* 51, 545, 2001.
25. Scott, J.M. et al. *Predicting Species Occurrences: Issue of Accuracy and Scale.* Island Press, Washington, DC, 2002.
26. Schimel, D.S. and C.S. Potter. Process modeling and spatial extrapolation. In P.A. Matson and R.C. Harriss (eds.), *Biogenic Trace Gases: Measuring Emissions from Soil and Water.* Blackwell Science Ltd., Cambridge, MA, 1995, p. 358.
27. Post, W.M. et al. The global carbon cycle. *Am. Sci.* 78, 310, 1990.
28. Eswaran, H. et al. Organic carbon in soils of the world. *Soil Sci. Soc. Am. J.* 57, 192, 1993.
29. Kern, J.S. Spatial patterns of soil organic carbon in the contiguous United States. *Soil Sci. Soc. Am. J.* 58, 439, 1994.
30. Guo, Y. et al. Quantity and spatial variability of soil carbon in the conterminous United States. *Soil Sci. Soc. Am. J.* 70, 590, 2006.
31. Amichev, B.Y. and J.M. Galbraith. A revised methodology for estimation of forest soil carbon from spatial soils and forest inventory data sets. *Environ. Manage.* 33(Suppl. 1), S74, 2004.
32. Tan, Z. et al. Taxonomic and geographic distribution of soil organic carbon pools in Ohio. *Soil Sci. Soc. Am. J.* 68, 1896, 2004.
33. Homann, P.S. et al. Regional soil organic carbon storage estimates for western Oregon by multiple approaches. *Soil Sci. Soc. Am. J.* 62, 789, 1998.
34. Yu, D.S. et al. Regional patterns of soil organic carbon stocks in China. *J. Environ. Manage.* 85, 680, 2007.
35. Meersmans, J. et al. A multiple regression approach to assess the spatial distribution of soil organic carbon (SOC) at the regional scale (Flanders, Belgium). *Geoderma* 143, 1, 2008.
36. Gessler, P.E. et al. Modeling soil-landscape and ecosystem properties using terrain attributes. *Soil Sci. Soc. Am. J.* 64, 2046, 2000.

37. Thompson, J.A. et al. Digital elevation model resolution: Effects on terrain attribute calculation and quantitative soil–landscape modeling. *Geoderma* 100, 67, 2001.
38. Thompson, J.A. and R.K. Kolka. Soil carbon storage estimation in a forested watershed using quantitative soil landscape modeling. *Soil Sci. Soc. Am. J.* 69, 1086, 2005.
39. McKenzie, N.J. and P.J. Ryan. Spatial prediction of soil properties using environmental correlation. *Geoderma* 89, 67, 1999.
40. Henderson, B.L. et al. Australia-wide predictions of soil properties using decision trees. *Geoderma* 124, 383, 2005.
41. Hastie, T. et al. *The Elements of Statistical Learning: Data Mining, Inference and Prediction*. Springer Series in Statistics. Springer Verlag, New York, 2001.
42. Gomez, C. et al. Soil organic carbon prediction by hyperspectral remote sensing and field vis-NIR spectroscopy: An Australian case study. *Geoderma* 146, 403, 2008.
43. Park, S.J. et al. Identification of the spatial distributions of soils using a process based terrain characterization. *Geoderma* 103, 249, 2001.
44. Moran, P.A.P. Notes on continuous stochastic phenomena. *Biometrika* 37, 17, 1950.
45. Burgess, T.M. and R. Webster. Optimal interpolation and isarithmic mapping of soil properties: The semi-variogram and punctual kriging. *J. Soil Sci.* 31, 315, 1980.
46. Burgess, T.M. and R. Webster. Optimal interpolation and isarithmic mapping of soil properties: Block kringing. *J. Soil Sci.* 31:331, 1980.
47. Goovaerts, P. Geostatistics in soil science: State of the art and perspectives. *Geoderma* 89, 1, 1999.
48. Hengl, T. et al. A generic framework for spatial prediction of soil variables based on regression kriging. *Geoderma* 120, 75, 2004.
49. Simbahan, G.C. et al. Fine resolution mapping of soil organic carbon based on multivariate secondary data. *Geoderma* 132, 471, 2006.
50. Mueller, T.G. and F.J. Pierce. Soil carbon maps: Enhancing spatial estimates with simple terrain attributes at multiple scales. *Soil Sci. Soc. Am. J.* 67, 258, 2003.
51. Lopez-Granados, F. et al. Using geostatistical and remote sensing approaches for mapping soil properties. *Eur. J. Agron.* 23, 3, 279, 2005.
52. DeAth, G. and K.E. Fabricius. Classificaiton and regression trees: A powerful yet simple technique for ecological data analysis. *Ecology* 81(11), 3178, 2000.
53. Holmberg, M. et al. An application of artificial neural networks to carbon, nitrogen, and phosphorus concentrations in three boreal streams and impacts of climate change. *Ecol. Model.* 195(1–2), 51, 2006.
54. Bricklemyer, R.S. et al. Monitoring and verifying agricultural practices related to soil carbon sequestration with satellite imagery. *Agr. Ecosyst. Environ.* 118, 201, 2007.
55. Breiman, L. Bagging predictors. *Mach. Learn.* 24, 123, 1996.
56. Freund, Y. and R.E. Schapire. Game theory, on-line prediction and boosting. *Proceedings of the Ninth Annual Conference on Computing and Learning Theory*, 1996, p. 325.
57. Grimm, R. et al. Soil organic carbon concentrations and stocks on Barro Colorado Island-Digital soil mapping using Random Forest Analysis. *Geoderma* 146, 102, 2008.
58. Jenny, H. *Factors of Soil Formation*. McGraw-Hill, New York, 1941.
59. Hudson, B.D. The soil survey as paradigm based science. *Soil Sci. Soc. Am. J.* 56, 836, 1992.
60. McBratney, A.B. et al. On digital soil mapping. *Geoderma* 117, 3, 2003.
61. Hutchinson, M.F. A locally adaptive approach to the interpolation of digital elevation models. *Proceedings of the Third International Conference Integrating GIS and Environmental Modelling*, Santa Fe, New Mexico, January 21–25, 1996.
62. Hall, C.A.S. and C.G. Olson. Predicting variability of soils from landscape models. In Spatial variability of soil and landforms. *Soil Sci. Soc. Am. Spec. Publ.* 28, 9, 1991.
63. McKenzie, N.J. et al. The role of terrain analysis in soil mapping. *Terrain Analysis—Principles and Applications*. John Wiley & Sons, New York, 2000, p. 245.

64. Moore, I.D. et al. Soil attribute prediction using terrain analysis. *Soil Sci. Soc. Am. J.* 57, 443, 1993.
65. Odeh, I.O.A. et al. Spatial prediction of soil properties from landform attributes derived from a digital elevation model. *Geoderma* 63, 197, 1994.
66. Skidmore, A.K. et al. An operational GIS expert system for mapping forest soil. *Photogramm. Eng. Remote Sens.* 62, 501, 1996.
67. Bell, J.C. et al. A soil terrain model for estimating spatial patterns of soil organic carbon. *Terrain Analysis—Principles and Applications*. John Wiley & Sons, New York, 2000, p. 295.
68. Agbu, P.A. Statistical comparison of SPOT spectral maps with field soil maps. *Soil Sci. Soc. Am. J.* 54, 818, 1990.
69. Irons, J.R. et al. Soil reflectance. *Theory and Applications of Optical Remote Sensing*. John Wiley & Sons, New York, NY, 1989, p. 66.
70. Chen, F. et al. Field scale mapping of surface soil organic carbon using remotely sensed imagery. *Soil Sci. Soc. Am. J.* 64, 746, 2000.
71. Sudduth, K.A. and J.W. Hummel. Evaluation of reflectance methods for soil organic matter sensing. *Trans. Am. Soc. Agric. Eng.* 34, 1900, 1991.
72. Henderson, T.L. et al. High dimensional reflectance analysis of soil organic matter. *Soil Sci. Soc. Am. J.* 56, 865, 1992.
73. Schulze, D.G. et al. Significance of organic matter in determining soil colors. *Soil Color*. SSSA Special Publication, 31. SSSA, Madison, WI, 1993, p. 71.
74. Fox, G.A. and G.J. Sabbagh. Estimation of soil organic matter from red and near infrared remotely sensed data using a soil line Euclidean distance technique. *Soil Sci. Soc. Am. J.* 66, 1922, 2002.
75. Harper, R.J. and R.J. Gilkes. Some factors affecting the distribution of carbon in soils of a dryland agricultural system in Southwestern Australia. *Assessment Methods for Soil Carbon*. Lewis Publishers, Boca Raton, FL, 2001, p. 577.
76. Sullivan, D.G. et al. IKONOS imagery to estimate surface soil property variability in two Albama physiographies. *Soil Sci. Soc. Am. J.* 69, 1789, 2005.
77. Barnes, E.M. et al. Remote- and ground-based sensor techniques to map soil properties. *Photogramm. Eng. Remote Sens.* 69, 6, 619, 2003.
78. Lagacherie, P. et al. Estimation of soil clay and calcium carbonate using laboratory, field and airborne hyperspectral measurements. *Remote Sens. Environ.* 112, 3, 825, 2008.
79. Ben-Dor, E. et al. Mapping of several soil properties using DAIS-7915 hyperspectral scanner data—A case study over clayey soils in Israel. *Int. J. Remote Sens.* 23, 1043, 2002.
80. Selige, T. et al. High resolution topsoil mapping using hyperspectral image and field data in multivariate regression modelling procedures. *Geoderma* 136, 235, 2006.
81. Stevens, A. et al. Detection of carbon stock change in agricultural soils using spectroscopic techniques. *Soil Sci. Soc. Am. J.* 70, 844, 2006.
82. Lal, R. Offsetting global CO_2 emissions by restoration of degraded soils and intensification of world agriculture and forestry. *Land Degrad. Dev.* 14, 309, 2003.
83. Singh, B.R. and R. Lal. The potential of soil carbon sequestration through improved management practices in Norway. *Environ. Dev. Sustainability* 7, 161, 2005.
84. Lal, R. and J.M. Kimble. Conservation tillage for carbon sequestration. *Nutr. Cycl. Agroecosyst.* 49, 243, 1997.
85. Shrestha, B.M. et al. Soil aggregate and partical associated organic carbon under different land uses in Nepal. *Soil Sci. Soc. Am. J.* 71, 1194, 2007.
86. Guo, Y. et al. Analysis of factors controlling soil carbon in the conterminous United States. *Soil Sci. Soc. Am. J.* 70, 601, 2006.
87. Mishra, U. et al. Assessing the value of using a remote sensing based evapotranspiration map in site specific management. *J. Plant Nutr.* 31, 1188, 2008.

88. Izaurralde, R.C. et al. Soil organic carbon dynamics: Measurement, simulation and site to region scale up. *Assessment Methods for Soil Carbon*. Lewis Publishers, Boca Raton, FL, 2001, p. 553.
89. Izaurralde, R.C. and W.B. McGill. How does landscape architecture affect the upscaling of soil processes to a regional level? Presented in *Frontiers in Soil Science Research*. National Academy of Sciences, Washington, DC, December 12–14, 2005.
90. Miller, J.R. et al. Spatial extrapolation: The science of predicting ecological patterns and processes. *Bioscience* 54, 310, 2004.
91. Heuvelink, G.B.M. and R. Webster. Modeling soil variation: Past, present, and future. *Geoderma* 100, 269, 2001.
92. Calhoun, F.G. et al. Predicting bulk density of Ohio soils from morphology, genetic principles, and laboratory characterization data. *Soil Sci. Soc. Am. J.* 65, 811, 2001.
93. USDA-NRCS, Soil characterization database. Available online at http://ssldata.nrcs.usda.gov/, verified at May 6, 2006.
94. Calhoun, F.G. et al. Intuitive user friendly approaches to entry and storage of soil characterization data in Ohio and Australia. Soil Resources: Their inventory, analysis, and interpretation for use in the 21st century. Poster No. 13 (absr.), University of Minnesota, Minneapolis, June 10–12, 1999.
95. USGS. National land cover database 2001 (NLCD 2001). Available online at http://seamless.usgs.gov/viewer.htm, verified at June 6, 2007.
96. Daly, C. et al. High quality spatial climate data sets for the United States and beyond. *Trans. ASAE* 43, 1957, 2001.
97. GLCF. http://glcf.umiacs.umd.edu, verified September 15, 2008.
98. Adams, W.A. The effect of organic matter on the bulk and true densities of some uncultivated podzolic soils. *J. Soil Sci.* 24, 10, 1973.
99. Fotheringham, A.S. et al. *Geographically Weighted Regression the Analysis of Spatially Varying Relationships*. John Wiley & Sons, Ltd., Chichester, West Sussex, Hoboken, NJ, 2002.
100. Charlton, M. et al. *GWR 3 Software for Geographically Weighted Regression*. University of Newcastle upon Tyne, England, 2003.
101. Frazier, B.E. and Y. Cheng. Remote sensing of soils in the eastern Palouse region with Landsat thematic mapper. *Remote Sens. Environ.* 28, 317, 1989.
102. Bhatti, A.U. et al. Estimation of soil properties and wheat yields on complex eroded hills using geostatistics and thematic mapper images. *Remote Sens. Environ.* 37, 181, 1991.
103. Arrouays, D. et al. Spatial analysis and modeling of topsoil carbon storage in temperate forest humic loamy soils of France. *Soil Sci.* 159, 191, 1995.
104. Cook, S.E. et al. A rule based system to map soil properties. *Soil Sci. Soc. Am. J.* 60, 1893, 1996.
105. Lilburne, L. et al. GIS-driven models of soil properties in the high country of the south island. *10th Colloquium of the Spatial Information Research Centre*. University of Otago, New Zealand, 1998, p. 173.
106. Chaplot, V. et al. Soil carbon storage prediction in temperate hydromorphic soils using a morphologic index and digital elevation model. *Soil Sci.* 166, 48, 2001.
107. Pennock, D.J. and M.D. Corre. Development and application of landform segmentation procedures. *Soil Till. Res.* 58, 151, 2001.
108. Florinsky, I.V. and R.G. Eilers. Prediction of the soil organic carbon content at micro-, meso- and macroscales by digital terrain modeling. *7th World Congress of Soil Science*. Bangkok, Thailand, August 14–21. Paper no. 24, 2002.
109. Hengl, T. et al. Mapping soil properties from an existing national soil data set using freely available ancillary data. *17th World Congress of Soil Science*, Bangkok, Thailand, August 14–21, Paper no. 1140, 2002.

110. Powers, J.S. and W.H. Schlesinger. Relationships among soil carbon distributions and biophysical factors at nested spatial scales in rain forests of northeastern Costa Rica. *Geoderma* 109, 165, 2002.
111. Kulmatiski, A. et al. Landscape determinants of soil carbon and nitrogen storage in southern New England. *Soil Sci. Soc. Am. J.* 68, 2014, 2004.
112. Thompson, J.A. et al. Soil-landscape modeling across a physiographic region: Topographic patterns and model transportability. *Geoderma* 133, 57, 2006.
113. Hu, K. et al. Spatial and temporal patterns of soil organic matter in the urban–rural transition zone of Beijing. *Geoderma* 141, 302, 2007.
114. Huang, X. et al. Total carbon mapping in glacial till soils using near infrared spectroscopy, Landsat imagery and topographical information. *Geoderma* 141, 34, 2007.
115. D'Acqui, L.P. et al. Use of ecosystem information to improve soil organic carbon mapping of a Mediterranean Island. *J. Environ. Qual.* 36, 262, 2007.
116. Chai, X. et al. Spatial prediction of soil organic matter in the presence of different external trends with REML-EBLUP. *Geoderma* 148, 159, 2008.
117. Mora-Vallejo, A. et al. Small scale digital soil mapping in Southeastern Kenya. *Catena* 76, 44, 2008.
118. Mishra, U. et al. Predicting soil organic carbon stock within different depth intervals using profile depth distribution functions and ordinary kriging. *Soil Sci. Soc. Am. J.* 73, 614, 2009.

14 Tillage and Crop Residue Effects on Soil Carbon Turnover Using the Michaelis–Menten Approach

Mahdi Al-Kaisi

CONTENTS

14.1 EXECUTIVE SUMMARY

One of the first steps in maximizing nutrient energy efficiency is maintaining soil organic carbon (SOC). Tillage and cropping systems play significant role in affecting soil carbon dynamics and carbon dioxide (CO_2) flux. The evaluation of various tillage systems can help determine the effectiveness of conservation practices in sustaining soil productivity and enhancing environment quality. Soil C storage and CO_2 emission respond to conservation tillage differently from conventional tillage because of their differential effects on soil properties. The purpose of this study was to determine the impact of tillage on soil C storage and CO_2 emission in Clarion–Nicollet–Webster soil association in a corn [*Zea mays* L.]–soybean [*Glycine max* (L.) Merr.] rotation in Iowa. Soil CO_2 emission was lower for less intensive tillage treatments compared with moldboard plow, with the greatest differences occurring immediately after tillage operations. Cumulative soil CO_2 emission was 19%–41% lower for less intensive tillage treatments than moldboard plow, and it was 24% less for no-tillage with residue than without residue during the 480 h measurement period. Findings suggest that adopting less intensive tillage such as no-tillage and strip-tillage, and better crop residue cover are effective in reducing CO_2 emission and thus improving soil C sequestration in a corn–soybean rotation. This case study demonstrates how mineralizable carbon and turnover can be calculated using the Lineweaver–Burk transformation of the Michaelis–Menten equation.

14.2 INTRODUCTION

The impact of soil carbon on soil gas emission, water retention, nutrient cycling, and plant growth makes the maintenance of soil carbon a crucial component in protecting the environment.[1–5] Increased atmospheric carbon dioxide (CO_2) has been considered a major contributor to global warming. Carbon loss from soil to the atmosphere as CO_2 or other gases has been enhanced due to inappropriate tillage practices.[6] However, soil can function as a net sink for sequestering atmospheric CO_2 through appropriate soil and crop management.[7,8]

From a long-term (≥10 years) perspective, soil can be managed to increase total soil organic C (TC) storage by implementing conservation tillage practices and annual cropping systems.[9–11] However, short-term (≤10 years) tillage effects on soil C dynamics are complex and often variable. Franzluebbers and Arshad[12] reported that there may be little to no increase in SOC in the first 2–5 years after changing to conservation management, but a large increase in TC occurred in the next 5–10 years. In addition, Duiker and Lal[13] found that there was a linearly positive response of SOC to residue application rate regardless of tillage system after 7 years.

Carbon dioxide is produced in the soil through the metabolism of plant roots, microflora, and fauna and, to a small extent, by chemical oxidation of carbon-bearing materials.[14] The rate of soil CO_2 emission is normally controlled by several factors, such as CO_2 concentration gradient between the soil and the atmosphere, soil temperature, soil moisture, pore size, and wind speed.[15] In addition, soil CO_2 emission is affected by agricultural practices such as tillage and residue management and varies with climatic conditions.[16–19] The measurement of soil CO_2 emission could provide a more sensitive indication of soil C sequestration than low-resolution data such as total or organic C values.[20,21]

Tillage accelerates soil CO_2 emission by improving soil aeration, increasing soil and crop residue contact, enhancing plant nutrient availability,[22,23] and increasing exposure of SOC in inter- and intra-aggregate zones to microbes for rapid oxidation.[24–26] The magnitude of CO_2 loss from the soil due to tillage practices is highly related to the frequency and intensity of soil disturbance caused by tillage. Although tillage effects on soil CO_2 emission are complex and often varied,[27,28] conservation tillage is regarded as one of the most effective agricultural practices for reducing soil CO_2 emission to the atmosphere from agricultural soils.[24,29,30]

Conservation tillage systems such as no-tillage, strip-tillage, and chisel plow are increasingly being used for crop production in the Midwest during the past decade due to their profitability and environmental advantages over moldboard plow. For example, no-tillage production in the Midwest was used in over 22% of all cropland area in 2002,[31] which almost doubled the amount in 1992. In contrast, conventional tillage systems accounted for 35% of all croplands in the Midwest. Deep rip is an effective and popular tool used to overcome soil compaction. Although deep rip is not a conservation tillage system, it still results in less soil disturbance and mixing and thus greater crop residue coverage on the soil surface than moldboard plow. There are few studies that quantify the effects of these main tillage alternatives with different intensities on soil CO_2 emission and C storage compared with more intensive tillage systems (i.e., moldboard plow) in the Midwest where a corn [*Zea mays* L.]–soybean [*Glycine max* (L.) Merr.] rotation has been the primary cropping system for decades. Even though moldboard plow use has been limited recently in the Corn Belt region, the inclusion of it in this study is to show the most extreme intensive tillage system effect on soil C dynamics as we evaluate a suite of tillage systems differing in their intensities in soil disturbance at different depths.

The objectives of this study are to evaluate (1) the short-term response of SOC pools to different tillage systems, (2) immediate and short-term effects of a suite of tillage systems with different intensities in soil disturbance on soil CO_2 emission, and (3) the influences of tillage systems on mineralizable C pools.

14.3 METHODS

The study was conducted on Clarion–Nicolet–Webster soil association that includes Canisteo (fine-loamy, mixed, calcareous, mesic Typic Haplaquolls) and Clarion (fine-loamy, mixed, mesic Typic Hapludolls) soil series. The tillage treatments included no-tillage, strip-tillage, deep rip, chisel plow, and moldboard plow, and that were established in a corn–soybean rotation in the fall of 1997. The study consisted of three replications with a randomized complete block design. Typically, no-tillage was defined as no preplant tillage. The crop in no-tillage was planted using a planter with a single coulter to cut through residues and loosen soil ahead of standard planter units. Tillage description, tillage depth, and width of disturbed soil zone due to tillage operations are presented in Table 14.1. The strip-tillage plots were tilled with a strip-tillage unit that consists of an anhydrous knife centered between two cover disks and a coulter for residue cleaning. The tilled zone was prepared in the proximity of previous season corn or soybean rows creating a tilled zone 10 cm high.

The chisel plow treatment was implemented using a commercially available model with straight shanks and twisted chisel plow sweeps at the bottom mounted on a tool bar. The shanks were mounted on four tool bars in a staggering order to ensure an effective spacing of 30 cm between shanks. The deep rip treatment was performed by using a commercially available deep ripper with four straight shanks

TABLE 14.1
Tillage System, Tillage Depth, and Width of Disturbed Soil Zone due to Tillage Operations

Tillage System	Shank or Disk Spacing (cm)	Tillage Depth (cm)	Disturbed Zone (cm)
Strip-tillage	76	20	20
Deep rip	76	40–46	30–40
Chisel plow	30	22–25	20
Moldboard plow	76	25	46

on a tool bar. The moldboard plow treatment resulted in a complete inversion of soil surface and nearly 100% incorporation of crop residue by using a commercially available model with four bottoms. In the spring prior to planting, all treatments except no-tillage and strip-tillage received one field-cultivation 10 cm deep. The field cultivator shovels were mounted on four tool bars in a staggering order to ensure an effective spacing between shovels of 30 cm. Tillage treatments were conducted during the fall immediately after harvest each season. The plot size for each treatment was 152 m wide × 272 m long.

14.3.1 Determination of Soil and Crop Residue Organic Carbon

Prior to the establishment of this study, an initial composite soil sample was taken across the site for soil texture, pH, and organic C analyses. In the fall of 2000 after corn harvest, soil sampling for TC, mineral fraction C (MFC), and particulate organic matter C (POMC) was conducted at soil depths of 0–5, 5–10, and 10–15 cm for each plot. Soil samples were collected from the tilled area in the strip-tillage treatment. Ten to twelve soil cores were randomly collected from each plot with a soil probe of 1.9 cm diameter after removing visible crop residue from the soil surface. Soil cores from the same depth in each plot were mixed and placed in a soil-sampling bag and stored in a cooler at 4°C. Soil samples were kept at workable wet conditions to pass through a 2 mm sieve and left to completely air dry afterward. TC equals the sum of MFC and POMC fractions. Soil POMC fractionation was conducted[32] to separate POMC associated with large, stable soil aggregates (>53 μm) from MFC associated with soil microaggregates (<53 μm or defined as silt + clay associated C fraction). Soil bulk density samples were taken at the same soil depth intervals as those used for SOC in the fall of 2000. Four samples per depth were taken to determine bulk density in each plot. Soil bulk density was determined using a core method with a copper cylinder of 5 cm in height and 5 cm in diameter similar to that used by Culley.[33] Bulk density was used to convert SOC concentrations (g kg^{-1}) to mass per soil volume (kg m^{-3}).

In the fall of 2001, a crop residue sample was collected from each plot after corn harvest before any tillage operations were performed for the determination of total C concentration in crop residue. A crop residue sample was taken by using a 1 m^2 frame thrown randomly on each plot three times to collect the above-ground crop residue. Residue samples were oven-dried at 64°C, cleaned from soil before weighing, and ground using a plant grinder with a 2 mm sieve (Wiley Mill, Model 2 pulverized carbon steel, Arthur H. Thomas Co., Philadelphia, Pennsylvania).

Soil organic C, MFC, POMC, and crop residue C concentrations were determined by dry combustion with a LECO CHN 2000 analyzer (LECO, St. Joseph, Michigan). Soil pH was measured using a 1:1 (soil:water) extraction. Prior to dry combustion, soil samples with pH greater than 7.1 were treated with 1 M HCl to eliminate any inorganic carbonate. On the other hand, TC was assumed to be equal to the soil total C if soil pH was not greater than 7.1.

Total C input from crop residue was estimated for the entire study period by including both corn and soybean seasons. Corn or soybean residue C input was estimated for each season separately by using crop grain yields that were measured each year for both corn and soybean. The total C input of each season was calculated as the quotient of grain yields by harvest index, then multiplied by the total C concentrations of crop residue (corn or soybean), respectively. The harvest index used in this computation was 0.59 for corn and 0.57 for soybean.[34] Total C concentrations of corn and soybean were determined by using the corn and soybean residue samples collected in 2001.

14.3.2 Soil Carbon Loss due to Tillage

The short-term, tillage-induced soil CO_2 emission was measured in the fall of 2001 with a LI-6400 CO_2 analyzer (LI-COR, Lincoln, Nebraska) immediately after tillage operations for each treatment. The LI-6400 CO_2 analyzer utilized a small chamber of PVC rings with 10 cm ID Immediately (1–2 min) after tillage operations in each tilled plot, five chambers were placed on the tilled soil surface and the measurements of CO_2 emission were taken along with soil moisture (with TRIME-TDR) and soil temperature. Soil temperature was measured by using a thermometer provided with the LI-6400 CO_2 analyzer. Soil moisture and temperature were measured in the top 5 cm. After completing the tilled soil CO_2 emission measurements, the same number of chambers was placed on the soil surface of the no-tillage plots, and CO_2 emission readings, along with soil moisture and soil temperature were taken. The no-tillage treatment was divided into two different treatments. One treatment was no-tillage with crop residue cover on the soil surface and the other one was no-tillage without corn residue cover on the soil surface, where the surface residues were completely removed by hand from inside the CO_2 measurement chambers. The CO_2 measurement chambers were installed on all plots immediately after tillage operations and kept in the same place for the entire duration of soil CO_2 measurements.

Measurements were taken using the same approach for all three replications. The tillage operations were conducted on November 7, 2001 and completed on the same day. The CO_2 emission measurements were repeated for the following time periods after tillage operations: 0, 2, 4, 8, 12, 24, 48, 96, 192, 288, and 480 h. The five chambers in each plot were kept at the same locations during the measurement period regardless of treatment.

Fertilizer rates were identical for all tillage treatments, but varied in placement methods. Prior to tillage operations in the fall, anhydrous ammonium was injected by using a mole knife with two cover disks in zones 76 cm apart at 135 kg N ha^{-1} for corn in all tillage treatments except no-tillage. For no-tillage treatment, anhydrous ammonium was injected in the fall by using a modified slot injector with minimum soil disturbance. No N fertilizer was applied for soybean regardless of treatment. Phosphorous and potassium were applied in the fall, as needed, according to soil test recommendations.

Weather data were collected from a nearby weather station on the same research farm. Daily measurements included solar radiation, air temperature, precipitation, and wind speed. There was no precipitation during the 20 day period of CO_2 emission measurements.

Strip-tillage in this study tilled 27% of the total soil surface area of each plot; the other 73% of soil surface area remained in no-tillage. There was minimum residue removal and disturbance in the no-tillage areas due to strip-tillage. Because CO_2 emission from the strip-tillage treatment was measured in the tilled zones only, these measurements were adjusted in order to reflect the true field conditions of the entire plot in the strip-tillage treatment. The following formula was used:

$$\begin{aligned}&\text{Adjusted soil } CO_2 \text{ emission}\\&\quad = 0.27 \times CO_2 \text{ emission from the tilled zones}\\&\qquad + 0.73 \times CO_2 \text{ emission from no-tillage treatment with residue}\end{aligned}$$

14.3.3 Quantifying Soil Carbon Loss

SOC loss due to tillage system can be quantified by using set of transformations and calculations based on the Michaelis–Menten[35] equation, where they reported the effect of substrate concentration on the velocity of enzyme-catalyzed reaction could be satisfactorily described in the following equation:

$$v = V_{max} \times \frac{S}{(K_m + S)} \tag{14.1}$$

where

v is the reaction velocity
V_{max} is the maximum reaction velocity
S is the substrate concentration
K_m is the Michaelis constant

Numerically, K_m is equal to the substrate concentration at half-maximum reaction velocity.

According to Equation 14.1, the cumulative mineralized S vs. distillation time reported in a previous study[36] that the Michaelis–Menten equation can also be used to describe the relationship between cumulative soil CO_2 emission and time.

Therefore, to calculate the amount of mineralizable C (i.e., the maximum cumulative soil CO_2 emission) due to different tillage systems, the Lineweaver–Burk transformation[37] of the Michaelis–Menten equation was used:

$$\frac{1}{c} = \frac{1}{C_{max}} + \frac{K_m}{C_{max}} \times \frac{1}{T} \tag{14.2}$$

where

c (kg CO_2 ha^{-1}) is cumulative soil CO_2 emission at a specific time after tillage operations
T is the time (h) after tillage operations
C_{max} (kg CO_2 ha^{-1}) is the potential maximum amount of cumulative soil CO_2 emission under a specific tillage system
K_m is the Michaelis constant, which equals to the time (h) at half-maximum cumulative soil CO_2 emission

Then, we took the inverse of cumulative soil CO_2 emission as the dependent variable and the inverse of time after tillage operations as the independent variable for each treatment for the entire 480h measurement periods according to the Lineweaver–Burk transformation of the Michaelis–Menten Equation 14.2 and used to plot the relationship between soil carbon loss (CO_2) and time for each tillage system (Figure 14.4).

The intercept of each line represents the total amount of soil carbon loss with each tillage system. To determine the actual amount of soil carbon loss for each tillage system, we took the inverse of each intercept value generated from Figure 14.4, which is shown in Table 14.5.

14.4 RESULTS AND DISCUSSION

14.4.1 TILLAGE DEPTH EFFECTS ON SOIL ORGANIC CARBON

A significant soil depth × tillage interaction effect ($p = 0.004$) on TC is observed (Table 14.2). At the 0–5 cm soil depth, TC is approximately 32% greater in no-tillage and strip-tillage treatments compared with chisel plow. Soil TC is 36%–41% greater with no-tillage and strip-tillage treatments compared with chisel plow at the 5–10 cm soil depth interval. However, no significant difference in TC at 10–15 cm is observed with no-tillage and strip-tillage compared with chisel plow. Our results generally suggest that reducing tillage intensity in a corn–soybean rotation can enhance TC at the 0–15 cm soil depth.

Similar to TC, the soil depth × tillage interaction effect ($p = 0.003$) on MFC is significant (Table 14.2). MFC at 0–5 cm is increased by 35%–42% for no-tillage and strip-tillage compared with other tillage systems. Soil MFC at 5–10 cm is approximately 25%–29% greater for no-tillage and strip-tillage, than chisel plow. At 0–15 cm, other tillage systems had similar MFC compared with chisel plow. No significant tillage or soil depth × tillage effect on POMC was observed (Table 14.2), although numerical increases were frequently detected under no-tillage, strip-tillage, and deep rip treatments relative to chisel plow.

TABLE 14.2
Tillage Effects on SOC in the Soil Profile after 3 Year Management in a Corn–Soybean Rotation

	TC				MFC				POMC			
	Soil Depth (cm)				Soil Depth (cm)				Soil Depth (cm)			
Tillage System	0–5 (Mg ha^{-1})	5–10 (Mg ha^{-1})	10–15 (Mg ha^{-1})	0–15 (Mg ha^{-1})	0–5 (Mg ha^{-1})	5–10 (Mg ha^{-1})	10–15 (Mg ha^{-1})	0–15 (Mg ha^{-1})	0–5 (Mg ha^{-1})	5–10 (Mg ha^{-1})	10–15 (Mg ha^{-1})	0–15 (Mg ha^{-1})
No-tillage	17.1a	16.9a	18.6a	52.6a	12.0a	13.3a	14.7a	40.0a	5.0a	3.6a	3.9a	12.5a
Strip-tillage	16.9a	17.6a	18.5a	53.0a	12.2a	13.7a	14.9a	40.8a	4.7a	3.9a	3.7a	12.2a
Deep rip	12.3b	13.0b	19.4a	44.7ab	8.7b	10.2b	13.7ab	32.6b	3.9a	2.8a	5.7a	12.1a
Chisel plow	12.9b	12.5b	18.3a	43.6ab	8.9b	10.7b	14.7a	34.3ab	3.9a	1.8a	3.6a	9.3a
Moldboard plow	11.4b	14.2ab	14.3b	39.9b	8.6b	11.9ab	13.0b	32.5b	2.7a	2.3a	2.4a	7.4a

TC, total soil organic C; MFC, mineral fraction C; POMC, particulate organic matter C.
Values in column followed by the same letter are not significantly different at $p < 0.05$.

Monreal and Janzen[38] reported that although SOC changes in response to management practices could be relatively rapid, it still took about 10 years to obtain stable management effects. By the time our soil C measurements were made, it had been only 3 years after no-tillage was initiated. Therefore, TC, MFC, and POMC levels likely have not reached a steady state in no-tillage, and the impact of no-tillage on increasing TC, MFC, and POMC would be greater with time.

14.4.2 Tillage Effects on Total Carbon Input from Crop Residue

Total C input from above-ground crop residue is significantly affected by tillage ($p = 0.02$) (Table 14.3). Total C input in 3 years was 10% lower with strip-tillage than chisel plow. All other tillage systems have similar total C input as chisel plow. Therefore, TC and MFC increases at 0–10 cm with no-tillage and strip-tillage over chisel plow (Table 14.2) could not be attributed to the total C input from above-ground crop residue in such a short period. Rather, it might relate to decreased mineralization of soil organic matter (SOM) due to less soil disturbance and cooler soil conditions in no-tillage and strip-tillage.

14.4.3 Tillage Effect on Soil Carbon Loss

Significant treatment effects on soil CO_2 emission are observed at almost all measuring times although CO_2 emission varied tremendously with time regardless of treatment (Figure 14.1). At the zeroth-hour measurement time, no-tillage with and without residue, strip-tillage, deep rip, and chisel plow treatments reduced CO_2 emission by 79%, 79%, 60%, 50%, and 14%, respectively, compared with moldboard plow immediately after tillage operations. Although all CO_2 emission reductions with less intensive tillage alternatives are statistically significant, the

TABLE 14.3
Tillage Effects on C Input from Above-Ground Corn and Soybean Residue during the First 3 Years in a Corn–Soybean Rotation

	C Input[a]		
Tillage System	Corn (Mg ha^{-1})	Soybean (Mg ha^{-1})	Total (Mg ha^{-1})
No-tillage	5.47a	2.76a	8.23ab
Strip-tillage	5.02a	2.44a	7.46b
Deep rip	5.86a	2.84a	8.70a
Chisel plow	5.65a	2.64a	8.29a
Moldboard plow	5.86a	2.90a	8.76a

Values in column followed by the same letter are not significantly different at $p < 0.05$.
[a] Carbon input under columns of corn, soybean, and total refers to C input from 1-year corn, 2-year soybean, and all 3 years, respectively.

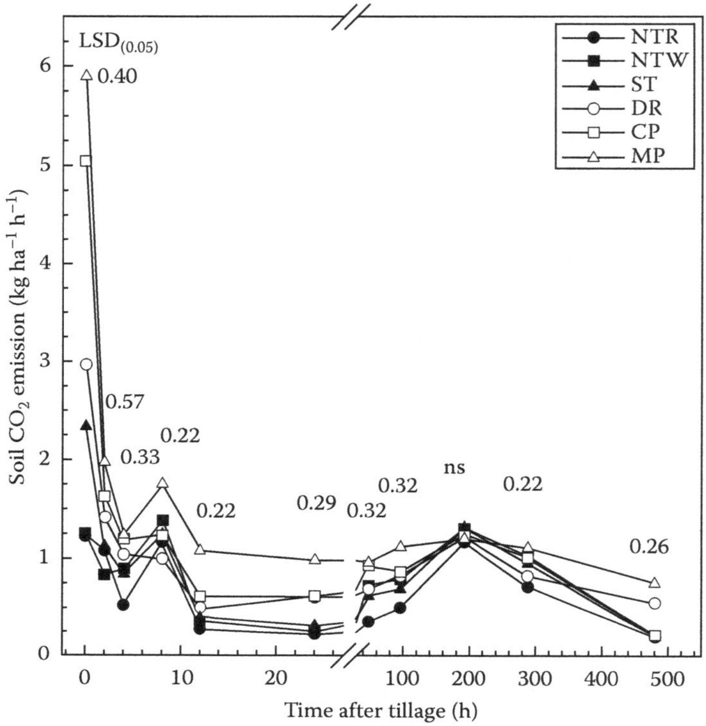

FIGURE 14.1 Tillage and crop residue effects on soil CO_2 emission after 3 years in a corn–soybean rotation. Least significant difference (LSD) at $p < 0.05$ for each measuring time is provided in the graph; ns, not significant at $p < 0.05$; NTR, no-tillage with residue; NTW, no-tillage without residue; ST, strip-tillage; DR, deep rip; CP, chisel plow; MP, moldboard plow.

greatest reductions in CO_2 emission are associated with those tillage systems having less soil disturbance, such as the two no-tillage treatments. Removal of crop residue from the soil surface under no-tillage did not alter CO_2 emission significantly compared with no-tillage with residue at the measurement time of zeroth hour.

At the second-hour measurement time, CO_2 emission from no-tillage with and without residue and strip-tillage are 43%–58% less than from moldboard plow (Figure 14.1), while chisel plow and deep rip treatments have statistically similar CO_2 emission as moldboard plow. The differences in CO_2 emission between the two no-tillage and moldboard plow treatments are significantly smaller at the second hour of measurement time than those at the zeroth hour according to a paired *t*-test.

Beyond second hour, following tillage operations, no-tillage with residue produces a significantly lower CO_2 emission than moldboard plow at all the measurement times except the 192nd hour (Figure 14.1). Compared with moldboard plow, CO_2 emission is less from no-tillage without residue at all the measurement times except at the 48th, 192nd, and 288th hour after tillage operations.

Similarly, strip-tillage produces lower CO_2 emission than moldboard plow at all measuring times except at the 192nd and 288th hour. Soil CO_2 emission is significantly less from deep rip than moldboard plow at the 8th-, 12th-, 24th-, and 288th-hour measurement times. Similarly, chisel plow has significantly lower CO_2 emission than moldboard plow at the 8th, 12th, 24th, and 480th hour after tillage operations.

In addition, no-tillage without residue results in greater CO_2 emission than no-tillage with residue at the measurement times of 4th, 48th, and 288th hour. Our results generally confirm the potential of reducing tillage intensity and increasing crop residue on the soil surface in reducing soil CO_2 emission to the atmosphere in a corn–soybean rotation.

The maximum CO_2 emission from all tilled treatments (strip-tillage, deep rip, chisel plow, and moldboard plow) is observed immediately after tillage operations (i.e., at the zeroth-hour measurement time) (Figure 14.1). However, CO_2 emission from these tilled treatments decreases sharply by 52%–68% within the first 2 h following tillage operations. In contrast, the two no-tillage treatments have only 12%–33% reduction during the same period (2 h after tillage operations). After the first 2 h, changes in CO_2 emission are much smaller regardless of treatment unless there is a sharp change in soil temperature or moisture. Reicosky et al.[6] reported that CO_2 emission decreased rapidly by 80% immediately after tillage operations on a harvested wheat field, which is greater than the reductions we observed in our study.

A sharp increase in CO_2 emission immediately after tillage operations may be attributed to the rapid increase in microbial activities in decomposing the labile SOM pool. However, Jackson et al.[39] and Roberts and Chan[40] concluded that the increase in soil CO_2 emission immediately after tillage operation was not due to the increase in microbial activities, but it was rather due to the increase in soil aeration that was induced by tillage disturbance. Reicosky and Lindstrom[24] also attributed greater CO_2 emission immediately after tillage practices to greater physical CO_2 emission from soil pores and solution.

Periodic soil CO_2 emission differs significantly among the treatments during most of these measurement periods (Table 14.4). For example, both no-tillage with residue and strip-tillage result in significantly less CO_2 emission than moldboard plow during all measurement periods except the 96th- to 192nd-, and 192nd- to 288th-hour periods after tillage operations. Cumulative CO_2 emissions for the entire 20 day period following tillage are 41%, 26%, 21%, and 19% lower for no-tillage with residue, strip-tillage, deep rip, and chisel plow than with moldboard plow, respectively (Table 14.4). Cumulative CO_2 emission from no-tillage without residue is 23% lower than that with moldboard plow, but 24% greater than the CO_2 emission from no-tillage with residue over the 20 day period.

In summary, lower CO_2 emission from no-tillage with residue than moldboard plow in this study could be partially attributed to slower decomposition of crop residue placed on the soil surface in no-tillage than when they were incorporated with moldboard plow.[41] Meanwhile, tillage operations may physically facilitate gas emission from the soil pores due to soil disturbance.[42] Others reported that soil temperatures primarily governed seasonal variations in soil CO_2 emission, with high soil

TABLE 14.4
Tillage and Crop Residue Effects on Soil CO_2 Emission after 3 Years in a Corn–Soybean Rotation

	Time Period after Tillage Operations										
Tillage System	0–2 h (kg ha^{-1})	2–4 h (kg ha^{-1})	4–8 h (kg ha^{-1})	8–12 h (kg ha^{-1})	12–24 h (kg ha^{-1})	1–2 Days (kg ha^{-1})	2–4 Days (kg ha^{-1})	4–8 Days (kg ha^{-1})	8–12 Days (kg ha^{-1})	12–20 Days (kg ha^{-1})	Total (kg ha^{-1})
No-tillage with residue	2.31e	1.61c	3.40c	2.89d	3.00d	6.89d	20.62c	80.60a	90.86a	88.19c	300.36c
No-tillage without residue	2.09e	1.74bc	4.56b	3.49bc	3.73d	11.79bcd	36.63ab	101.12a	112.27a	118.94bc	396.36b
Strip-tillage	3.46d	1.97bc	4.16bc	3.28bcd	4.32cd	11.35cd	31.96bc	96.45a	108.89a	111.73bc	377.57bc
Deep rip	4.37c	2.46ab	4.08bc	2.99cd	6.68bc	15.73bc	36.50ab	98.28a	98.47a	132.66b	402.20b
Chisel plow	6.68b	2.82a	4.87b	3.70b	7.44b	18.63ab	43.06ab	101.51a	108.54a	118.61bc	415.83b
Moldboard plow	7.87a	3.20a	5.96a	5.66a	12.41a	23.43a	50.09a	111.94a	111.36a	179.43a	511.34a

Values in column followed by the same letter are not significantly different at $p < 0.05$.

CO_2 emission during the summer when soil moisture and substrate C were adequate and low CO_2 emission during the winter when soil biological activity was minimal due to near-freezing soil temperatures.[43,44] Therefore, given the timing of this study, soil CO_2 emission is most likely due to CO_2 exchange through soil pores rather than by microbial activity. These findings show the effects of tillage that is normally conducted in the fall on soil C loss in the Corn Belt region.

The results show trends similar to others.[45–47] Reicosky et al.[45] reported that cumulative soil CO_2 emission from conventional tillage at the end of 80 h was nearly three times larger than from no-tillage. Additionally, crop residue on the soil surface with no-tillage contributes to the reduction of soil CO_2 emission by serving as a barrier for CO_2 emission from soil to the atmosphere, having a lower crop residue decomposition rate due to minimum residue–soil contact, and lowering soil temperature.[45] Soil temperature at 5 cm during the 20 day measurement period ranged from −9°C to 6°C (Figure 14.2). Soil temperature changes from one measurement time to another are similar for all treatments. Soil temperature is below zero at the 24th-, 48th-, 96th-, 288th-, and 480th-hour measurement times regardless of treatment. Accordingly, CO_2 emission is also low at these measurement times regardless of treatment (Figure 14.1). This observation confirms that soil CO_2 emission is affected by soil temperature. Furthermore, tillage and crop residue effects on soil temperature are significant at 4 of 11 measurement times during the 20 day period (Figure 14.2). Our soil temperature results generally agree with those by Reicosky et al.,[6] who observed that the relatively small temperature differences among tillage treatments probably had less influence on soil CO_2 emission than the differences in tillage-induced soil disruption.

Compared with soil temperature, fluctuations in soil moisture are much lower during the 20 day measurement period (Figure 14.2). In general, the two no-tillage treatments result in significantly higher soil moisture than moldboard plow (data not presented). Chisel plow and strip-tillage treatments had similar moisture content during the first 12 h after tillage operations, but significantly higher moisture when compared with moldboard plow. Soil moisture is similar or higher under deep rip relative to moldboard plow. A significant difference in soil moisture is observed between the two no-tillage treatments only at the 192nd- and 288th-hour measurement times, when no-tillage with residue has greater soil moisture. Overall, soil moisture effects on CO_2 emission seem to be minor during such a short period of time.

14.4.4 Soil Organic Carbon Pool Loss

No significant linear or quadratic relationship between cumulative soil CO_2 emission and TC, MFC, or POMC is observed over the 480 h measurement period regardless of treatment (data not presented). This finding indicates that SOC substrate is not the limiting factor to soil CO_2 emission. Rather, soil CO_2 emission in such a short-term experiment may be governed by soil structural pore changes due to tillage and microbial community population and its activity. This is reasonable because our measurements were taken for only a 3-week period from middle November to early December, when farmers normally conducted the tillage operations. In addition,

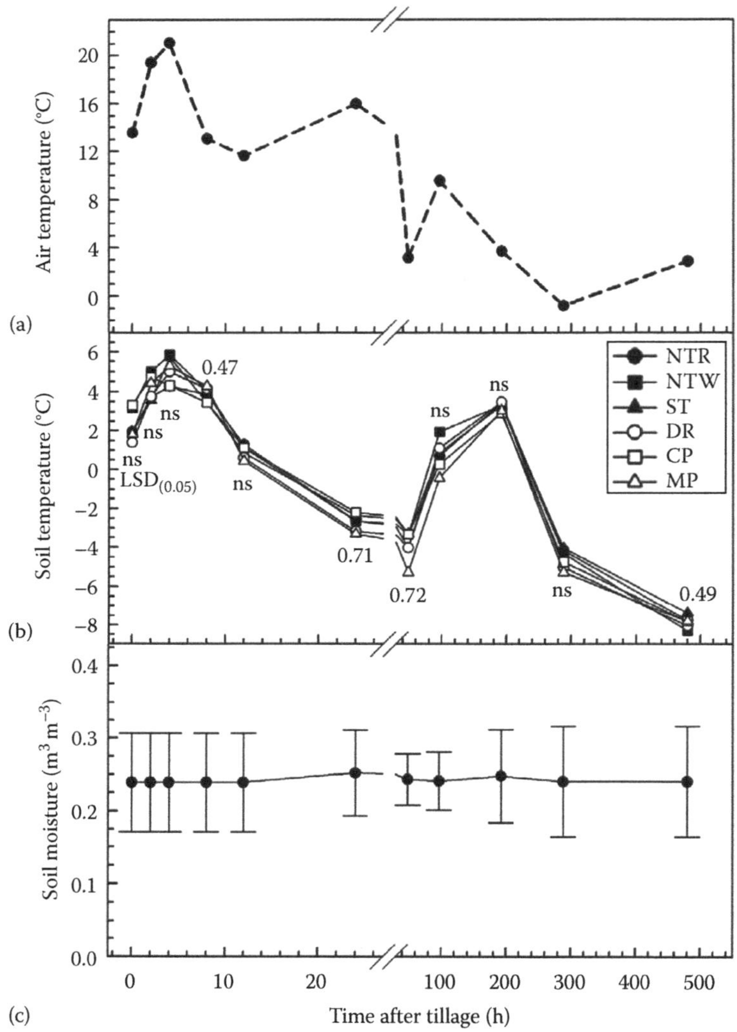

FIGURE 14.2 (a) Air temperature, (b) soil temperature, and (c) soil moisture in the top 5 cm of different tillage systems and residue covers from 7 to 26 November, 2001, after 3 yr in a corn–soybean rotation. For soil temperature, least significant difference (LSD) at $p < 0.05$ for each measurement time is provided; NS, not significant at $p < 0.05$. For soil moisture, an error bar for each measurement time is provided. NTR, no-tillage with residue; NTW, no-tillage without residue; ST, strip-tillage; DR, deep rip; CP, chisel plow; MP, moldboard plow.

low soil temperatures and stable soil moisture during the measurement period may cause a decrease in microbial activity, thus reducing the significance of relationship between cumulative soil CO_2 emission and TC, MFC, or POMC. Overall, our results suggest that short-term measurement of CO_2 emission during the dormant season can be considered as an indicator of soil physical changes due to tillage disturbance and microbial activity rather than SOC pool levels.

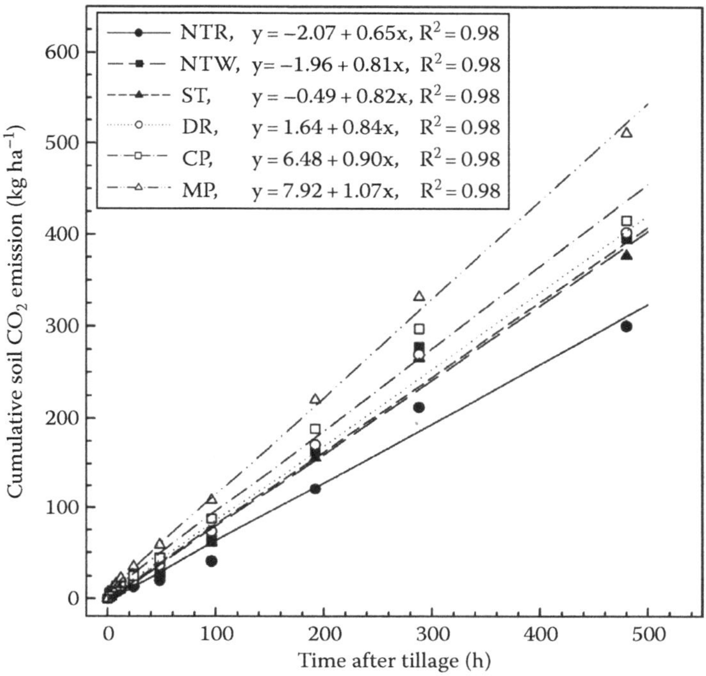

FIGURE 14.3 Tillage and crop residue effects on cumulative soil CO_2 emission after 3 years in a corn–soybean rotation. NTR, no-tillage with residue; NTW, no-tillage without residue; ST, strip-tillage; DR, deep rip; CP, chisel plow; MP, moldboard plow.

Cumulative soil CO_2 emission and time after tillage operations is linearly related regardless of treatment (Figure 14.3). Slope of the equation decreases as tillage intensity is reduced. Moldboard plow has the greatest slope and no-tillage with residue has the lowest slope. This finding suggests that cumulative soil CO_2 emission to the atmosphere can be lowered by adopting less intensive tillage systems compared with moldboard plow.

14.4.5 Tillage Impact on Mineralizable C

The Lineweaver–Burk transformation of the Michaelis–Menten equation applied to cumulative soil CO_2 emission vs. time after tillage operations is shown in Figure 14.4 for each tillage system. There is a linear relationship between the inverse of cumulative soil CO_2 emission and the inverse of time after tillage operations regardless of tillage system. Inverse of the intercept of each linear relationship represents the size of potentially mineralizable C pool (C_{max}) due to the effect of each tillage system. The C_{max} value is lower with less intensive tillage systems compared with moldboard plow (Table 14.5). No-tillage with and without residue, strip-tillage, deep rip, and chisel plow reduced the size of mineralizable portion of the maximum C pool (C_{max}) by 66%, 40%, 51%, 28%, and 22% relative to moldboard plow, respectively. This trend suggests adopting less intensive tillage systems and maximizing residue coverage on the soil surface can reduce the amount of C mineralized due to tillage.

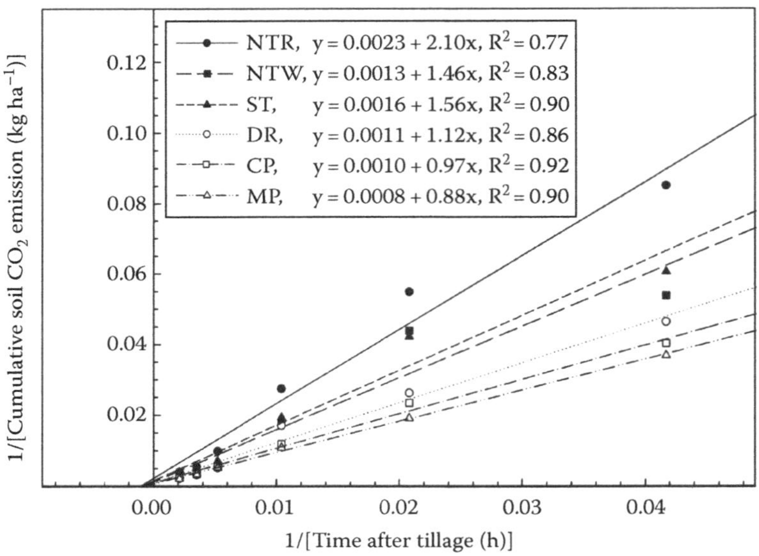

FIGURE 14.4 Linear plot of the Michaelis–Menten equation for cumulative soil CO_2 emission. NTR, no-tillage with residue; NTW, no-tillage without residue; ST, strip-tillage; DR, deep rip; CP, chisel plow; MP, moldboard plow.

TABLE 14.5
Maximum Mineralizable Soil C under Different Tillage Systems Estimated by the Lineweaver–Burk Transformation of the Michaelis–Menten Equation

Tillage System	C_{max} (kg ha^{-1})
No-tillage with residue	435
No-tillage without residue	775
Strip-tillage	629
Deep rip	935
Chisel plow	1010
Moldboard plow	1294

14.5 CONCLUSIONS

Reducing the intensity of tillage operations could increase soil C storage in a corn–soybean rotation from a short-term perspective. No-tillage with residue and strip-tillage significantly increase TC and MFC at the 0–5 and 5–10 cm soil depths compared with chisel plow after 3 years of tillage practices. This short-term tillage effect is not attributed to the annual C input from above-ground crop residue but most likely is related to decreased mineralization rate of SOM with no-tillage.

Soil CO_2 emission is generally lower with less intensive tillage alternatives relative to moldboard plow, with the greatest differences occurring at the time immediately following tillage operations. Over the 480 h measurement period, less intensive tillage alternatives lower cumulative soil CO_2 emission by 19%–41% compared with moldboard plow. Carbon dioxide emission is 24% less with no-tillage with residue than without residue covers.

Relationship of soil CO_2 emission with TC, MFC, POMC, soil temperature, or moisture content is not observed or was weak over the short period of measurement. However, a positive linear relationship between cumulative CO_2 emission and time is observed. Estimated mineralizable C pool is reduced by 22%–66% with less intensive tillage alternatives relative to moldboard plow. The decrease in mineralizable C pool may be partially responsible for the reduced soil CO_2 emission especially from the no-tillage treatments.

The results suggest that from a short-term perspective, adopting less intensive tillage alternatives, such as no-tillage and strip-tillage, and leaving more crop residue cover on the soil surface are effective in reducing soil CO_2 emission, and thus improving soil C sequestration in a corn–soybean rotation in Corn Belt soils. The benefit of less soil CO_2 emission or C loss along with other economic and environmental advantages, such as higher production profitability and less soil erosion, associated with less intensive tillage and better crop residue management systems should be taken into account when soil management decisions are made for conservation planning.

14.6 CALCULATING MINERALIZABLE C AND THE MICHAELIS–MENTEN CONSTANT

1. Enter the following hypothetical data in Microsoft Excel.

	A	B	C	D	E	F	G	H	I	J
1	Time (h)	1/t	CO_2–C rate	CO_2–C rate		Accumulative CO_2–C			Accumulative CO_2–C	
2			Soil 1 (kg ha^{-1} h^{-1})	Soil 2 (kg ha^{-1} h^{-1})		Soil 1 (kg ha^{-1})	Soil 2 (kg ha^{-1})		Soil 1 (1 kg^{-1} ha^{-1})	Soil 2 (1 kg^{-1} ha^{-1})
3										
4	2	0.5	2.31	7.87		2.31	7.87		0.4329	0.127065
5	4	0.25	1.61	3.2		3.92	11.07		0.255102	0.090334
6	8	0.125	3	5.96		6.92	17.03		0.144509	0.05872
7	12	0.083333	2.89	5.66		9.81	22.69		0.101937	0.044072
8	24	0.041667	3	12.41		12.81	35.1		0.078064	0.02849
9	48	0.020833	6.89	23.43		19.7	58.53		0.050761	0.017085
10	96	0.010417	20.62	50.09		40.32	108.62		0.024802	0.009206

2. Conduct the regression analysis for column I vs. B and J vs. B, select Tools, Data analysis, Regression, and I4:I10 in the Y range and B4:B10 in the X range. Click OK.

   The regression equations are

   Soil 1: $y = 0.0354 + 0.81X$

   Soil 2: $y = 0.0189 + 0.235X$
3. Calculate C_{max} and K_m
   a. C_{max} is 1/intercept, and therefore for Soil 1 is $1/(0.0354) = 28.24\,\text{kg}\ CO_2\ \text{ha}^{-1}$

   And for Soil 2 C_{max} is $1/0.0189 = 52.91\,\text{kg}\ CO_2\ \text{ha}^{-1}$

   b. $K_m = \text{slope} \times C_{max}$
      i. Soil 1 K_m is $0.81 \times 28.24 = 22.9\,\text{h}$
      ii. Soil 2 K_m is $0.235 \times 52.91 = 12.43\,\text{h}$
4. These values mean Soil 1 has a much lower mineralization potential and the midpoint of reaching the potential is much slower in Soil 1 than Soil 2. Findings from the study above report that tillage impacts the soils mineralization. Similar measurements can be collected across landscapes. By understanding both tillage and landscape impacts on carbon mineralization, recommendations that improve management can be developed.
5. Once the rate constants are known, they can be integrated into geographic information systems (GIS) program that predicts carbon turnover.

ACKNOWLEDGMENTS

The materials are based on published work supported by the Department of Agronomy Endowment Fund at Iowa State University. The author would like to acknowledge Dr. Xinhua Yin, University of Tennessee, and Mark Licht, Iowa State University Extension, for extending their assistance.

REFERENCES

1. Bauer, A. and Black, A.L. Quantification of the effect of soil organic matter content on soil productivity. *Soil Sci. Soc. Am. J.* 58, 185, 1994.
2. Doran, J.W. and Parkin, T.B. Defining and assessing soil quality. In J.W. Doran et al. (eds). *Defining Soil Quality for a Sustainable Environment.* SSSA Special Publication No. 35. SSSA, Madison, WI, 1994, pp. 3–21.
3. Robinson, C.A., Cruse, R.M., and Ghaffarzadeh, M. Cropping system and nitrogen effect on Mollisol organic carbon. *Soil Sci. Soc. Am. J.* 60, 264, 1996.
4. Sainju, U.M. and Good, R.E. Vertical root distribution in relation to soil properties in New Jersey Pinelands forest. *Plant Soil.* 150, 87–97, 1993.
5. Sainju, U.M. and Kalisz, P.J. Characteristics of "coal bloom" horizons in undisturbed forest soils in eastern Kentucky. *Soil Sci. Soc. Am. J.* 54, 879, 1990.
6. Reicosky, D.C., Dugas, W.A., and Torbert, H.A. Tillage-induced soil carbon dioxide loss from different cropping systems. *Soil Till. Res.* 41, 105, 1997.
7. Lal, R., Fausey, N.R., and Eckert, D.J. Land use and soil management effects on emissions of relatively active gases from two Ohio soils. In R. Lal et al. (eds.). *Soils and Global Climate Change. Advances in Soil Science.* CRC Press, Boca Raton, FL, 1995, pp. 41–59.

8. Paustian, K., Robertson, G.P., and Elliott, E.T. Management impacts on carbon storage and gas fluxes (CO_2, CH_4) in mid-latitudes cropland. In R. Lal et al. (eds.). *Soils and Global Climate Change. Advances in Soil Science*. CRC Press, Boca Raton, FL, 1995, pp. 69–83.
9. Havlin, J.L., Kissel, D.E., Maddux, L.D., Claasen, M.M., and Long, J.H. Crop rotation and tillage effects on soil organic carbon and nitrogen. *Soil Sci. Soc. Am. J.* 54, 448, 1990.
10. Franzluebbers, A.J., Hons, F.M., and Zuberer, D.A. Tillage and crop effects on seasonal soil carbon and nitrogen dynamics. *Soil Sci. Soc. Am. J.* 59, 1618, 1995.
11. Halvorson, A.D., Wienhold, B.J., and Black, A.L. Tillage, nitrogen, and cropping system effects on soil carbon sequestration. *Soil Sci. Soc. Am. J.* 66, 906–912, 2002.
12. Franzluebbers, A.J. and Arshad, M.A. Soil organic matter pools during early adoption of conservation tillage in northwestern Canada. *Soil Sci. Soc. Am. J.* 60, 1422, 1996.
13. Duiker, S.W. and Lal, R. Crop residue and tillage effects on carbon sequestration in a Luvisol in central Ohio. *Soil Till. Res.* 52, 73, 1999.
14. Lundegardh, H. Carbon dioxide evolution and crop growth. *Soil Sci.* 23, 417, 1927.
15. Raich, J.W. and Schlensinger, W.H. The global carbon dioxide flux in soil respiration and its relationship to vegetation and climate. *Tellus* 44B, 81, 1992.
16. Fernandez, I.J., Son, Y., Kraske, C.R., Rustad, L.E., and David, M.B. Soil carbon dioxide characteristics under different forest types and after harvest. *Soil Sci. Soc. Am. J.* 57, 1115, 1993.
17. Burton, D.L. and Beauchamp, E.G. Profile nitrous oxide and carbon dioxide concentrations in a soil subject to freezing. *Soil Sci. Soc. Am. J.* 58, 115, 1994.
18. Osozawa, S. and Hasegawa, S. Daily and seasonal changes in soil carbon dioxide concentration and flux in an Andisol. *Soil Sci.* 160, 117, 1995.
19. Yavitt, J.B., Fahey, T.J., and Simmons, J.A. Methane and carbon dioxide dynamics in a northern hardwood ecosystem. *Soil Sci. Soc. Am. J.* 59, 796, 1995.
20. Fortin, M.C., Rochette, P., and Pattey, E. Soil carbon dioxide fluxes from conventional and no-tillage small-grain cropping systems. *Soil Sci. Soc. Am. J.* 60, 1541, 1996.
21. Grant, R.F. Changes in soil organic matter under different tillage and rotation: Mathematical modeling in ecosystem. *Soil Sci. Soc. Am. J.* 61, 1159, 1997.
22. Logan, T.J., Lal, R., and Dick, W.A. Tillage systems and soil properties in North America. *Soil Till. Res.* 20, 241–270, 1991.
23. Angers, D.A., N'dayegamiye, A., and Cote, D. Tillage induced difference in organic matter of particle-size fractions and microbial biomass. *Soil Sci. Soc. Am. J.* 57, 512, 1993.
24. Reicosky, D.C. and Lindstrom, M.J. Fall tillage method: Effect on short-term carbon dioxide flux from soil. *Agron. J.* 85, 1237, 1993.
25. Beare, M.H., Cabrera, M.L., Hendrix, P.F., and Coleman, D.C. Water-stable aggregates and organic matter fractions in conventional and no-tillage soils. *Soil Sci. Soc. Am. J.* 58, 777, 1994.
26. Jastrow, J.D., Boutton, T.W., and Miller, R.M. Carbon dynamics of aggregate-associated organic matter estimated by carbon-13 natural abundance. *Soil Sci. Soc. Am. J.* 60, 801, 1996.
27. Mosier, A.R., Schimel, D., Valentine, D., Bronson, K., and Parton, W. Methane and nitrous oxide fluxes in native, fertilized, and cultivated grasslands. *Nature (London)*. 350, 330, 1991.
28. Lauren, J. and Duxbury, J.M. Methane emissions from flooded rice emended with a green manure. In E. Rolston et al. (eds.), *Agricultural Ecosystems Effects on Trace Gases and Global Climate Change*. ASA Special Publication No. 55. ASA, CSSA, and SSSA, Madison, WI, 1993, pp. 183–192.
29. Kern, J.S. and Johnson, M.G. Conservation tillage impacts on national soil and atmospheric carbon levels. *Soil Sci. Soc. Am. J.* 57, 200, 1993.

30. Lal, R. and Kimble, J.M. Conservation tillage for carbon sequestration. *Nutr. Cycl. Agroecosyst.* 49, 243, 1997.
31. Conservation Technology Information Center. *2002 National Crop Residue Management Survey*. Conservation Technology Information Center, West Lafayette, IN, 2003. http://www.ctic.purdue.edu/CTIC/CRM.html, May 28, 2004.
32. Cambardella, C.A. and Elliott, E.T. Particulate soil organic matter changes across a grassland cultivation sequence. *Soil Sci. Soc. Am. J.* 56, 777, 1992.
33. Culley, J.L.B. Density and compressibility. In M.R. Carter (ed.). *Soil Sampling and Methods of Analysis*. CRC Press, Boca Raton, FL, 1993, pp. 529–539.
34. Licht, M.A. Comparative influence of tillage systems and nutrient timing on the soil environment and crop response in Iowa soils. MS thesis, Library of Iowa State University, Ames, IA, 2003.
35. Michaelis, L. and Menten, M.L. Die kinetik der Invertinwirtung. *Biochem. Z.* 49, 333, 1913.
36. Pirela, H.J. and Tabatabai, M.A. Reduction of organic sulfur in soils with tin and phosphoric acid. *Soil Sci. Soc. Am. J.* 52, 959, 1988.
37. Tabatabai, M.A. Soil enzymes. In D.L. Sparks et al. (eds.). *Methods of Soil Analysis, Part 2. Microbiological and Biochemical Properties*. SSSA Book Series, No. 35. SSSA, Madison, WI, 1994, pp. 775–833.
38. Monreal, C.M. and Janzen, H.H. Soil organic-carbon dynamics after 80 years of cropping a Dark Brown Chernozem. *Can. J. Soil Sci.* 73, 133, 1993.
39. Jackson, L.E., Calderon, F.J., Steenwerth, K.L., Scow, K.M., and Rolston, D.E. Responses of soil microbial processes and community structure to tillage events and implications for soil quality. *Geoderma* 114, 305, 2003.
40. Roberts, W.P. and Chan, K.Y. Tillage-induced increases in carbon dioxide loss from soil. *Soil Till. Res.* 17, 143, 1990.
41. Curtin, D., Wang, H., Selles, F., McConkey, B.G., and Campbell, C.A. Tillage effects on carbon fluxes in continuous wheat and fallow-wheat rotations. *Soil Sci. Soc. Am. J.* 64, 2080, 2000.
42. Ellert, B.H. and Janzen, H.H. Short-term influence of tillage on CO_2 fluxes from a semi-arid soil on the Canadian Prairies. *Soil Till. Res.* 50, 21, 1999.
43. Kirschbaum, M.U.F. The temperature dependence of soil organic matter decomposition, and the effect of global warming on soil organic C storage. *Soil Biol. Biochem.* 27, 753, 1995.
44. Follett, R.F. CRP and microbial biomass dynamics in temperate climates. In R. Lal et al. (eds.). *Management of Carbon Sequestration in Soil*. CRC Press, Boca Raton, FL, 1997, pp. 305–322.
45. Reicosky, D.C., Reeves, D.W., Prior, S.A., Runion, G.B., Rogers, H.H., and Raper, R.L. Effects of residue management and controlled traffic on carbon dioxide and water loss. *Soil Till. Res.* 52, 153, 1999.
46. Dao, T.H. Tillage and crop residue effects on carbon dioxide evaluation and carbon storage in a Paleustoll. *Soil Sci. Soc. Am. J.* 62, 250, 1998.
47. Jacinthe, P.A., Lal, R., and Kimble, J.M. Carbon budget and seasonal carbon dioxide emission from a central Ohio Luvisol as influenced by wheat residue amendment. *Soil Till. Res.* 67, 147, 2002.

15 Geospatial Management of Andean Technology by the Inca Empire

F. Mamani Pati, David E. Clay, and H. Smeltekop

CONTENTS

15.1 EXECUTIVE SUMMARY

For the past 10,000 years, agriculture has provided much of the food needed by urban and rural communities. The ability of a community to manage climate variability and soil sustainability has influenced its long-term success. The inability to manage climate variability of the adoption of non-sustainable practices can result in societal collapse. Precision farming is one tool that can increase energy efficiency as well as maintain productivity and the soil resource. The concepts of precision farming and efficiency are not new and have been used to manage variability and resource allocations since antiquity. One of the most successful peoples for managing variability were the Incas in South America. Over 500 years ago, the Incas, indigenous peoples of the Andes region in the South America, developed agricultural management practices for one of the world's most extremely variable climates. These peoples conducted agriculture activities from sea level to over 4400 m. Extreme climatic variability in rainfall and temperatures were observed across these elevations. The Incas relied on grain storage capacity and research, most likely conducted at Moray agricultural station (located near present-day Cuzco, Peru), to help them manage climatic uncertainty. Many of the activities they implemented could be viewed as precision farming. For example, they used production zones and the development of site-specific management techniques designed to protect crops from low temperatures. They also fertilized their fields with bird guano harvested from islands near the coast of Peru. We believe that time-tested lessons learned by the Incas as well as other peoples from antiquity may help us develop low-technology-based solutions that will improve energy efficiency and manage climate and soil variability today. The purpose of this chapter is to review these successes and discuss the consequences of adopting non-sustainable practices.

15.2 HISTORICAL PERSPECTIVES OF GEOSPATIAL NUTRIENT MANAGEMENT

A perception exists that precision agriculture is a new invention and that it requires the use of global position systems (GPS), geographic information systems (GIS), and computer programs. This perception is not correct. Throughout history, ancient cultures have used precision farming concepts (site-specific management and on-farm research) as tools to increase productivity. However, when non-sustainable agricultural practices were adopted, societal collapse often followed. There are numerous examples where this occurred.[1,2] One of the causes of agriculture failure is climatic variability[3–6] that reduces productivity, which in turn increases agricultural intensification and the loss of soil resources through saltation and erosion.[7–9]

Different peoples developed different solutions to similar questions. Peoples within the Andean environment developed solutions to problems associated with farming in an environment of extreme climatic variability. In this extremely hostile environment, one approach to increase food production was the use of vertical management zones. The zones were created by moving up and down the topographic relief. This technique matches crops to ecological niches. In addition, within a niche, many different types of plants were cultivated. It is believed that these people had over 70 different domesticated crops to choose from. The advantage of this approach was that food diversity was increased and food security against unpredictable weather was increased. We believe that rediscovering knowledge learned by ancient peoples can help us develop low-technology management systems that will increase energy efficiency, food production, and security issues today. This chapter explores some of the approaches the Andean peoples improved food security in one of the harshest environments in the world.

15.2.1 Inca Agricultural System

The Andes mountain region of South America is an incredibly difficult place to produce a sustainable food supply. Peoples in this region adopted and continue to adopt many of the concepts and practices relied on in precision agriculture including site-specific management techniques and on-farm field research. Two basic management zone approaches were used to overcome climatic variability. The first approach was vertical zones and the second approach was climatic modification within a zone. Peoples in the Andes are rediscovering techniques developed by the region's indigenous peoples.[10,11]

FIGURE 15.1 Agricultural research at Moray.

In the Inca Empire, agricultural activities were based on research conducted at both the local and empire scales. At the community and individual scales, research was conducted near their homes or in specific production zones where production fields were located. Observations were made on the behavior of different crops under different climatic conditions, the resistance of crops to frost, and crop response to excessive or deficient rainfall, extreme hot and cold climate, and exposure to pests including diseases and insects. Based on these observations, landscapes were separated into production zones.[12]

At the empire scale, many people believe that the Incas conducted agricultural research at Moray (Peru).[12] Moray has an infrastructure of concentric circular terraces (Figure 15.1). Each terrace produces a microclimate where different treatments can be tested. Modern-day analysis of this system suggests that this terrace system can produce up to 20 different types of microclimates. Moray is believed to have served as a model for calculating the Empire's agricultural production capacity, domesticating new plants, and testing the environment tolerances of domesticated plants. Information learned at the state scale was transferred to local communities by experts.

15.2.2 Andean Foods Produced

Producing a reliable food supply in an environment of extreme climatic and topographic variability requires an attention to detail. The use of sustainable management practices was partially responsible for the success of the Incas.[13,14] It is estimated that the Inca cultivated about 70 crop species. This large number of crops was needed to provide a sustainable food supply in an area with extreme climatic variability. Some of the crops were more resistant to climatic stresses than others. Major crops domesticated by the Incas included potatoes (*Solanum* spp.), sweet potatoes (*Ipomoea batatas*), corn (*Zea mays*), chili peppers (*Capsicum* spp.), zapallo (*Cucubita maxima*), coca (*Erythroxylon coca*), peanuts (*Arachis hypogaea*), tarwi (*Lupinus mutabilis*), edible roots called oca (*Axalis tuberosa*), maca (*Lepidium meyenii*), Achira (*Canna adulis*), Arracacha (*Arracacia xanthorrhiza*), isano (*Tropaeolum tuberusum*), ulluku (*Ullucus tuberosus*), beans, and grains known as quinoa (*Chenopodium quinoa*), canihua (*C. pallidicauli*), and amaranth (*Amaranthus caudatus*).[5,13,15–17] The most important crops to the Incas were potato, corn, and quinoa. Chili peppers are warm-season plants sensitive to freezing temperatures, while quinoa is drought tolerant, resists brutal weather, and can be grown at high altitudes.[18] By domesticating crops with different responses to climatic variability, crops can be matched to the likely environmental conditions. Using this approach, food security can be maintained even under variable conditions. In addition to producing crops, the Inca also raised llamas and alpacas for their wool and meat, as well as, helping in transportation and travel. The manure from these animals was used to fertilizer crop areas. Wild vicunas were captured, and their fine hair was used for clothing and blankets.

15.2.3 Managing Erosion and Increasing Soil Temperatures: Terraces

In mountainous Andean environments, terraces, also known as andenes (Figure 15.2), were constructed by the people at great expense. The terraces, which are still in existence today, are flat-stepped areas that resist erosion and landslides, maintain soil fertility, save water, and create microclimates where crops are grown.[14] Terraces consist of three parts, the retaining wall and two distinct layers of earth that fill the space behind the wall (Figure 15.2). The underlying stratum, an artificial subsoil, is composed of coarse stones and clay to a thickness that depends upon the height of the retaining wall; this stratum was covered by a layer of rich soil up to 1 m deep (Figure 15.3). Often the top soil was carried up from rich river bottoms to provide areas with better soil texture and nutrient content than soils present in the area. The terrace platform serves as a raised bed for crop cultivation and improves cultivating conditions through enhanced infiltration, recycling of soil nutrients, and biological processes necessary for agronomic production. Solar radiation is captured and retained in the soil as a result of the enhanced water

FIGURE 15.2 Terraces or Andenes built by the Incas at Charazani, La Paz. (Courtesy of Bolivian Catholic University, UAC Carmen Pampa, Pa Paz, Bolivia.)

FIGURE 15.3 Terraces are still be constructed today by Bolivian farmers. (Photo courtesy of Public University of El Alto (UPEA), La Paz, Bolivia.)

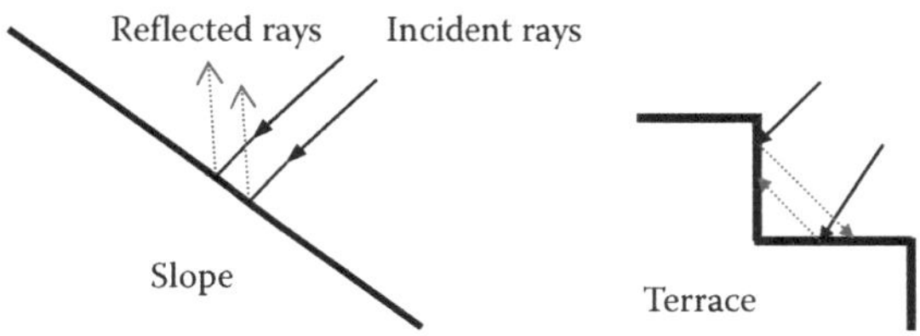

FIGURE 15.4 Use of terraces to provide shelter to crops, support their development, and protect from frost.

levels, which protects the terrace soil from the effects of frost.[19] An added benefit of the terraces is that they reduce heat loss by storing more heat from the solar radiation than a flat surface or slope because part of the reflected rays reflect again on the terrace wall that then acts like thermoregulator of the microclimate within the platform (Figure 15.4). The terrace length ranges from 4 to 100 m and width from 2 to 20 m; it is dependent on slope. Exceptionally, terraces can be found 1 × 1.5 m on steep slopes and 30 × 150 m or more in places almost flat. Other practices the Incas implemented included an effective gravity-based irrigation system, the use of digging stick to minimize soil disturbance, and constructing effective drainage systems.

15.2.4 Protecting Crops from Frost Damage at High Elevation: Camellones

The Incas used camellones (ridges) or suka kollus (waru waru in Quechua and sukka qolla in Aymara) in flat high terrains.[10,20–27] Camellones is a system where water is used to create microclimate that reduces the risk of frost. In a camellon, rainwater is channeled around areas that are used to grow crops. The water has a dual purpose. First, the water tempers temperature changes, making it possible to grow crops at high elevations. Second, the water provides irrigation water for the crops (Figure 15.5). Other benefits of the camellones include reducing erosion and increasing relative humidity and solar energy storage.

FIGURE 15.5 Waru Waru or suka kollus was first developed in the year 300 BC.

15.2.5 Nutrient Management

The Incas used manure from humans and domesticated llamas and alpacas and fallen tree leaves to fertilize their crops. Soils near coastal regions were fertilized with bird guano as a source of P collected from islands off the coast of Peru that are inhabited by thousands of fish-eating birds. The Incas valued this resource so highly that the punishment for anyone caught killing a single guano-producing bird was death.

In the upland valleys and basins, human or llama manure was applied as a source of plant nutrient and also served as a soil amendment. Manure was applied to the maize field at the time of sowing.[15] On the high mountain slope fields, however, most llama dung was used as a fuel. Fields were often rested after 1 or 2 years of cultivation to sustain the soil resource.[15] Today, Bolivian farmers have been improving their efficiency by combining historical techniques with modern knowledge.[28–30] For example, in the suka kollus area, people are improving their ability to produce food by using the recurrent floods to reestablish camellones and terraces (http://www.sas.upenn.edu/~cerickso/applied3.html).

15.3 CONSEQUENCES OF NOT ADOPTING SUSTAINABLE PRACTICES

In the past, climatic variability and the adoption of non-sustainable practices has influenced the ability of civilizations to produce a reliable food supply. Changes in temperature and rainfall patterns can have profound impacts on a civilization's ability to produce food. The inability to provide a sustainable food supply, resulting from climate change, has resulted in the collapse of many civilizations.

Worldwide, between AD 900 and AD 1300, climatic conditions were highly variable. In Europe, the period is known as the Medieval Warm Period. During this time, the Norse founded colonies in Iceland and Southern Greenland, while the Anasazi colonized the Colorado Plateau in North America, the Wari Empire located in the South American Andes failed, while the Inca Empire expanded.[31–34] During the Medieval Warm Period, peoples of the South America Andes built terraces, planted trees in the mountains to reduce erosion, and irrigated crops with water from melting glaciers.[14] These activities help the Incas' expand their food production capacity and their ability to support a larger population.

The Medieval Warm Period was also a time where prolong droughts occurred. For example, a drought that occurred around AD 1100 lowered the water level of Lake Titicaca by 12–17 m. This drought may partially be responsible for the collapse of the South America Wari Empire. Following the collapse of the Wari, the Incas expanded. The Incas success most likely was related to their ability to maximize agricultural productivity (precision farming) by adapting to regional environments. To assist them maximize agricultural productivity, they likely invested in agricultural research. The outgrowth of their science-based activities was improved agricultural practices, a redesigned landscape, and food surpluses that contributed to population increases.[33] We believe that in response to climate variability, the peoples of the Andes made a critical agricultural research investment that helped them produce a sustainable food supply.

Civilizations in other parts of the world were less successful at surviving climate change. For example, droughts in Central America and North America may have resulted in the collapse of the Maya civilization and the Anasazi,[31–38] while the Norse colonies in Southern Greenland did not survive the Little Ice Age that occurred between AD 1400 and AD 1850.

Climate change and the need to increase agricultural energy efficiency will undoubtedly be the most devastating environmental problems of the next century. Lessons from history tell us investments into agriculture research are critical for providing a sustainable food supply and reducing the potential for societal collapse. Many people believe that over 700 years ago, the Incas made this investment. Other important lessons for overcoming climate variability are diversity and adaptability. As in the past, precision farming is a tool that can help increase energy efficiency and food security.

15.4 SUMMARY

Even though ancient peoples did not have GIS or GPS, they used precision farming techniques to help produce a reliable food supply. In the Andes, the Incan people most likely invested resources in agricultural research. In addition, they engineered and built highly effective systems that produced microclimates required for food security. In many areas of the Andes, agricultural knowledge of the indigenous peoples is being rediscovered. Adoption of these techniques may provide low-technology solutions to maintain food security and improve resistance to the many inherent regional climate problems. Critical historical lessons for preparing for an uncertain future are maintaining investments into agricultural research, encouraging producer adaptability and creativity, and the maintenance of crop diversity.

ACKNOWLEDGMENTS

Funding provided by NC SARE and the South Dakota State Experiment Station.

REFERENCES

1. Erickson, C.L. Neo-environmental determinism and agrarian collapse in Andean prehistory. *Antiquity* 73, 634, 1999.
2. Willians, P.R. Rethinking disaster-induced collapse in the Demise of the Andean Highland States: Wari and Tiwanaku. *World Archaeology* 33, 361, 2002.
3. Binford, M.W., Kolata, A.L., Janusek, J.W., Seddon, M.T., Abboti, M., and Curtis, J.H. Climate variation and the rise and fall of an Andean Civilization. *Quaternary Research* 47, 235, 1997.
4. Kolata, A. and Ortloff, C. Agroecological perspectives on the decline of the Tiwanaku State. In Kolata, A. (ed.). *Tiwanaku and its Hinterland: Archaeology and Paleoecology of the Andean Civilization*. Smithsonian Institution Press, Washington, DC, pp. 181–202, 1996.
5. Brush, S.B. The natural and human environment of the central Andes. *Mountain Research and Development* 2, 19, 1982.
6. Abbott, M.B., Wolfe, B.B., Wolfe, A.P., Seltzer, G.O., Aravena, R., Mark, B.G., Polissar, P.J., Rodbell, D.T., Rowe, H.D., and Vuille, M. Holocene paleohydrology and glacial history of the central Andes using multiproxy lake sediment studies. *Palaeogeography, Palaeoclimatology, Palaeoecology* 194, 123, 2003.
7. Dollfus, O. Development of land-use patterns in the Central Andes. *Mountain Research and Development* 2, 39, 1982.
8. Branch, N.P., Kemp, R.A., Silva, B., Meddens, F.M., Williams, A., Kendall, A., and Pomacanchari, C.V. Testing the sustainability and sensitivity to climatic change of terrace agricultural systems in the Peruvian Andes: A pilot study. *Journal of Archaeological Science* 34, 1, 2007.
9. Erickson, C.L. Intensification, political economy, and the farming community; in defense of a bottom-up perspective of the past. In Marcus, J. and Stanish, C. (eds.). *Agricultural Strategy*. Cotsen Institute, Los Angeles, CA, pp. 233–265, 2006.
10. Erickson, C.L. Prehistoric landscape management in the Andean Highlands: Raised field agriculture and its environmental impact. *Population and Environment: A Journal at Interdisciplinary Studies* 13, 285, 1992.
11. Erickson, C.L. and Candler, K.L. Raised fields and sustainable agriculture in the Lake Titicaca Basin of Peru. In Browder, J.O. (ed.). *Fragile Lands in Latin America: Strategies for Sustainable Development*. West view Press, Boulder, CO, pp. 230–248, 1989.
12. Earls, J. The character of Inca and Andean agriculture. Essay available at http://macareo.pucp.edu.pe/~jearls/documentosPDF/theCharacter.PDF, 1998.
13. Morris, A. The agricultural base of the pre-Incan Andean civilization. *Geographical Journal* 65, 286, 1999.
14. Chepstow-Lusty, A.J., Frogley, M.R., Bauer, B.S., Leng, M.J., Boessenkool, K.P., Carcaillet, C., Ali, A.A., and Gioda, A. Putting the rise of the Inca Empire within a climatic and land management context, *Climate of the Past* 5, 375, 2009.
15. Kelly, K. Land-use regions in the central and northern portions of the Inca Empire. *Annals of the Association of American Geographers* 55, 327, 1965.
16. Flores, H.E., Walker, T.S., Guimarães, R.L., Bais, H.P., and Vivanco, J.M. Andean root and tuber crops: Underground rainbows. *Hortscience* 38, 161, 2003.
17. Ochoa, C. *The Potatoes of South America: Bolivia*. D. Ugent (trans.). Cambridge University Press, Cambridge, U.K., 1991.
18. PROPINPA (Promoción y Investigación de los Productos Andinos). Study on the social, environmental and economic impacts of quinoa promotion in Bolivia. La Paz, 80 p., 2004.

19. UNEP-International Environmental Technology Centre United Nations Environment Programme. *Source Book of Alternative Technologies for Freshwater Augmentation in Latin America and the Caribbean.* Unit of Sustainable Development and Environment General Secretariat, Organization of American States Washington, DC. Available at http://www.oas.org/dsd/publications/Unit/oea59e/begin.htm#Contents, 1997.
20. Erickson, C.L. El valor actual de los Camellones de cultivo precolombinos: Experiencias del Perú y Bolivia, Antropologo. Department of Anthropology, University of Pennsylvania, Philadelphia, PA, 2006.
21. Erickson, C.L. Raised field agriculture in the Lake Titicaca basin: Putting ancient agriculture back to work. *Expedition* 30, 8, 1988.
22. Erickson, C.L. Applied archaeology and rural development: Archaeology's potential contribution to the future. *Journal of the Steward Anthropological Society* 20, 1, 1992.
23. Smith, C.T., Denevan, W.M., and Hamilton, P. Ancient ridged fields in the region of Lake Titicaca. *The Geographical Journal* 134, 353, 1968.
24. Lhomme, J.P. and Vacher, J.J. Frost mitigation in the raised fields of the Andean Altiplano. Institut français d'Études andines, Av. Arequipa 4595, Miraflores Lima, Perú, 32, 377, 2002.
25. Graffam, G. Raised fields without bureaucracy: An archaeological examination of intensive wetland cultivation in the Pampa Koani Zone, Lake Titicaca, Bolivia, PhD dissertation, University of Toronto, Toronto, 1990.
26. Biesboer, D.D. Nitrogen fixation in soils and canals of rehabilitated raised-fields of the Bolivian Altiplano, *Biotropica* 31, 255, 1999.
27. Carney, H.J., Binford, M.W., Kolata, A.L., Marin, R.R., and Goldman, C.R. Nutrient and sediment retention in Andean raised-field agriculture. *Nature* 364,131, 1993.
28. Bandy, M.S. Energetic efficiency and political expediency in Titicaca Basin raised field agriculture. *Journal of Anthropological Archaeology* 24, 271, 2005.
29. Sanchez de Lozada, D., Baveye, P., and Riha, S. Heat and moisture dynamics in raised fields of the Lake Titicaca region (Bolivia). *Agricultural and Forest Meteorology* 92, 251, 1998.
30. Sanchez de Lozada, D., Baveye, P., Lucey, R.F., Mamani, R., and Fernandez, W. Potential limitations for potato yields in raised soil field systems near Lake Titicaca. *Scientia Agricola* 63, 444, 2006.
31. Chepstow-Lusty, J., Bennett, K.D., Fjeldsa, J., Kendall, A., Galiano, W., and Herrera, A.T. Tracing 4,000 years of environmental history in the Cuzco Area, Peru, from the pollen record. *Mountain Research and Development* 18, 159, 1998.
32. Ortloff, C.R. and Kolata, A.L. Climate and collapse: Agro-ecological perspectives on the Decline of the Tiwanaku state. *Journal of Archaeological Sciences* 20, 195, 1993.
33. Zaro, G., Builth, H., Rivera, C., Roldan, J., and Suvires, G. Landscape evolution and human agency: Archaeological case studies from drylands in western South America and Australia. *Revista de Antropología Chilena* 40, 261, 2008.
34. Larson, D.O. and Michaelsen, J. Impacts of climatic variability and population growth on Virgin Branch Anasazi cultural developments. *American Antiquity* 55, 227, 1990.
35. Benson, L., Petersen, K., and Steins, J. Anasazi (Pre-Columbian Native American) migrations during the middle 12th and late 13th centuries—Where they drought induced? *Climatic Change* 83, 187, 2007.
36. Hodell, D.A., Curtis, J.H., and Brenner, M. Possible role of climate in the collapse of the Classic Maya civilization. *Nature* 375, 391, 1995.
37. Haug, G.H., Gunther, D., Peterson, L.C., Sigman, D.M., Hughen, K.A., and Aeschlimann, B. Climate and the collapse of Maya civilization. *Science* 299, 1731, 2003.
38. Peterson, L.C. and Haug, G.H. Climate and the collapse of Maya civilization: A series of multi-year droughts helped to doom an ancient culture. *American Scientist* 93, 322, 2005.

16 Calculating Energy Efficiency of Applying Fresh and Composted Manure to Soil

R.J. Wiederholt, Shafiqur Rahman, and A. Ehni

CONTENTS

16.1 EXECUTIVE SUMMARY

Composted livestock manure is a popular approach for reducing the amount of material required for land spreading, providing weed seed and pathogen control, developing a product that is easy to spread uniformly, and developing a product that can be sold separately. The treatment system adopted to handle manure, either composted or applied fresh, impacts the energy and nutrient efficiency of the system. Little information is available that outlines differences in time and energy required to treat manure via composting vs. handling and land applying it without treatment. This chapter provides a review of the composting process and an energy budget of the composting process. Data from this study was obtained from a 2 year case study that was conducted in Central North Dakota. In this pilot study, producers were paid an incentive to adopt composting. Data was collected on various parameters, including volume reduction of the manure and time spent making the compost. Based on data collected from the case studies, the energy requirement for the fresh manure relative to the composted manure was 1.56:1.00. This analysis does not consider N that may have been lost during the composting process.

16.2 INTRODUCTION

All livestock operations need an efficient system for managing manure. The types of systems that are routinely used include handling fresh, stockpiling and applying at a later date, or composting. Even though each operation may use different practices, there are similarities in manure handling components. Each system may include collection, treatment, storage, and utilization. However, not all components are in every system nor is there a specific order.[1] Composting livestock manure is one treatment component that is attractive to livestock producers wishing to increase their efficiency of manure handling. Composting is a very effective method of decreasing manure volume.[2] The decrease in volume can reduce energy spent on transportation costs of handling and land applying manure. Additionally, composting manure can open up alternative markets for manure.[3,4] There are other benefits of composting manure including elimination of pathogens,[5] human parasites,[6] decreased weed seed viability,[7] and odor.[8] The benefits of composting do not come without a cost. Time, money, equipment, and labor requirements need to be met during the composting process.[9] This chapter will use information from a case study of a mechanically turned windrow composting project conducted in Wells County, North Dakota (ND), to determine the energy efficiency of composting manure vs. handling fresh manure.

As with all manure management systems, a portion of the nutrients contained within the manure can be lost. In the Great Plains, several of the easiest approaches to increase energy gain is to move from an open to a closed nutrient management system. In open systems, nutrients are removed from the system in the harvested grain, while in closed systems, harvested nutrients are returned to the soil in the manure. The total amount of fertilizer can be substantially reduced by the application of manure. During the composting process, P and K are generally conserved, while a portion of the N can be lost. Hao et al.[10] reported that 41.4% and 11.8% of the total N in manure can be lost during composting straw and wood chip-bedded (WBM) manure, respectively. Others have reported similar losses for non-composted manure. Munox et al.[11] reported that in Wisconsin, 36% of the ^{15}N contained in ^{15}N-labeled manure was either volatilized or denitrified within 20 h of being disked. Sharpe et al.[12] reported that when swine manure was applied through irrigation, about 12% of the ammonium was lost during application and an additional 23% was volatilized from the soil surface within 48 h of application. Bittman et al.[13] reported that ammonia volatilization in the first 2 weeks after application ranged from 36% to 61% for broadcast manure and 17%–32% for surface-banded manure.

N lost through volatilization can reduce the energy efficiency of the system. For straw-bedded manure, the loss of 8.3 kg/Mg (43 MJ/kg N) reported by Hao et al.[10] was equivalent to 356 MJ of energy.[14]

16.2.1 Composting Process

Composting is a natural decomposition process conducted under controlled conditions relying on microorganisms found in soils.[15] Composting takes raw organic products containing nitrogen (N), carbon (C), and water and degrades them into

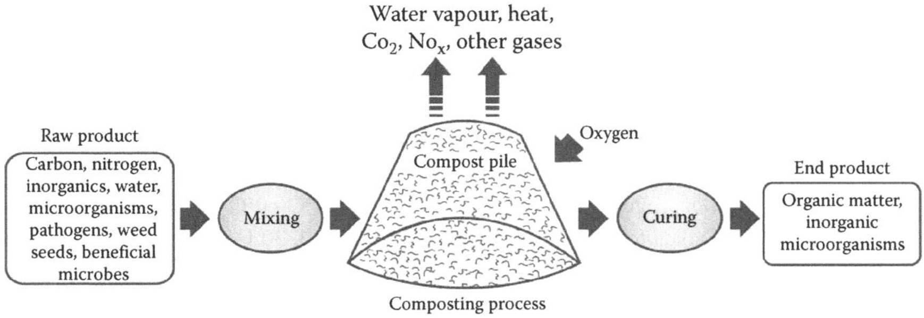

FIGURE 16.1 Material flow for the conventional composting process. (Adapted from Manure Composting Manual, Agdex 400/27-1, Alberta Agriculture and Rural Development, Alberta, Canada, 2005. With permission of Alberta Agriculture and Rural Development. With permission.)

stable organic matter by incorporating oxygen (O) through a mixing procedure. During the active phase of composting, microbial activity raises the temperature of the windrow to 50°C–60°C (Figure 16.1). These temperatures are held for several weeks while degradable compounds are being consumed. To keep the compost in the active phase, turning the windrow incorporates oxygen allowing the bacteria to thrive. Once the easily degradable compounds are degraded, turning the windrow will no longer increase compost temperature and the active phase is complete. Following the active phase is the curing phase. During the curing phase, temperatures are around 40°C and will become similar to ambient conditions when the curing phase ends. Besides C, N, O, and water, there are other factors that must be acceptable for composting to be successful such as ambient air temperature, carbon-to-nitrogen ratio, manure pH. Depending on the ambient environmental conditions, the composting process can take from 20 to 60 days.

There are five methods that can be utilized to make compost. Bin, passive windrow, active windrow, aerated static windrow, or in-vessel.[16] When composting livestock manure, all of these methods will work but bin composting is probably the least desirable because the volume of manure composted is so great. In-vessel composting is very effective but the most capital intensive since a specialized composting vessel must be purchased and maintained. Windrow composting is the most popular method. During active windrow composting, windrows are made and then turned with a loader bucket on a tractor or a mechanical turner. Of the windrow methods, active is the most efficient since you are turning the manure consistently, thereby speeding up the process. Passive windrow composting consists of making a windrow and allowing the manure to compost naturally without any mechanical turning. Passive windrow composting is the slowest method but costs very little in time spent managing the process. Aerated static windrow is the least popular windrow method because aeration equipment needs to be installed into the windrow and removed before the finished compost is utilized. Active windrow composting was utilized in this case study. In this case study, a mechanical turner that was capable of building and turning windrows was used to simplify the process (Table 16.1) (Figure 16.2).

TABLE 16.1
Factors Affecting the Composting Process and Acceptable Ranges

Factor	Acceptable Range
Temperature	54°C–60°C
Carbon-to-nitrogen ratio (C:N)	25:1–30:1
Aeration, percent oxygen	>5%
Moisture content	50%–60%
Porosity	30–36
pH	6.5–7.5

FIGURE 16.2 Tractor-mounted compost windrow turner. (Photo courtesy of Ron Wiederholt, NDSU.)

16.3 MATERIALS AND METHODS

In 2007, the Wells County, ND, Soil Conservation District (SCD) received a Natural Resource Conservation Service Conservation Innovation Grant to help them enhance the adoption of livestock manure composting by livestock producers. The grant funds were used to pay an incentive to producers who participated. The incentive paid was calculated based on the tons of manure each participant provided to the project and was available for only the first 2 years of the project. The Wells County SCD provided a person, tractor, and compost turner at a set charge per ton of manure processed. The incentive was then used to offset this charge.

In 2007, three livestock producers enrolled in the program and in 2008, 10 producers enrolled in the program. Information was collected and included time spent windrowing and turning the manure, temperatures of the manure during the process, and volume reduction during the composting process. This allowed for documentation of the effectiveness of composting. All of the participants in the project raised beef cattle in open air feedlots and used straw bedding in their livestock pens. The inclusion of the bedding in the manure pack added sufficient C for a favorable C:N ratio of the manure that was composted. The manure was windrowed for composting

FIGURE 16.3 Manure windrow turned with mechanical compost turner. (Photo courtesy of Brown Bear Corp., Corning, IA.)

in spring or early summer each year. The finished compost product was analyzed for nutrients and land applied in the fall.

Following established guidelines, the windrowed manure was turned with a mechanical turner (Figure 16.3) immediately after windrow construction and each time the temperature of the compost crested (about 65°C) and dropped below 50°C. Under summer environmental conditions of ND, the increment between turning of the windrows was typically 10 days. Since the final use of the compost was as an agronomic application to crop fields, the windrows were only turned during the active phase. The active phase was completed after three turns of the windrows. The manure was at a sufficient moisture level that no added water was needed to facilitate the composting process.

At the end of each project year, a group review was conducted with the participants to gauge the producer's satisfaction. All three of the 2007 participants reenrolled in the 2008 project indicating their satisfaction with composting manure. At the end of the 2008 project, all of participants wanted to continue composting even without the incentive payments. Each participant had a specific set of reasons why they adopted composting, but there were several commonalities among all the participants. The significant reduction in volume and resultant decreased number of loads required to land apply the finished compost was unanimously favored among participants. There was also universal agreement as to the importance of weed seed and pathogen reduction in the manure as a result of composting. Anecdotally, several participants stated that herbicide cost savings due to weed seed reductions in the compost offset the cost of composting.

16.4 ENERGY REQUIREMENTS OF COMPOSTING VS. HANDLING FRESH MANURE

According to the data collected during the case study, there was a 65% reduction in volume of final product when the manure was composted. It was also determined that the total time required to turn the manure during the composting process was 6 min/head. The average herd size for each participant was 180 animals. The following information summarizes and goes through the step-by-step process of determining an energy balance between composting or handling fresh manure.

Step 1. Summarize Known Information

Average herd size	= 180 animals
Total pile turning during composting	= 3 times
Tractor power	= 100 hp
Time to turn pile and windrow preparation	= 6 min/head
Volume reduction	= 65%
Total bedding materials used	= 150 lb/head

Step 2. Identify Assumptions

Animal weight	= 1000 lb
Manure produced	= 60 lb/day-animal
Animals are on feeds	= 120 days
Density of bedding materials	= 2.5 lb/ft^3
Manure density	= 60 lb/ft^3
Final compost density	= 25.35 lb/ft^3
Assume transport energy	= 2.9 kcal/km kg[17]
Box spreader size	= 180 ft^3
Hauling energy requirement	= 2.9 kcal/km kg[17]
Average gasoline consumption (gal/jr) Q_{ave} for power take-off (PTO)	= $0.044 \times P_{pto}$
Labor energy requirement	= 2500 kcal/h

In the use of the bedding material, the bedding material is compressed to ½ of its original volume

The total volume of material is reduced by 65% during the composting process

Time required to windrow and turn compost = 6 min/head

Step 3. Determine the Amount of Manure Produced

a. Calculate the volume of manure produced

1. Volume of manure produced/animal

$$= \frac{60\,\text{lb manure}}{\text{animal day}} \cdot \frac{1\ \text{ft}^3}{60\ \text{lb}} = \frac{1\ \text{ft}^3}{\text{day animal}}$$

2. Total manure produced for 180 animals over 120 days

$$= \frac{1\,\text{ft}^3}{\text{day animal}} \times 120\ \text{days} \times 180\ \text{animals} = 21{,}600\ \text{ft}^3$$

b. Amount of bedding materials

1. Bedding materials used	= 150 lb/1000 lb for 120 days
	= 150/120 = 1.25 lb/1000 lb animal/day
2. Total bedding materials	= 1.25 (lb/animal-day) × 120 days × 180 animals
	= 27,000 lb
3. Volume of bedding	= 27,000 lb/2.5 (lb/ft^3) = 10,800 ft^3
4. Volume of bedding materials + manure	$= \text{manure volume} + \frac{1}{2}\text{ bedding volume}$
	$= 21{,}600\text{ ft}^3 + \frac{1}{2}10{,}800\text{ ft}^3$
5. Total material volume	= 27,000 ft^3

Step 4. Calculate the Number of Loads

a. Calculate the amount remaining after composting

$$= \text{Initial compost} \times \text{compost remaining after processing}$$
$$= 27{,}000\text{ ft}^3 \times 0.65 = 17{,}550\text{ ft}^3$$

b. Calculate the number of loads

$$= 17{,}550\text{ ft}^3 \times \frac{\text{load}}{180\text{ ft}^3} = 98\text{ loads}$$

c. Calculate the time required to windrow and turn compost

$$= \frac{6\text{ min}}{\text{animal}} \times 180\text{ animals} = 1080\text{ min}$$

Step 5. Calculate the Energy Use during Windrowing and Composting

a. Tractor fuel consumption per hour for PTO

$$= \frac{0.044\text{ gal}}{\text{h} \times \text{hp}} \times 100\text{ hp} = \frac{4.4\text{ gal}}{\text{h}}$$

b. Total tractor fuel consumption for turning

$$= \frac{4.4\text{ gal}}{\text{h}} \times \frac{1\text{ h}}{60\text{ min}} \times 1080\text{ min} = 79.2\text{ gal}$$

c. Convert gal to kcal

$$= 79.2 \text{ gal} \times \frac{11{,}404 \text{ kcal}}{\text{L}} \times \frac{3.78 \text{ L}}{\text{gal}} = 3{,}417{,}078 \text{ kcal}$$

Step 6. Calculate the Hauling Energy from Source/Load

a. Calculate the energy per mile load

$$= \frac{180 \text{ ft}^3}{\text{load}} \times \frac{25.35 \text{ lb}}{\text{ft}^3} \times \frac{1 \text{ kg}}{2.24 \text{ lb}} \times \frac{2.9 \text{ kcal}}{\text{kg km}} \times \frac{1.61 \text{ km}}{\text{mi}} = 9511 \text{ kcal/mi load}$$

b. Calculate the total hauling energy
$= 9511 \times 98$
$= 932{,}078 \text{ kcal}$

Step 7. Energy Associated with Labor

$$= \frac{2{,}500 \text{ kcal}}{\text{h}} \times \frac{1 \text{ h}}{60 \text{ min}} \times 1080 \text{ min} = 45{,}000 \text{ kcal}$$

Step 8. Calculate the Total Energy Required for Composting to Hauling

= (machinery energy + hauling energy + labor energy)
= (3,417,078 + 932,078 + 45,000) kcal
= 4,394,156 kcal
= 4.39 Mcal

Step 9. Calculate the Scraping Pens, Loading Spreader, and Hauling Energy for Fresh Manure

a. Amount of fresh manure

$$= \text{manure volume} + \frac{1}{2}\text{bedding volume}$$

$$= 21{,}600 \text{ ft}^3 + \frac{1}{2} 10{,}800 \text{ ft}^3 = 27{,}000 \text{ ft}^3$$

b. Calculate the Number of Loads

$$= 27{,}000 \text{ ft}^3 \times \frac{\text{load}}{180 \text{ ft}^3} = 150 \text{ loads}$$

c. Tractor fuel consumption per hour for scraping pens and loading spreader

$$= \frac{0.044\text{ gal}}{\text{h} \times \text{hp}} \times 100\text{ hp} = \frac{4.4\text{ gal}}{\text{h}}$$

d. Total tractor fuel consumption for scraping and loading

$$= \frac{4.4\text{ gal}}{\text{h}} \times \frac{1\text{ h}}{60\text{ min}} \times 1080\text{ min} = 79.2\text{ gal}$$

e. Convert gal to kcal

$$= 79.2\text{ gal} \times \frac{11{,}404\text{ kcal}}{\text{L}} \times \frac{3.78\text{ L}}{\text{gal}} = 3{,}417{,}078\text{ kcal}$$

f. Calculate the hauling energy for fresh manure

$$= \frac{180\text{ ft}^3}{\text{load}} \times \frac{60\text{ lb}}{\text{ft}^3} \times \frac{1\text{ kg}}{2.24\text{ lb}} \times \frac{2.9\text{ kcal}}{\text{kg km}} \times \frac{1.61\text{ km}}{\text{mi}} = 22{,}511\text{ kcal/mi load}$$

g. Calculate the total hauling energy

$$= \frac{22{,}511\text{ kcal}}{\text{mi load}} \times 150\text{ loads} = 3{,}376{,}650\text{ kcal/mi}$$

Step 10. Calculate the Labor Energy for Fresh Manure

a. Let us assume that manure handling time is equal to compost pile preparation and turning time, i.e., 1080 min. Then, labor energy

$$= \frac{2{,}500\text{ kcal}}{\text{h}} \times \frac{1\text{ h}}{60\text{ min}} \times 1080\text{ min} = 45{,}000\text{ kcal}$$

b. Total energy required for hauling fresh manure
= (machinery energy + manure hauling energy + labor energy)
= (3,417,078 + 3,376,650 + 45,000) kcal
= 6,838,728 kcal
= 6.84 Mcal

c. Fresh manure vs. compost energy requirement ratio
= 6.84/4.39 Mcal
= 1.56:1.00

16.5 SUMMARY AND CONCLUSIONS

This study suggests that for these producers, it took less total energy to windrow, turn, haul, and apply composted beef manure than it did to scrape, haul, and apply fresh beef manure. Even though there is some time and energy invested in making and turning windrows of manure during the composting process, there is still a net energy savings based on fewer loads of final product that need to be hauled and field applied. The results of this case study point to the decrease in volume of final product during the composting process as the main contributor to the energy savings.

However, when composted livestock manure is utilized as an agronomic soil fertility amendment, there is a decrease in the amount of readily plant available nitrogen (N). To overcome this nutrient supply deficiency, commercial N may be supplemented on acres that receive composted manure to meet the total annual crop N needs. Hao et al.[10] reported that 41.4% and 11.8% of the total N in manure can be lost during composting straw and WBM, respectively. These N losses are similar to the losses that occur following manure applications of fresh manure. These results suggest that (1) more research is needed to identify the best bedding material for energy efficiency and (2) determine which crop rotation systems respond best to composted manure. This case study does show that composting the manure can improve energy efficiency.

REFERENCES

1. Moore, J.A. and Hart, J.M. Manure management design strategies: How and why. *J. Dairy Sci.* 80, 2655–2658, 1997.
2. Larney, F.J., Olson, A.F., Carcamo, A.A., and Chang, C. Physical changes during active and passive composting of beef feedlot manure in winter and summer. *Bioresour. Technol.* 75, 139–148, 2003.
3. Rynk, R. Status of dairy manure composting in North America. *Compost Sci. Util.* 2, 20–26, 1994.
4. Sharpley, A.N., Meisinger, J.J., Breeuwsma, A., Sims, T., Daniel, T.C., and Schepers, J.S. Impacts of animal manure management on ground and surface water quality. In: Hatfield, J. (ed.). *Effective Management of Animal Waste as a Soil Resource*. Ann Arbor Press, Chelsea, MI, 1998, pp. 173–242.
5. Larney, F.J., Yanke, L.J., Miller, J.J., and McAllister, T.A. Fate of coliform bacteria in composted beef cattle feedlot manure. *J. Environ. Qual.* 32, 1508–1515, 2003.
6. Van Herk, F.H., Cockwill, C.L., Gusselle, N., Larney, F.J., Olson, M.E., and McAllister, T.A. Elimination of Giardia cysts and Cryptosporidium oocysts in beef feedlot manure compost. *Compost Sci. Util.* 12, 235–241, 2004.
7. Larney, F.J. and Blackshaw, R.E. Weed seed viability in composted beef cattle feedlot manure. *J. Environ. Qual.* 32, 1105–1113, 2003.
8. Sweeten, J.M. Composting manure and sludge. In: *National Poultry Waste Management Symposium*, Columbus, OH, April 18–19. Department of Poultry Science, Ohio State University, Columbus, OH, 1998, pp. 38–44.
9. Rynk, R., van de Kamp, M., Willson, G.B., Singley, M.E., Richard, T.L., and Kolega, J.J. On farm composting. *Northeast Regional Agricultural Engineering Service*, Ithaca, NY, 1992.
10. Hao, X., Chang, C., and Larney, F.J. Carbon, nitrogen balance, and greenhouse gas emissions during cattle feedlot manure composting. *J. Environ. Qual.* 33, 37–44, 2004.

11. Munox, G.R., Owell, J.M., and Kelling, K.K. Nitrogen budget and soil N dynamics after multiple applications of unlabeled or ^{15}Nitrogen-enriched dairy manure. *Soil Sci. Soc. Am. J.* 67, 817–825, 2003.
12. Sharpe, R.R., Schomberg, H.H., Harper, L.A., Endale, D.M., Jenkins, M.B., and Franzebbers, A.J. Ammonia volatilization from surface applied poultry litter under conservation tillage management practices. *J. Environ. Qual.* 33, 1183–1188, 2002.
13. Bittman, S., van Vliet, V.J.P., Kowalenko, C.G., McGinn, S., Hunt, D.E., and Bounaix, F. Surface-banding liquid manure over aeration slots: A new low disturbance method for reducing ammonia emissions and improving yield of perennial grasses. *Agron. J.* 97, 1304–1313, 2005.
14. Shapouri, H., Duffield, J.A., and Wang, M. The energy balance of corn ethanol: An update. *Agricultural Economic Report 813*, U.S. Department of Agriculture, Washington, DC, 2002, 14 p.
15. Cooperband, L. *The Art and Science of Composting: A Resource for Farmers and Compost Producers*. Center for Integrated Agricultural Systems. University of WI-Madison, Madison, WI, 2002. http://www.cias.wisc.edu/wp-content/uploads/2008/07/artofcompost.pdf
16. Manure Composting Manual. Alberta Agriculture, Food and Rural Development. http://www1.agric.gov.ab.ca/$department/deptdocs.nsf/all/agdex8875/$file/400_27-1.pdf?OpenElement, 2005.
17. Pimentel, D. *Handbook of Energy Utilization in Agriculture*. CRC Press, Inc., Boca Raton, FL, 1980.

17 Quantifying Greenhouse Gas (CO_2, CH_4, and N_2O) Fluxes from Soil in a Pasture

Nsalambi V. Nkongolo

CONTENTS

17.1 EXECUTIVE SUMMARY

Greenhouse gas (GHG) fluxes exhibit strong spatial and temporal variability across agricultural fields. However, because of limitations in the number of samples one can collect, measurements of fluxes across agricultural fields are often limited to a few points. The average value of point measurements is later used to calculate the total flux for the entire field. This approach may result in an over- or underestimation of the total flux. The objective of this study was to assess how geographic information systems (GIS) could improve the estimation of N_2O, CH_4, and CO_2 total field fluxes from soil in a pasture in central Missouri. We sampled for N_2O, CH_4, and CO_2 fluxes in a pasture, fitted variogram models to fluxes data, predicted fluxes at un-sampled locations by inverse distance weighting (IDW), produced fluxes maps, and classified them according to fluxes distribution zones. Thereafter, we first calculated a total flux for the entire field by a traditional approach (TA) consisting in multiplying field minimum and maximum flux values obtained from points sampling to the total area of the pasture. Second, we also computed a GIS-based total field flux as the sum of "total flux" for each flux distribution zone for each gas. Results showed that "TA" method underestimated (up to 1000%) the minimum and overestimated the maximum flux of N_2O, CH_4, and CO_2 for the entire pasture. Our approach provides an improved quantification of greenhouse flux. The approach can be extended to other soil and environmental parameters.

17.2 INTRODUCTION

Quantifying soil GHG emissions is an essential task for developing management practices that increase energy efficiency. In agricultural fields, GHG emissions contain a large amount of spatial and temporal variability.[1,2] In fact, differences in soil types, moisture, temperature, season, crop type, fertilization, and other agricultural practices apparently all play a part in emissions from soils. Estimates of GHG emissions are influenced by the data processing approach. Differences between the calculation approaches lead to uncertainty. Techniques for developing better estimates are needed.

GIS, global positioning system (GPS), and computer mapping and geostatistics are technologies that can be used to better understand systems that contain large amounts of spatial and temporal variability. Verhagen[3] generated fertilizer maps using simulation and identified three pattern types. Guy and Levine[4] created a GIS-based deterministic model to assess the potential impact of different land-use strategies for mitigating Ohio's carbon dioxide (CO_2) emissions and lowering its total CO_2 budget. Using ARC/INFO, Bausch et al.[5] mapped near-infrared index to show plant-deficient areas for two different days. Kristensen and Olesen[6] analyzed aerial photography to map soil moisture content in the root zone using kriging, co-kriging, and inverse distance algorithms. Their observations indicated that even when they included soil texture, there was no improvement in prediction accuracy among the algorithms employed. Anderson and Yang[7] conducted a study on site-specific farm management. They used ArcView to visualize and query spatial data, generate statistics for each management zone, and create charts. These researchers concluded that the integration of aerial photography, GPS, and GIS provided an effective way to collect, process, and analyze information. The objective of this study was to demonstrate a GIS-based approach for quantifying CO_2, CH_4, and N_2O fluxes from a pasture system. Two approaches of quantifying the flux for the entire pasture (minimum and maximum) for each gas were compared: the first is a TA and consisted in computing the flux for the entire pasture as the product of the pasture area multiplied by the flux value (minimum and maximum) obtained from points measurements. The second is a GIS-based approach and consisted in producing an interpolated map (from points measurements) portraying flux distribution zones across the pasture, classifying the flux map to determine the area covered by each zone, and computing the entire pasture flux or total flux (minimum and maximum) as the sum of individual total flux for each classified zone.

17.3 METHODS

17.3.1 Study Area

The experiment was conducted on a pasture at George Washington Carver farm at Lincoln University in Jefferson City, MO. The experimental field and sampling sites is shown in Figure 17.1. The latitude and longitude coordinates of the site are 38°31′45″N and 92°08′07″W, respectively. The study area is a 1.42 ha area that is dominated by Brome grass. Brome grass is a cool season grass that is very popular in the production of hay. The soil type of this site is Elk Silt Loam (*Ultic hapludalts*). In 2006, this area experienced a drought during the spring and summer months and only received an annual precipitation of 990 mm. The total rainfall from June through November was 310 mm with an average temperature of 27°C.

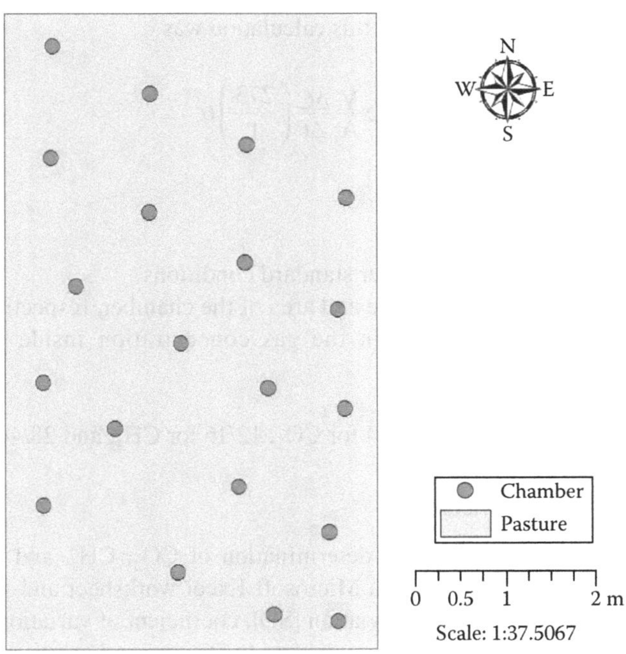

FIGURE 17.1 Experimental field (pasture) used in the case study.

17.3.2 Air Sampling for Determination of CO_2, CH_4, and N_2O Emissions

Twenty cylindrical polyvinylchloride chambers of 0.30 m length and 0.20 m in diameter were permanently inserted into the soil to a depth of 0.03 m since summer of 2003. The sampling chambers were arranged in five rows and each row had four chambers separated each other by about 30 m. The design of the sampling chamber is a modified version of Hutchinson and Mosier[8] and Robertson.[9] The chambers were constructed with two ventilation holes on the sides. They had circular tops made from Plexiglas, which contained two additional holes. One of the holes was covered by a stopper for the extraction of gases and while the other served for ventilation. Installation of these chambers permanently since 2003 kept soil undisturbed. Before air sampling, an assessment of the chambers was done to assure that all chambers were in perfect condition. During air sampling, the groove was filled with Dow Corning high-vacuum grease and a lid was placed over the top. Atmospheric samples in the chambers were collected by sealing the chamber holes with rubber stoppers, sealing the surface of the chamber, and collecting a sample with a 50 mL syringe placed in a 200 mL Tedlar bag after 30 min. An air sample was also collected at about 1.5 m above the chamber for the determination of CO_2, CH_4, and N_2O concentrations in the ambient air. Samples were analyzed for CO_2, CH_4, and N_2O at the Lincoln University's Dickinson Research Laboratory within 2 h of collection. The concentration of each GHG was measured using a gas chromatograph (GC) with an electron capture detector. Three standards with known concentrations of CO_2, CH_4, and N_2O were first analyzed before injecting the samples into the GC. The data was then transferred into an excel data sheet where the gas fluxes were calculated. A positive value represents gas emission from the soil, while a negative value represents gas uptake. Fluxes were calculated according to Ginting et al.[10] The equation used in this calculation was

$$F = \rho \frac{V}{A} \frac{\Delta C}{\Delta t} \left(\frac{273}{T} \right) \alpha$$

where

F is the gas production rate
ρ is the gas density (kg m^{-3}) under standard conditions
V (m^3) and A (m^2) are the volume and area of the chamber, respectively
$\Delta C/\Delta t$ is the ratio of change in the gas concentration inside the chamber (10^{-6} m^3 m^{-3} h^{-1})
T is the absolute temperature
α is the transfer coefficient (12/44 for CO_2, 12/16 for CH_4, and 28/44 for N_2O)

17.3.3 Geospatial Analysis

After air samples analyses for the determination of CO_2, CH_4, and N_2O concentrations, the data was entered into a Microsoft Excel worksheet and summaries of simple statistics (mean, standard deviation [SD], coefficient of variation [CV], minimum, median, and maximum) were calculated. These simple statistics (minimum and maximum) were used to calculate the total flux for the entire pasture using the TA.

This approach consisted in calculating the lower end (minimum) and upper end (maximum) flux for the entire pasture as the product of the pasture area (14,200 m^2) multiplied by the flux value (minimum and maximum) obtained from points measurements. In the second approach, interpolated maps (from points measurements) portraying the distribution zones (classes) of CO_2, CH_4, and N_2O fluxes across the pasture for each sampling period were produced, classified, and their areas determined. Then, a total flux was calculated as the sum of individual total flux for each classified zone. The all process consisted in transferring the pasture map into ARCGIS 9.3 (ArcMap) and joining the flux data to ArcMap, producing interpolated maps of CO_2, CH_4, and N_2O using ARCGIS 9.3 Spatial Analyst Extension with the IDW as the interpolation method. The default map classification method was equal interval and the number of classes (zones) was limited to nine. Interpolated flux maps were exported in JPEG format but later converted into TIFF format using Adobe Photoshop 7.0. Converted flux maps were then opened and classified using Multispec 3.2 software. The maximum likelihood method was used as the classification method and the classification accuracy was assessed using the kappa coefficient. Nine zones (classes) corresponding to the map produced with ARCGIS were classified in each flux map (treated as a satellite image). All classified maps had an overall classification accuracy (kappa coefficient) above 95%. After classification, maps fluxes were retrieved in ArcView–GIS 3.2 for compatibility reason, then transferred in ARCGIS 9.3. The entire approach is summarized in Nkongolo et al.[11] This process is summarized in Figure 17.2.

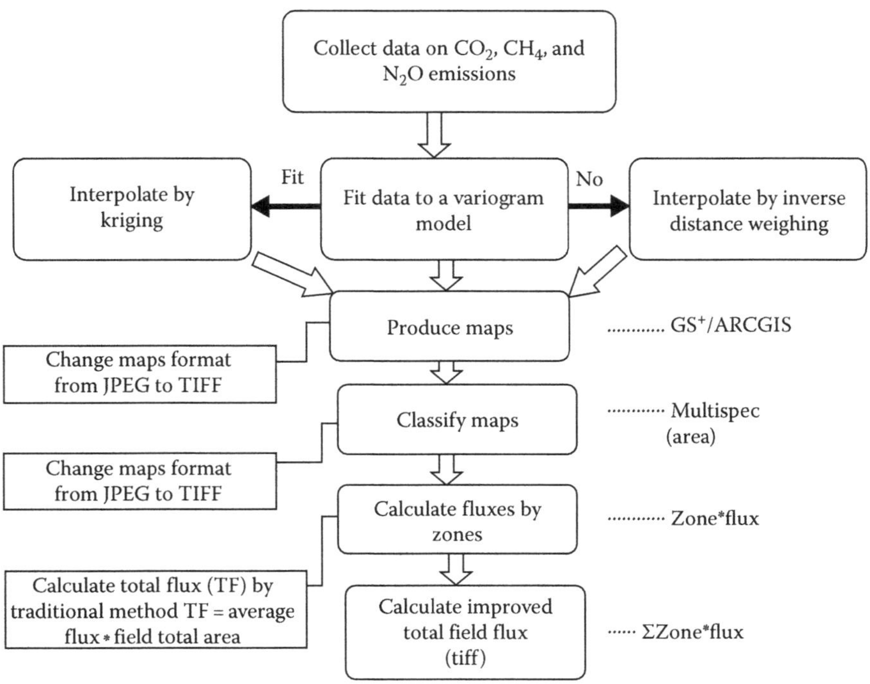

FIGURE 17.2 A schematic showing project activities.

17.4 RESULTS

17.4.1 Fluctuations of CO_2, CH_4, and N_2O in the Pasture

Summaries of simple statistics for GHG fluxes measured in June, July, and August are shown in Table 17.1. To demonstrate the type of data that was collected, interpolated, and classified, maps for CO_2, N_2O, CH_4, and CH_4 classification procedure for June 2006 are shown in Figures 17.3 through 17.6. Means CO_2 emissions decreased from June to August and were 87.11, 48.04, and 47.43 mg CO_2–C m^2 h^{-1}, respectively. The variability in CO_2 emissions also decreased from June to August as shown by low coefficients of variation and SD. Overall, CO_2 emissions data approached normality as shown by the means that are almost the same with medians (June data) and low skewness and kurtosis values. Methane (CH_4) showed a different trend, shifting from emissions in June to July and August. The means were 7.45, −14.15, and −12.29 µg C–CH_4 m^2 h^{-1}, respectively. Similarly to CO_2 emissions, the variability in CH_4 fluxes decreased nearly 100 times from June to August as shown by low CV and SD in July and August. In addition, CH_4 uptake (negative emissions) means in July and August were closer to their medians, implying that this data approached normality. Nitrous oxide (N_2O) emissions tripled from June to July and decreased 10 times from July to August. The means were 13.47, 40.64, and 4.06 N_2O–N m^2 h^{-1}, respectively. SD and CV also increased then decreased in the same pattern as did the means. CH_4 data of August approached normality as shown by the mean that was closer to its median.

17.4.2 Calculation of Total Flux for the Entire Pasture by the Traditional Approach

CO_2, CH_4, and N_2O fluxes (minimum and maximum) for the entire pasture (14,200 m^2) calculated using the TA are shown in Table 17.2. The minimum CO_2 emission for the entire pasture was 164.58 g CO_2–C h^{-1} in June, then increased three times in

TABLE 17.1
Summary of Simple Statistics for Greenhouse Gases (CO_2, CH_4, and N_2O) in a Pasture in Missouri in 2006

	CO_2 (mg CO_2–C m^{-2} h^{-1})			CH_4 (µg CH_4–C m^{-2} h^{-1})			N_2O (µg N_2O–N m^{-2} h^{-1})		
	June	July	August	June	July	August	June	July	August
Mean	87.18	48.04	47.43	7.45	−14.15	−12.29	13.47	40.64	4.06
SD	33.81	11.71	14.47	44.91	8.67	11.49	14.54	59.41	4.59
CV	38.78	24.37	30.51	602.85	61.29	93.45	107.90	146.19	113.05
Min	11.59	33.74	20.44	−62.22	−28.85	−31.65	−4.01	3.92	−2.92
Med	87.14	46.45	46.01	−0.25	−15.22	−16.56	8.60	18.64	3.91
Max	173.62	83.53	74.23	108.93	0.16	11.91	58.57	238.26	15.33
Skewness	0.21	1.36	0.01	0.64	0.20	0.32	2.09	2.31	0.58
Kurtosis	1.24	2.21	−0.28	−0.31	−0.85	−0.74	3.78	4.54	0.12

Min, minimum; Max, maximum; Med, median; SD, standard deviation; CV, coefficient of variation.

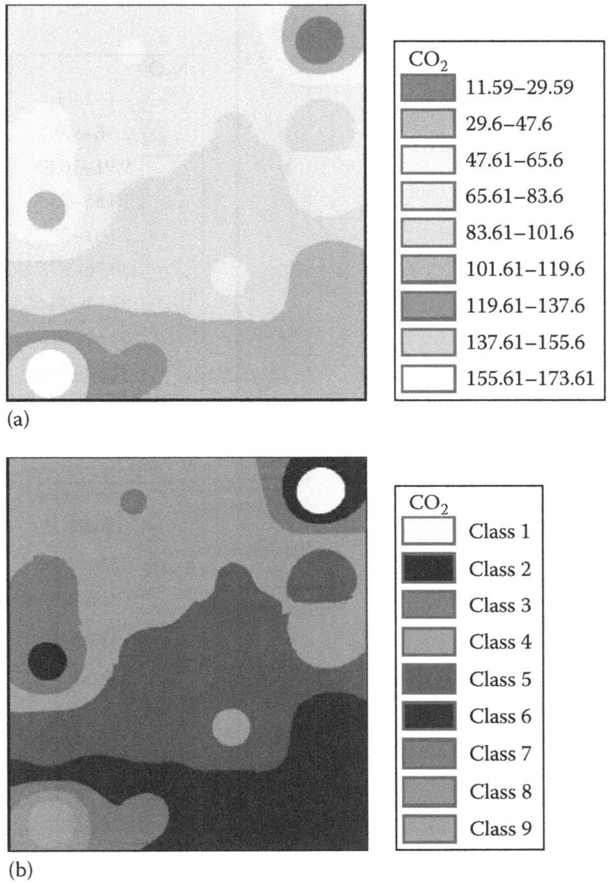

FIGURE 17.3 Interpolated (a) and classified (b) maps of CO_2 emissions in June 2006.

July (479.11 g CO_2–C h^{-1}), and finally decreased about 1.5 times in August (290.25 g CO_2–C h^{-1}). The maximum emissions of CO_2 were higher (as expected), but followed the same trend as the minimum emissions, i.e., increasing from June to July and decreasing from July to August. The ratios between the maximum and the minimum CO_2 flux were 15, 2.5, and 3.6, for June, July, and August, respectively. For methane (CH_4), the lower-end fluxes in the pasture resulted in an uptake of 883.52, 409.67, and 449.43 mg CH_4–C h^{-1} in June, July, and August, but emissions of 1546.81, 2.27, and 169.12 mg CH_4–C h^{-1} for the same months, respectively. Similarly to CH_4, the minimum fluxes of N_2O for the pasture resulted in an uptake of 56.94 and 41.46 mg N_2O–N h^{-1} in June and August, but an emission of 55.66 mg N_2O–N h^{-1} in July.

17.4.3 Calculation of Total Flux for the Entire Pasture by the GIS Approach

CO_2, CH_4, and N_2O fluxes (minimum and maximum) for the entire pasture (14,200 m^2) calculated using the GIS approach are shown in Tables 17.3 through 17.5,

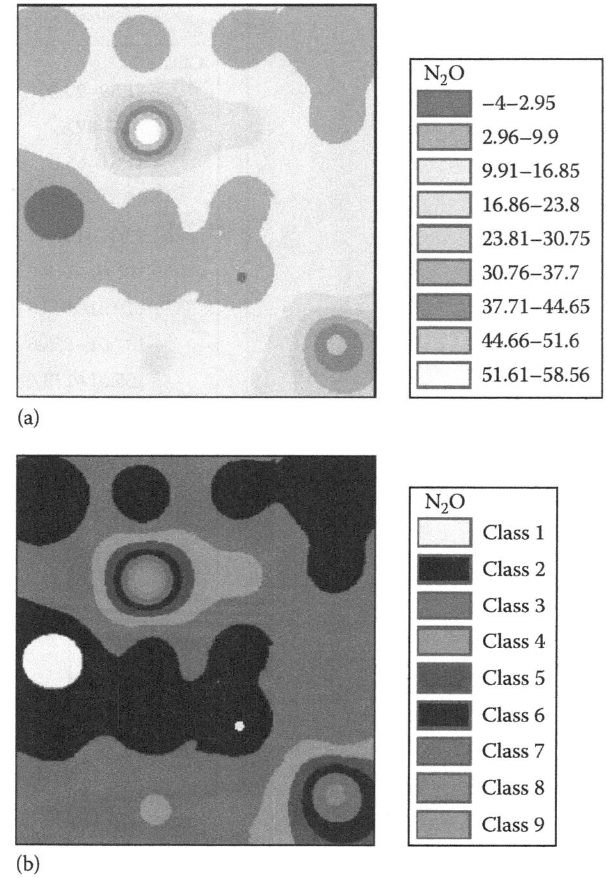

FIGURE 17.4 Interpolated (a) and classified (b) maps of N_2O fluxes in June 2006.

respectively, for June, July, and August. Table 17.3 shows that both the minimum and maximum emissions of CO_2 emissions in the pasture were highest in June and similar for July and August. The ratios between the maximum and minimum were small and close to one. Methane (CH_4) exhibited similar monthly trends with highest values of either uptake or emissions in June and lower values in July and September. Finally, the minimum emissions for N_2O in the pasture were 138.76 mg N_2O–N h^{-1} in June and doubled to 379.18 mg N_2O–N h^{-1} in July and decreased eight times to 44.08 mg N_2O–N h^{-1} in August. The maximum emissions followed similar monthly trends, almost tripling from June to July and decreasing 10 times from July to August.

17.4.4 Comparison between TA and GIS Approaches in Calculating Total Flux in the Pasture

Table 17.6 shows the minimum and maximum fluxes of CO_2, CH_4, and N_2O calculated using both the TA and GIS approaches. Overall, the TA approach underestimated the minimum flux while it overestimated the maximum flux of CO_2, CH_4, and N_2O for the

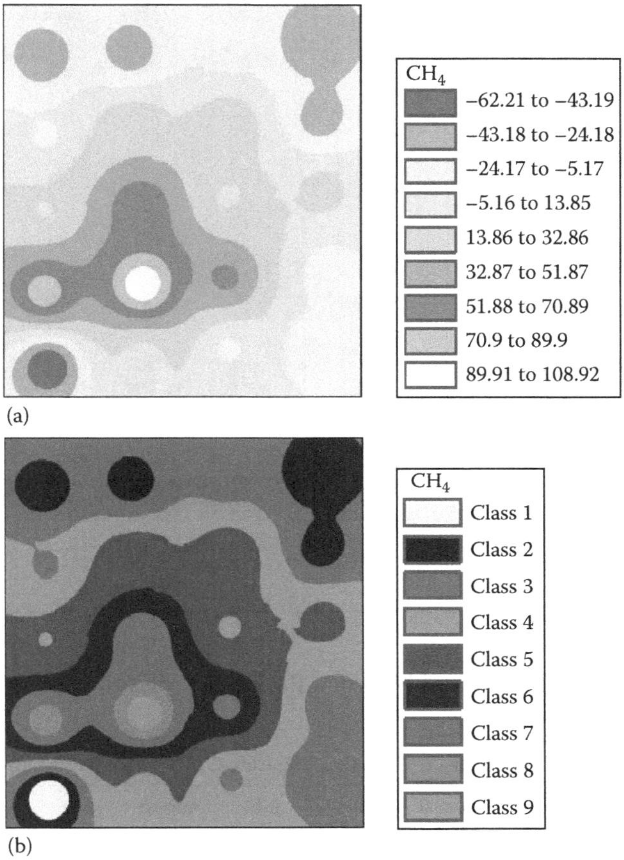

FIGURE 17.5 Interpolated (a) and classified (b) maps of CH_4 fluxes in June 2006.

entire pasture. For CO_2, the minimum flux was underestimated by 25% and 54% in July and August while the maximum flux was overestimated by 80%, 66%, and 48% during the same months. CH_4 minimum uptake was underestimated by 140141, 84%, and 100% in June, July, and August while the maximum flux was overestimated by 474%, 101%, and 208% in the same months. Finally, N_2O minimum flux was underestimated by 141%, 85%, and 194% and the maximum flux overestimated by 250%, 351%, and 199%.

17.5 CASE STUDY

17.5.1 What You Need: Software and Data

You can use a variety of GIS programs, such as ArcView 3.xx, ARCGIS 8.xx, ARCGIS 9.xx, or another GIS software, to help you produce interpolated maps. If you are using ArcView or ARCGIS, make sure that you have a corresponding version of spatial analyst extension. You will also need a software that will help you classify your map such as Erdas Imagine, Multispec, Scion. In this case study, we used ARCGIS 9.2 for mapping and MultispecW32 for classifying our map. MultispecW32

FIGURE 17.6 Classification process of CH_4 fluxes map in Multispec 3.2.

TABLE 17.2
Total Flux of CO_2, CH_4, and N_2O in a Pasture in Missouri in 2006: Calculation-Based TA[a]

	CO_2 (g CO_2–C h^{-1})			CH_4 (mg CH_4–C h^{-1})			N_2O (mg N_2O–N h^{-1})		
	June	July	August	June	July	August	June	July	August
Min	164.58	479.11	290.25	−883.52	−409.67	−449.43	−56.94	55.66	−41.46
Max	2465.40	1186.13	1054.07	1546.81	2.27	169.12	831.69	3383.29	217.69
Range	2300.83	707.02	763.82	2430.33	411.94	618.55	888.64	3327.63	259.15

[a] The field area (14,200 m^2) is multiplied by the min (minimum) or the max (maximum) flux value found in Table 17.1 to obtain the min and max flux for the entire field.

is available freely on the Internet. Download MultispecW32 at http://cobweb.ecn.purdue.edu/~biehl/MultiSpec/download_win.html. Make sure that you choose a version compatible with your hardware and software. Save it in your computer and follow the directions given on the website to install the software. Before classifying your map (if you will use MultispecW32), you will need to preprocess your map: clip the map body and save it in "TIFF" format. If you are using ARCGIS 9.2, you may complete these operations within the same software and transfer your clipped map to MultispecW32 directly for classification. However, in this case study, even though we are using ARCGIS 9.2, we used Adobe Photoshop 7.0 to clip our map and change its format before classification as this may also be helpful to those producing their interpolated maps in GS+, Surfer, MapViewer, or other software that might not allow clipping, change of format, and direct transfer to MultispecW32 or the map classification software being used.

TABLE 17.3
Total Flux of CO_2 in a Pasture in Missouri in 2006: Calculation from Classified Flux Maps—GIS-Based Approach[a]

	Zones	Min Flux (mg C–CO_2 m^{-2} h^{-1})	Max Flux (mg C–CO_2 m^{-2} h^{-1})	Area (%)	Area (m^2)	Total Min (g C–CO_2 h^{-1})	Total Max (g C–CO_2 h^{-1})
June	1	11.59	29.59	1.30	184.60	2.14	5.46
	2	29.59	47.59	3.10	440.20	13.02	20.95
	3	47.59	65.59	8.40	1,192.80	56.77	78.24
	4	65.59	83.59	34.40	4,884.80	320.41	408.32
	5	83.59	101.60	27.10	3,848.20	321.65	390.98
	6	101.60	119.60	18.60	2,641.20	268.33	315.89
	7	119.60	137.60	3.60	511.20	61.12	70.34
	8	137.60	155.60	1.70	241.40	33.16	37.56
	9	155.60	173.60	1.60	227.20	35.32	39.44
	Total			99.89	14,171.60	1,111.92	1,367.18
July	1	33.75	39.28	5.87	833.54	28.13	32.74
	2	39.28	44.81	40.05	5687.10	223.39	254.84
	3	44.81	50.34	27.52	3,907.84	175.11	196.72
	4	50.34	55.87	14.16	2,010.72	101.22	112.34
	5	55.87	61.4	6.89	978.38	54.66	60.07
	6	61.4	66.93	1.56	221.52	13.60	14.83
	7	66.93	72.46	1.79	254.18	17.01	18.42
	8	72.46	78	1.39	197.38	14.30	15.40
	9	78	83.53	0.79	112.18	8.75	9.37
	Total			100.02	14,202.84	636.18	714.72
August	1	20.44	26.41	1.14	161.88	3.31	4.28
	2	26.41	32.39	3.35	475.70	12.56	15.41
	3	32.39	38.36	4.8	681.60	22.08	26.15
	4	38.36	44.31	25.83	3,667.86	140.70	162.52
	5	44.31	50.31	34.38	4,881.96	216.32	245.61
	6	50.31	56.29	17.29	2,455.18	123.52	138.20
	7	56.29	62.26	8.41	1,194.22	67.22	74.35
	8	62.26	68.24	3.84	545.28	33.95	37.21
	9	68.24	74.22	0.96	136.32	9.30	10.12
	Total			100	14,200	628.96	713.85

[a] Flux maps produced with ARCGIS spatial analyst extension are classified using Multispec to determine the area covered by each zone.

To start, go to the folder of this chapter and copy the following files: "chamber.shp, chamber.dbf, chamber.shx, chamber.prj, pasture.shp, pasture.dbf, pasture.shx, pasture.prj, and Table 7.xls." Pasture.shp is the 14,200 m^2 plot where this study was conducted. Chamber.shp contains the 20 locations where chambers were installed in order to collect soil air and measure its concentration of CO_2, CH_4, and N_2O.

TABLE 17.4
Total Flux of C–CH_4 in a Pasture in Missouri in 2006: Calculation from Classified Flux Maps—GIS-Based Approach[a]

	Zones	Min Flux (mg C–CH_4 m^{-2} h^{-1})	Max Flux (mg C–CH_4 m^{-2} h^{-1})	Area (%)	Area (m^2)	Total Min (mg C–CH_4 h^{-1})	Total Max (mg C–CH_4 h^{-1})
June	1	−62.2	−43.19	0.9	127.8	−7.95	−5.52
	2	−43.19	−24.17	9.3	1,320.6	−57.04	−31.92
	3	−24.17	−5.16	25.38	3,603.96	−87.11	−18.60
	4	−5.16	13.84	23.69	3,363.98	−17.36	46.56
	5	13.84	32.86	21	2,982	41.27	97.99
	6	32.86	51.87	9.77	1,387.34	45.59	71.96
	7	51.87	70.88	7.49	1,063.58	55.17	75.39
	8	70.88	89.9	1.8	255.6	18.12	22.98
	9	89.9	108.91	0.68	96.56	8.68	10.52
	Total			100.01	14,201.42	−0.63	269.35
July	1	−28.85	−25.63	1.60	227.20	−6.55	−5.82
	2	−25.63	−22.4	4.00	568.00	−14.56	−12.72
	3	−22.4	−19.18	10.30	1,462.60	−32.76	−28.05
	4	−19.18	−15.96	13.70	1,945.40	−37.31	−31.05
	5	−15.96	−12.73	32.90	4,671.80	−74.56	−59.47
	6	−12.73	−9.51	22.90	3,251.80	−41.40	−30.92
	7	−9.51	−6.29	8.00	1,136.00	−10.80	−7.15
	8	−6.29	−3.07	4.70	667.40	−4.20	−2.05
	9	−3.07	−0.16	1.80	255.60	−0.78	−0.04
	Total			99.90	14,185.80	−222.93	−177.28
August	1	−31.64	−26.80	8.70	1,235.40	−39.09	−33.11
	2	−26.80	−21.96	4.80	681.60	−18.27	−14.97
	3	−21.96	−17.12	17.10	2,428.20	−53.32	−41.57
	4	−17.12	−12.28	26.80	3,805.60	−65.15	−46.73
	5	−12.28	−7.45	17.60	2,499.20	−30.69	−18.62
	6	−7.45	−2.61	16.90	2,399.80	−17.88	−6.26
	7	−2.61	2.22	5.30	752.60	−1.96	1.67
	8	2.22	7.06	1.80	255.60	0.57	1.80
	9	7.06	11.90	1.00	142.00	1.00	1.69
	Total			100.00	14,200.00	−224.79	−156.10

[a] Flux maps produced with ARCGIS spatial analyst extension are classified using Multispec to determine the area covered by each zone.

Both chamber and pasture files are projected into "Geographic Coordinates," that is why files with extension "prj" are included. Table 7.xls contains the longitude and latitude of each chamber as well as data on CO_2 (mg C–CO_2 m^2 h^{-1}), CH_4 (μg C–CH_4 m^2 h^{-1}), and N_2O (μg N–N_2O m^2 h^{-1}) collected at each point. Temp is the soil temperature.

TABLE 17.5
Total Flux of N_2O in a Pasture in Missouri in 2006: Calculation from Classified Flux Maps—GIS-Based Approach[a]

	Zones	Min Flux (mg $N-N_2O$ m^{-2} h^{-1})	Max Flux (mg $N-N_2O$ m^{-2} h^{-1})	Area (%)	Area (m^2)	Total Min (mg $N-N_2O$ h^{-1})	Total Max (mg $N-N_2O$ h^{-1})
June	1	−4	2.95	1.94	275.48	−1.10	0.81
	2	2.95	9.89	34.4	4,884.8	14.41	48.31
	3	9.89	16.85	46.33	6,578.86	65.06	110.85
	4	16.85	23.8	8.23	1,168.66	19.69	27.81
	5	23.8	30.75	3.3	468.6	11.15	14.41
	6	30.75	37.7	2.94	417.48	12.84	15.74
	7	37.7	44.66	1.81	257.02	9.69	11.48
	8	44.66	51.6	0.69	97.98	4.38	5.06
	9	51.6	58.56	0.36	51.12	2.64	2.99
	Total			100	14,200	138.76	237.47
July	1	3.94	29.97	53.55	7,604.10	29.96	227.89
	2	29.97	56.01	29.49	4,187.58	125.50	234.55
	3	56.01	82.04	7.27	1,032.34	57.82	84.69
	4	82.04	108.07	3.82	542.44	44.50	58.62
	5	108.07	134.11	1.90	269.80	29.16	36.18
	6	134.11	160.14	1.48	210.16	28.18	33.66
	7	160.14	186.17	1.00	142.00	22.74	26.44
	8	186.17	212.21	0.97	137.74	25.64	29.23
	9	212.21	238.24	0.52	73.84	15.67	17.59
	Total			100.00	14,200.00	379.18	748.85
August	1	−2.91	−0.89	2.11	299.62	−0.87	−0.27
	2	−0.89	1.13	9.48	1,346.16	−1.20	1.52
	3	1.13	3.16	28.93	4,108.06	4.64	12.98
	4	3.16	5.19	25.41	3,608.22	11.40	18.73
	5	5.19	7.21	24.65	3,500.30	18.17	25.24
	6	7.21	9.24	4.65	660.30	4.76	6.10
	7	9.24	11.27	2.48	352.16	3.25	3.97
	8	11.27	13.29	1.39	197.38	2.22	2.62
	9	13.29	15.32	0.9	127.80	1.70	1.96
	Total			100	14,200.00	44.08	72.85

[a] Flux maps produced with ARCGIS spatial analyst extension are classified using Multispec to determine the area covered by each zone.

TABLE 17.6
Comparison between TA and GIS for Calculating CO_2, CH_4, and N_2O Fluxes for the Entire Pasture

		CO_2 (g CO_2–C h^{-1})		CH_4 (µg CH_4–C h^{-1})		N_2O (g N_2O–N h^{-1})	
		TA	GIS	TA	GIS	TA	GIS
June	Min	164.58	1111.19	−883.52	−0.63	−56.94	138.76
	Max	2465.40	1367.18	1546.81	269.35	831.69	237.47
July	Min	479.11	636.18	−409.67	−222.93	55.66	379.18
	Max	1186.13	714.72	2.27	−177.28	3383.29	748.85
August	Min	290.25	628.96	−449.34	−224.79	−41.46	44.08
	Max	1054.07	713.85	169.12	−156.10	217.69	72.85

17.5.2 Production of Interpolated Maps in ARCGIS 9.2

This exercise uses data stored in the data disk under Table 17 data tables. To start, open ARCGIS and launch ArcMap. On "Start using ArcMap with" tab, select "A new empty map," then click "Ok." Click on "add data" (the + button) and navigate to the folder where you saved the data. Click on chamber, then press and hold the "ctrl" key and click also on "pasture" and release "ctrl" and click "Ok." Both chamber and pasture will be displayed. Right click on "chamber," then "open attribute table," the table shows only "FID" (0–19), "shape" (point), and "ID" (1–20). To add the data, right click on "chamber" → "join and relate" → "join." In the "join table," make sure that "join attributes from a table" is displayed under "what do you want to join to this layer?" Choose "ID" for "1. Choose the field in this layer that the join will be based on," for "2. Choose the table to join, or load the table from disk," browse to the folder where you saved the data and click on "Table 7.txt." Table 7.txt will appear on (2). If it does not work, try one of these two options, either "add" Table 7.txt to your layers as you did for "chamber.shp" before joining, or try to join "Table 7.xls" instead of "Table 7.txt." For joining "Table 7.xls," browse to your folder and click "Table 7.xls." On the "Add" table that will appear, double click on "sheet 1$." You will see "sheet 1$" under (2). For step "3. Choose the field in the table to base join on," choose "ID," then click "Ok." Make also sure that other fields such as "longitude, latitude, CO_2, CH_4..." appear when you click on ID. If not, your join was not successful. To make sure the table was joined correctly, right click on "chamber" again, then "open attribute table." The data has been joined but "sheet 1$" (Figure 17.7) has been added to fields such as longitude, latitude... to tell that the source of the data is "sheet 1$."

Now you are ready to start making your maps. You need to activate your "spatial analyst extension." To do so, on ArcMap file menu, go to "tools" → extensions and in the extensions table, check "Spatial Analyst." The extension is activated but may not show up on the toolbars. Therefore, you need to go to "View → Toolbars" → then check "Spatial Analyst." The extension is now displayed. Go to "Spatial Analyst → Options"

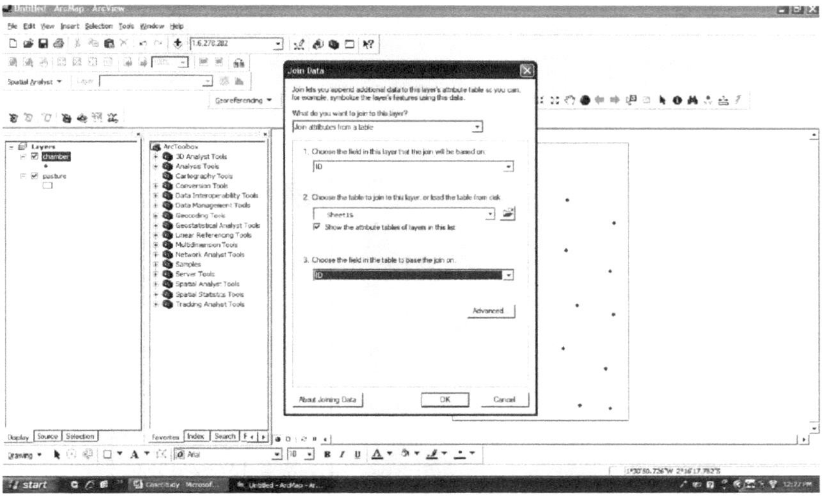

FIGURE 17.7 Joining Table 7.xls to chamber.shp in ArcMap.

and in the option table, click the "general tab," for the "working directory," browse and select your working folder, for "analysis mask," select "pasture.shp" in your working folder. Then, click on "extent tab," for "analysis extent," select "same as layer" "pasture.shp," then move to "cell size tab" and for "analysis cell size," select "maximum of inputs" and click "Ok." You are done setting the analysis mask. To make your first map for "CO_2" for example, go to "Spatial Analyst"→Interpolate to raster→Inverse distance weighted. On the "inverse distance weighted" table, use the following parameters: Inputs point: Chamber; Z value field: "Sheet 1\$.CO_2" and leave the other parameters to default, but for the "output raster," browse to your working folder and on the "save as" table, give a name such as "CO_2" and click "Ok." An interpolated map of CO_2 is displayed. If you cannot see it, uncheck the chamber layer or simply move the new "CO_2" layer at the top of your layers. To make maps of CH_4 and N_2O, repeat the process as you did for CO_2. Now, let's go back to our CO_2 map. Right click on CO_2 layer (top left of your screen), then "properties" and the "layer properties" is displayed. On the "layer properties," right click on the data below "label"-format level, the category is "numeric" and below "rounding," check "number of decimal places" and set this number to "2." The rest is left to default. Note that the classification method is "equal interval" and the number of class is left to 9 (default). Double click on the color of Class 1 (11.62–29.62) in the legend and a box with colors appears. Point your mouse to any of the colors, the color's name appears. In fact, you need to have a map with clear distinction between the colors of each class. If the distinction is not clear enough, you may have hard time classifying the map and the best option is to change the color as we did in this study case. For this case study, the following colors were chosen: Class 1: Black; Class 2: Lemon grass; Class 3: Fir green, Class 4: Light vert, Class 5: Cherry cola, Class 6: Solar yellow, Class 7: Lapio lazuli, Class 8: Mars red, and Class 9: Arctic white. Uncheck

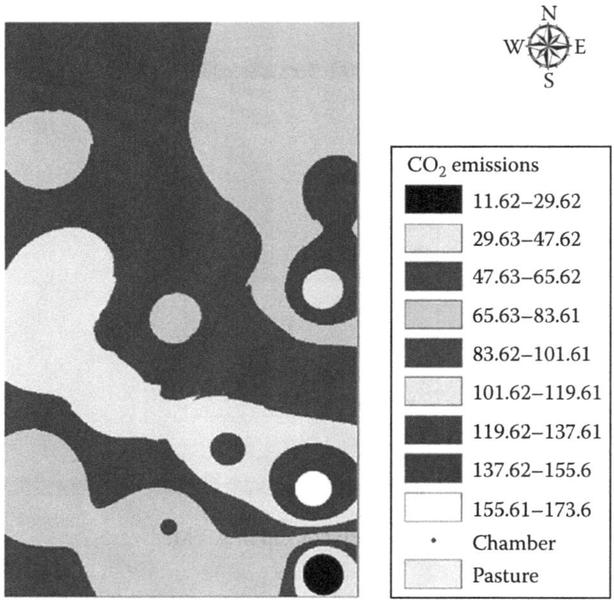

FIGURE 17.8 Interpolated map of CO_2 emissions in the pasture.

layers "chamber" and "pasture," change the display to "layout view" to create a map layout—then insert the legend as follows: go to Insert→legend and on the "legend wizard" table, just continue to click next until you reach the end (finish) and the legend is displayed in the layout view. Go to file→page and print setup and in "page and print setup table," change the orientation to "landscape" so that it can easily fit the CO_2 map and its legend inside the box. Add other map elements (north arrow, title) as needed. When you finish, print a copy (preferably with a color printer) of your map, then go to File→Export on the "Export map tab," change the file name to "CO_2_Emissions" and change the "save as" to "JPEG" from the dropdown list. Save your map (Figure 17.8) to your folder and repeat the same process for CH_4 and N_2O.

Make sure that you uncheck the CO_2 layers when making CH_4 map and uncheck CH_4 layer when you make N_2O map. When you are done, save your work: File→save as. This will be a map document with an extension "mxd." If you need to come back and redo your maps, you will not need to go through the entire process again, but after launching ArcMap, "open an existing map" to have this screen again (as far as you keep all your files in the same directory). Exit ArcMap and take your hard copy map. Open excel and create a file and write in column A: Class (1–9, see above), column B: the color of each class (see above), column C: "Point Min CO_2" which is the minimum value of CO_2 emission of the class as given in the legend (this value is 11.62 for Class 1), and column D: "Point Max CO": or the maximum value of CO_2 emission of the class as given in the legend (this value is 29.62 for Class 1), then save your work (Table 8.xls). Repeat the operations for CH_4 and N_2O. When you are done with recording your classes and the corresponding minimum and maximum values, close excel and open Adobe Photoshop.

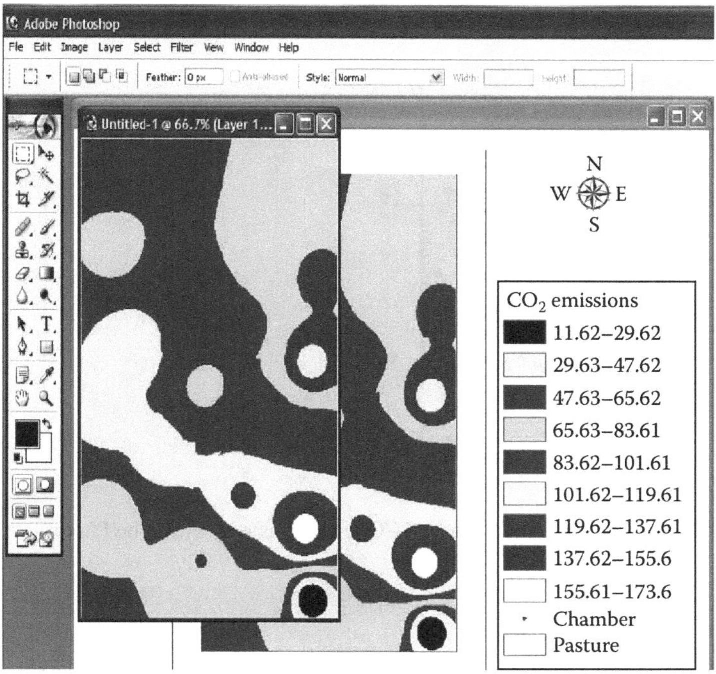

FIGURE 17.9 Clipping the map of CO_2 emissions in Adobe Photoshop.

17.5.3 Reformatting Maps Using Adobe Photoshop 7.0

In Adobe Photoshop, open "CO_2_Emissions.JPEG" that you exported earlier from ArcMap. Using the "Rectangular Marquee" tool (M), select only the body of your map, avoiding the legend, north arrow, and white areas, then Edit→Copy→File→New→Ok (Figure 17.9).

On the "New" tab, go to→Edit→paste. Then File→Save as, in the "save as" table, make sure that your folder is the one shown in "save in," change the file name to "CO_2_Em" and change the format to "TIF (*.TIF; *.TIFF)" from the dropdown list (Figure 17.10). Below "Save options," make sure that "Layers" and "Use Lower Case Extension" are checked.

Then, click on "save." In the "TIF Options" that will appear after you click save, make sure that "None" is selected under "Image Compression," under "Byte Order," select "IBM PC" and for "Layer Compression," "RLE (faster saves, bigger files)" should be checked. Then, click Ok→Ok (Figure 17.11).

Repeat these operations for CH_4 and N_2O maps. You are now ready to classify your map using MultispecW32.

17.5.4 Classification of Maps Using MultispecW32

Open "MultispecW32" that you downloaded from the Internet and go to File→Open Image, browse to your folder, click on "CO_2_Em.TIFF," then click

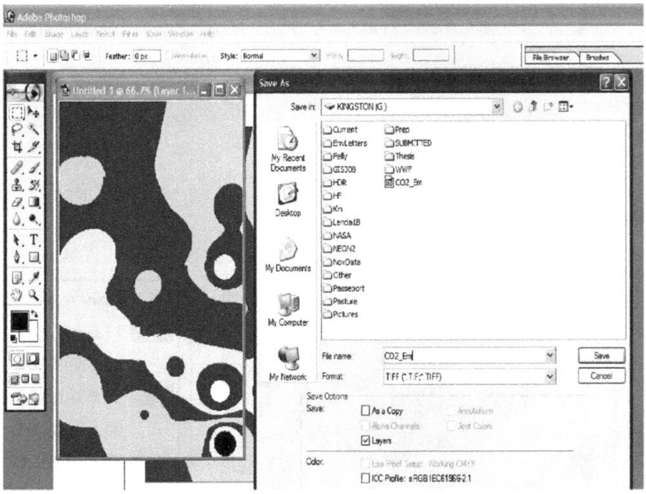

FIGURE 17.10 Changing the format of CO_2 emissions map in Adobe Photoshop.

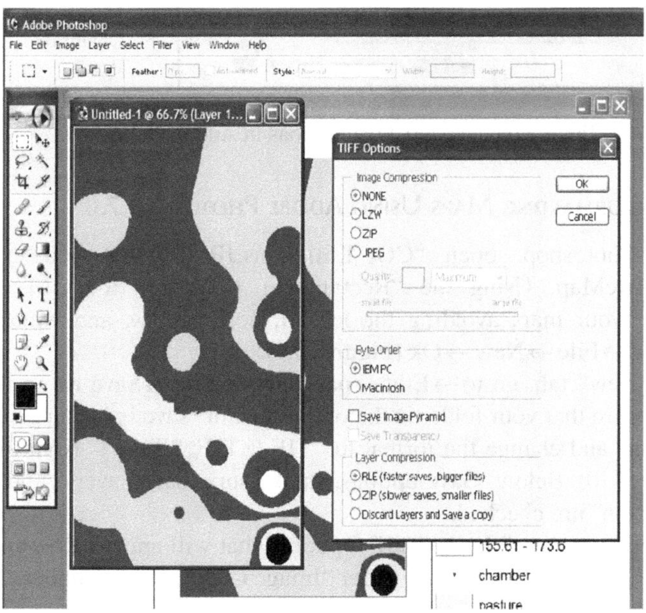

FIGURE 17.11 Saving the final map of CO_2 emissions in Adobe Photoshop.

"Ok" on the "Set Display Specifications for" tab that is added to your view and the CO_2 map is displayed. Click on "Processor → Statistics," and a "Set Project Options" tab is added to your view. In the "Set Project Options" tab, make sure that "training fields, test fields, show classes names, and show fields names" are all checked (Figure 17.12), then click "Ok" and another tab named "Project" is added to your view.

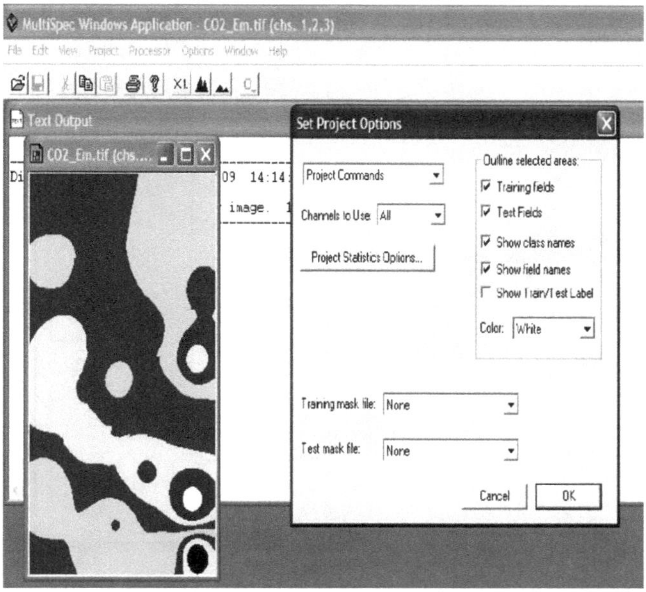

FIGURE 17.12 Opening CO_2 emissions clipped map in MultispecW32.

Take the hard copy of the map you printed earlier and use it as reference. Assuming that your Class 1 corresponds to "black," Class 2 to "Lemon grass," Class 3 to "Fir green," and so on (see Section 17.5.3), start your classification with Class 1. Using your mouse, draw a small square or rectangle in Class 1 (inside the dark circle), then click on "Add to List" in the "Project" tab. In the "Define Class and/or Field Description" tab that appears, the "Class" is set to "New" and "Enter Class Name" is automatically set to "Class 1." If you did not start with Class 1 (or black color), write the appropriate class number. Below Class 1, the number of pixels is given for the sample you took (the square or rectangle you drew). It is advised to take many samples per class, depending on your map (simulated to a satellite image), but in this case study, you took only one sample per class. Also, in the "Enter Field Identifier," "Field 1" is automatically given and will continue to increase as you sample. On "Area Type," make sure that "Training Field" is checked when you are ready to classify your map. However, if you are not ready to classify your map, you can first practice by checking "Test Field." Click "Ok" and "Class 1, Field 1" appear inside the black color in your map and move to "Class 2" (color Lemon grass). Draw a small rectangle in area of Class 2, then click "Add to List." In the "Define Class and/or Field Description," change "Class 1" to "New" using the dropdown list and "Class 2" is displayed under "Enter Class Name." Do not worry about the field, which is automatically changed to field 2. Click "Ok" and move on to Class 3 (color: Fir green), Class 4 (color: Light vert), 5, 6, 7, 8, and 9 (color: Arctic red) (Figure 17.13).

When you finish, go to "Processor"→"Classify." In the "Set Classification Specifications" tab that appears, make sure that the "Procedure" is "Maximum Likelihood" and check "Disk File" under "Write Classification Results To." Click

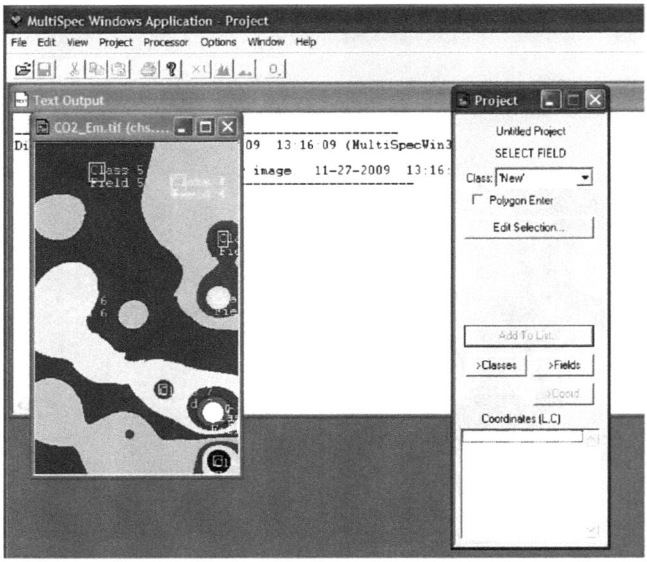

FIGURE 17.13 Classifying the map of CO_2 emissions in MultispecW2.

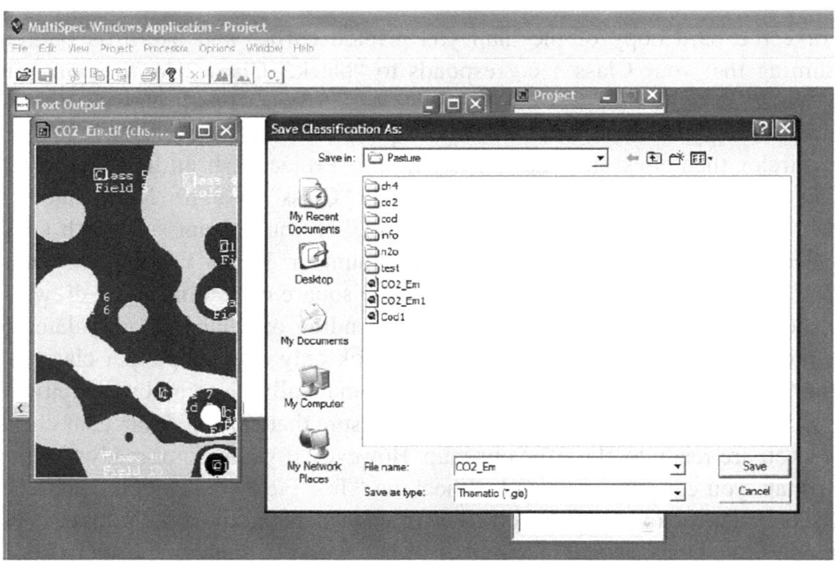

FIGURE 17.14 Saving the classified map of CO_2 emissions in MultispecW32.

"Ok," then on "Multispec Windows Application," click "Ok" and save the "classification" as "CO_2_Em." On the dropdown list, the file type should be "Thematic.gis" (this is your classified map that you can open in ArcMap) (Figure 17.14).

The classification results (text output) are displayed behind the map on your view. You can minimize your map, then copy the text output and save it in MS word

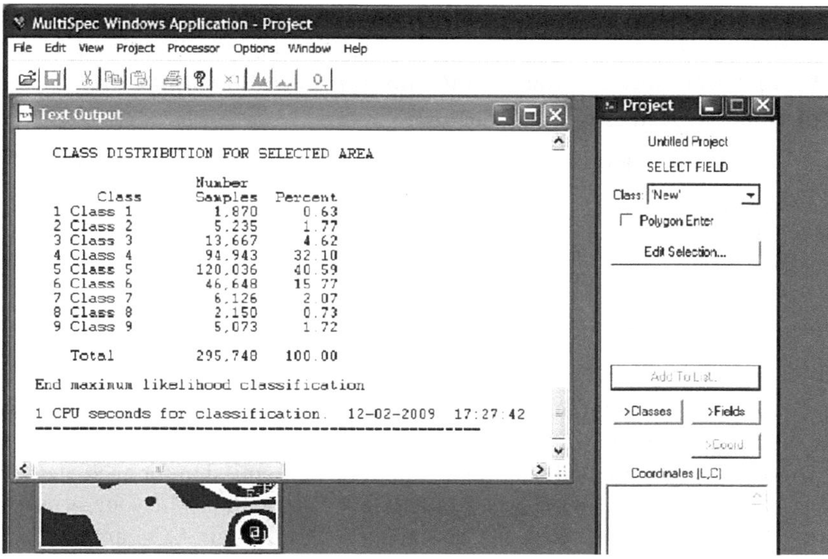

FIGURE 17.15 Classification performance and class distribution for CO_2 emissions map.

(Table 9.doc). You can also directly save the text output as a txt file and later open it with MS word or excel. This output contains information on all the operations, but you only need the last page showing both your "Kappa Statistic" and the "Class Distribution" (percent of each class). If your kappa coefficient is between 90% and 100%, you did a good job in your classification (Figure 17.15).

However, verify that the percentage given is proportional to the size of each color in the map. For example, Class 1 cannot be more than Class 3. Print your "Class Distribution" and save your output. Repeat these operations to classify CH_4 and N_2O maps (note these are maps in TIFF format prepared in Adobe Photoshop).

17.5.5 Calculation of Gas Fluxes for the Field

17.5.5.1 Traditional Approach

Tables 17.7 through 17.9 are available in Chapter 17 data disk. Open Table 7 on the data disk (Chapter 17 data) that you downloaded at the beginning of this project and generate a summary of simple statistics, but you are interested only in the minimum and maximum values of CO_2, CH_4, and N_2O from point measurements across the pasture (Table 17.1). For CO_2, the minimum emission is 11.59 mg and the maximum emission is 173.62 mg C–CO_2 m^2 h^{-1}. The total area of the pasture is 1.42 ha or 14,200 m^2. Therefore, the minimum CO_2 emission for the entire pasture (Pasture Min CO_2) = 11.59 mg C–CO_2 m^2 h^{-1} × 14,200 m^2 = 164.58 g of C–CO_2 h^{-1}. The maximum CO_2 emission for the pasture (Pasture Max CO_2) = 173.62 mg C–CO_2 m^2 h^{-1} × 14,200 m^2 = 2,465.40 g C–CO_2 h^{-1}. Repeat the same operations to calculate the pasture minimum and maximum emissions (or uptake) for CH_4 and N_2O.

TABLE 17.10
Comparison between the TA and GIS Approach for Estimation of Total CO_2 Emissions in a Pasture

	Traditional Approach (TA)	GIS Approach (GIS)	Ratio (TA/GIS)
Pasture min CO_2 (g C–CO_2 h^{-1})	164.58	1135.85	0.15
Pasture max CO_2 (g C–CO_2 h^{-1})	2465.40	1391.92	1.77

17.5.5.2 GIS Approach

Take your "class distribution" printout from Section 17.5.4. and open your "Table 8.xls" you made in Section 17.5.2 after making your interpolated maps and in which you wrote the minimum and maximum values of CO_2 emission for each class. Add new columns: E. Class % (values from "Class Distribution" in your printout in Section 17.5.4, see Table 9.doc), F. Class area, G. Pasture min CO_2, and H. Pasture max CO_2. Then, calculate columns F, G, and H as follows: $F = (E \times 14{,}200)/100$, $G = F \times C$, and $H = F \times D$. At the bottom of Table 17.2, calculate the total for F, G, and H. Verify that F = 14,200 (or closer to) for the total area.

17.5.5.3 Comparison of Both Approaches

Compare values of pasture min CO_2 and pasture max CO_2 obtained in Sections 17.5.5.1 and 17.5.5.2 (Table 17.10). For the GIS approach (Section 17.5.5.2), these values are given by the totals in column G for pasture min CO_2 and column H for pasture min CO_2. Calculate the ratio TA/GIS to decide whether there is an overestimation (TA/GIS > 1) or an underestimation (TA/GIS < 1) of total CO_2 emissions by the TA approach. The TA underestimated six times the minimum CO_2 emissions from the pasture and overestimated about two times the maximum CO_2 emissions from the entire pasture.

17.6 CONCLUSION

We evaluated two approaches to calculate the total flux (minimum and maximum) of CO_2, CH_4, and N_2O emitted from a 14,200 m^2 pasture. The first approach, named as the traditional approach (TA), consisted in multiplying the minimum and maximum flux values obtained from 20 point measurements across the pasture by the pasture area. The second approach, named "GIS," consisted in producing an interpolated flux map from the 20 points, treating the map as a satellite image, classifying the map to determine the area covered by each flux zone, and finally, calculating a total flux as the sum of flux in each classified zone. Our results showed that the TA did underestimate the minimum flux while it overestimated the maximum flux for the entire pasture. This approach is a promising tool for quantifying GHG fluxes in agricultural fields. It can also be used to quantify the amount of nutrients to apply to soil during fertilization campaign.

ACKNOWLEDGMENTS

This material is based upon work supported by the Cooperative State Research, Education and Service (CSREES), United States Department of Agriculture (USDA), USDA/CSREES, Project type: Evans-Allens, ACCESSION#: 0208180.

REFERENCES

1. Robertson, G. and Grace, P. Greenhouse gas fluxes in tropical and temperate agriculture: The need for a full-cost accounting of global warming potentials. *Environ. Dev. Sustain.* 6, 51, 2004.
2. Robertson, G.P., Paul, E.A., and Harwood, R.R. Greenhouse gases in intensive agriculture: Contributions of individual gases to the radiative forcing of the atmosphere. *Science* 289, 1922–1925, 2000.
3. Verhagen, J. Modeling soil and crop responses in a spatially variable field. In: Stafford, V.J. (ed.). *Precision Agriculture*, vol. 1. Bios Scientific Publishers, Oxford, U.K., 1997, pp. 197–204.
4. Guy, E.D. and Levine, N.S. GIS modeling and analysis of Ohio's CO_2 budget: Mitigating CO_2 emissions through reforestation. *Ohio J. Sci.* 101, 34, 2001.
5. Bausch, W.C., Duke, H.R., and Porter, L.K. Remote sensing of plant nitrogen status in corn. *ASAE Summer Meetings*, Kansas City, MO, June 19–22, 1994.
6. Kristensen, K. and Olesen, S.E. Mapping root zone capacity by co-kriging areal photographs and point measurements of available soil moisture. In: Stafford, V.J. (ed.). *Precision Agriculture*, vol. 1. Bios Scientific Publishers, Oxford, U.K., 1997, pp. 197–204.
7. Anderson, G.L. and Yang, C. Multispectral videography and geographic information systems for site-specific farm management. In: *Proceedings of the 3rd International Conference,* Minneapolis, MN, June 23–26, 1996, pp. 681–692.
8. Hutchinson, G.L. and Mosier, A.R. Improved soil cover method for field measurement on nitrous oxide fluxes. *Soil Sci. Soc. Am. J.* 45, 311, 1981.
9. Robertson, G.P. Nitrification and denitrification in humid tropical ecosystems: Potential controls on nitrogen retention. In: Procter, J (ed.). *Mineral Nutrients in Tropical Forest and Savana Ecosystems.* Blackwell Scientific Publications, Oxford, U.K., 1989, pp. 55–59.
10. Ginting, D., Kessavalou, A., Eghball, B., and Doran, J.W. Greenhouse gas emissions and soil indicators four years after manure and compost applications. *J. Environ. Qual.* 32, 23, 2003.
11. Nkongolo, N.V., Schmidt, K., Paro, R.M., Hoilett, N.O., Adisa S.J., and Johnson S.S. Improved quantification of CO_2, CH_4, and N_2O fluxes from soil in agricultural fields in central Missouri. *J. Environ. Mon. Rest.* 5, 179, 2008.

18 Improved Nitrogen and Energy-Use Efficiency Using NIR-Estimated Soil Organic Carbon and N Simulation Modeling

Christopher J. Graham, Harold M. van Es, Jeffrey J. Melkonian, and David A. Laird

CONTENTS

18.1 EXECUTIVE SUMMARY

In production agriculture, energy efficiency can be significantly improved by matching the crop's N fertilizer needs and requirements. Current approaches for estimating optimum nitrogen fertilizer rates for maize (*Zea mays* L.) are generally based on regionalized mass-balance equations or expected economic returns. However, N losses occur from dynamic and complex interactions among weather, soil organic matter (SOM) mineralization and hydrology, crop water and N uptake, and management practices. This results in spatially and temporally variable fertilizer N needs. Studies have documented that early-season weather impacts changes between the inorganic and organic soil N pools, which contributes to variability in calculated maize economic optimum in-season N rate values. These interacting and complex spatiotemporal processes can be simulated by well-calibrated models. This chapter discusses the integration of multiple data sources for improved estimates of maize N fertilizer needs. Data from an 11 ha field located in Iowa, United States, will be used in this study. Equipment-mounted near-infrared (NIR) spectroscopy was used to estimate soil organic carbon (SOC) content at 1342 locations and kriged for 136 blocks. Soil texture and hydrology were determined from soil survey information. Using daily weather data and soil information, 24-year simulations were conducted using the precision nitrogen management (PNM) model to estimate late spring rootzone inorganic N content. This was combined with information on crop N uptake potential, mid-season N mineralization, and price ratio corrections to determine optimum sidedress N rates. Spatial and temporal variability in optimum N rate had a range of 60 kg ha^{-1} and field-scale maps were derived for 10th and 90th percentile climate scenarios. This approach provides a framework for integration of relevant spatiotemporal processes to create more precise and locally adapted N fertilizer recommendations for maize.

18.2 INTRODUCTION

18.2.1 NITROGEN CONCERNS

Applying unnecessary N fertilizer can reduce energy efficiency and profitability and increase agricultural impacts on the environment.[1] Maize, a C_4 plant, is physiologically more efficient at utilizing N (more yield per unit N accumulation) than most other major crops, which are generally C_3 plants.[2] But paradoxically, maize production systems as a whole have low fertilizer N uptake and recovery efficiencies (RE). Through on-farm experiments in six north-central U.S. states, average RE was determined to be 37% with a standard deviation of 30%.[3] This suggests low nutrient-use efficiency and both high and variable N losses to the environment.

Bergstrom[4] and Randall et al.[5] reported that in lysimeter studies, maize had higher nitrate-N concentrations in leachate than less-fertilized crops such as soybean (*Glycine max* L.), wheat (*Triticum aestivum* L.), and perennial crops (e.g., alfalfa [*Medicago sativa* L.] and grasses). This was attributed to different fertilizer rates, fertilizer application schedules, timing of crop water and N uptake, and rooting depth. Intensive maize production areas, therefore, pose a risk for N losses to surface and groundwater systems and have become the focus of policy debates on addressing eutrophication and hypoxia concerns.[6]

Energy consumption for N fertilizer production through the Haber–Bosch process and subsequent conversion of NH_3 to other forms of fertilizer ranges from 45 to 65 MJ per kg of N.[7] Energy consumption associated with maize grain production can often account for up to 50% of the total energy use. Nitrous oxide losses from U.S. agriculture are estimated at 377 pG CO_2 equivalent, which accounts for 57% of the total agricultural greenhouse warming potential (GWP) associated with agriculture.[8] Based on several sources and assuming a mixture of N fertilizers, Snyder et al.[9] estimated an overall GWP of 4 kg CO_2 kg^{-1} N.

The environmental impacts of N fertilizer become a larger concern when N is applied in excess of the plant requirements. Studies by van Es et al.[10,11] and Randall[12] indicate that high nitrate leaching is primarily of concern when N is applied in excess of plant uptake. Similarly, Bouwman et al.[13] determined that N_2O emissions remain relatively constant in the range below the crop demand level (i.e., conservative fertilizer rates), but increase significantly when crop demand is exceeded. The precise estimation of optimum N fertilizer rates is therefore critical for both agronomic and environmental reasons,[14] but maize response to applied N is often highly variable and economically optimal N rates (EONR) may range from 0 to 250 kg N ha^{-1}.[15] Therefore, the estimation of the true EONR for a given specific location and growing season has remained elusive.

18.2.2 Estimating Optimum N Rates for Maize

In recent decades, the mass-balance approach has been the most widely used method for making N fertilizer recommendations.[16] It is generally based on a yield goal and associated N uptake, minus credits given for non-fertilizer N sources such as mineralized N from SOM, preceding crops, and organic amendments. Several studies have documented, however, that the relationship between yield and EONR is very weak or nonexistent for humid regions.[17–21] For dryer regions, N response is more influenced by water limitations and associated unattained yield potential.[22] The increased use of yield monitors for site-specific yield measurement and grid-based soil sampling with SOM assessment have generated renewed interest in combining spatial yield data with the mass-balance approach to develop variable-rate fertilizer application technology.[23,24] But Scharf et al.[15] found that the EONR was more strongly related to spatial variability of soil N factors than crop N requirements. In humid regions, yield patterns themselves are highly variable from year to year depending on weather conditions,[20,25] and any mass-balance approach to N fertilizer recommendations would pose the challenging task of predicting yields in the early growing season.

Several leading U.S. maize-producing states have adopted the maximum return to N (MRTN) approach,[21] which largely abandons the mass-balance method. It provides highly generalized recommendations based on multiyear and multilocation field trials, curve-fitting, and economic analyses.[19] Adjustments based on realistic yield expectation are sometimes encouraged. However, owing to its generalization over large areas, soil types and across seasons, it is not adaptive to local conditions and does not address or account for spatial and temporal processes that affect N availability to maize.

18.2.3 Temporal and Spatial N Dynamics

EONR is affected by spatial and temporal processes, and multiple N sources may contribute to the pool of N available for maize N uptake. Approximately 190 kg N ha^{-1} is needed to produce 10 Mg ha^{-1} of maize grain in Nebraska, according to Cassman et al.[3] Mineralization of SOM in central United States can range from 50 to 250 kg N ha^{-1}. In eastern United States, mineralization is slightly lower and averages 80 kg N ha^{-1}.[26] Lobell[27] reported that more precise N management and potentially significant long-term reductions in N rates for maize production with no yield penalty could be obtained if site-specific estimates of SOM mineralization can be made. The difference between the crop requirement (which itself is affected by seasonal environmental stresses) and the soil supply is ideally provided by fertilizer supplements. But the precise estimation of this difference and the associated fertilizer-use efficiency remains a challenge due to numerous sources of variability.

Early season weather, particularly precipitation, has been highly correlated with seasonal variation in optimum fertilizer N rates and nitrate-N losses from crop fields.[28,29] Sogbedji et al.[30] found that growing seasons with excessive wetness in late spring showed lower maize yields and higher EONRs than other years, with an estimated range of 90 kg ha^{-1}. In normal years in humid temperate climates, accumulation of mineral N in the rootzone from SOM mineralization may contribute about half of the required maize N (Figure 18.1). The crop N uptake curve lags behind the organic N mineralization curve until the rapid uptake phase during the mid-vegetative period. During the late spring, high quantities of soil mineral N (SMN) reside in the soil profile, mostly in the nitrate form that is subject to loss. If excessive rainfall occurs during this critical period, significant N losses may occur from leaching or

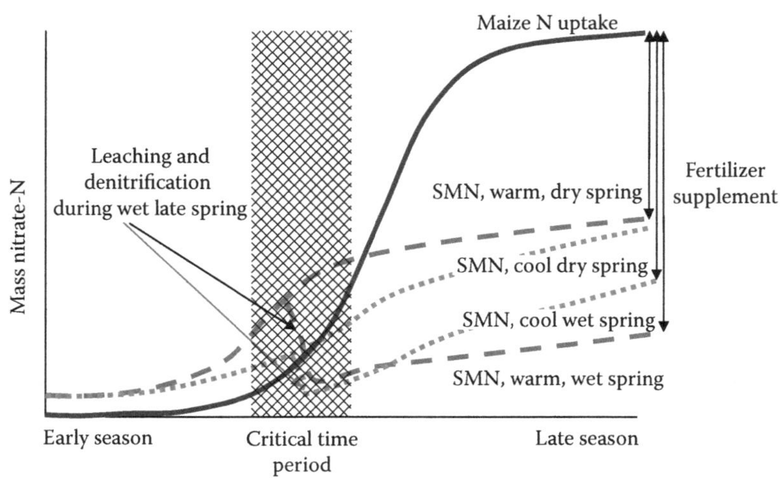

FIGURE 18.1 Conceptual gains and losses of SMN and crop N over a growing season for a low soil N system. Broken and dotted lines represent SMN accumulations; solid line depicts maize N uptake. (Modified from van Es, H.M. et al., Nitrogen management under maize in humid regions: Case for a dynamic approach, in Bruulsema, T., ed., *Managing Crop Nutrition for Weather*, International Plant Nutrition Institute Publ., Norcross, GA, 2007, pp. 6–13.)

denitrification (with warm soil, Figure 18.1). SMN accumulation is generally higher for soils high in organic matter, but this may be subject to losses as well.[31] Losses are also affected by the accumulation of heat units over the first months of a growing season. In cool springs, N mineralization is slow, and the accumulation and subsequent loss of SMN is smaller when excessive wetness occurs (Figure 18.1). Also, the date of planting affects the length of the critical period before the rapid growth period.

Early-season weather appears to be the strongest determinant for seasonal N availability for the crop, but *mid*- and *late*-season weather, especially the occurrence of drought, may also affect yields and create unattained yield potential. When the crop is well fertilized, this tends to result in high residual N levels at the end of the growing season, which are subject to environmental losses.

Several spatial factors impact EONR as well. Sogbedji et al.[32] determined that spatial effects were not consistent from year to year and that they interacted with weather. That is, poorly drained areas had a wider range of annual EONR values (high in wet springs and low in dry springs) than well-drained areas. Similarly, Kay et al.[31] determined that organic matter content (OMC) (a spatially variable N source) and early-season rainfall (driving force for N losses) were the main predictors of seasonal N availability, but they strongly interacted in a nonlinear manner.

In all, optimum N fertilizer rates vary greatly and primarily depend on (1) the amount of readily mineralizable nitrogen in the soil and the mineralization pattern during the growing season, (2) early growing season N losses related to the occurrence and timing of excess wetness and high soil temperatures during those times of saturation, and (3) the occurrence of drought during the mid and late season, resulting in unattained yield potential. Static methods for determining fertilizer rates, including current mass-balance and MRTN approaches, have limited success because they neglect temporal and spatial dynamics in soil N. Use of static fertilizer rates generally results in excessive fertilization in years with dry springs and summers, and inadequate fertilization in years with high N losses from wet spring soil conditions. In many cases, especially when fertilizer to crop price ratios are low, farmers opt to use higher N fertilizer rates (insurance fertilizer) to avoid the risk of yield loss in the event of a wet season. In the majority of years, this results in excessive fertilizer application, unnecessary expense, and increased N losses that adversely impact the environment.[12,32]

Improving the current in-season N recommendations for maize is critical to the credibility of fertility recommendation systems, and increased N use efficiency is expected to reduce residual soil N that may be lost to the environment.[10] This chapter demonstrates a locally adaptive approach that accounts for several sources of spatial and temporal variability, notably variations in weather and SOM, to obtain more precise and site-specific N fertilizer recommendations.

18.3 OBJECTIVES, APPROACH, AND TECHNOLOGIES

18.3.1 Objectives and Approach

Our objective is to establish a framework for site-specific and season-specific N recommendations for maize production in Midwestern United States. We are accomplishing this through the integration of several technologies, including a vehicle-mounted NIR spectrometer that is providing information for estimating soil organic carbon, soil survey information, weather (climate) data, a deterministic N simulation model, and a geographic information system (GIS). Our study site is situated in east-central Iowa (Bremer County; 42° 45′ 25″N, 92° 32′ 13″ W), and consists of gently sloping loam soils across approximately 11 ha. We used a site- and season-specific mass-balance approach in that the various components of the N mass balance were estimated in a spatially and temporally explicit manner and were allowed to interact.

18.3.2 Estimating Soil Organic Carbon with Near-Infrared Spectroscopy

Mineralized organic N is an important source of N for a maize crop, but may have high spatial variability across single fields due to soil forming and erosion patterns. Mapping soil fertility indicators and quantifying soil parameters that control soil processes are important for site-specific soil management. Large numbers of samples must be collected and analyzed in order to capture this spatial variability and adequately estimate soil properties. Conventional methods may be expensive and require large amounts of labor and chemicals for performing these tasks.[33] NIR reflectance spectroscopy is a low-cost method that can be used to substitute or complement traditional soil characterization methods. It measures soil constituents that have unique absorption features in this wavelength region due to overtones related to stretching and bending vibrations in molecular bonds such as C–C, C–H, N–H, and O–H.[34] Once calibrated, the methodology can be used to predict multiple soil characteristics simultaneously and explain within-field spatial variability. Chang et al.[35] and Reeves et al.[36] reported successful predictions ($R^2 > 0.80$) for several properties including total organic carbon and N (g kg^{-1}), which are important properties for precise N management.

18.3.3 Using Models for N Management

More precise management of N under maize in humid regions requires the explicit consideration of interacting factors that vary in both space (site-specific) and time (primarily, as defined by variation in weather conditions). In humid regions, crop N requirements for maize cannot be accurately predicted at the beginning of the growing season (even less so during the previous fall), because one of the main determining factors (spring weather) is still undetermined at that time. Even slow-release or nitrification-inhibition technology and early season soil testing can only achieve limited accuracy, because a large part of the maize N needs are derived from organic N mineralization that is affected by early-season weather factors. Environmental information systems and simulation models effectively allow for incorporation of both temporal and spatial processes related to N dynamics. This approach can take advantage of increasingly sophisticated environmental databases (e.g., radar-based precipitation estimates)[37]; that can be accessed as input information for dynamic soil and plant models to estimate crop growth and soil N dynamics, and provide more precise estimates of seasonal crop N needs.[38,39] If these models are well calibrated and tested, they can provide information to growers to adjust in-season N applications to more precisely match crop N demand.[40,41]

The PNM model[37,42] was developed to track soil and crop N flows in maize cropping systems. Critical outputs of the PNM model are simulated values of mineralized N and losses through leaching, denitrification, and volatilization, as well as crop N uptake and biomass (vegetative and grain) accumulation.

The model has two components: LEACHN, the N (and phosphorus) module of LEACHM[43] and a maize N uptake, growth, and yield model.[44] LEACHN is a process-based, one-dimensional (1D) model that simulates water and solute transport, and chemical and biological N transformations in the unsaturated soil zone.[43] Flows between different pools of C and N are simulated in each soil segment as well as on the soil surface. LEACHN is well suited for simulating soil N processes and has been extensively used and tested in several studies.[32,45–50] The rate constants in the equations describing nitrification, denitrification, manure mineralization, and plant residue mineralization were calibrated based on multiyear, replicated field experiments.[50,51] In the PNM model, SOM mineralization is simulated using two rate constants instead of one as in LEACHN: a higher rate constant for early season SOM mineralization (up to July 15) and a lower rate constant for later in the season, based on studies by Dharmakeerthi et al.[52]

The crop component of the PNM model is based on a maize N uptake, growth, and yield model.[44] The subroutines of the maize N uptake, growth, and yield model incorporate the effects of temperature, solar radiation, water supply, and parameters influencing the crop N budget during the three major phases of maize development: vegetative growth, anthesis, and grain fill.[44,53–55]

18.3.4 Methodology

18.3.4.1 Mass-Balance Inputs

The database for the following project is available in the data disk provided with this book. In this file, each point, a northing and easting, represents a point on the 30 × 30 m grid in ArcGIS. The N recommendations are based on the mass-balance model:

$$\text{N recommendation} = 21.4 \times \text{Yield Goal} - N_s \text{ (indigenous soil N)}$$

The recommendations are in columns L–M and are defined as

$$\text{Column J} = N - H - I - E$$

$$\text{Column K} = N - H - I - F$$

$$\text{Column L} = N - G - I - E$$

$$\text{Column M} = N - G - I - F$$

where column N contains the value for the yield goal calculated as 21.4 × 11 (Mg ha^{-1}), which equals 235.4 kg ha^{-1}. This value essentially represents the estimated N needed to attain the desired/predicted yield. The columns J–M represent the simulated amounts of recommended fertilizer for the 10th and 90th percentiles (Figure 18.7) at two price ratios.

To calculate the site- and season-specific N-fertilizer recommendations, we used the basic mass-balance approach.[16,17] As Stanford[16] notes, the mass-balance approach as it was originally formulated does not necessarily encompass the dynamic nature of the ecological system controlling N. We postulate that the concerns with the mass-balance approach are primarily the result of unaccounted for site- and season-specific conditions (especially seasonal variations in rainfall), thereby masking the inherent scientific validity of the methodology. In the approach discussed below, we explicitly incorporate the spatial and temporal sources of variability (soil types, drainage class, OMC, weather, and crop management) as well as their interaction in the estimation of maize N fertilizer needs, assuming that the timing of fertilizer addition is optimized and occurs as sidedress or topdress in the late spring—for our case presumed to be soon after June 15. A modified version of the mass-balance equation was used:

$$\text{Nrec}_i = \text{Nupt}_i - \text{Nrz\_pre15jun}_i - \text{Nrz\_post15jun}_i - \text{Npr\_corr}_i \qquad (18.1)$$

where

Nrec_i is the sidedress N recommendation for a management unit i
Nupt_i is the total crop N uptake
Nrz_pre15jun_i is the total rootzone N on June 15
Nrz_post15jun_i is the estimated N mineralized from June 15 to harvest
Npr_corr_I is a correction between the estimated maximum yield and the economically optimum yield based on the fertilizer-to-grain price ratio

These recommendations were made for each of 136 blocks within the field for two climate years and two price ratios. The following sections describe how the input data for the mass-balance equation were obtained and the recommendations derived.

18.3.4.2 Yield Goal and Estimated N Uptake

The yield goal was fixed at 11.0 Mg ha^{-1} for the purpose of this research, which was in line with the estimates from the PNM model (Table 18.1). Nupt_i (kg ha^{-1}) was estimated by multiplying the yield goal by 21.4, which is a standard assumption for maize production.[22]

18.3.4.3 Soil Survey Data

The PNM model requires basic soil information including textural class, drainage class, and organic C (or organic matter) content by depth to 1 m. Other than organic C, basic soils data were derived from the natural resources conservation service

TABLE 18.1
Maximum Yield and Optimum N Fertilizer Rates at Maximum Yield and 5.6 and 11.2 Fertilizer-to-Grain Price Ratios

	Max Yield (Mg ha^{-1})	N Rate at Maximum Yield (kg ha^{-1})	EONR at 5.6 Price Ratio (kg ha^{-1})	EONR at 11.2 Price Ratio
Floyd loam—1.5% OC	10.92	190	175	163
Floyd loam—2% OC	10.98	180	170	154
Floyd loam—2.5% OC	11.22	181	168	151
Floyd loam—3% OC	11.12	174	162	144

soil survey database through the web soil survey portal (http://websoilsurvey.nrcs.usda.gov). Six soil types were mapped for the field (Figure 18.2), mostly Floyd loam (fine-loamy, mixed, superactive, mesic Aquic Pachic Hapludoll) and geographically associated soils (Clyde silty clay loam, Kenyon loam, Ostrander loam, Waukee loam, and Dickinson fine sandy loam). Slope classes included A: 0%–2%; B: 2%–5%; and C: 5%–9%. Soil profile data including drainage class and textural class by horizon were derived from the soil survey report.[18] Soil survey data were adapted to quantitative data by depth layer as close as possible, which in some cases involved interpolation. The soil survey information was digitized and entered into ArcGIS (ESRI, Redlands, CA) software.

18.3.4.4 Soil Organic Carbon from NIR Reflectance Spectroscopy

The field distribution of soil organic C content was obtained through the use of a vehicle-mounted visible and near infrared (VIS-NIR) reflectance spectroscopy unit that was located in a soil-engaged shank (Veris Technologies, Salina, KS). Data from this field and seven nearby fields were used to calibrate the NIR estimates (with visible wavelengths deleted) with laboratory measured samples using partial least squares regression with an optimum number of principal components (5) using The Unscrambler v 8.0 software (CAMO Software AS, Oslo, Norway). Approximately 20 GPS (Global Positioning System) registered soil samples were collected from each field immediately after the field mobile NIR spectrometer passed the sample location. Thus, spectra, collected real time in the field, were available for each of the calibration samples. Conservative leave-one-field-out calibrations were based on truncated first difference spectra in the wavelength range of 1002–1681 nm. This yielded a calibration curve with a coefficient of determination of 0.75 (Figure 18.3). The field mobile NIR system was used to estimate organic C for 1302 locations within the field, which were consolidated into one hundred and thirty-six 30 × 30 m management units using block kriging based on spherical variograms and examination of quantile–quantile

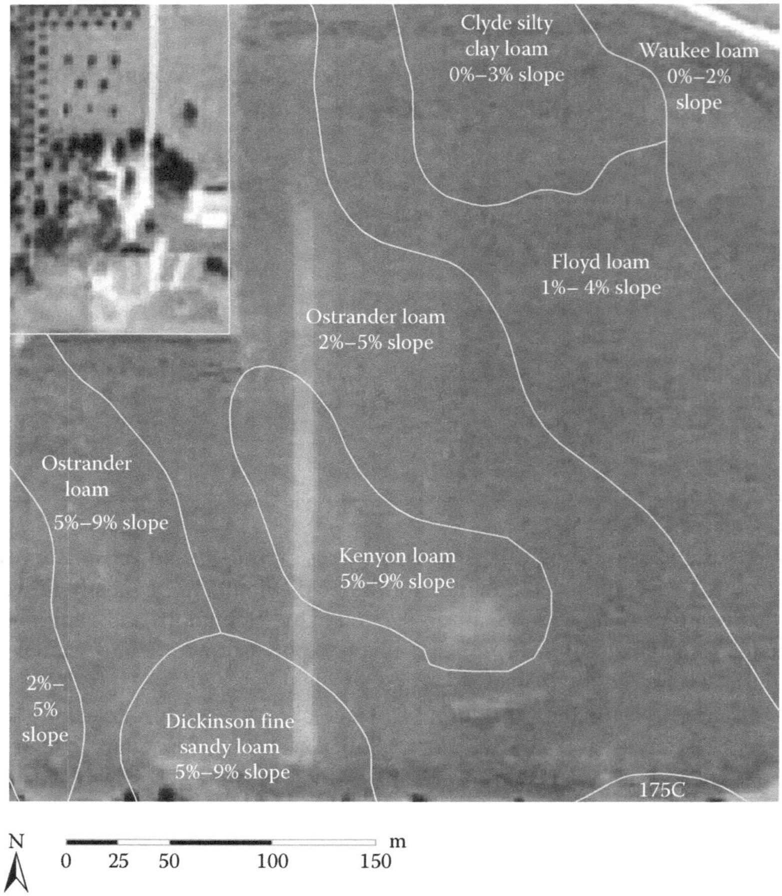

FIGURE 18.2 Study site and soil map. (From USDA-NRCS Web Soil Survey. Available at, http://websoilsurvey.nrcs.usda.gov.)

plots in the ArcGIS Geostatistical_Analyst tool (Figure 18.4). These organic C estimates, combined with the soil survey information, provided the basic soil inputs for the N simulations.

18.3.4.5 PNM Model Simulations

After the soil data were entered into the model, multiyear PNM simulations of soil and crop N dynamics were performed for 24 climate years (1985–2008) based on observations from a nearby weather station (Dumont, IA). We simulated a continuous maize cropping system under a plow-till system. N was applied as a 32% solution of urea and ammonium nitrate (UAN32) with 22.42 kg ha^{-1} applied at planting on May 1 each year and an additional sidedress application of 145.73 kg ha^{-1} (130 lb ac^{-1}) applied on June 15. Simulations were executed at the Cornell Center for Advanced Computing for each of the 136 management units of the field based on the site-specific soil information.

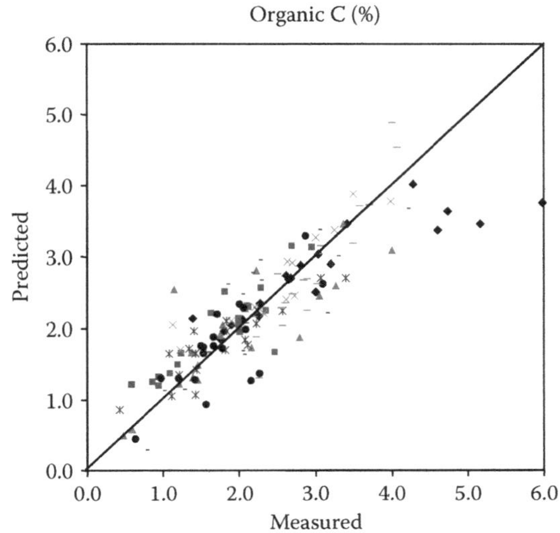

FIGURE 18.3 Correlation between laboratory-measured and NIR-estimated organic C contents obtained using leave-one-field-out cross-validation for eight fields in east-central Iowa ($R^2 = 0.78$; RPD = 2.1). Different symbols represent different fields.

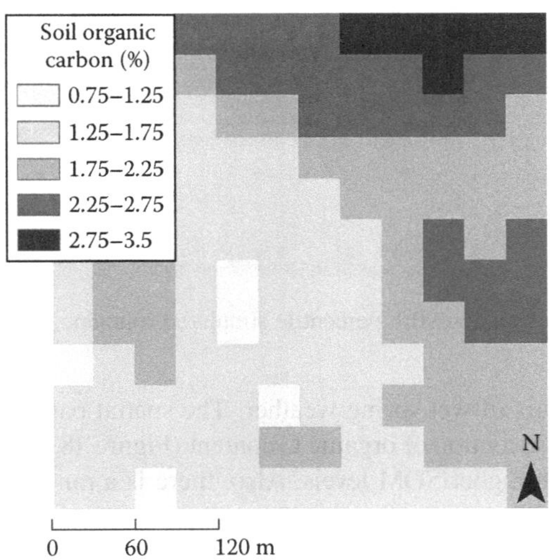

FIGURE 18.4 Soil organic C content distribution across the study site.

18.4 RESULTS

18.4.1 Rootzone N in Early Season

The PNM simulations yielded estimates of the amount of mineral rootzone N on June 15 ($Nrz\_pre15jun_i$) for each of the 136 management units and each of the 24 simulation years. Figure 18.5 shows the distribution of the rootzone N on June 15 for the 10th and 90th percentile of the 24 simulation years, the former generally

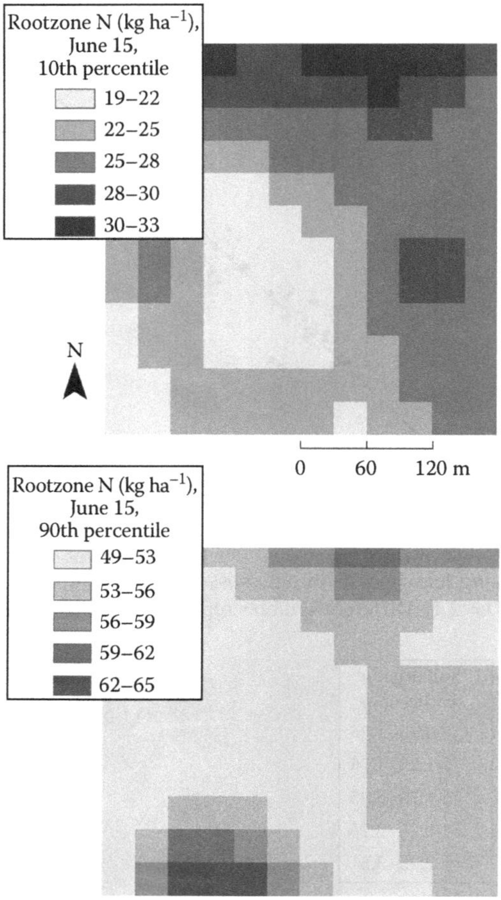

FIGURE 18.5 The 10th and 90th percentile simulated rootzone N at June 15 based on 24 climate years.

associated with years of wet spring weather. The spatial patterns in both years are correlated to the distribution of organic C content (Figure 18.4), indicating greater N mineralization with higher SOM levels. Also, there is a range of 40 kg ha^{-1} among the locations and years.

18.4.2 Rootzone N Mineralization in Late Season

The model simulations also provided estimates for post-sidedress mineralization amounts (June 16 to harvest; $Nrz\_post15jun_i$). This represents the soil N that is mineralized in the mid- and late-growing season that contributes to crop N availability. Since this contribution occurs after the time of fertilizer application, it cannot be estimated based on the conditions for a specific growing season. However, it can be assumed that this N release is not subject to leaching or denitrification losses due to high crop transpiration amounts during this time period (primarily

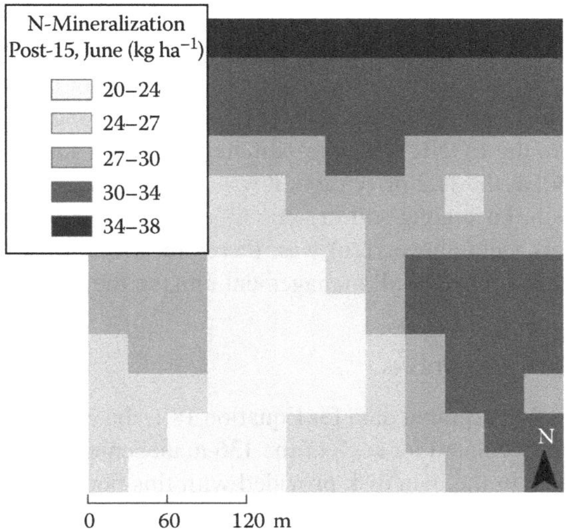

FIGURE 18.6 Average simulated N contributed during mid and late season (June 15 and harvest).

due to the large leaf area in mid and late season). We, therefore, generated site-specific estimates for each of the 136 management units based on the average mineralization during the post-June 15 period (Figure 18.6). The N contributions for mid- and late-season mineralization were estimated to range from 20 to 38 kg ha^{-1} for the different locations in the field, following mostly the distribution in soil organic C content (Figure 18.4).

18.4.3 Price Ratio Correction

Maize response to nitrogen is a nonlinear process, and the maximum yield is generally not the economically optimum yield. In his original assessment, Stanford[16] noted that the final 10% of yield required up to half the total N applied. The EONR accounts for this fact by calculating the marginal return of N,[56] which effectively calculates the rate at which fertilizer application no longer returns a profit. This generally depends on the cost of fertilizer relative to the return rate (price) of the grain, which is expressed as the price ratio (\$ kg^{-1} of N per \$ kg^{-1} of maize grain). These price ratios vary with commodity prices, but have nevertheless not changed much over the long term, because maize grain and fertilizer prices tend to rise and fall together.[57]

Estimating the EONR was performed using the PNM model. Naturally, the ratio of the price of fertilizer to the price of maize will alter the rate at which maximum profit can be attained where the larger ratio (i.e., high fertilizer cost relative to grain price) decreases the economic optimum rate. We conducted a separate set of 24-year PNM simulations to estimate appropriate price ratio corrections, Npr_corr_i in Equation 18.1, for the dominant Floyd loam soil at organic C levels of 1.5%, 2.0%, 2.5%, and 3.0%. Simulations were performed for N sidedress rates of 0, 25, 50, 75,

100, 125, 150, 175, and 200 kg N ha^{-1} to define the N response curve, which was subsequently modeled with a quadratic equation using Microsoft Excel (Microsoft Corp., Redmond, WA). This allowed for the quantification of the maximum yield as well as the economic optimum yield at two price ratios, 5.6 and 11.2 (Table 18.1). For the 5.6 price ratio, the EONR was approximately 10–15 kg ha^{-1} lower than the rate at maximum yield; at the 11.2 price ratio, it was 26–30 kg ha^{-1} lower. The price ratio effect was very similar among soil organic C levels and a soil-averaged Npr_corr_i value (corrected as yield above zero) was, therefore, used for each price ratio, and each was uniformly applied to all management units in the field.

18.4.4 N Recommendations

After having obtained all input data for Equation 18.1, the N fertilizer recommendation, $Nrec_i$ can be calculated for each of the 136 management units in the field. This data set is available in the data disk provided with this book. Estimates were made for 2 × 2 factorial combinations: the 10th and 90th percentile of the distribution of simulated June 15 N mineralization for the 24 climate years, and the 5.6 and 11.2 fertilizer-to-maize grain price ratios. Figure 18.7 shows the maps of recommended N fertilizer rates. The higher recommended rates are in the areas with low organic C content (Figure 18.4) and N mineralization potential (Figures 18.6 and 18.7). The recommended rates ranged from 149 to 179 kg ha^{-1} for the 10th percentile at the 5.6 price ratio, and were reduced by approximately 30 kg ha^{-1} (119–146) for the 90th percentile, which represents years with dry, early growing seasons that experience little N loss from excessive rainfall. The effect of the price ratio reduces the recommended N rate by 17 kg ha^{-1} for the 11.2 compared to the 5.6 price ratio.

18.5 DISCUSSION

The above process describes a framework for improved (more precise and locally adapted) estimation of economically optimum fertilizer N rates for late spring applications. We used 24-year climate data to gain a better understanding of the effects of early-season weather and we used NIR-based estimates of soil organic C and soil survey information as inputs for assessing the spatial variability component. These technologies are currently available and implementation and testing of this approach is therefore feasible. For a typical management implementation, the simulations would be done for only the present growing season.

In our analysis, the ranges associated with spatial (primarily soil-related) variability and temporal (primarily weather-related) variability were estimated to be approximately the same, 30 kg ha^{-1}, with little interaction. This implies that EONR for any location and any year may vary by up to 60 kg N ha^{-1} (Figure 18.7) for a given price ratio. This could provide significant savings to a farmer and reductions in environmental losses. The benefits of using seasonally adapted N rates will be enhanced if the supplemental N applications occur somewhat later (e.g., early July) in the growing and can incorporate longer weather records. The approach can also be used with conservative at-planting N fertilizer applications and model-based supplemental rates in the years with wet springs.

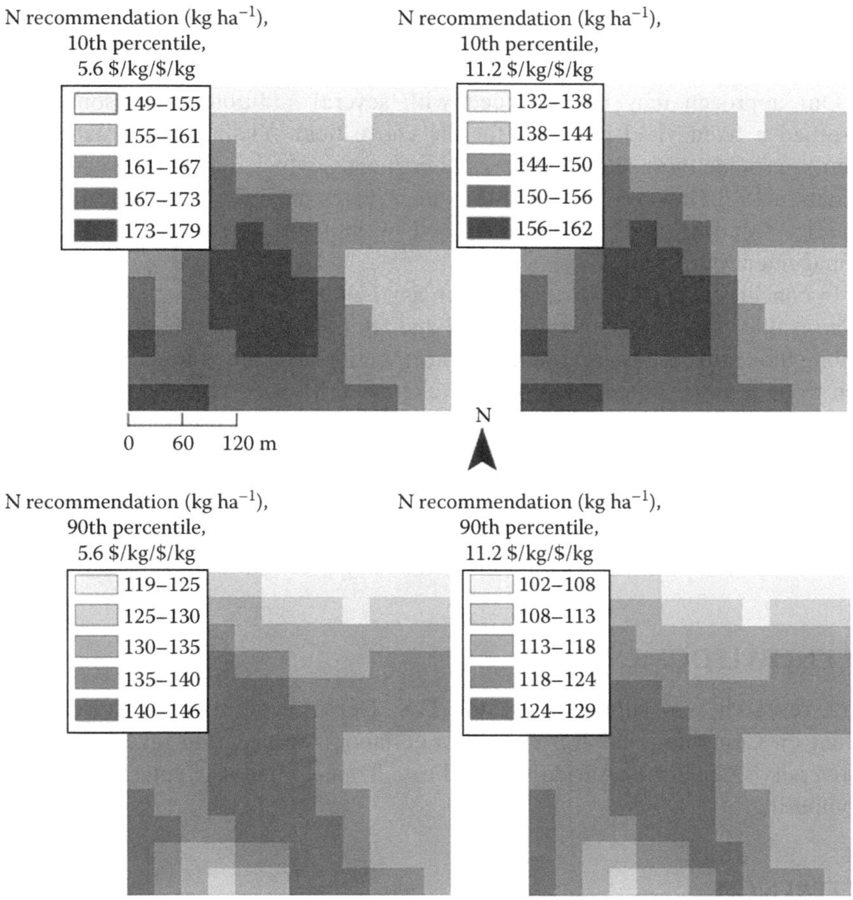

FIGURE 18.7 Recommended nitrogen fertilizer rates at 5.6 and 11.2 price ratios for the 10th and 90th percentiles of the distribution of rootzone N on June 15.

The spatial variability may be somewhat higher than predicted by our model. The PNM model performs 1D simulations of the soil-crop system, and cannot account for three-dimensional redistribution of water in the landscape. This means that imperfect drainage in lower landscape positions may be accentuated and the N losses are greater in those locations in wet, early growing seasons.

The effort involved in developing the N fertilizer recommendations may appear excessive, but the framework mostly involves one-time investments. With the increased availability of NIR spectroscopy for field scale soil assessment—or alternatively, the more conventional grid-based soil analyses or Order 1 soil surveys[58]—many farmers are willing to invest in the characterization of soils at the subfield scale if they can more profitably manage crop inputs. This one-time investment can then be entered into a GIS and used for multiple years of soil and crop management applications. The N recommendation system can be built into a GIS, but requires an infrastructure to perform dynamic simulations in real time. The Adapt-N tool (http://adapt-N.eas.cornell.edu) is currently available for such purposes, where the PNM model can be run based on stakeholder inputs and up-to-date high-resolution climate data.[37]

Our approach may be enhanced with several additional data sources. We assumed a fixed yield potential for the entire field. Yield monitor data can be used to provide more site-specific yield estimates, which can then be entered into Equation 18.1. However, yield patterns themselves may vary among growing seasons[20,25,58] and are still mostly undefined by late spring, posing a challenge for management purposes.

In conclusion, the EONR for maize in any field is not a fixed quantity, but varies as a result of several interacting spatial and temporal factors. The most significant among those are early-season weather (precipitation and temperature), N mineralization from organic sources, and crop development. Most current N fertilizer recommendation systems ignore these dynamic processes, which limits their capacity to precisely manage N. We presented a framework for improved estimation of EONR for maize, which accounts for these spatial and temporal processes. Gains from this approach appear to be considerable, but field verification will be required for future adoption.

ACKNOWLEDGMENTS

This research was supported by the U.S. Department of Agriculture Special Grant on Computational Agriculture, a collaborative program involving Cornell University's College of Agriculture and Life Sciences and its Center for Advanced Computing.

REFERENCES

1. McIsaac, G.F., David, M.B., Gertner, G.Z., and Goolsby, D.A. Relating net nitrogen input in the Mississippi River basin to nitrate flux in the lower Mississippi River: A comparison of approaches. *J. Environ. Qual.* 31, 1610, 2002.
2. Greenwood, D.J., Lemaire, G., Gosse, G., Cruz, P., Draycott, A., and Neetson, J.T. Decline in the percentage of N of C3 and C4 crops with increasing plant mass. *Ann. Bot.* 66, 425, 1990.
3. Cassman, K.G., Dobermann, A., and Walters, D.T. Agroecosystems, nitrogen-use efficiency, and nitrogen management. *AMBIO* 31, 132, 2002.
4. Bergstrom, L. Nitrate leaching and drainage from annual and perennial crops in tile-drained plots and lysimeters. *J. Environ. Qual.* 16, 11, 1987.
5. Randall, G.W., Huggins, D.R., Russelle, M.P., Fuchs, D.J., Nelson, W.W., and Anderson, J.L. Nitrate loss through subsurface tile drainage in conservation reserve program, alfalfa, and row crop systems. *J. Environ. Qual.* 26, 1240, 1997.
6. Diaz, R.J. and Rosenberg, R. Spreading dead zones and consequences for marine ecosystems. *Science* 321, 926, 2008.
7. Wang, M. *The Greenhouse Gases, Regulated Emissions, and Energy Use in Transportation (GREET) Model.* Argonne National Laboratory, Chicago, IL. http://www.transportation.anl.gov/software/GREET/index.html (verified January 8, 2010), 2007.

8. U.S. EPA. *Inventory of U.S. Greenhouse Gas Emissions and Sinks: 1990–2005*. EPA 430-R-07-002. U.S. Environmental Protection Agency, N.W., Washington, DC, 2007.
9. Snyder, C.S., Bruulsema, T.W, Jensen, T.L., and Fixen, P.E. Review of greenhouse gas emissions from crop production systems and fertilizer management effects. *Agric. Ecosyst. Environ*. 133, 247, 2009.
10. van Es, H.M., Czymmek, K.J., and Ketterings, Q.M. Management effects on N leaching and guidelines for an N leaching index in New York. *J. Soil Water Conserv*. 57, 499, 2002.
11. van Es, H.M., Kay, B.D., Melkonian, J.J., and Sogbedji, J.M. Nitrogen management under maize in humid regions: Case for a dynamic approach. In: Bruulsema, T. (ed.). *Managing Crop Nutrition for Weather*. International Plant Nutrition Institute Publ., Norcross, GA, 2007, pp. 6–13.
12. Randall, G. Risks associated with nitrogen rate decisions. In: Sawyer, J., Nafziger, E., Randall, G., Bundy, L., Rehm, G., and Joern, B. *Concepts and Rationale for Regional Nitrogen Guidelines for Corn*. Iowa State University Extension Publication, PM2015, 2006, 27 p.
13. Bouwman, A.F., Boumans, L.J.M., and Batjes, N.H. Emissions of N_2O and NO from fertilized fields: Summary of available measurement data. *Global Biogeochem. Cycl*. 16, 1058, 13 p., 2002.
14. Ostergaard, H.S. Agronomic consequences of variable N fertilization. In: Stafford, J.V. (ed.). *Precision Agriculture '97, Vol. I, Spatial Variability in Soil and Crop*. BIOS Scientific Publishers, Oxford, U.K., 1997, pp. 315–320.
15. Scharf, P.C., Kitchen, N.R., Sudduth, K.A., and Davis, J.G. Spatially variable corn yield is a weak predictor of optimum nitrogen rate. *Soil Sci. Soc. Am. J*.70, 2154, 2006.
16. Stanford, G. Rationale for optimum nitrogen fertilization in corn production. *J. Environ. Qual*. 2, 159, 1973.
17. Lory, J.A. and Scharf, P.C. Yield goal versus delta yield for predicting nitrogen fertilizer need in corn. *Agron. J*. 95, 994, 2003.
18. Camp, L.D. *Soil Survey of Bremer County, Iowa*. U.S. Department of Agriculture. Natural Resources Conservation Service, 2009.
19. Vanotti, M.B. and Bundy, L.G. Corn nitrogen recommendations based on yield response data. *J. Prod. Agric*. 7, 249, 1994.
20. Katsvairo, T., Cox, W.J., van Es, H.M., and Glos, M.A. Spatial yield responses of two corn hybrids to two N levels. *Agron. J*. 95, 1012, 2003.
21. Sawyer, J., Nafziger, E., Randall, G., Bundy, L., Rehm, G., and Joern, B. *Concepts and Rationale for Regional Nitrogen Guidelines for Corn*. Iowa State University Extension Publication. PM2015, 2006, 27 p.
22. Kim, K.I., Clay, D.E., Carlson, C.G., Clay, S.A., and Trooien, T. Do synergistic relationships between nitrogen and water influence the ability of corn to use nitrogen derived from fertilizer and soil. *Agron. J*. 100, 551–556, 2008.
23. Ferguson, R.B., Hergert, G.W., Schepers, J.S., Gotway, C.A., Cahoon, J.E., and Peterson, T.A. Site-specific nitrogen management of irrigated maize: Yield and soil residual nitrate effects. *Soil Sci. Soc. Am. J*. 66, 544, 2002.
24. Khosla, R., Fleming, K., Delgado, J.A., Shaver, T., and Westfall, D.G. Use of site-specific management zones to improve nitrogen management for precision agriculture. *J. Soil Water Conserv*. 57, 513, 2002.
25. Kahabka, J.E., van Es, H.M., McClenahan, E.J., and Cox, W.J. Spatial analysis of maize response to N fertilizer in Central New York. *Precis. Agric*. 5, 463, 2004.
26. Ketterings, Q.M., Klausner, S.D., and Czymmek, K.J. Nitrogen guidelines for field crops in New York. Second Release. *Department of Crop and Soil Extension Series E03-16*. Cornell University, Ithaca, NY, 2003, p. 70.

27. Lobell, D.B. The cost of uncertainty for nitrogen fertilizer management: A sensitivity analysis. *Field Crop. Res.* 100, 210, 2007.
28. Balkcom, K.S., Blackmer, A.M., Hansen, D.J., Morris, T.F., and Mallarino, A.P. Testing soils and cornstalks to evaluate nitrogen management on the watershed scale. *J. Environ. Qual.* 32, 1015, 2003.
29. Mitsch, W.J., Day, J.W., Gilliam, J.W., Groffman, P.M., Hey, D.L., Randall, G.W., and Wang, N. Reducing nitrogen loading to the Gulf of Mexico from the Mississippi River basin: Strategies to counter a persistent ecological problem. *BioScience* 51, 373, 2001.
30. Sogbedji, J.M., van Es, H.M., Klausner, S.D., Bouldin, D.R., and Cox, W.J. Spatial and temporal processes affecting nitrogen availability at the landscape scale. *Soil Till. Res.* 58, 233–244, 2001.
31. Kay, B.D., Mahboubi, A.A., Beauchamps, E.G., and Dharmakeerthi, R.S. Integrating soil and weather data to describe variability in plant available nitrogen. *Soil Sci. Soc. Am. J.* 70, 1210, 2006.
32. Sogbedji, J.M., van Es, H.M., and Hutson, J.L. N fate and transport under variable cropping history and fertilizer rate on loamy sand and clay loam soils: I. Calibration of the LEACHMN model. *Plant Soil.* 229, 57, 2001.
33. Viscarra Rossel, R.A. and McBratney, A.B. Soil chemical analytical accuracy and costs: Implications from precision agriculture. *Aust. J. Exp. Agr.* 38, 765, 1998.
34. Dalal, R.C. and Henry, R.J. Simultaneous determination of moisture, organic carbon, and total nitrogen by near infrared reflectance spectrophotometry. *Soil Sci. Soc. Am. J.* 50, 120, 1986.
35. Chang, C.W., Laird, D.A., Mausbach, M.J., Maurice, J., and Hurburgh, J.R. Near-Infrared reflectance spectroscopy—Principal components regression analyses of soil properties. *Soil Sci. Soc. Am. J.* 65, 480, 2001.
36. Reeves, J., McCarty, G., and Mimmo, T. The potential of diffuse reflectance spectroscopy for the determination of carbon inventories in soils. *Environ. Pollut.* 116, 277, 2002.
37. Melkonian, J., van Es, H.M., DeGaetano, A.T., Sogbedji, J.M., and Joseph, L. Application of dynamic simulation modeling for nitrogen management in maize. In: Bruulsema, T. (ed.). *Managing Crop Nutrition for Weather*. International Plant Nutrition Institute, 2007, pp. 14–22.
38. Olness, A.E., Lopez, D., Cordes, J., Sweeney, C., and Voorhees, W.B. Predicting nitrogen fertilizer needs using soil and climatic data. In: Vermoesen, (ed.). *Proceedings of the 11th World Fertilizer Congress*, Gent, Belgium, September 7–13, 1997. International Centre of Fertilizers, Gent, Belgium, 1998, pp. 356–364.
39. van Alphen, B.J. and Stoorvogel, J.J. A Methodology for precision nitrogen fertilization in high-input farming systems. *Precis. Agric.* 2, 319, 2000.
40. Kersebaum, K.C. Application of a simple management model to simulate water and nitrogen dynamics. *Ecol. Modell.* 81,145, 1995.
41. Smith, J.U., Dailey, A.G., Glendining, M.J., Bradbury, N.J., Addiscott, T.J., Smith, P., Bide, A. et al. Constructing a nitrogen fertilizer recommendation system using a dynamic model: What do farmers want? *Soil Use Manage.* 13, 225, 1997.
42. Melkonian, J., van Es, H.M., and Joseph, L. *Precision Nitrogen Management Model: Simulation of Nitrogen and Water Fluxes in the Soil-Crop-Atmosphere Continuum in Maize (Zea mays* L.*) Production Systems. Version 1.0.* Department of Crop and Soil Sciences, Research Series No. R05-2. Cornell University, Ithaca, NY, 2005.
43. Hutson, J.L. *Leaching Estimation and Chemistry Model: A Process-Based Model of Water and Solute Movement, Transformations, Plant Uptake, and Chemical Reactions in the Unsaturated Zone. Version 4.* Department of Crop and Soil Sciences, Research series No. R03-1. Cornell University, Ithaca, NY, 2003.
44. Sinclair, T.R. and Muchow, R.C. Effect of nitrogen supply on maize yield: I. Modeling physiological responses. *Agron. J.* 87, 632, 1995.

45. Jabro, J.D., Lotse, J., Fritton, D.D., and Baker, D.E. Estimation of preferential movement of bromide tracer under field conditions. *J. Hydrol.* 156, 61, 1994.
46. Jemison, J.M. Jr., Jabro, J.D., and Fox, F.H. Evaluation of LEACHM: Simulation of cumulative drainage, bromide leaching, and corn bromide uptake. *Agron. J.* 86, 843, 1994.
47. Jemison, J.M. Jr., Jabro, J.D., and Fox, F.H. Evaluation of LEACHM: Simulation of nitrate leaching from nitrogen-fertilized and manured corn. *Agron. J.* 86, 852, 1994.
48. Lotse, E.G., Jabro, J.D., Simmons, K.E., and Baker, D.E. Simulation of nitrogen dynamics and leaching from arable soils. *J. Contam. Hydrol.* 10, 183, 1992.
49. Sogbedji, J.M., van Es, H.M., Hutson, J.L., and Geohring, L.D. N fate and transport under variable cropping history and fertilizer rate on loamy sand and clay loam soils: II. Performance of LEACHMN using different calibration scenarios. *Plant Soil.* 229, 71, 2001.
50. Sogbedji, J.M., van Es, H.M., Melkonian, J., and Schindelbeck, R.R. Evaluation of the PNM model for simulating drain flow nitrate-N concentration under manure-fertilized maize. *Plant Soil.* 282, 343, 2006.
51. van Es, H.M., Sogbedji, J.M., and Schindelbeck, R.R. Effect of manure application timing, crop, and soil type on nitrate leaching. *J. Environ. Qual.* 35, 670, 2006.
52. Dharmakeerthi, R.S., Kay, B.D., and Beauchamp, E.G. Factors contributing to changes in plant available nitrogen across a variable landscape. *Soil Sci. Soc. Am. J.* 69, 453, 2005.
53. Muchow, R.C. and Sinclair, T.R. Water deficit effects on maize yields modeled under current and "greenhouse" climates. *Agron. J.* 83, 1052, 1991.
54. Muchow, R.C., Sinclair, T.R., and Bennett, J.M. Temperature and solar radiation effects on potential maize yield across locations. *Agron. J.* 82, 338, 1990.
55. Sinclair, T.R. and Amir, J. A model to assess nitrogen limitations on the growth and yield of spring wheat. *Field Crop. Res.* 30, 63, 1992.
56. Nafziger, E.D., Sawyer, J.E., and Hoeft, R.G. Formulating N recommendations for corn in the cornbelt using recent data. In: *Proceedings of the North Central Extension-Industry Soil Fertility Conference*, Des Moines, IA, 20, 5, 2004.
57. Bruulsema, T.W. and Murrell, T.S. Corn fertilizer decisions in a high-priced market. *Better Crops* 92, 16, 2008.
58. Chang, J., Clay, D.E., Carlson, C.G., Reese, C.L., Clay, S.A., and Ellsbury, M.M. Defining yield goals and management zones to minimize nitrogen and phosphorus fertilizer recommendation errors. *Agron. J.* 96, 825, 2004.

19 Computing Wheat Nitrogen Requirements from Grain Yield and Protein Maps

Daniel S. Long and R.E. Engel

CONTENTS

19.1 EXECUTIVE SUMMARY

Optical protein sensors and mass-flow yield monitors provide the opportunity to continuously measure grain quality and quantity during harvesting. This chapter illustrates how yield monitor and grain protein measurements may provide useful postharvest information for evaluating water or nitrogen (N) limitations in wheat. The surface-mapping software Surfer is used to create yield and protein maps that share a common grid, and then calculate maps of critically low protein, N removed in grain, and N management zones. Analysis of a critical spring wheat protein level provided site-specific information needed to assess where N had been adequate or deficient within an irrigated northern Montana production field. Where N was adequate (≥132 g of protein kg^{-1} of grain), variable-rate N management could be based on replacing N at the rate at which it was removed in grain. Where it was deficient (<132 g kg^{-1}), N management was based on university recommendations that involve yield potentials and soil nitrate-N test values.

19.2 INTRODUCTION

On-combine yield monitors are essential tools for characterizing the spatiotemporal variability of crop yields within farm fields. Yield maps produced from yield monitors provide important feedback for determining the effectiveness of management and gauging the influence of weather, soil properties, pests, and other environmental factors on crop productivity. In addition, producers who have recorded multiple years of yield data are able to identify the most consistently productive (or unproductive) portions of their fields.[1,2] This information can serve as a basis for identifying management zones and implementing variable-rate application.[3,4] Though yield maps document yield spatial variability, they do not reveal the cause of the variation. If yield maps are to be valuable, they must be properly interpreted for their incorporation into enterprise analysis, decision making, and overall farm planning.

One way to better understand a yield map is to study the relationship between yield and grain protein concentration. For example, water most likely limited yields in areas with high grain protein concentration.[5–7] An inverse yield–protein relationship is caused by the dilution of grain nitrogen (N) by a much greater grain biomass accumulation.[8] On the other hand, N fertility most likely limited yields in areas where the relationship is positive (i.e., grain yield and protein increase together). Whelan et al.[9] utilized these relationships to assess spring wheat yield variability within farm fields in eastern Australia. Locally positive relationships represented conditions where N availability was limited by soil/landscape conditions, whereas water stress was indicated where the local relationship was negative. These results were useful for improving N fertilizer management within farm fields.

In addition, the protein concentration of grain has been proposed for use as a qualitative indicator of a grower's N fertilizer program when contrasted with an established critical level.[10] The rationale is that crops are excellent indicators of soil conditions in the root zone and that spatial patterns in grain protein are correlated with patterns in soil N fertility. In Montana, Engel et al.[10] measured the grain protein of hard-red spring wheat grown under varying N and water regimes. Grain yield was reduced when grain protein fell below a critical concentration of 132 g kg^{-1}. Based on this critical level, a field could be separated into two groups: one where N fertility was most likely inadequate and another where N most likely was adequate. Similarly, in Colorado, Goos et al.[11] found that protein at 115 g kg^{-1} indicated the transition between N-sufficient and N-deficient winter wheat. A critical level of 88 g kg^{-1} was determined necessary for N sufficiency in soft-white winter wheat in Oregon.[12]

Nitrogen management maps can be derived at the scale of fertilizer application using maps of grain yield and grain protein, a geographic information system and simple algorithms relating N removed in grain to wheat yield and grain protein.[13] By adjusting N rates in accordance with N removal and other N factors that varied spatially across a farm field in Montana, the spatially variable N placement improved the uniformity of the protein concentration of hard-red spring wheat. Similarly, Bonfil et al.[14] computed N removed in grain from yield and protein data, and applied this information into precision N management of spring wheat in Israel. They were able to increase grain yield and dollar returns over conventional uniform management based on yield potentials and soil tests.

Optical, near-infrared sensors are commercially available for measuring and mapping the grain protein concentration of wheat from a combine harvester during harvest. Field validation experiments reported in the literature reveal on-combine measurement accuracies of 5.7 g kg^{-1} for wheat in Belgium,[15] 4.9 g kg^{-1} for hard-red spring wheat in Montana,[16] and 3.1 g kg^{-1} for soft-white winter wheat in Oregon.[17] These results are sufficiently promising to support the use of on-combine optical sensing for measuring and mapping the protein concentration during harvest. Together with yield monitors, this information can be applied into N fertility management based on critical protein levels and N removal in grain.

Combine harvesters equipped with a mass-flow yield monitor and optical near-infrared grain analyzer were used to acquire site-specific measurements of grain yield and grain protein concentration. These data were then arithmetically combined using the surface-mapping program Surfer (Golden Software, Inc., Golden, Colorado). In this chapter, approaches to apply this information into precision N management are demonstrated with results from an irrigated wheat field in northern Montana. The objectives of this chapter are to (1) describe the use of Surfer for mapping N removed in grain, (2) provide georeferenced, edited, and gridded wheat yield and grain protein data, and (3) provide an exercise in identifying N management zones using Surfer.

19.3 METHODS

19.3.1 Grain Yield and Protein Mapping

Data for this chapter are from a 32.4 ha (80 ac) field near Malta, Montana (48.3842°N, −107.7663°W) that was planted to hard-red spring wheat (*Triticum aestivum* L.) in 2004 (Figure 19.1). The climate is continental, semiarid with 32.7 cm (12.9 in.) of average annual rainfall, which is typical of the drier portions of the northern Great Plains. More than half of the rainfall (58%) is received during the growing season (April–July). Each year, natural rainfall is augmented with about 15.2 cm (6 in.) of water from a center pivot irrigation system during vegetative growth. The landscape is characterized by moderate relief with uneven rolling topography. Soils are derived from glacial till and comprise Telstad (fine loamy, mixed, frigid Aridic Ariustolls) and Joplin (fine loamy, mixed, frigid Aridic Argiustolls) series with clay loam surface texture.

Site-specific wheat yield data were acquired from a 700 m × 360 m (25.2 ha) portion of this field using a Case-IH 1660 combine equipped with a 4.57 m (15 ft) wide header, AgLeader YM2000 yield monitor, and Trimble Ag132 receiver with ±1 m of horizontal accuracy and ±2 m of vertical accuracy. The yield monitor was calibrated to measure grain yield with ±1% accuracy. Yield was sensed at a rate of 1 Hz, while the combine's ground speed was held at approximately 1.56 m s^{-1} (3.5 mph). The number

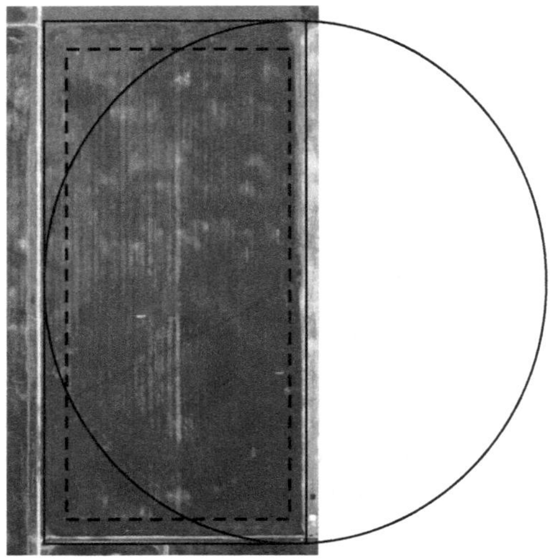

FIGURE 19.1 Aerial image of rectangular field with outline of main circular area of center pivot irrigation system. Dashed line bounds region where yield monitor data were obtained.

of yield measurements in the field was approximately 33,000, with each observation representing a 7.1 m^2 harvest area (1.6 m × 4.6 m). Instead of recording onto an SRAM card, the yield data stream was logged to an external computer by means of the AgLeader YM2000 export serial port.

A Zeltex AccuHarvest grain analyzer and the Trimble GPS receiver were used to independently map and record the grain protein measurements. The AccuHarvest sensor operates within the near-infrared portion (893–1045 nm) of the electromagnetic spectrum. The instrument relies upon 14 light-emitting diodes and narrow band filters to generate 14 channels of transmission spectra. Sample presentation is by means of a gravity-filled, quartz-windowed chamber and a mechanical sampling system of inlet and outlet ports that permit the grain to flow into and out of the sensor. Light is scattered and absorbed within the grain sample, and the instrument records the reflectance spectra by means of an inexpensive silicon detector. Protein measurements are collected at a rate of approximately 0.0833 Hz. Diodes, filters, the detector, and sampling mechanism are integrated into one unit that is designed to be mounted to the middle of the clean grain elevator. Instrument control and data transfer/storage are accomplished by means of an RS-232 serial port for interface with a palmtop computer.

19.3.2 Preparing Data for Analysis

The time lag between cutting at the header and sensing of the grain at the top of the elevator was determined to be 11 s. This time difference was used to offset the yield position along the travel path of the combine based on its ground speed. To do this, the ASCII yield data file that had been logged by the external computer was imported into MS-Excel. Each record in that file had been tagged with its unique coordinated universal time (UTC) of collection in intervals of 1 s. Therefore, the time lag could be easily implemented by physically repositioning (i.e., offsetting) the column of yield data backward in time by 11 s, or 11 records. As an aside, later models of the AgLeader monitor utilize a specialized routine within the proprietary SMS software for this purpose. Thus, it is not always necessary to import these yield data into a spreadsheet. Protein data recorded onto the palmtop computer from the Zeltex instrument were offset in the same manner with the use of the UTC tags and MS-Excel. Each record was separated by a time interval of approximately 12 s. Therefore, offsetting the protein points was accomplished by repositioning the column containing its data series by 12 s, which was equivalent to one record.

To minimize potential errors, the combine was operated at a ground speed that prevented surging and loss of grain in the transport and threshing mechanisms. Despite these measures, it was necessary to remove erroneous values from the data file. Data points not considered reasonable often include zeros and extremely large values. Zeros can be generated during unloading and when the combine travels outside of the field. Extremely large values can be generated when the combine slows or stops. These were identified by creating a histogram of the data series (Figure 19.2) and inspecting the histogram to determine the cutoff values for outliers. In this case study, yields less than 800 kg ha^{-1} or greater than 6800 kg ha^{-1} were deleted.

By measuring both yield and protein, our ability to compute the N removed in grain and assess the yield-limiting factors is improved. However, combining these data arithmetically is not straightforward. The AgLeader yield monitor operated at 1 Hz while the Zeltex protein sensor operated at 0.0833 Hz, resulting in 12 yield points for every protein point. Differences in the spatial resolution complicate their statistical correlation and use in arithmetic calculations.

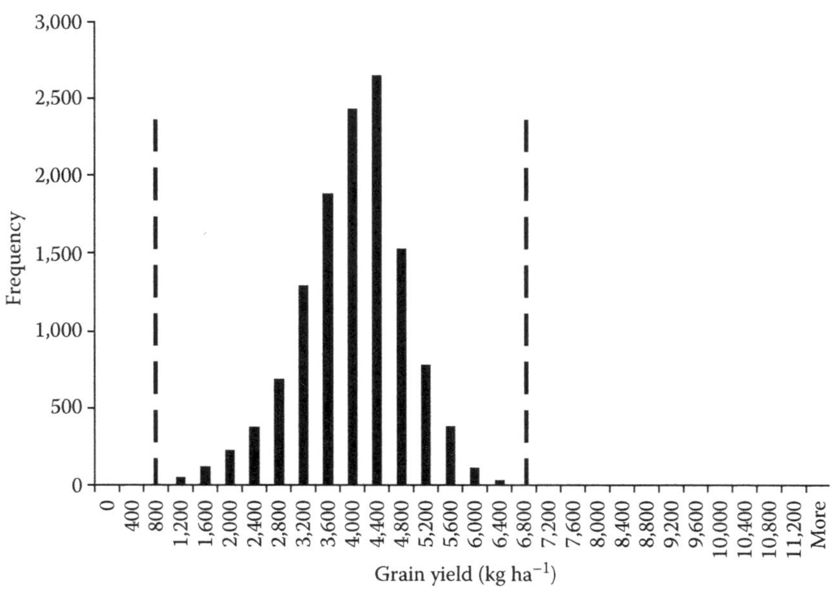

FIGURE 19.2 Frequency of observations versus yield as recorded with AgLeader 2000 yield monitor with cutoff values for excluding values <800 and >6800 kg ha^{-1}.

Fortunately, the value of any point on a regular grid can be estimated, or interpolated, from a limited number of spatially referenced samples. An edited and lagged data set was input to version 8.09 of Surfer, and the geostatistical interpolation procedure of kriging was used to superimpose a common 1.6 m × 1.6 m estimation grid upon the original yield and protein data. In kriging, an average value is computed for each unit of this grid using a spatially weighted combination of the original values available for each unit. The spatial weights are deduced through the regionalized variable theory and implemented by means of semivariograms that can be fit to pairs of point of different distance classes. The Grid Math feature of Surfer could then be used to arithmetically combine the interpolated values of yield and protein together for the computation of maps of N sufficiency based on critical protein level and N removed in grain.

19.3.3 Producing a Map of Critically Low Protein

For Montana conditions, the critical protein level for hard-red spring wheat is 132 g kg^{-1}.[10] Based on this value, two categories:

Category 0: Protein <132 g kg^{-1}
Category 1: Protein ≥132 g kg^{-1}

were identified. Cells with protein <132 g kg^{-1} (N deficient for yield) were categorized with a value of 0, while the cells with values ≥132 g kg^{-1} were categorized with 1 (N sufficient). The resulting binary class map was used for assessing the spatial pattern in N sufficiency.

Field areas that had sufficient plant-available N (PAN), as indicated by protein ≥132 g kg^{-1}, would receive N applications equal to the rates at which N was removed in the harvested grain. The N replacement approach, however, would not be satisfactory for portions of this irrigated field indicated to be deficient in N fertility (protein <132 g kg^{-1}) because not enough fertilizer N would be applied to satisfy the requirement for yield before protein can be increased. Instead, fertilizer applications for these areas were based on university recommendation methods that rely upon yield potential and soil test N information.[18] Methods for computing the N removed in grain are described in the following.

19.3.4 Computing Nitrogen Removed in Grain

The N removed in grain reflects the amount of N that must be applied to a future crop in the following year to replace that N removed by the harvested crop in the current year. A "maintenance" approach to N fertility management is inferred in which the N need of the following crop equals the N removed by the crop in the previous year.[10] In this study, the N removed in the grain was computed by multiplying a map of grain yield by a map of grain protein and dividing the result by the theoretical fraction of N in protein (5.7):

$$\frac{\text{kg N}}{\text{ha}} = \frac{\text{kg grain}}{\text{ha}} \times \frac{\text{g protein}}{\text{kg grain}} \times \frac{1\,\text{kg}}{1000\,\text{g}} \times \frac{1}{5.7} \quad (19.1)$$

This maintenance approach to N management may be modified if wheat straw is harvested for livestock feed or as a feedstock for bioenergy production. Under this scenario, N removed in straw can be estimated from known relationships between grain N and straw N. For example, in Israel[14], the total N removed in grain and straw is computed by adding the N contained in the grain and straw together, using the following equation:

$$\text{Total N removed (kg ha}^{-1}) = \text{N removed in grain (kg ha}^{-1}) + \text{N removed in straw (kg ha}^{-1}) \quad (19.2)$$

where N removed in grain is derived from Equation 19.1 and N removed in straw is computed by multiplying N removed in grain by 0.01. The N content of 1% for straw is based on extensive laboratory testing, a harvest index of 33%, and postharvest incorporation of half of the straw into the soil.

In Australia, an N budget inventory approach has been proposed that considers the amount of soil N converted from organic matter to PAN by mineralization as well as any fertilizer N that is applied prior to planting.[19] Accordingly, the amounts of PAN and applied N is used to offset the N removed in grain using the equation:

$$\text{N budget} = \text{PAN} + \text{applied N} - \text{N removed in grain} \quad (19.3)$$

Currently, the N budget approach is constrained by a lack of soil sensors to obtain site-specific maps of PAN prior to planting. Crude maps of PAN may be obtained economically by means of composite sampling and testing of soils within individual management zones, but with considerably less spatial resolution than associated grain yield or grain protein maps.

19.3.5 Computing Local Neighborhood Correlation

Correlation analysis was used to determine the association within relatively small neighborhoods between the georegistered yield and protein maps. The Pearson product moment correlation coefficient (r) is calculated using the following equations:

$$r = \frac{\text{cov}_{12}}{s_1 s_2} \quad (19.4)$$

where

$$\text{cov}_{12} = \frac{\sum_{i=1}^{n} (\text{GY}_{i1} - \mu_1)(\text{GP}_{i2} - \mu_2)}{n-1} \quad (19.5)$$

and cov_{12} is the covariance between yield and protein values in the neighborhood, s_1 is the standard deviation of the yield values in the neighborhood, and s_2 is the standard deviation of the protein values in the neighborhood. In addition, GY_{i1} is the ith value of elements found in the neighborhood of the yield map, GP_{i2} is the ith value of the elements found in the neighborhood of the protein map, n is the total number of values found in the neighborhood, and μ_1 and μ_2 are the means of the series of values found in the neighborhood of each map. In this study, the correlation coefficient was computed for a neighborhood of size 11 × 11 cells and was expected to indicate the association between the two maps on a cell by cell basis. Local neighborhood correlation analysis cannot be performed in Surfer and MS-Excel, and instead was undertaken with the numerical computation software MATLAB®. Procedures with MATLAB are beyond the scope of this chapter. Nevertheless, the results are illustrated for contemplation by readers.

19.4 RESULTS

19.4.1 Maps of Grain Yield and Grain Protein

The northwest and southwest corners of the field are outside of the main circular area of the center pivot irrigation system and thus were limited to 17 cm (89% of long term) of rain that fell during the 2004 growing season. Grain yield ranged from 1500 to 3000 kg ha^{-1} in the field's corners and from 3000 to 6000 kg ha^{-1} in the main circular area (Figure 19.3A). Protein concentration of grain varied from 150 to 180 g kg^{-1} in the corners and from 120 to 170 g kg^{-1} in the rest of the field (Figure 19.3B).

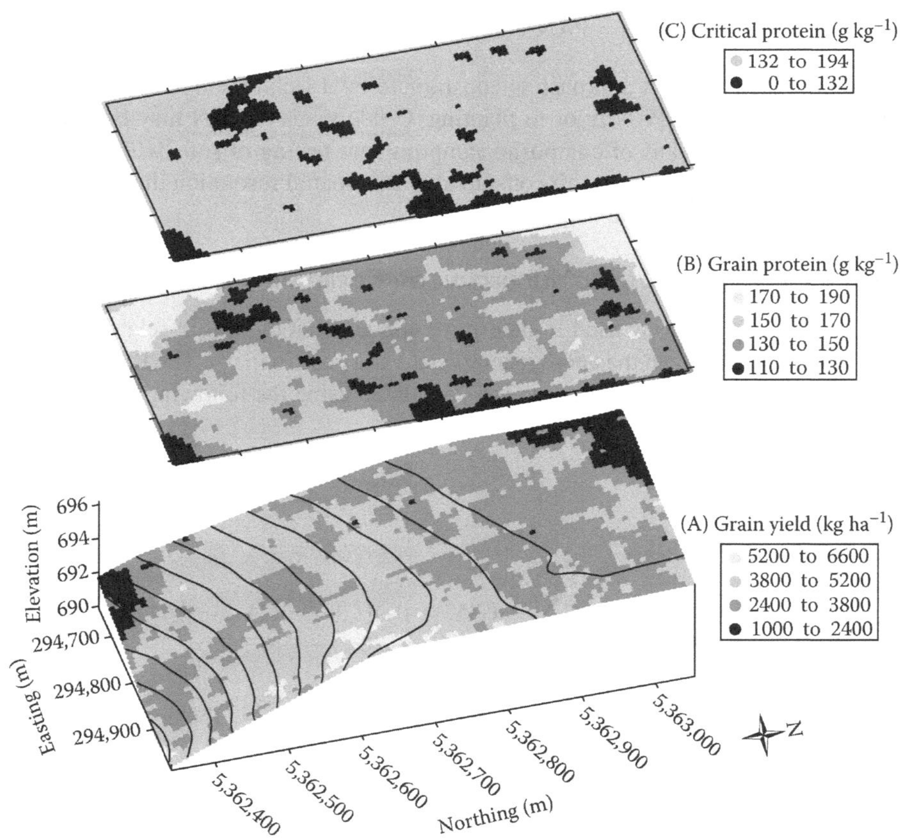

FIGURE 19.3 Map of grain yield (A), map of grain protein concentration (B), and map of critically low protein, indicating areas where nitrogen could be deficient for yield (C).

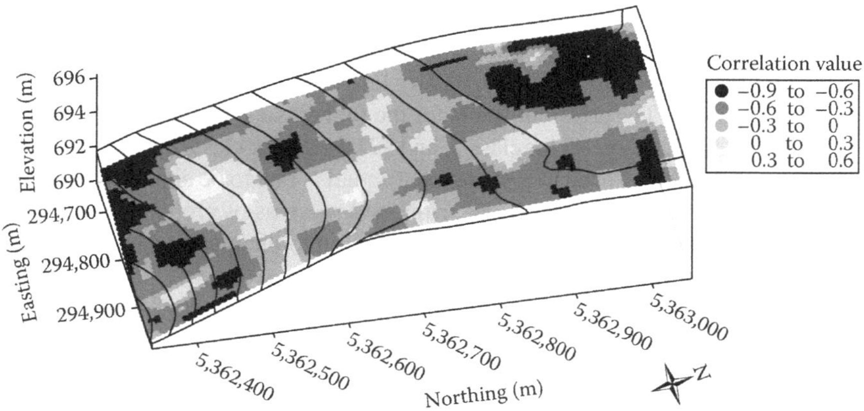

FIGURE 19.4 Correlation between grain yield and grain protein computed from values within an 11 × 11 kernel for each cell shared by the two maps.

The local correlation between grain yield and grain protein is strongly negative in the corners of the field where the crop likely was water stressed (Figure 19.4). In contrast, the irrigated portion comprises several positively correlated areas where crop productivity was likely determined by available N rather than water. Regions where the correlation is near-zero indicate transitional zones between water stress and N stress.

19.4.2 Map of Critically Low Protein

In hard-red spring wheat grown in Montana, protein above 132 g kg^{-1} is indicative of N sufficiency. Therefore, the map of critical protein (Figure 19.3C) gives a grower, or crop consultant, knowledge that available N was sufficient for yield where protein exceeded the critical level. The majority of the field exceeded this critical protein level, thus suggesting that PAN had been adequate for meeting production goals.

Proteins less than the critical level of 132 g kg^{-1} also exist within this irrigated field. Unfortunately, available N would be deficient for yield in these areas if fertilizer N requirements were based on the N removed in grain alone. A solution is to use the critical protein map to direct soil sampling and testing into the areas indicated to be N deficient. The resulting nitrate-N values would be used with university fertilizer guidelines to determine the N required for these areas. For example, 164 kg of N ha^{-1} would be required if: (1) the soil contained 50 kg nitrate-N ha^{-1}, (2) the yield potential is 4031 kg grain ha^{-1}, and (3) 0.053 kg of applied N kg^{-1} grain is needed to produce wheat with a protein concentration of 140 g kg^{-1}.

19.4.3 Map of Nitrogen Removed

The amount of N removed in grain ranged between 21 and 155 kg ha^{-1} throughout the field (Figure 19.5A). Nitrogen removed corresponds with spatial patterns in grain yield, thus indicating that this variable was responsible for most absorption of N. Accordingly, the high-yielding, irrigated area absorbed the greatest amounts of available N and the low-yielding, dryland corners the least.

Boolean operators in the Grid Math routine of Surfer were used to classify the ratio-interval values of N removed into four classes. The class interval was computed by dividing the range by four (134/4 = 33.5 kg ha^{-1}), which was the desired number of management zones. The apparent limits for the four classes were 21–54.5 for class 1, 54.5–88 for class 2, 88–121.5 for class 3, and 121.5–155 for class 4. The mid-value in each class could then be used to represent the actual N rate to be applied for variable-rate fertilizer application. In addition, critically low areas of protein (Figure 19.5B) formed a fifth management zone after their addition to the map of N removed using Grid Math (Figure 19.5C).

Therefore, the variable-rate strategy that is illustrated consists of five N rates: 38, 71, 105, 138, and 165 kg ha^{-1} and would represent five settings that a task controller on a tractor would use to implement variable-rate fertilizer N application. In general, the largest amounts to be applied are indicated in the irrigated field area where grain yield had been greatest and grain protein had been least. Surfer is capable of exporting the final grid map to ESRI ArcGIS where a Shapefile could be constructed, which is the data format read by a variable-rate controller.

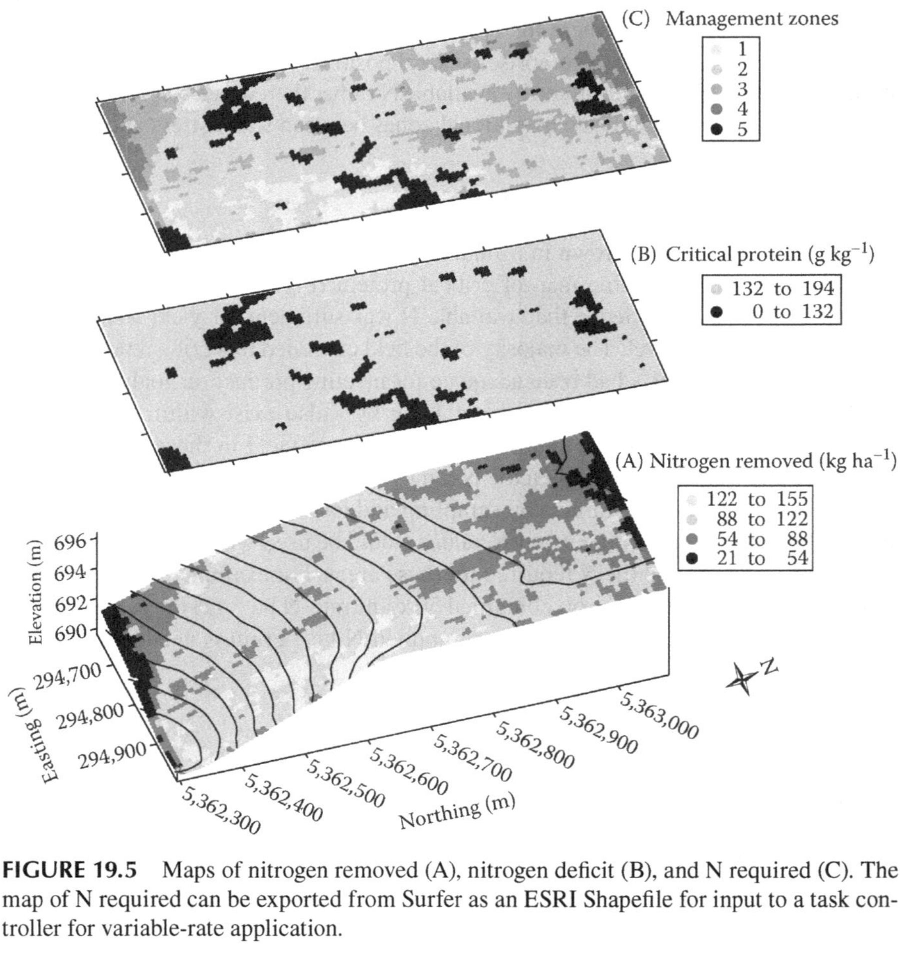

FIGURE 19.5 Maps of nitrogen removed (A), nitrogen deficit (B), and N required (C). The map of N required can be exported from Surfer as an ESRI Shapefile for input to a task controller for variable-rate application.

19.5 SUMMARY

The advent of on-combine optical sensing gives rise to the opportunity for measuring and mapping the protein concentration of grain within farm fields. When combined with grain yield, this information can be utilized to compute a variety of crop N-related attributes that have application in N fertility management. Mapping software such as Surfer can be used to estimate values for a regular grid and arithmetically combine the gridded values of grain yield and grain protein to compute the N removed in grain. The ratio-interval values of the N removed can be then generalized into three or four classes representing the management zones for variable-rate N application.

A grain protein map, when compared to an established critical protein level, can be used to indicate where available N had been sufficient, or deficient, for yield within a field. Plus, maps of the neighborhood correlation between grain yield and grain protein can be used to identify field areas where the crop had been N-limited versus water-limited. Together, this postharvest information can be utilized to assess the effectiveness of a grower's N fertility management program as well as to identify field areas where soil testing and corrective N treatments are needed.

There are drawbacks to the use of crop yield parameters to diagnose N nutrition adequacy and derive N recommendations. First, N removed in grain is confounded by severe drought that abnormally elevates grain protein, causes yield reductions, and influences N availability and plant N uptake.[8] Second, wheat cultivars sometimes do not experience yield loss when protein is below the critical level and thus may be insensitive to this protein index.[20] Chief advantages include the ability to rapidly index variability in crop N-related attributes at the scale of fertilizer application, forecast N fertilizer requirements, and direct future soil and plant sampling efforts.

19.6 STEP-BY-STEP EXERCISE WITH SURFER

In this exercise, you will use Surfer to generate maps of critically low protein, N removed in grain, and an N management map. Grain yield and protein data are from the field near Malta, Montana. To expedite, these data have been prepared for analysis by adjusting for the time lag in data collection and removing extraneous values.

1. To create a map of critically low protein, open Surfer to a new, empty plot.
2. Click Grid > Data. In the Open dialog box, navigate to Chapter 22\Exercise\ on the CD accompanying this book and open the file *grain yield.txt*. This is a tab delimited text file of grain yield with 12,502 yield records referenced in UTM coordinates.
3. In the Grid Data dialog box, select Kriging as the gridding method. For Grid Line Geometry, set the Minimum and Maximum values for X Direction to 294,635.45 and 295,006, and for Y Direction to 5,362,305.27 and 5,363,070.48. The Grid Data dialog box should look like Figure 19.6. Once it does, click OK to initiate Kriging and create the grid file *grain yield.grd* with 4900 records (49 × 100 cells).

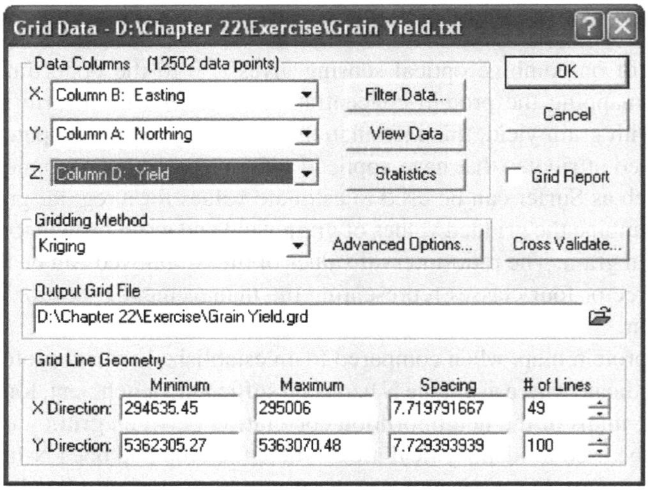

FIGURE 19.6 Grid Data dialog box with settings for Kriging the grain yield data.

4. Repeat step 3 for the file *grain protein.txt*, which is a tab delimited text file of grain protein with 696 records. The Minimum and Maximum values for X Direction and Y Direction need to be identical to those used with the grain yield file. These actions will create the grid file *grain protein.grd* with 4900 records.
5. Click Grid > Math. In the Open Grid dialog box, select the newly created grid file *grain protein.grd* and click Open to open the Grid Math dialog box. In the Grid Math dialog box, the grid file *grain protein.grd* is revealed as Input Grid File A. Input Grid File B should have an entry of None because a Boolean operation will be performed only on *grain protein.grd*. Click on the folder icon for Output Grid File C and the Save Grid As dialog box will open. Enter the name of the output grid file to be created (i.e., *critical protein.grd*). Click Save to return to the Grid Math dialog box. Enter the following Boolean function into the opening at the bottom of the Grid Math dialog box: C = if(A < 132, 5, 0), which will return a value of 5 if protein is less than the critical value of 132 g kg^{-1} and 0 otherwise. The Grid Math dialog box should resemble Figure 19.7.
6. Click OK. The grid file *critical protein.grd* is created having a value of 5 where A < 132 is true and 0 otherwise. Surfer returns to a screen with an empty plot.
7. To create a map of N removed, click Grid > Math. In the Open Grid dialog box, select the grid file *grain yield.grd* and click Open to open the Grid Math dialog box. In the Grid Math dialog box, grid file *grain yield.grd* becomes Input Grid File A. Click on the folder icon for Input Grid File B, select *grain protein.grd*, and click Open. To create a file for Output Grid File C, click on its folder icon, and enter the name of the output grid file to be created (i.e., *N removed.grd*).

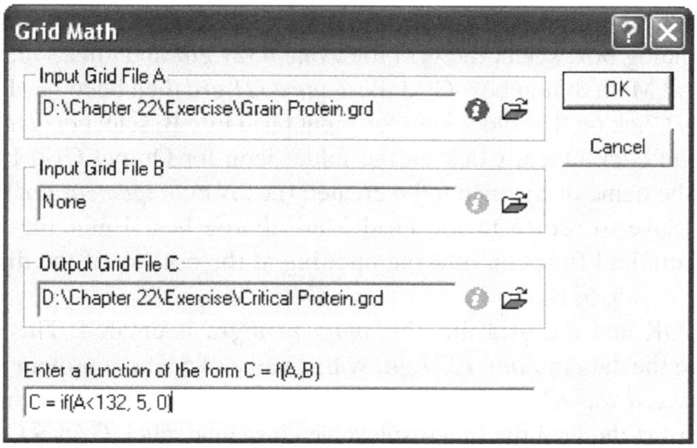

FIGURE 19.7 Grid Math dialog box with settings for creating a map of critically low protein.

8. Click Save to return to the Grid Math dialog box. Enter the following mathematical function into the opening at the bottom of the Grid Math dialog box: C = A*B/1000/5.7.
9. Click OK. The grid file *N removed.grid* is created with ratio-interval values from 20.8 to 154.8 kg ha^{-1}. Surfer returns to a screen with an empty plot.
10. To create a map of N management, click Grid > Math. In the Open Grid dialog box, select the grid file *N removed.grd* and click Open to open the Grid Math dialog box. Grid file *N removed.grd* then becomes Input Grid File A. Leave the entry for Input Grid File B as None. Click on the folder icon for Output Grid File C and the Save Grid As dialog box will open. Enter the name of the output grid file to be created (i.e., *Zone 4.grd*).
11. Click Save to return to the Grid Math dialog box. Enter the following Boolean function into the opening at the bottom of the Grid Math dialog box: C = if(A > 121.5, 4, A) where values are coded with *4* where A > 121.5 is true and original values of A otherwise.
12. Click OK. The grid file Zone 4.grd is created. Surfer returns to a screen with an empty plot.
13. Repeat step 11 except that *Zone 4.grd* is used as Input Grid File A instead of *N removed.grd*. Enter the name of the output grid file to be created as *Zone 43.grd*.
14. Click Save to return to the Grid Math dialog box. Enter the following Boolean function into the opening at the bottom of the Grid Math dialog box: C = if(A > 88, 3, A) where values are coded with 3 where A > 88 is true and original values of A otherwise.
15. Click OK. The grid file *Zone 43.grd* is created. Surfer returns to a screen with an empty plot.
16. Repeat step 11 for the remaining lower class limits (54.5 and 21 kg ha^{-1}) remembering to use the previous grid file as Input Grid File A until output grid file *Zone 4321.grd* is created containing class values of 1, 2, 3, and 4.

17. To create the final N management map, click Grid > Math. In the Open Grid dialog box, select the grid file *Zone 4321.grd* and click Open to open the Grid Math dialog box. Grid file *Zone 4321.grd* then becomes Input Grid File A. Click on the folder icon for Input Grid File B, select *critical protein.grd*, and click Open. Click on the folder icon for Output Grid File C and enter the name of the map to be created (i.e., *N management.grd*).
18. Click Save to return to the Grid Math dialog box. Enter the following mathematical function into the opening at the bottom of the dialog box: C = if(A = 5, 5, B).
19. Click OK and the grid file *N management.grd* is created. This grid will replace the data in *Zone 1234.grd* with a value of 5 where protein in *critical protein.grd* was <132 g kg^{-1}. Surfer returns to a screen with an empty plot.
20. To convert the grid file to a displayable class map, click Grid > Convert. In the Open Grid dialog box, select *N management.grd* and click Open. In the Save Grid As dialog box, enter the file name as *N management* and save it as type ASCII XYZ. Click Save. Surfer returns to a screen with an empty plot.
21. To display the ASCII *N management.dat* file as a class map, click Map > Post Map > New Classed Post Map. In the Open dialog box, select *N management.dat*. The map appears as a plot on the main screen. Double-click on the map to open the Map: Classes Post Properties dialog box to adjust the number of classes, class limits, and color of symbols representing the classes.

ACKNOWLEDGMENTS

The authors gratefully acknowledge Karl Mavencamp for providing field access and permission to release the data used in this exercise. Funding was provided by USDA-NIFA Initiative for Future Agricultural Systems Grant 2001-52103-11321.

REFERENCES

1. Flowers, M., Weisz, R., and White, J.G. Yield-based management zones and grid sampling strategies: Describing soil test and nutrient variability. *Agron. J.* 97, 968, 2005.
2. Kleinjan, J., Clay, D.E., Carlson, C.G., and Clay, S.A. Developing productivity zones from multiple years of yield monitor data. In *GIS Applications in Agriculture*, CRC Press, Boca Raton, FL, p. 65, 2007.
3. Kleinjan, J., Clay, D.E., Carlson, C.G., and Clay, S.A. Developing productivity zones from multiple years of yield monitor data. Site-specific management guidelines SSMG-45. *International Plant Nutrition Institute*. Available at http://www.ppi-ppic.org/ppiweb/ppibase.nsf/$webindex/06E070534CA50AB2052569DB0033F992 (verified December 21, 2009), 2006.
4. Ping, J.L. and Dobermann, A. Processing of yield map data. *Precis. Agric.* 6, 193, 2005.
5. Smika, D.E. and Greb, B.W. Protein content of winter wheat grain as related to soil and climatic factors in the semiarid Central Great Plains. *Agron. J.* 65, 433, 1973.
6. Simmonds, N.W. The relation between yield and protein in cereal grain. *J. Sci. Food Agric.* 67, 309, 1995.
7. Terman, G.L., Ramig, R.E., Dreier, A.F., and Olson, R.A. Yield protein relationships in wheat grain, as affected by nitrogen and water. *Agron. J.* 61, 755, 1969.

8. Selles, F. and Zentner, R.P. Grain protein as a postharvest index of N sufficiency for hard red spring wheat in the semiarid prairies. *Can. J. Plant Sci.* 81, 631, 2001.
9. Whelan, B.M., Taylor, J.A., and Hassall, J.A. Site-specific variation in wheat grain protein concentration and wheat grain yield measured on an Australian farm using harvester-mounted on-the-go sensors. *Crop Pasture Sci.* 60, 808, 2009.
10. Engel, R.E., Long, D.S., Carlson, G.R., and Meier, C. Method for precision nitrogen management in spring wheat: I. Fundamental relationships. *Precis. Agric.* 1, 327, 1999.
11. Goos, R.J., Westfall, D.G., Ludwick, A.E., and Goris, J.E. Grain protein content as an indicator of N sufficiency for winter wheat. *Agron. J.* 74, 103, 1982.
12. Glenn, D.M., Carey, A., Bolton, F.E., and Vavra, M. Effect of N fertilizer on protein content of grain, straw, and chaff tissues in soft white winter wheat. *Agron. J.* 77, 229, 1985.
13. Long, D.S., Engel, R.E., and Carlson, G.R. Method for precision nitrogen management in spring wheat: II. Implementation. *Precis. Agric.* 2, 25, 2000.
14. Bonfil, D.J., Mufradi, I., Asido, S., and Long, D.S. Precision nitrogen management based on nitrogen removal in rainfed wheat [CD-ROM computer file]. In Khosla, R., ed. *Proceedings of the 9th International Conference on Precision Agriculture*, Colorado State University, Fort Collins, CO, 2008.
15. Maertens, K., Reyns, P., and De Baerdemaeker, J. On-line measurement of grain quality with NIR technology. *Trans. ASAE.* 46, 1135, 2004.
16. Long, D.S. and Rosenthal, T. Evaluation of an on-combine wheat protein analyzer on Montana hard red spring wheat. In Stafford, J.V., ed. *Proceedings of the 5th European Conference on Precision Agriculture*. Wageningen Academic Publishers, Wageningen, the Netherlands, p. 385, 2005.
17. Long, D.S., Engel, R.E., and Siemens, M.C. Measuring grain protein concentration with in-line near infrared reflectance spectroscopy. *Agron. J.* 100, 247, 2008.
18. Jones, C. and Jacobsen, J. Nitrogen cycling, testing and fertilizer recommendations. *Nutrient Management Module No. 3*. Montana State University Extension Service, 2001.
19. Taylor, J.A., Whelan, B.M., Thylén, L., Gilbertsson, M., and Hassall, J. Monitoring wheat protein content on-harvester: Australian experiences. In Stafford, J.V., ed. *Precision Agriculture, 2005: Proceedings of the 5th European Conference on Precision Agriculture.*, Wageningen Academic Publishers, Wageningen, the Netherlands, p. 369, 2005.
20. Fowler, D.B. Crop nitrogen demand and grain protein concentration of spring and winter wheat. *Agron. J.* 95, 260, 2003.

20 Review of Low- and High-Technology Nitrogen Management Approaches for Improved Nitrogen Use Efficiency

Daryl B. Arnall and Robert W. Mullen

CONTENTS

20.1 EXECUTIVE SUMMARY

The relatively low (33%) nitrogen use efficiency (NUE) of many cereal crops grown in the United States suggests that by adopting improved management practices agricultural, economic, and energy efficiency could be improved.

Techniques designed to improve energy efficiency can be separated into low- and high-technology techniques. Low-technology approaches include fertilizer timing, the application of slow release and inhibitor enhanced fertilizer, the use of yield goals, and soil and plant testing to improve N recommendations. High-technology approaches include the management zones, chlorophyll meter (CM) or other optical sensors, and soil testing designed to predict N mineralization. This chapter discusses and reviews the various low- and high-technology approaches that can be used to improve energy and NUE. Case studies that discuss how the adoption of improved N management techniques increased profitability are provided.

20.2 INTRODUCTION

NUE is a commonly used term to determine the relative efficiency of the N fertilizer. This value is calculated using a number of different approaches. The most common approach is to divide the difference in the amount of N taken up fertilized and unfertilized control plots by the N rate. A low NUE suggests that the N fertilizer was relatively ineffective. Many cereal crop nitrogen (N) fertilizer management strategies have relatively low (33%) NUE.[1] Using management strategies with low NUE can result in low energy use efficiency, environmental contamination, and economic losses. Development of alternative N management strategies that increase productivity and energy efficiency while reducing the environmental consequences of agriculture is an important goal worldwide. This chapter provides a brief overview of the N cycle as a framework for discussing the cause of low NUE, and then using this framework presents various strategies that can be implemented to improve NUE and achieve the goal of sustainable crop production.

The N cycle depicted in Figure 20.1 portrays the movement of N in all its different forms in both natural and managed agricultural ecosystems between the atmosphere, biosphere, and geosphere. Similar to the carbon cycle, the nitrogen cycle consists of various storage pools. In the N cycle, there are four primary pathways for potential loss: (1) leaching, (2) denitrification, (3) ammonia volatilization, and (4) plant loss.

NUE can be improved by preventing loss once in the soil system and supplying the crop with just enough nutrients to satisfy demand. The approach advocated by the International Plant Nutrition Institute (IPNI) is to adopt the concept using the right source, at the right rate, at the right time, and in the right place, also known as the 4Rs. Adopting the 4Rs, has beneficial impacts on environmental, economic, and social concerns. The 4R nutrient stewardship concept involves crop producers and their advisers selecting the right source-rate-time-place combination from practices validated by research conducted by agronomic scientists. Goals for economic, environmental, and social progress are set by—and are reflected in performance indicators chosen by—the stakeholders of crop production systems (http://www.ipni.net/). Implementing a 4R program should result in selecting the

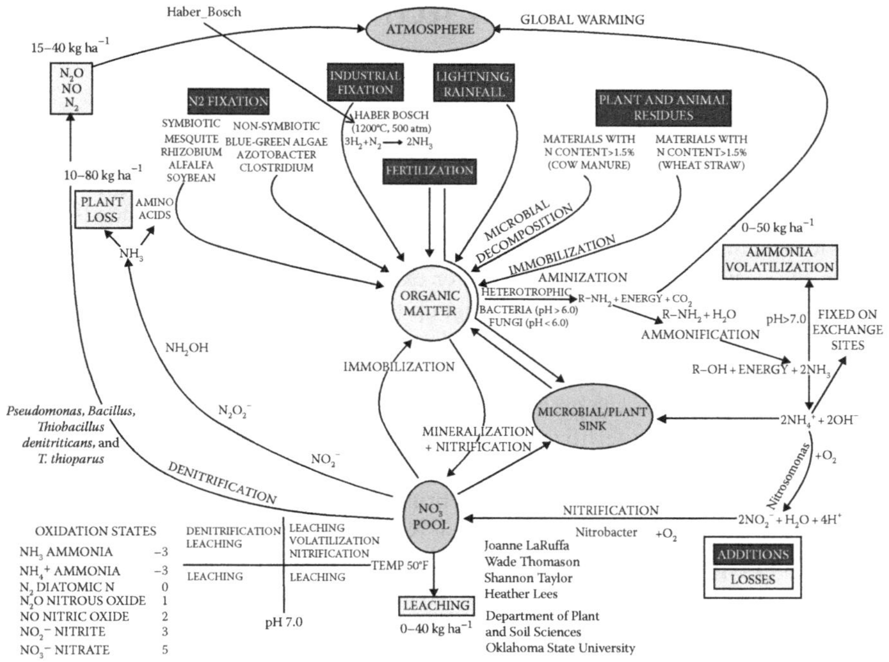

FIGURE 20.1 The nitrogen cycle. Pathways of loss are in light grey boxes while the addition pathways are dark grey. The circles represent pools or reservoirs for nitrogen.

correct N form for a given problem. For example, avoiding the use of urea in no-till or pasture environments when irrigation is not available; rain is not forecasted in the near future; or the weather is hot, dry, and windy. Under these conditions, significant loss of N via ammonia volatilization can occur.[2,3] The right time suggests applying fertilizer as close to the time of crop need as possible, preventing extended periods of time where inorganic-N (specifically nitrate-N) is available in the soil system.[4] The right place is primarily focused on the immobile nutrients (phosphorus and potassium) where techniques such as banding can improve nutrient use efficiency.[5] The right place can still be considered in N management especially in larger scale zone management and variable-rate (VR) N management.[6,7] The fourth R of the 4Rs, right rate, minimizes inorganic-N loss by limiting the amount of residual N carried over from 1 year to the next and reducing the risk of loss to the environment by not applying more than is needed.[1] The majority of the NUE improvement methods are focused on the right rate. Many of the methods discussed in this chapter would fit into multiple categories, such as slow-release fertilizers which are aimed at providing the correct form of nutrient to be released at the proper period in the crops life cycle to the use of advanced in-season (right time) sensing techniques to identify where (right place) and how much (right rate) N should be applied.

20.3 METHODS OF NUE: LOW TECH

20.3.1 TIMING OF N APPLICATION

The timing of N application has been extensively studied. The general conclusion is that N should be applied to coincide with the time it is needed by the crop. Plants generally have the highest demand for N following emergence, during periods of rapid growth, and during grain filling. To account for these different needs, N can be split applied.[8,9] The goal of split application is to provide in-season N to the crop during the period of greatest demand. For example, in winter wheat production, N was historically applied prior to planting, up to 2 months before the seed was sown in September to October (in the Central Plains). Winter wheat germinates and grows during the fall and then lays dormant during the cold winter months. It then greens up in the spring and has the greatest need for N from February to May. This means that the majority of the fertilizer N had been in the soil for up to 7 months. Split application strategies apply a low percentage of the total N as preplant, with the remainder being applied in January or February. Olson and Swallow[10] observed that in 4 out of 5 year's in-season application of N led to an increase in fertilizer use efficiency over preplant treatments.

When considering in-season N management, foliar applications is also a possibility. Finney et al.[11] recorded an increase in wheat grain protein by as much as 4.4% when foliar applications of urea solutions were made at flowering. Wuest and Cassman[12] found that the recovery of N applied at anthesis was 55%–80%, while the recovery of preplant N ranged from 30% to 55%. While damage of the leaf surface is a concern with urea-ammonium nitrate solutions when temperatures exceed 65°F, there are many low-salt N solutions that minimize this injury such as ureaformaldehyde (UF) and isobutylidene diurea (IBDU). Supplementing N later in the growing season when N demand is higher can result in improved NUE for most cereal crops.

20.3.2 SLOW OR CONTROLLED RELEASE

Slow-release or controlled-release fertilizers (CRF) are either specially formulated or coated with a less soluble compound to alter how N is released into the soil system. For the purposes of this discussion, the two terms will be considered synonymous. These fertilizers are very popular in the horticulture and turf grass industries, as well as in agricultural production regions with high rainfall. Slow-release fertilizers have advantages in their ability to increase plant nutrient recovery by lowering denitrification and NO_3 leaching.[13–16]

There are many forms of slow-release fertilizers, but most can be separated into two groups: coated and uncoated. To achieve a slower release of N, fertilizer prills can be coated with sulfur or polymers and are made of an alternate form of N such as UF, methylene urea, or IBDU uncoated fertilizers.

The premise of coated fertilizers is that a less soluble compound should surround the soluble N source. The availability of N from urea with a sulfur and/or protective overcoat is dependent upon physical breakage, biological oxidation, and diffusion. The release is primarily controlled by moisture, temperature, and thickness of coating. Generally the release of sulfur-coated material is slow but uncontrolled. For polymer-coated N, release is controlled by polymer chemistry, manufacturing process, thickness, temperature, and moisture. The release is very controlled and can be matched with expected plant uptake.

During the 1960s and 1970s, the technology to coat urea with sulfur was developed by the National Fertilizer Development Center. Sulfur was chosen as the principle coating material because of its low cost and its value as a secondary nutrient. Sulfur-coated ureas are comprised of particles of urea coated with a layer of sulfur and usually a sealant plus a conditioner.[17] They are typically brown to tan or yellow, depending on the source of urea and whether a sealant is used. Nitrogen is released from the prill when water penetrates through micropores, imperfections, and microbial degradation of the coating. Once penetration occurs, it is released rapidly. Wax seals can be used, which must be degraded by microbes prior to release. Polymer-coated fertilizers (PCFs) have a different release mechanism than coated materials. These products release nutrients by diffusion through a semipermeable polymer membrane, and the release rate can be controlled by varying the composition and thickness of the coating. The type of fertilizer source also may influence the rate of N release.

Meister programmed release fertilizer products are produced by using thermoplastic resins as coating materials. The resin coating can be applied to most nutrient sources including N, P, or K. Release-controlling agents such as ethylene-vinyl acetate and surfactants are added to the coating to obtain the desired diffusion characteristics, while coating thicknesses remain similar for most products. The nutrients are released by diffusion through the coating, which are the common methods of release for most of the PCFs. Because of this, the release is largely controlled by temperature and moisture.

Reactive layer coating (RLC) is a technology that combines two reactive monomers as they are simultaneously applied to the fertilizer source. By doing this, an ultra-thin membrane coating controls nutrient release by osmotic diffusion. Just as with Meister products, RLC can be applied to most solid nutrient sources. The thickness of the coating will control the rate of diffusion and the duration of nutrient release.

Polymer- and sulfur-coated fertilizers (PSCFs) are hybrid products that utilize both a primary coating of sulfur and a secondary polymer coat. These fertilizers were developed to achieve the same or near similar performance of PCFs but at a much reduced cost. The combination of the two coatings permits a positive cost/benefit value over products with single coatings of either sulfur or polymer.

Uncoated slowly available compounds protect N by delaying the availability to the soil environment. For the CRF chemistries, N release relies upon biochemical decomposition, and the length of protection lasts weeks to months. The release rate of each compound is determined by the chemical structure's resistance to breakdown, the molecular weight and degree of polymerization, and the environmental conditions. The release of N is considered slow, yet generally uncontrolled. Another characteristic of uncoated products is that they are homogenous, that is, their composition is the same throughout the particle. By contrast, coated products, which consist of a fertilizer core surrounded by a coating, are not homogenous.

We will now discuss two uncoated slow-release fertilizers. UF reaction products represent one of the oldest controlled-release N technologies, having been first produced in 1936 and commercialized in 1955. IBDU is a white crystalline solid available in a wide range of particle sizes that contains a minimum of 30% N with 90% of the N in water-insoluble form.

For uncoated fertilizers, the release of N is performed through a process of converting the N of the product into a plant-available form of N. For UF products, this is a multistep process involving dissolution of the compound and then microbial decomposition into plant-available N. Since microbial decomposition is the primary process involved in the conversion, environmental conditions play a critical role in the rate of nutrient release. Nitrogen from IBDU becomes available to plants through hydrolysis. In the presence of water, the compound will hydrolyze (break down) into urea and isobutyraldehyde. The low water solubility of IBDU controls the transport of the product into the soil solution. The particle size and water availability have an impact on IBDU dissolution. Once in the soil solution, the hydrolysis rate is affected by soil pH and temperature. Because of IBDU independence of microbial interaction, N release can readily occur at low soil temperatures that may hinder microbial activity. This is why IBDU products are preferred in cool-season applications.

It is important to know the mechanisms of N release for both slow- and controlled-release fertilizers in order to choose the correct product for the given environment. For example, it would not be advised to use a sulfur-coated urea in a highly acidic soil, as the sulfur lowers soil pH,[18] while coatings that require microbial degradation might not be the best choice for cool soils. A limiting factor in the use of these fertilizers is the high price per unit of N. In many cases, the price for slow-release fertilizer is two to six times higher per pound of actual N.

20.3.3 Inhibitors: Urease and Nitrification

Urease and nitrification inhibitors are products that are added to commercial N fertilizers. Urease inhibitors such as NBPT [N-(n-butyl)-thiophosporic triamide] are added to urea-based fertilizers to delay/inhibit the activity of the urease enzyme. Urease is an enzyme that catalyzes the hydrolysis of urea into carbon dioxide and ammonia (NH_3). In the soil system, the NH_3 is then quickly converted to ammonium (NH_4). Urea hydrolysis has to take place for the N in urea to be plant available. However, preventing urea hydrolysis is beneficial in situations where the urea is not incorporated or injected into soil. When urea sits on the surface, the urease enzyme breaks down urea into the volatile form of NH_3. If urea is not incorporated with rain or irrigation within 2–4 days of a surface application, even faster on high pH soils,[19] the NH_3 can be volatilized (Figure 20.1). The addition of the inhibitor delays this process up to 14 days (Agrotain, St. Louis, MO). This product is especially useful in no-till systems and pasture systems.[20] This product is only useful for urea-type fertilizers and should not be used when the fertilizer is injected into the soil. It may not increase NUE when liquid urea ammonia nitrate (UAN) is applied with flat fan nozzles, UAN banded with streamer nozzles, or rainfall occurs soon after applications.

Nitrification inhibitors slow down the process of NH_4 being converted to nitrate (NO_3) by killing or interfering with the metabolism of *Nitrosomonas* bacteria (Figure 20.1).

The benefit of preventing this conversion is that that NO_3^- having a negative charge is highly mobile. Nitrate will travel with the soil water so leaching is a particular problem, especially in coarse-textured soils. High rainfall or irrigation in naturally well-drained or tile-drained soils increases the likelihood that NO_3^- will leach out of the rooting zone. The loss of N from the rooting zone can be minimized by lengthening the time that the N is in the ammonium form.

In addition, nitrification inhibitors can also reduce denitrification (the microbiological conversion of nitrate and nitrite to gaseous forms of N). Denitrification is a major pathway of N loss from most fine-textured soils. Denitrification can occur in waterlogged soils. High soil pH, high carbon, warm soil temperatures (>60°F), and an abundance of nitrate, and a carbon source will create a maximum rate of denitrification. In waterlogged soils, more than 100 kg NO_3 ha^{-1} can be denitrified within a 5-day period. However, in cold (40°F) or low pH (5 or below) soils, denitrification can be slow. Nelson and Huber[21] reported that the application of nitrification inhibitors delayed the conversion of NH_4 to NO_3 for 4–10 weeks, depending upon soil pH and temperature. With fall applications of N fertilizers, nitrification inhibitors minimize nitrification until low soil temperatures (40°F) stop the process. With spring applications, nitrification inhibitors prevent the formation of nitrate during the late spring when rainfall is high and uptake of N by crops is low.

20.3.4 Yield Goal

Yield goal is the "yield per unit area you hope to grow."[22] A yield goal is the desired outcome from which the N rate can be calculated. However, Dahnke et al.[22] noted that desired and actual are two different things. Yield goals can be defined using a variety of approaches. For example, yield goals can be defined as the fields past average yield, the past average yield over the past 10 years plus 10%, the average yield plus 10% minus outliers, and potential yield. The potential yield is the highest possible yield that can be obtained with ideal management, soil conditions, and weather. Hoeft and Zafziger[23] and Zhang and Raun[24] recommend that the yield goal is usually 30%–33% higher than the average yield. When yield goals are applied, it explicitly places the risk of predicting the environment (good or bad year) on the producer.

Many researchers in the United States have shown that yield goals are poorly correlated with optimum N rates.[25–27] For example, in data from a long-term Oklahoma winter wheat study (initiated in 1971), the average optimum N rate over a 36-year span was 55 kg ha^{-1} (Figure 20.2a). The error associated with this average was ±44 kg N ha^{-1}, and the maximum average yield was 2863 kg ha^{-1} (43 bu ac^{-1}) with an error of ±870 kg ha^{-1} (13 bu ac^{-1}) (Figure 20.2b). The maximum yield during the study was 5936 kg ha^{-1} (89 bu ac^{-1}). In this study, the optimum N rate ranged from 0 to 157 kg N ha^{-1}. Predicted yield goal-based N recommendations were poorly corrected with measured values. Poor correlations between measured and predicted N rates have been recorded in many crops all across the world. This is attributed to a wide range of N needs from one year to the next. University extension, fertilizer dealers, and private consulting organizations have historically used yield goals due to the lack of a better alternative.

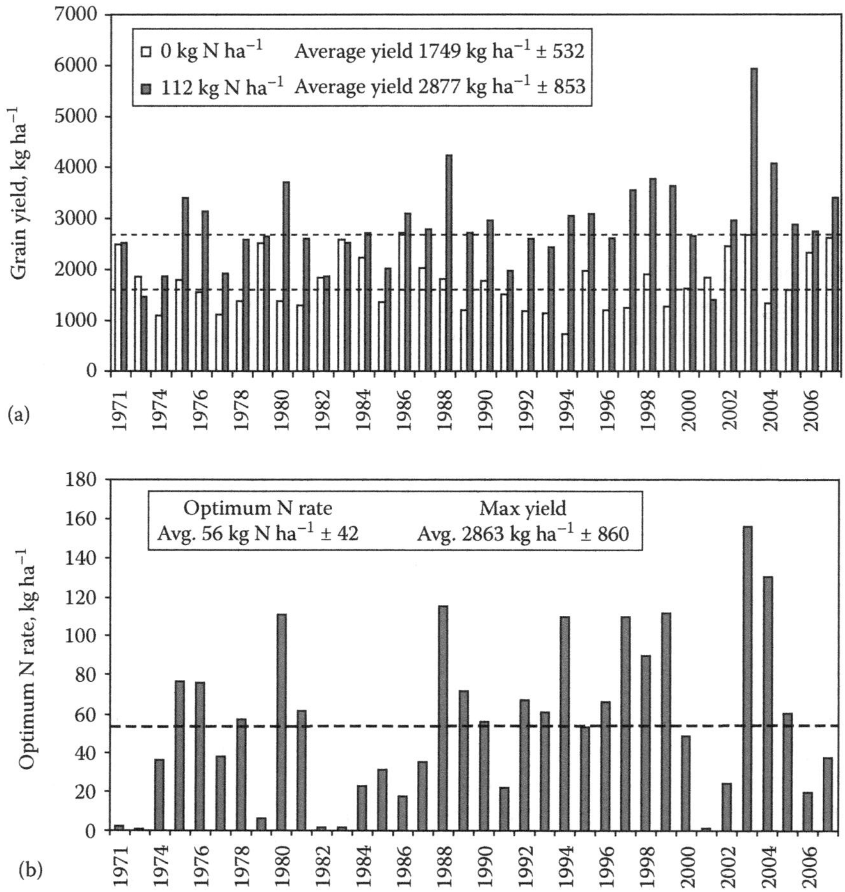

FIGURE 20.2 (a) Winter wheat grain yield of two fertility treatments (0 and 112 kg N ha) from 1971 to 2007. Over the 26-year period, the yield ranged from the 1422 to 5936 kg ha^{-1}. Long-term fertility study located at Central Oklahoma. (b) Optimum N fertilizer rate to reach maximum winter wheat grain yield from 1971 to 2007. Over the 26-year period, optimum rate ranged from 0 to 156.8 kg N ha^{-1}. Long-term fertility study #502 located at Lahoma, Oklahoma.

20.3.5 Soil Testing: Preplant and Pre-Sidedress Nitrate

Soil testing can be used to determine the amount of inorganic and organic N contained in the soil. Many N recommendation protocols use some form of soil testing. Oklahoma State University generally recommends that farmers apply 33 kg N ha^{-1} for every 1 Mg of wheat (2 lb N ac^{-1} for every bushel of wheat) they hope to produce, minus the amount of NO3-N in the surface (0–15 cm) soil profile. 24 Nitrogen's dynamic nature leads to the downfall of soil test N recommendations. Any change in environment, whether it is temperature, rainfall, or plant growth, after the sample is collected affects N concentrations in soil.

The depth at which a soil sample is taken can affect the efficiency at which fertilizer N is used. The soil sampling depth should match the rooting depth.

Pre-sidedress soil test nitrate (PSNT) is collected prior to a sidedress N application.[28] This method has predominantly been used in the production of corn. For PSNT, it is recommended that soil samples be collected when the crop is 20–30 cm in height, and the sample is collected to a depth of 30 cm. In most cases, a concentration of 20–30 mg kg^{-1} is considered to be nonresponsive to additional N inputs.[29–34] The PSNT has been found to be an adequate method of separating responsive and nonresponsive fields.[35,36] Guillard et al.,[34] Magdoff,[37] and Durieux et al.[38] found that the PSNT reduced overfertilization. It is most useful in fields that have received an application of animal waste or have a legume crop included within its rotation.

20.3.6 Plant Tissue Testing and Tiller Density

Plant analysis is the quantitative determination of the elements in plant tissue. Plant analysis consists of procedures that quantify laboratory analysis, while tissue testing refers to semiquantitative "quick" tests that analyze plant sap and can be conducted on-site for troubleshooting purposes.[39] Both laboratory and in-field tests can be performed to evaluate N concentrations in the plant samples. Research suggests that plant N concentrations at various growth stages provide good predictive values of corn or wheat yields and N responsiveness.[39–46] Vaughan et al.[47] found that a combination of spring soil inorganic-N and plant N concentrations provided a better prediction of wheat N response than either soil or plant N measurements alone. Unfortunately, Vaughan et al.[47] reported that stem and whole-plant NO_3-N concentrations were too variable for reliable use in wheat N recommendations.

An N management strategy for winter wheat, which both involved a soil NO_3-N test, plant analysis, and tiller density estimation, was developed by Scharf and Alley.[45] Tillers are the lateral shoots that form out of auxiliary buds in the wheat plant. The method determined the whole-field tiller density at GS 25.[48] If the tiller density was low (<1000 tillers m^{-2}), applying N at GS 25 improved grain yields and economic returns. At GS 30, they found that a tissue test could be used to optimize further N applications. Weisz et al.[49] examined the critical range for GS 25 N application for a wide range of tiller densities. They found that when tiller density was <540 tillers m^{-2}, applying all of the spring N at GS 25 produced grain yields that were greater than those obtained by applying all of the N at GS 30. However, when tiller density was >540 tillers m^2, applying all of the spring N at GS 25 produced grain yields that were less than those obtained when all of the N was applied at GS 30. These studies show that to obtain maximum grain yield, wheat growers must know the tiller density at GS 25 and be able to respond quickly by applying N when needed. This method was shown to increase the producer income by an average of \$73 ha^{-1}.[45] Flowers et al.[50] reported that in spite of the benefits of timing N based on tiller density, growers are reluctant to use this system. Lack of adoption is attributed to the fact that tiller counts are difficult and laborious to obtain, and tiller density is highly variable across a field.

20.4 METHODS OF NUE IMPROVEMENT: HIGH TECH

20.4.1 Management Zones

In traditional N management, a single rate is based on the average perceived need of an entire field, which often consists of large areas of land. Because of field spatial variability, this method inherently places excessive amounts of N in some areas, while others may receive an inadequate amount. The spatial variability of a field has long been recognized as having an impact on crop yield.[51] The advent of the combine yield monitor and its widespread use has revealed to many agronomic producers just how variable yield is across agricultural landscapes. Solie et al.[52] reported that variability in soil test values was determined to exist at resolutions of less than 1 m^2, illustrating that coarse resolution approaches (>1 ha) may not be the optimal agronomic resolution.

Fields can be separated into management zones, and these zones can be differentiated by a multitude of characteristics. Zones can be delineated by historic yield levels (i.e., zones of low, average, and high yields), soil type or texture, slope, bare soil imagery, remotely sensed canopy reflectance, farmer experience, soil electrical conductivity, or depth to limiting layer.[53–56] Reducing the size of the field into zones can improve energy and NUE.[57,58] A challenge with this approach is defining the management zones.[59,60] For example, in a dryland field with a coarse-textured soil type in one well-defined area and a fine-textured soil in a different area can be separated into two zones. Managing these two areas is difficult because the sandy soil might have very high yields some years and low yields other years. Solie et al.[61] stated that, in order to capitalize on any potential N fertilizer savings and increased energy and NUE, management decisions need to be made at the appropriate field element size. Field element size is defined as that area or resolution that provides the most precise measure of the available nutrient where the level of that nutrient changes with distance. With the current state of fertilizer applicators and global positioning system technologies, the field element size can be reduced to the width of the applicator and a similar distance in the direction travel.

Research has shown that site-specific management zones and VR application (VRA) can increase NUE.[62–64] Mulla and Bhatti[65] showed that in eastern Washington, N application for winter wheat could be reduced by as much as 42 kg ha^{-1} using VRA. In irrigated corn production, Koch et al.[66] observed that a VR yield goal zone management strategy produced net returns ranging from \$18.21 to \$29.57 ha^{-1} over the uniform N management strategy. Khosla et al.[67] found that using VRA on a 20.4 ha field located in Virginia reduced total N applied by 22 kg N ha^{-1} without a reduction in grain yield when compared with a uniform N treatment.

Combining management zones (right place) with the concept of the right source can result in improved NUE. Noellsch et al.[68] demonstrated that the use of controlled-release N fertilizer (polymer-coated urea) or anhydrous ammonia in specific management zones can be beneficial over urea. Within their study, they utilized landscape position as a factor to determine where polymer-coated urea or anhydrous ammonia should be used. They concluded that under certain environmental conditions, low-lying areas benefited from the use of polymer-coated urea or anhydrous ammonia compared to urea.

20.4.2 Reference Strips

The use of reference strips (N-rich strip) as a tool to improve energy and NUE has potential in both low- and high-tech applications. Reference strips are a simple technique for identifying areas that may respond to additional N.[60] N strips are placed in fields by applying a relatively high rate in a strip across the field. This can be accomplished by doubling or tripling the preplant N rate on one pass, by returning after preplant application, or by planting with a single pass through the field with a high rate of N. It is recommended that the reference strip contain at least 125% of the recommended total N need for the crop.[69] This strip should contain enough N to carry the crop through the entire season, minimizing the risk of N deficiency. By comparing the N strip to other areas in the field, N recommendations can be fine-tuned. CM and remote sensing have been used to quantify differences. If there is a visible difference between the reference strip and the rest of the field, then there is a definite need for added N. Conversely, if differences are not detected, then additional may not be needed.

Variations of the N-rich strip have been evaluated and used, much of which employ a sequence of N rates.[70,71] In essence, an N rate study is applied in each field to determine optimum N rate. The "Ramped" reference strip is a strip where different N rates are applied to different areas across the strip. One approach is the start with a high N rate which is sequentially decreased every 3–9 m until the zero rate is achieved. The ramps would measure from 60 to 150 m in total length.[71] In a field with high spatial variability, the ramp strip may not provide reliable prescription.

20.4.3 Illinois Soil N Test

The Illinois soil N test (ISNT), developed by Mulvaney and Khan,[72] determines different forms of N in the soil hydrolysates.[73] The ISNT values are strongly correlated to amino-sugar N value[72] which is directly related to the soil N mineralization potential and inversely related to a crop potential to respond to additional N.[72] This suggests that a critical level may be determined to indicate whether or not crops will respond to fertilizer N in a particular field. Since the test's introduction, it has been evaluated on many different soils and environments. Unfortunately, like numerous other N mineralization indexes, it has had mixed results.[73]

20.4.4 Light: CM and Optical Sensors

Chlorophyll has been shown to be highly correlated with leaf N concentration.[74–76] It is most useful when it is used in combination with a reference strip. Instruments such as the SPAD meter (Figure 20.3) measure relative chlorophyll concentration and report the results as arbitrary units. CM work by being clamped down on the leaf so that light may be transmitted through the leaf and transmittance determined at the 650 and 940 nm wavelengths. Research conducted across the corn belt with this tool has shown that it can improve in-season N recommendations.[9,77–80]

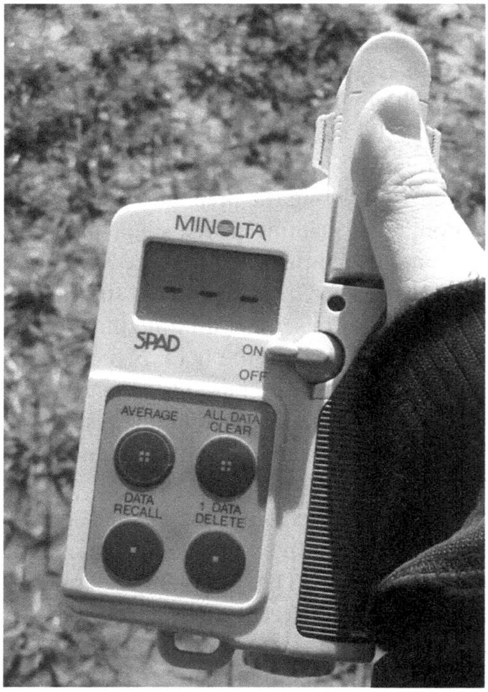

FIGURE 20.3 Image of the Minolta SPAD 502CM. (Courtesy of Konica Minolta Sensing, Inc. Tokyo, Japan.)

Sims et al.[81] indicated that the leaf CM could be an alternative to the PSNT for refining in-season fertilizer N requirements in corn. A problem with the CM is the length of time required to collect field data. Optical sensors and aerial imagery has been proposed as a technique to overcome this problem.

Canopy reflectance has evolved as a promising high-tech approach to improving NUE. Crop health can be assessed using a variety of remote-sensing indexes. The most common indexes used are normalized difference vegetation index (NDVI) and Green band NDVI (GNDVI).[82–87] NDVI utilizes visible (670nm) and near-infrared (NIR) light, approximately 780nm, and is calculated as (NIR − Red)/(NIR + Red), while GNDVI is calculated with (NIR − Green)/(NIR + Green). Different remote-sensing indexes may be better at predicting one stress than another.[88]

Remote-sensing information can be collected from satellite imagery, aerial photography, and ground-based sensors, and passive and active sensing. All satellite and aerial imagery and some ground-based systems use passive sensing, which involves the measurement of natural reflected light as well as irradiated, ambient light from the sun. Active sensors generate their own light source. NDVI has been used as an index of plant health because the red band is absorbed by chlorophyll and the NIR band is reflected by healthy vegetation. Plants with a higher leaf area and chlorophyll contents reflect less red and more NIR light. The NIR wavelength is able to penetrate the inner layer plant cells. The healthier or more vigorous a plant that has a denser inner cell reflects more NIR while the weaker cells allow greater penetration of the

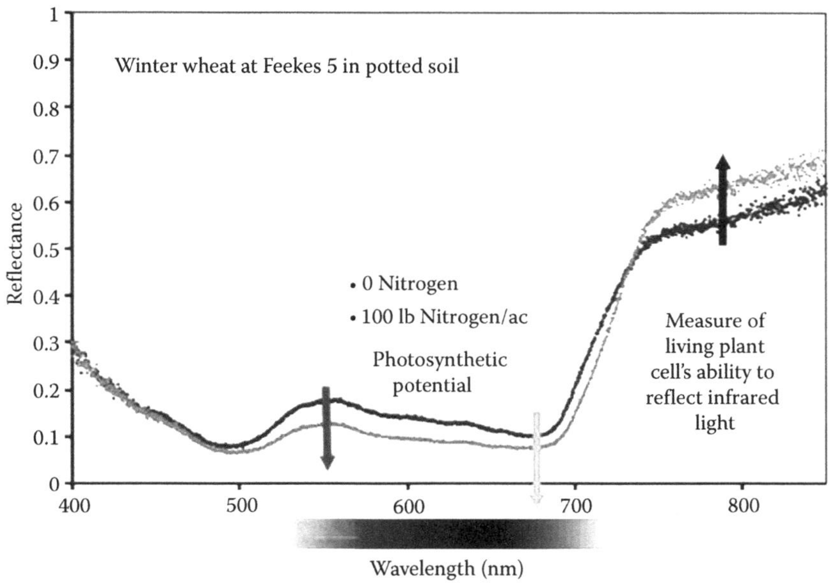

FIGURE 20.4 Spectral reflectance of two potted plants: one healthy and vigorously growing (100 lb N ac^{-1}) and one stunted and smaller due to nutrient deficiencies (0 nitrogen). (Courtesy of Dr. Marvin Stone, Oklahoma State University, Stillwater, Oklahoma.)

inner wall, as seen in Figure 20.4. Past work has shown that NDVI measurements in winter wheat between Feekes physiological growth stages 4 and 5 can provide a reliable prediction of both N uptake and biomass.[61,89]

Two terms often used with optical sensor/plant reflectance technology are sufficiency and response. While the technological meaning of the two terms is different, they are actually the mathematical reciprocal of each other. This is also where reference strips play a part in reflectance-based N rate prescriptions. The reference strip is just that, a reference of what the crop looks like when supplied with adequate amounts of N. Both the sufficiency and response indices compare the N-rich reference (NR) with the rest of the field, which we will refer to as the farmer practice.

Sufficiency index (SI) is calculated by dividing the NDVI of the farmer practice by the NDVI of the NR, giving you a percent of total. For example, if the NDVI in field equals 0.67 and the NDVI of the N-rich strip was 0.83, then the SI would be 0.81 (0.67/0.83). Using the SI concept, the rest of the field only has 81% of the total N required for maximum yield. Using a predetermined economic optimum N rate (EONR), a sidedress N prescription can be made. For this example, if the EONR for the areas yield potential and soil type is 150 kg N ha^{-1}, the calculation for sidedress N rate would be simply 150 * (1.0 – 0.81), and the prescription for the field would be 28.5 kg N ha^{-1}.

The response index, as mentioned before, is the reciprocal of the sufficiency calculation. The NDVI of the N-rich strip is divided by the NDVI in the field. So, using the same NDVI values in the previous example, response index would be 1.23. Using this concept, additional N would be needed to optimize yields.

An alternative approach is to combine crop reflectance with growing degree days (GDDs).[90,91] GDDs are calculated by taking the daily minimum temperature plus maximum temperature and dividing by 2. The sum of the GDDs accumulated from planting to sensing is cumulative GDD. The use of GDDs in predicting grain yield was a successful method of normalizing the NDVI reading for variable environments, growing conditions, planting dates, and stage of growth at the time of sensing. A method of determining N rate when using response index and yield potential was outlined by Lukina et al.[92] The N rate is calculated with the equation:

$$\text{N rate} = (\text{YP0} * \text{RI} - \text{YP0}) * \%\text{N}/\text{NUE}$$

where

YP0 is the yield potential of the farmers practice
RI is the response index and is calculated by dividing the NDVI of the N-rich strip by the NDVI of the farmer practice
%N is the percent of N in the grain

NUE is the expected N use efficiency of the fertilizer application In essence, this method determines the difference in yield between the farmer's field and well-fertilized reference strip, and the algorithm recommends enough N fertilizer to correct for this difference. The amount of N is determined through a back calculation of the average grain N content and amount of yield removed. Because yield potential is a part of the equation, two fields that have the same measured response indices may have very different N rate recommendations if yield potentials are different. If both fields share a response index of 1.23 but the first field farmer practice has a yield potential of 2000 kg ha^{-1}, then the recommendation would be to fertilize for a 460 kg ha^{-1} yield increase, but if the second field has a farmer practice yield potential of 4000 kg ha^{-1}, then the N rate required would be for a 920 kg ha^{-1} yield increase. The amount of actual N required to achieve a 920 kg ha^{-1} increase in yield would be (920 * 0.0239)/0.50 = 43.9 kg N ha^{-1}, where 0.0239 represents the average percent N in winter wheat grain (2.39%) and 0.50 is the expected efficiency of the N when applied topdress. The expected efficiency can change depending on application conditions, crop growth stage, and a range of other factors that must be identified by the producer.

20.5 CASE STUDIES: CALCULATING N RATES IN AREAS WITH DIFFERENT YIELDS

Case Study 1: Reference Strip Identifies In-Field Variability

Location: North East Oklahoma in a 32 ha corn field, the previous crop was winter wheat following soybeans.

The producer had applied an N-rich strip that ran the length of the field. At the time of sensing, eight-leaf stage, the N-rich strip stood out from the rest of the field. However, there was a significant soil texture change, and the eastern third of the field consisted of a much more coarse-textured soil. Viewing the N-rich strip, this change was very evident. In the finer soil, the field was much less responsive and the N-rich strip did not present the same significant difference as it did in the eastern one-third. The sensor readings were individually collected from the N-rich strip and farmer practice within the fine and coarse soils. The NDVIs indicated an RI of 1.15 from the fine soil and an RI of 1.3 from the coarse-textured soil with recommended N sidedress rates 22 and 66 kg N ha^{-1}, respectively. With this knowledge, the producer could have applied the N according to management zones as the sensor recommended. However, this producer decided to apply a flat rate of 44 kg N ha^{-1} across the field, acknowledging that he was overapplying to two-third's of the field while underapplying and losing yield on the additional one-third.

Case Study 2: Optical Sensor Identifies a Difference in Past Management Practices

Location: North Central Oklahoma in a 65 ha wheat field, previous crop was unknown.

The producer had recently purchased the 65 ha, and this was the first crop, previously the field was farmed as two independent fields, each with a different history. The producer owns a commercially available sensor-based VR applicator. At planting, the producer placed an N-rich strip in the field and took NDVI measurement prior to topdress. The producer had expected to apply approximately 55 kg ha^{-1}; the actual applied average N rate was 52 kg ha^{-1}. When the application map created by the software was reviewed, there was a significant difference in average N rate and a definite break that occurred in the middle of the field. On the west 32.5 ha, only 26 kg ha^{-1} was applied while the east half received 78 kg ha^{-1}. Because of this difference, the history of the land was further investigated with the conclusion that the east 32.5 ha had historically been under mono culture winter wheat production for many years, while the western 32.5 ha had been recently taken out of alfalfa hay production. When the yield monitor data were reviewed after harvest, the entire 65 ha field had a relatively homogenous grain yield of 3900 kg ha^{-1}. Without the benefit of the reference strip in optical sensor technology, the producer would have likely overapplied N to 32.5 ha and lost a significant amount of yield on the remaining 32.5 ha.

Case Study 3: Optical Sensor Identifies a Growing Season with High-Fertilizer N Demand

Location: Central Oklahoma in an 81 ha continuous cultivated wheat field.

The producer had agreed to support a large-scale replicated study evaluating sensor-based N management strategies. Three treatments were evaluated—preplant only, producers practice, and sensor-based nitrogen rate calculator (SBNRC) flat rate—on 0.2 ha plots with three replications. The SBNRC utilizes an optical NDVI sensor and a reference strip to calculate potential yield, response index, and the N fertilizer recommendation. The producer had collected soil samples during the summer; soil test NO_3 results recommended a total of 39 kg N ha^{-1} to be applied to reach the 3360 kg ha^{-1} yield goal. The producer applied 20 kg N ha^{-1} preplant and planned to apply the remainder as topdress. At the time of sensing, Feekes 5, the N-rich strips that were applied at sowing were very visible after a very dry and cold winter (immobilization and lack of N contributed from rainfall, Figure 20.1). Because of the dramatic visual difference, the producer applied 30 kg N ha^{-1} to the rest of his field (11 kg higher than soil test recommendation), based on the NDVI and GDDs the SBNRC recommended 66 kg N ha^{-1}. The wheat was harvested by the producer and the yield was recorded using a weigh wagon. The SBNRC flat rate out yielded the producers practice by 389.74 kg grain ha^{-1}. This increase in grain yield resulted in an increase in profit of \$33.42 ha^{-1}, assuming a grain price of \$0.206 and \$1.23 kg^{-1} for the N fertilizer. The use of reference strips and optical sensors provided an opportunity to increase yields by identifying a situation where the crop would be highly responsive to fertilizer N.

Case Study 4: Optical Sensor Identifies a Growing Season with Low-Fertilizer N Demand

Location: East Central Missouri a 32.4 ha cornfield, previous crop was also corn.

The producer had agreed to support a large-scale replicated study evaluating sensor-based N management strategies. Three treatments were evaluated—producers practice, flat rate SBNRC, and sensor-based variable-rate treatment (VRT)—on 0.1 ha plots with three replications. At the time of sensing, nine-leaf stage, the N-rich strip had slightly higher NDVI values; however, it was not apparent visually, with the exception of a slight difference in height. The producer had 17 kg N ha^{-1} preplant to the entire field while the N-rich strip had received 392 kg N ha^{-1}. A good N response at this location was expected considering it was third-year corn on corn production with very little preplant N. However, it was a very moist spring and early summer and a soil with 2% organic matter (mineralization and additions from rainfall, Figure 20.1). At the time, the SBNRC calculated yield potentials to be near 13,180 kg ha^{-1}. The producers rate, and the rest of the field, received 246 kg N ha^{-1}. The SBNRC recommended 112 kg N ha^{-1}, and an average rate of 76 kg N ha^{-1} was applied to the VRT. The plots were harvested by the producers and yield data collected with a yield monitor. The farmer practice had the highest yield with 12,260 kg ha^{-1} and lowest gross profit at \$1726.40 ha^{-1}, assuming \$0.157 kg^{-1} grain and \$0.77 kg^{-1} N fertilizer. The SBNRC flat rate resulted in a yield of 11,910 kg ha^{-1} grain yield and a profit of \$1774.52 ha^{-1}. The VRT rate resulted in the highest profit of \$1806.4 ha^{-1}; this is not taking into account cost of the equipment. The VRT did yield slightly lower than the producers practice with an average grain yield of 11,940 kg ha^{-1}. In this case, the use of reference strips and optical sensors provided an opportunity to increase profits by identifying a situation where the crop would be minimally responsive to fertilizer N and the producers traditional practice would apply N at an excessive rate.

20.6 SUMMARY

Among the many methods of NUE improvement, all are valid. Some, however, may be more regionally or system selective than others. Every step taken to improve NUE in agricultural production is a step forward. While considering these different methods, you should not ask "Is this the perfect method or solution?" The question that must be asked should be "Does this method improve upon the previous practice?" It is essential to recognize that there is currently no perfect answer, and likely there will not be one any time in the near future. Nitrogen's dynamic relationship with the environment adds to our inability to control the fate of N. A worldwide solution for improved N management will ultimately be tied to better incorporation of climatological inputs that define the exact conditions encountered from planting to the time mid-season N decisions are made. The development of new innovations, testing of new concepts, and cooperation of agricultural producers, industry, and the scientific community on a global scale will assist in realizing change that is sorely needed in world N management. Today, high-tech methods use real-time sensor measurements to make recommendations at the field or zone level. Tomorrow's high-tech solutions will undoubtedly incorporate the ability to economically and efficiently apply N to individual plants.

REFERENCES

1. Raun, W.R. and Johnson, G.V. Improving nitrogen use efficiency for cereal production. *Agron. J.* 91, 357–363, 1999.
2. Viswakumar, A., Mullen, R.W., Sundermeier, A., and Dygert, C.E. Tillage and nitrogen application methodology impacts on corn grain yield in a poorly drained soil. *J. Plant Nutr.* 31, 1963–1974, 2008.
3. Ferguson, R.B. and Kissel, D.E. Effects of soil drying on ammonia volatilization from surface applied urea. *Soil Sci. Soc. Am. J.* 50, 485–549, 1986.
4. Keeney, D.R. Nitrogen management for maximum efficiency and minimum pollution. In Stevenson, F.J. (ed.). *Nitrogen in Agricultural Soils*. Agronomy Monograph 22. ASA, CSSA, and SSSA, Madison, WI, 1982, pp. 605–649.
5. Randall, G.W. and Hoeft, R.G. Placement methods for improved efficiency of P and K fertilizers: A review. *J. Prod. Agric.*1, 70–79, 1988.
6. Mamo, M., Malzer, G.L., Mulla, D.J., Huggins, D.R., and Strock, J. Spatial and temporal variation in economically optimum nitrogen rate for corn. *Agron. J.* 95, 958–964, 2003.
7. Schmidt, J.P., Dejoia, A.J., Ferguson, R.B., Taylor, R.K., Young, R.K., and Havlin, J.L. Corn yield response to nitrogen at multiple in-field locations. *Agron. J.* 94, 798–806, 2002.
8. Gauer, L.E., Grant, C.A., Gehl, D.T., and Bailer, L.D. Effects of nitrogen fertilization on grain protein content, nitrogen uptake, and nitrogen use efficiency of six spring wheat (*Triticum aestivum*, L.) cultivars, in relation to estimate moisture supply. *Can. J. Plant Sci.* 72, 235–241, 1992.
9. Vetsch, J.A. and Randall, G.W. Corn production as affected by nitrogen application timing and tillage. *Agron. J.* 96, 502–509, 2004.
10. Olson, R.V. and Swallow, C.W. Fate of labeled nitrogen fertilizer applied to winter wheat for five years. *Soil Sci. Soc. Am. J.* 48, 583–586, 1984.
11. Finney, K.F., Meyer, J.W., Smith, F.W., and Fryer, H.C. Effect of foliar spraying on Pawnee wheat with urea solutions on yield, protein content, and protein quality. *Agron. J.* 49, 341–347, 1957.
12. Wuest, S.B. and Cassman, K.G. Fertilizer-nitrogen use efficiency of irrigated wheat: I. Uptake efficiency of preplant vs. late season application. *Agron. J.* 84, 682–688, 1992.
13. Delgado, J.A. and Mosier, A.R. Mitigation alternatives to decrease nitrous oxides emissions and urea-nitrogen loss and their effect on methane flux. *J. Environ. Qual.* 5, 1105–1111, 1996.
14. Hauck, R.D. Slow-release and bioinhibitor-amended nitrogen fertilizers. In: Engelstad O.P. (ed.). *Fertilizer Technology and Use*, 3rd edn. SSSA, Madison, WI, 1985, pp. 293–322.

15. Peoples, M.B., Freney, J.R., and Mosier, A.R. Minimizing gaseous losses of nitrogen. In: Bacon, P.E. (ed.). *Nitrogen Fertilization in the Environment*. Marcel Dekker, Inc., New York, 1995, pp. 565–602.
16. Prasad, R. and Power, J.F. Nitrification inhibitors for agriculture, health and the environment. *Adv. Agron*. 54, 1995, pp. 233–281.
17. Allen, S.E. Slow release nitrogen fertilizers. In: Hauck, R.D. (ed.). *Nitrogen in Crop Production*. ASA-CSSA-SSSA-Madison, WI, 1984, pp. 195–206.
18. Camberato, J.J. Nitrogen in soils and fertilizers. In: *South Carolina Turfgrass Foundation News*, 6–10, 2001.
19. Overdahl, J.C., Rehm, G.W., and Meredith, H.L. *Fertilizer Urea FO-00636-GO*. University of Minnesota Extension, Minneapolis, MN, 1991.
20. Schlegel, A.J., Nelson, D.W., and Sommers, L.E. Field evaluation of urease inhibitors for corn production. *Agron. J*. 78, 1007–1012, 1986.
21. Nelson, D.W. and Huber, D.H. Nitrification inhibitors for corn production. In: *National Corn Handbook, Crop Fertilization* NCH-55. Cooperative Extension Service, University of Illinois, Urbana-Champaign, IL, 1988.
22. Dahnke, W.C., Swenson, L.J., Goos, R.J., and Leholm, A.G. *Choosing a Crop Yield Goal*. SF-822. North Dakota State Extension Service, Fargo, ND, 1988.
23. Hoeft, R.G. and Nafziger, E.D. *Illinois Agronomy Handbook*, 23rd edn. University of Illinois, Urbana-Champaign, IL, 2002.
24. Zhang, H. and Raun, W.R. *Soil Fertility Handbook*. Oklahoma Agricultural Experiment Station, Stillwater, OK, 2006.
25. Vanotti, M.B. and Bundy, L.G. An alternative rationale for corn nitrogen fertilizer recommendations. *J. Prod. Agric*. 7, 249–256, 1994.
26. Doerge, T.A. Variable rate nitrogen management creates opportunities and challenges for corn producers. Available at http://www.plantmanagmentnetwork.org/pub/cm/review/variable%2Dn/ (accessed July 1, 2009). *Crop Management*, doi:101094/cm-2002-0905-01-RS, 2002.
27. Nafziger, E.D., Sawyer, J.E., and Hoeft, R.G. Formulating N recommendations for corn in the corn belt using recent data. In: *Proceedings of the North Central Extension Industry Soil Fertility Conference*, Des Moines, I., November 17–18, 2004. Potash and Phosphate Institute, Brookings, SD, 2004, pp. 5–11.
28. Magdoff, F.R., Ross, D., and Amadon, J.A. Soil test for nitrogen availability to corn. *Soil Sci. Soc. Am. J*. 48, 1301–1304, 1984.
29. Blackmer, A.M., Pottker, D., Cerrato, M.E., and Wedd, J. Correlations between soil nitrate concentrations in late spring and corn yields in Iowa. *J. Prod. Agric*. 2, 103–109, 1989.
30. Fox, R.H., Roth, G.W., Iverson, K.V., and Piekielek, W.P. Soil and tissue nitrate tests compared for predicting soil nitrogen available to corn. *Agron. J*. 81, 971–974, 1989.
31. Magdoff, F.R., Jokela, W.E., Fox, R.H., and Griffin, G.F. A soil test for N availability in the northeastern U.S. *Commun. Soil Sci. Plant Anal*. 21, 1103–1115, 1990.
32. Meisinger, J.J., Bandel, V.A., Angle, J.S., O'Keefe, B.E., and Reynolds, C.M. Presidedress soil nitrate test evaluation in Maryland. *Soil Sci. Soc. Am. J*. 56, 1527–1532, 1992.
33. Klausner, S.D., Reid, W.S., and Bouldin, D.R. Relationship between late spring soil nitrate concentration and corn yields in New York. *J. Prod. Agric*. 6, 350–354, 1993.
34. Guillard, K., Morris, T.F., and Kopp, K.L. The pre-sidedress soil nitrate test and nitrate leaching from corn. *J. Environ. Qual*. 28, 1845–1852, 1999.
35. Fixen, P.E. and Grove, J.H. Testing soils for phosphorus. In: Westerman, R.L. (ed.). *Soil Testing and Plant Analysis*, 3rd edn. SSSA Book Series 3, Madison, WI, 1990, pp. 141–180.

36. Haby, V.A., Russelle, M.P., and Skogley, E.O. Testing soils for potassium, calcium, and magnesium. In: Westerman, R.L. (ed.). *Soil Testing and Plant Analysis*, 3rd edn. SSSA Book Series, Madison, WI, 1990, pp. 181–228.
37. Magdoff, F.R. Understanding the Magdoff pre-sidedress nitrate test for corn. *J. Prod. Agric.* 4, 297–305, 1991.
38. Durieux, R.P., Brown, H.J., Stewart, E.J., Zhao, J.Q., Jokela, W.E., and Magdoff, F.R. Implications of nitrogen management strategies for nitrate leaching potential: Roles of nitrogen source and fertilizer recommendation systems. *Agron. J.* 87, 884–887, 1995.
39. Schulte, E.E. and Kelling, K.A. Plant analysis: A diagnostic tool. In: *National Corn Handbook, Crop Fertilization* NCH-46. Cooperative Extension Service, University of Illinois, Urbana-Champaign, IL, 1988.
40. Donohue, S.J. and Brann, D.E. Optimum N concentration in winter wheat grown in the coastal plain region of Virginia. *Commun. Soil Sci. Plant Anal.* 15, 651–661, 1984.
41. Baethgen, W.E. and Alley, M.M. Optimizing soil and fertilizer nitrogen use by intensively managed winter wheat: I. Crop nitrogen uptake. *Agron. J.* 81, 116–120, 1989.
42. Baethgen, W.E. and Alley, M.M. Optimizing soil and fertilizer nitrogen use by intensively managed winter wheat: II. Critical levels and optimum rates of nitrogen fertilizer. *Agron. J.* 81, 120–125, 1989.
43. Roth, G.W., Fox, R.H., and Marshall, H.G. Plant tissue tests for predicting nitrogen fertilizer requirements of winter wheat. *Agron. J.* 81, 502–507, 1989.
44. Vaughan, B., Barbarick, K.A., Westfall, D.G., and Chapman, P.L. Tissue nitrogen levels for dryland hard red winter wheat. *Agron. J.* 82, 561–565, 1990.
45. Scharf, P.C. and Alley, M.M. Spring nitrogen on winter wheat: II. A flexible multicomponent rate recommendation system. *Agron. J.* 85, 1186–1192, 1993.
46. Scharf, P.C., Alley, M.M., and Lei, Y.Z. Spring nitrogen on winter wheat: I. Farmer-field validation of tissue test-based rate recommendations. *Agron. J.* 85, 1181–1186, 1993.
47. Vaughan, B., Westfall, D.G., Barbarick, K.A., and Chapman, P.L. Spring nitrogen fertilizer recommendation models for dryland hard red winter wheat. *Agron. J.* 82, 565–57, 1990.
48. Zadok, J.C., Chang, T.T., and Konzak, C.F. A decimal code for the growth stages of cereals. *Eucarpia Bulletin* 7, 42–52, 1974.
49. Weisz, R., Crozier, C.R., and Heiniger, R.W. Optimizing nitrogen application timing in no-till soft red winter wheat. *Agron. J.* 93, 435–442, 2001.
50. Flowers, M., Weisz, R., and Heiniger, R. Remote sensing of winter wheat tiller density for early nitrogen application decisions. *Agron. J.* 93, 783–789, 2001.
51. Johnson, C.K., Mortensen, D.A., Wienhold, B.J., Shanahan, J.F., and Doran, J.W. Site-specific management zones based on soil electrical conductivity in a semiarid cropping system. *Agron. J.* 95, 303–315, 2003.
52. Solie, J.B., Raun, W.R., and Stone, M.L. Submeter spatial variability of selected soil and bermudagrass production variables. *Soil Sci. Soc. Am. J.* 63, 1724–1733, 1999.
53. Lark, R.M. and Stafford, J.V. Classification as a first step in the interpolation of temporal and spatial variation of crop yield. *Ann. Appl. Biol.* 130, 111–121, 1997.
54. Kitchen, N.R., Sudduth, K.A., and Drummond, S.T. Soil electrical conductivity as a crop productivity measure for claypan soils. *J. Prod. Agric.* 12, 607–617, 1999.
55. Fleming, K.L., Westfall, D.G., Wiens, D.W., and Brodah, M.C. Evaluating farmer developed management zone maps for variable rate fertilizer application. *Precis. Agric.* 2, 201–215, 2000.
56. Fridgen, J.J., Kitchen, N.R., Sudduth, K.A., Drummond, S.T., Wiebold, W.J., and Fraisse, C.W. Management zone analyst (MZA): Software for subfield management zone delineation. *Agron. J.* 96, 100–108, 2004.

57. Raun, W.R., Solie, J.B., Johnson, G.V., Stone, M.L., Mullen, R.W., Freeman, K.W., Thomason, W.E., and Lukina, E.V. Improving nitrogen use efficiency in cereal grain production with optical sensing and variable rate application. *Agron. J.* 94, 815–820, 2002.
58. Shanahan, J.F., Kitchen, N.R., Raun, W.R., and Schepers, J.S. Responsive in-season nitrogen management for cereals. *Comput. Electron. Agric.* 61, 51–62, 2007.
59. Jaynes, D.B. and Colvin, T.S. Spatiotemporal variability of corn and soybean yield. *Agron. J.* 89, 30–37, 1997.
60. Johnson, G.V. and Raun, W.R. Nitrogen response index as a guide to fertilizer management. *J. Plant Nutr.* 26, 249–262, 2003.
61. Solie, J.B., Raun, W.R., Whitney, R.W., Stone, M.L., and Ringer, J.D. Optical sensor based field element size and sensing strategy for nitrogen application. *Trans. ASAE.* 39, 1983–1992, 1996.
62. Khosla, R., Fleming, K., Delagado, J.A., Shaver, T.M., and Westfall, D.G. Use of site-specific management zones to improve nitrogen management for precision agriculture. *J. Soil Water Conserv.* 57, 513–518, 2002.
63. Hornung, A., Khosla, R., Reich, R., and Westfall, D.G. Evaluation of site-specific management zones: Grain yield and nitrogen use efficiency. In: Stafford, J. and Werner, A. (eds.). *Precision Agriculture.* Wageningen Academic Publishers, Wageningen, the Netherlands, 2003, pp. 297–302.
64. Inman, D., Khosla, R., Westfall, D.G., and Reich, R. Nitrogen uptake across site specific management zones in irrigated corn production systems. *Agron. J.* 97, 169–176, 2005.
65. Mulla, D.J. and Bhatti, A.U. An evaluation of indicator properties affecting spatial patterns in N and P requirements for winter wheat yield. In: Stafford, J.V. (ed.). *Precision Agriculture* '97, First European Conference on Precision Agriculture, Warwick University, U.K. September 7–10, 1997. Bios Scientific Publishers, Oxford, Herndon, VA, 1997, pp. 145–153.
66. Koch, B., Khosla, R., Frasier, W.M., Westfall, D.G., and Inman, D. Economic feasibility of variable-rate nitrogen application utilizing site-specific management zones. *Agron. J.* 96, 1572–1580, 2004.
67. Khosla, R., Alley, M.M., and Griffith, W.K. Soil-specific nitrogen management on mid-Atlantic coastal plain soils. *Better Crops* 83, 6–7, 1999.
68. Noellsch, A.J., Motavalli, P.P, Nelson, K.A., and Kitchen, N.R. Corn response to conventional and slow-release nitrogen fertilizers across a claypan landscape. *Agron. J.* 101, 607–614, 2009.
69. Arnall, D.B., Edwards, J.T., and Godsey, C.B. *Reference Strip Series: Applying your Nitrogen-Rich and Ramp Calibration Strips.* CR 2255. Oklahoma Cooperative Extension Service. Oklahoma State University, Stillwater, OK, 2008.
70. Raun, W.R., Solie, J.B., Stone, M.L., Zavodny, D.L., Martin, K.L., and Freeman, K.W. Automated calibration stamp technology for improved in-season nitrogen fertilization. *Agron. J.* 97, 338–342, 2005.
71. Raun, W.R., Solie, J.B., Taylor, R.K., Arnall, D.B., Mack, C.J., and Edmonds, D.E. Ramp calibration strip technology for determining midseason nitrogen rates in corn and wheat. *Agron. J.* 100, 1088–1093, 2008.
72. Mulvaney, R.L. and Khan, S.A. Diffusion methods to determine different forms of nitrogen in soil hydrolysates. *Soil Sci. Soc. Am. J.* 65, 1284–1292, 2001.
73. Laboski, C.A.M., Sawyer, J.E., Walters, D.T., Bundy, L.G., Hoeft, R.G., Randall, G.W., and Andraski, T.W. Evaluation of the Illinois soil nitrogen test in the north central region of the United States. *Agron. J.* 100, 1070–1076, 2008.
74. Wolfe, D.W., Henderson, D.W., Hsiao, T.C., and Alvino, A. Interactive water and nitrogen effects on senescence of maize: II. Photosynthetic decline and longevity of individual leaves. *Agron. J.* 80, 865–870, 1988.

75. Lohry, R.D. Effect of N fertilizer rate and nitrapyrin on leaf chlorophyll, leaf N concentration, and yield of three irrigated maize hybrids in Nebraska, Lincoln. Dissertation Abstract 9013612, 1989.
76. Schepers, J.S., Francis, D.D., Vigil, M.F., and Below, F.E. Comparisons of corn leaf nitrogen and chlorophyll meter readings. *Commun. Soil Sci. Plant Anal.* 23, 2173–2187, 1992.
77. Blackmer, T.M. and Schepers, J.S. Techniques for monitoring crop nitrogen status in corn. *Commun. Soil Sci. Plant Anal.* 25, 1791–1800, 1994.
78. Varvel, G.E., Schepers, J.S., and Francis, D.D. Ability for in-season correction of nitrogen deficiency in corn using chlorophyll meters. *Soil Sci. Soc. Am. J.* 61, 1233–1239, 1997.
79. Scharf, P.C., Brouder, S.M., and Hoeft, R.G. Chlorophyll meter readings can predict nitrogen need and yield response of corn in the north-central USA. *Agron. J.* 98, 655–665, 2006.
80. Ruiz Diaz, D.A., Hawkins, J.A., Sawyer, J.E., and Lundvall, J.P. Evaluation of in-season nitrogen management strategies for corn production. *Agron. J.* 100, 1711–1719, 2008.
81. Sims, J.T., Vasilas, B.L., Gartley, K.L., Milliken, B., and Green, V. Evaluation of soil and plant nitrogen test for maize on manured soils of the Atlantic coastal plain. *Agron. J.* 87, 213–222, 1995.
82. Bartholome, E. Radiometric measurements and crop yield forecasting: Some observations over millet and sorghum experimental plots in Mali. *Int. J. Remote Sens.* 9, 1539–1552, 1988.
83. Reeves, D.W., Mask, P.L., Wood, C.W., and Delaney, D.P. Determination of wheat nitrogen status with a hand-held chlorophyll meter: Influence of management practices. *J. Plant Nutr.* 16, 781–796, 1993.
84. Lillesand, T.M. and Kiefer, R.W. *Remote Sensing and Image Interpretation*, 3rd edn. John Wiley & Sons, Inc., New York, 1994.
85. Moran, M.S., Inoue, Y., and Barnes, E.M. Opportunities and limitations for image-based remote sensing in precision crop management. *Remote Sens. Environ.* 61, 319–346, 1997.
86. Tucker, C.J., Elgin, J.H., Jr., and McMurtrey III, J.E. Relationship of spectral data to grain yield variation. *Photogramm. Eng. Remote Sens.* 46, 657–666, 1980.
87. Pinter, P.J., Jr., Jackson, R.D., Idso, S.B., and Reginato, R.J. Multidate spectral reflectance as predictors of yield in water stressed wheat and barley. *Int. J. Remote Sens.* 2, 43–48, 1981.
88. Clay, D.E., Kim, K.I., Chang, J., Clay, S.A., and Dalsted, K. Characterizing water and N stress in corn using remote sensing. *Agron. J.* 98, 579–587, 2006.
89. Stone, M.L., Solie, J.B., Raun, W.R., Whitney, R.W., Taylor, S.L., and Ringer, J.D. Use of spectral radiance for correcting in-season fertilizer nitrogen deficiencies in winter wheat. *Trans. ASAE.* 39, 1623–1631, 1996.
90. Raun, W.R., Solie, J.B., Johnson, G.V., Stone, M.L., Lukina, E.V., Thomason, W.E., and Schepers, J.S. In-season prediction of potential grain yield in winter wheat using canopy reflectance. *Agron. J.* 93, 131–138, 2001.
91. Raun, W.R., Solie, J.B., Stone, M.L., Martin, K.L., Freeman, K.W., Mullen, R.W., Zhang, H., Schepers, J.S., and Johnson, G.V. Optical sensor-based algorithm for crop nitrogen fertilization. *Commun. Soil Sci. Plant Anal.* 36, 2759–2781, 2005.
92. Lukina, E.V., Freeman, K.W., Wynn, K.J., Thomason, W.E., Mullen, R.W., Stone, M.L., Solie, J.B., et al. Nitrogen fertilization optimization algorithm based on in-season estimates of yield and plant nitrogen uptake. *J. Plant Nutr.* 24, 885–898, 2001.

21 Use of GIS-Based Site-Specific Nitrogen Management for Improving Energy Efficiency

Kevin F. Bronson, Peter C. Scharf, and Newell R. Kitchen

CONTENTS

21.1 EXECUTIVE SUMMARY

To our knowledge, geographical information system (GIS)-based site-specific nitrogen management (SSNM) techniques have not been used to assess agricultural energy costs and efficiency. This chapter uses SSNM case studies for corn (*Zea mays* L.) grown in Missouri and cotton (*Gossypium hirsutum* L.) grown in Texas. In five case studies, the impact of SSNM will be compared with blanket N fertilizer recommendations. The five case studies are investigating (1) the impact N on energy produced in cotton production, (2) the impact of variable-rate N for cotton production based on soil nitrate and crop reflectance, (3) the feasibility of variable-rate N based on corn crop reflectance, (4) the use of corn management zones and crop reflectance for improving N recommendations and energy efficiency, and (5) the ability of using aerial photographs to improve N recommendations in corn.

21.2 BACKGROUND

In production agriculture, nitrogen (N) is a nutrient that often limits crop growth and when applied at rates that are sufficient to optimize yield, represents one of the single largest energy investments. Nitrogen fertilizer use, which has increased 80 times since the 1920s has contributed to worldwide yield increases.[1–3] In the United States, corn, wheat, and cotton use 70% of total fertilizer used, with corn accounting for 50% of the N.[4] Asia is one of the areas in the world where it is used, and resulting yields are expanding rapidly. Higher yields are needed to feed an expanding population that desires more meat in their diets.

One of the primary energy costs of cropping systems is associated with N fertilizer.[5] Most commercially available N fertilizer is made from nitrogen gas (N_2) which makes up 70% of the atmosphere. To convert the N_2 molecule to a biological active form requires a large amount of energy (Figures 21.1 and 21.2). Not all N sources have the same energy production requirements (Table 21.1). Of the commonly used

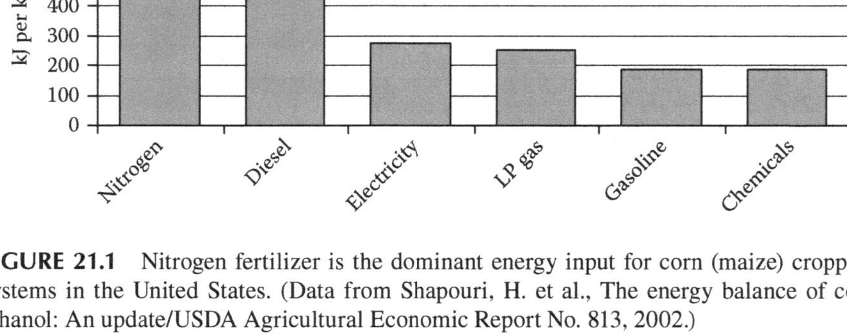

FIGURE 21.1 Nitrogen fertilizer is the dominant energy input for corn (maize) cropping systems in the United States. (Data from Shapouri, H. et al., The energy balance of corn ethanol: An update/USDA Agricultural Economic Report No. 813, 2002.)

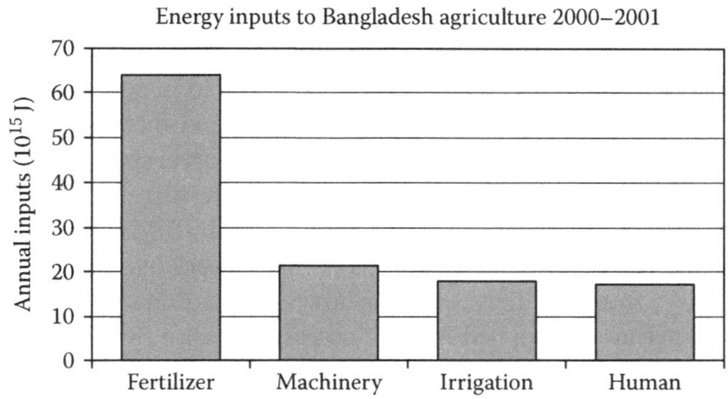

FIGURE 21.2 Fertilizer is the dominant energy input for cropping systems (mainly rice) in Bangladesh. (Data from Alam, M.S. et al., *Am. J. Environ. Sci.*, 1, 213, 2005.) Breakdown of fertilizer energy into NPK is not given, but is dominantly for N.

TABLE 21.1
Energy Needed for the Production of Common N Fertilizers

N Fertilizer Source	N Concentration (%)	Energy Production Requirement (MJ kg^{-1} N)
Ammonia	82	55
Ammonium sulfate	21	58
Liquid UAN	32	65
Ammonium nitrate	34	67
Urea	46	70

Source: Adapted from Hood, C.F. and Kidder, G., Fertilizers and energy, Fact Sheet EES-58, November 1992, Florida Cooperative Extension Service, University of Florida, 1992.

N fertilizers, anhydrous ammonia has the lowest amount of energy associated with its production and transport. Most of the energy cost is in the production of N fertilizer, and only a small proportion of energy is expended for transport and application.[6] Kuesters and Lammel[7] reported that the energy requirement for transporting a kilogram of urea 8000 km by sea was 5.56 MJ kg^{-1} urea-N, while the energy requirement for producing the urea was 8400 MJ. Natural gas is the main energy input into the production of N fertilizer.[3]

Although N fertilizer has increased crop yields, the overapplication of N can have unintended negative economic and environmental consequences. Both environmental and efficiency-related concerns have fueled thousands of field studies of N fertilizer management, cycling, export, and balances in various cropping systems. Nitrogen behavior in soil turns out to be remarkably complex. The fates of N fertilizer not utilized by plants include NO_3^- leaching, ammonia volatilization, immobilization into soil organic matter, fixation in clay particles, and denitrification. Nitrous oxide is a product of denitrification and nitrification and is a potent greenhouse gas. Emissions of N_2O increase several fold in soils following N fertilization.[8–10] Management strategies to reduce N loss and increase crop N recovery have been studied extensively, including N fertilizer source, application method, timing of fertilizer application, tillage, N loss inhibitors (fertilizer additives), and, more recently, site-specific management to account for within-field crop N needs.[11,12]

Nitrogen use efficiency (NUE) in terms of its recovery by row crops is generally less than 50%.[13,14] In spite of these relatively low values, agronomic NUE (i.e., increase in grain yield per unit of applied N) has increased in corn 36% since 1980.[14] Much of this improvement is due to cultivar development, but higher plant populations and improved soil management practices, such as conservation tillage, contribute as well. Improved N management practices include less fall-applied N fertilizer and more split N applications.[14] Plant breeding has clearly made major contributions to corn and wheat yield gains the last several decades. Studies in wheat[15] and corn[16] have compared historical cultivars with modern ones and found an increase in agronomic NUE. With wheat, agronomic NUE was reported to have increased between 1950 and 1985 by 1% $year^{-1}$, evenly divided between gains in N recovery and physiological NUE.[15]

Reducing trade deficits and improving energy independence are also rationale for improving NUE in agriculture. Using the United States as an example, most (52% in 2007) of N fertilizer used is imported,[17] whereas in 1992, only 25% of N fertilizer was imported.[4] This reflects a trend seen over the last 15 years of decreasing U.S. N fertilizer production, and an increased reliance on imported N fertilizer from Russia, Ukraine, Egypt, and Trinidad. This change is the result of higher natural gas costs in many developed nations.

Nitrogen best management practices include NO_3^-–N soil testing, considering all sources that provide N to the crop, proper timing of application, sound water management and fertigation, proper calibration/operation of equipment, and realistic yield goals.[3,18–20] Over the past 15 years, these practices have been tested in a large number of research projects.[19]

Precision agriculture is an area that has only recently been explored. In precision agriculture, site-specific inputs are based on locally derived soils, soil test results, yield goals, and landscape positions. The interest in SSNM is driven by decreases in the costs of obtaining spatial information using GIS and increasing fertilizer costs. Site specific of variable-rate N fertilizer management strategies include: grid soil sampling management zone-based soil sampling, yield map/yield goal approach, and canopy reflectance-based N management.[21–30] Variable-rate N management can also reduce NO_3^- runoff and leaching losses and NO_3 leaching.[31–33]

SSNM is one approach that might improve agricultural energy efficiency. With respect to N fertilizer inputs, SSNM can increase the net energy output if

1. N fertilizer use is reduced, without reducing yields
2. N fertilizer use is maintained or increased, but that yield responses to N are greater

Among most SSNM studies, only a few have demonstrated improved NUE by both processes. Inman et al.[34] classified irrigated cornfields in Colorado into low-, medium-, and high-productivity management zones, based on bare soil imagery and farmers' input. Nitrogen uptake and N fertilizer response varied by zone, suggesting that SSNM can be implemented based on management zones. Historically, yield goals have been part of U.S. corn N recommendation algorithms. However, Scharf et al.[25] reported that in humid environments, corn yield spatial variation is a weak predictor of economically optimum N rates (EONR). Spatial variation in the soil N supply is often more important. Plant spectral reflectance may provide the information needed to assess N supply.

Scharf and Lory[28] and Kitchen et al.[30] in Missouri, and Schmidt et al.[29] in Pennsylvania, estimated EONR in corn using spectral reflectance. They achieved the best predictions by using reflectance ratios of the area of interest relative to a well-fertilized area. Yabaji et al.[20] reported that basing in-season SSNM of drip-irrigated cotton on canopy reflectance resulted in 17–28 kg N ha^{-1} savings in N fertilizer compared to regional N recommendations, without reducing lint and seed yields. Bronson et al.[24] compared variable-rate N applications based on grid soil sampling to blanket regional N management in a 3-year study of center-pivot irrigated cotton. In just 1 year out of 3 years, variable-rate N resulted in a greater lint yield response than blanket N. The average variable-rate N fertilizer application rate was nearly identical with the blanket N fertilizer rate all 3 years.

Life-cycle analysis and greenhouse gas budgets are increasingly being used to determine agricultural energy and system efficiency.[35–38] Tilman et al.[37] reported that the energy output to input ratios were marginally positive for corn ethanol, and that perennial grasses for cellulosic conversion of biomass to ethanol have a relatively high energy ratio. This ratio is very sensitive to energy inputs and generally decreases with increasing N. For example, soybean, which does not require N fertilizer, has a relatively high-energy output to input ratio. However, different recommendations can result if energy yields are the selection criteria. Kuesters and Lammel[7] reported that N fertilization resulted in a fivefold gain in energy in wheat and sugar beets grown in Germany. This was despite the fact that the optimal N fertilizer rates (160 kg N ha^{-1} in wheat and 120 kg N ha^{-1} in sugar beet) were 40% of the total energy input. Hülsbergen et al.[39] had similar results and reported that N fertilizer rates required to optimize energy yields were higher than the N needed to maximize the ratio for wheat, sugar beets, potatoes, and barley.

Many studies have assessed the net energy return to ethanol production from corn production, considering the energy from N fertilizer production. Shapouri et al.[40] reported positive energy values for just 6 of 10 studies that assessed the energy efficiency of producing ethanol from corn grain. However, in several studies, positive energy yield was only possible by considering co-products such as gluten meal, gluten feed, and corn oil.[37,41,42] These studies did not address the impact of N fertilizer on ethanol and energy yields.[7,39] The purpose of this chapter is to use five case studies to demonstrate how GIS-based SSNM approaches can be used to improve energy costs and efficiencies.

21.3 CASE STUDY 1: N IMPACTS ON ENERGY PRODUCED IN COTTON

21.3.1 METHODS

The description and results of this 3-year study were published in Bronson et al.[24]. The study site is near Lamesa, Texas, approximately 100 km south of Lubbock, Texas and consists of 14 ha under a 48 ha center-pivot irrigation system. The soil at this site is an Amarillo fine sandy loam (fine-loamy, mixed, superactive, thermic, Aridic Paleustalf). The experimental design was a randomized complete block with three replicates.

The experiment consisted of three N treatments (zero-N, blanket-rate N, and variable-rate N). The N management plots were eight rows wide, and since the rows were planted in a circular fashion, plot lengths ranged from 500 to 1000 m (Figure 21.3). In March of each year, soil samples were taken at differential global positioning system (DGPS)-referenced points within the 14 ha experimental area on a 0.2 ha grid. Two subsamples were taken of the 15–30, 30–60, and 60–90 cm depths with a Giddings soil sampling machine (Giddings Machine Co., Windsor, Colorado), within 3 m of the DGPS point.

Soils from all depths were analyzed for KCl-extractable NO_3–N.[43] The N fertilizer rate for both the blanket-rate N and variable-rate N treatments was calculated using an N supply requirement of 134 kg N ha^{-1} for a constant yield goal of 1100 kg lint ha^{-1},[44] minus extractable soil NO_3–N in 0–60 cm soil. Nitrogen was applied as urea ammonium nitrate (UAN) (320 g N kg^{-1}) with a liquid fertilizer system fitted with spoke applicators. Half of the N fertilizer was applied at 3 weeks after planting and half was applied at 5–6 weeks after planting (early fruit set or squaring). The blanket rate of N fertilizer was based on the average 0–60 cm soil NO_3–N content of the nine blanket-N plots. Inverse distance interpolation of 0–60 cm

FIGURE 21.3 Blanket-rate, variable-rate, and zero-N fertilizer strip plots in center-pivot cotton of case study 1, Lamesa, Texas, 2002.

NO_3–N values from all DGPS points was used to create variable-rate application maps in 2002.

In May of each year, Roundup Ready® cotton was planted into glyphosate-(isoprophylamine salt of N-phosphomonomethyl glycine) terminated rye in 1 m rows at a seeding rate of 18 kg ha^{-1}. Hand harvesting of lint and seed were done on 8 m of row at each DGPS-referenced point in October of each year. The hand samples were ginned on a one-saw plot gin equipped with a one-stage lint cleaner to give a unique percentage turnout of lint for each DGPS point.

Energy from N fertilizer was calculated by multiplying the N rate by 65 MJ kg^{-1} (Table 21.1). Gross beef cattle maintenance (GBCM) energy was calculated by first calculating total digestible nutrients (TDN) and then metabolizable energy with the following equations:

$$\text{TDN} = 40.26 + (0.1969 * \text{CP}) + (0.422 * \text{NFE}) + (1.19 * \text{Fat}) - (0.1379 * \text{CF}).^{45}$$

$$\text{ME (MJ kg}^{-1}) = 0.1516 * \text{TDN}^{45}$$

$$\text{GBCM} = -0.508 + (1.37 * \text{ME}) - 0.3042 * \text{ME} * \text{ME}) + (0.051 * \text{ME} * \text{ME} * \text{ME}).^{45}$$

$$\text{Net energy}_{\text{fertilizer}} = \text{gross energy} - \text{N fertilizer energy (Table 21.2)}.$$

where

CP is crude protein (%)
NFE is nitrogen-free extract (%)
Fat is in %
CF is crude fiber (%)
ME is metabolized energy
TDN is total digestible nutrients
Net $energy_{fertilizer}$ is the net return of cottonseed energy to N fertilizer application

21.3.2 Results

Cotton lint and seed yields responded to N fertilizer in all 3 years of the study.[24] The delta yields of the SSNM treatment improved each year, such that by the third season, yields with variable-rate N were significantly greater than the blanket-N treatment. Averaged across the 3 years, N fertilizer responses in cottonseed yield and protein above the zero-N treatment were observed (Table 21.2). There was no difference between blanket-rate and variable-rate N in seed yield, protein, or fat. Nitrogen fertilizer rates were similar between the two N-fertilized treatments in all 3 years of the study. Fat yield averaged 383 kg ha^{-1} and was not affected by N. Multiplying fat yields by 45.2 MJ kg^{-1} (higher heating value of cottonseed methyl

TABLE 21.2
CottonSeed, Protein, Fat, and Energy Yields as Affected by Variable-Rate N Fertilizer Management with Center-Pivot Irrigation, Lamesa, Texas, 2002–2004

Nitrogen Treatment	Nitrogen Applied (kg ha^{-1})	Fat (kg ha^{-1})	Protein (kg ha^{-1})	Crude Fiber (kg ha^{-1})	NFE (kg ha^{-1})	Seed Yield (kg ha^{-1})	Energy from N Fertilizer Production (MJ ha^{-1})	Gross Energy (MJ ha^{-1})	Net Energy (MJ ha^{-1})
Blanket	89a	386a	386a	395a	523a	1,757a	6,118a	23,380a	17,263b
Variable	85a	383a	389a	387ab	516ab	1,744a	5,838a	23,210ab	17,371b
Zero	0b	380a	307b	361b	487b	1,599b	0b	21,921b	21,921a
LSD	34	42	51	32	36	108	2,060	1,399	1,341

Numbers in a column followed with the same letter are not significantly different from each other ($p = 0.05$).

esters produced with 97% yield,[46]) gives energy from cottonseed oil of 17,311 MJ ha^{-1}, averaged across N rates. This value reflects energy from potential biodiesel yields and is 75% of the total energy value, which includes fat energy and feed value of the meal. Gross energy from cottonseed was significantly greater with blanket-rate N than the zero-N. However, when the energy from N fertilizer production was subtracted to give net energy yields, the two N-fertilized treatments resulted in 21% less energy than the nonfertilized plots (Table 21.2). This result is very different from the large net energy returns to N fertilizer in the Missouri corn case studies. The main reason for this negative return to N fertilizer in Texas cotton is that the "delta yield" or cottonseed response to N was only 10% or about 151 kg ha^{-1}. However, profitable lint returns to N fertilizer of \$15–25 ha^{-1} were observed in 2003 and in 2004.[24]

21.4 CASE STUDY 2: SSNM BASED ON NO_3–N OR CROP REFLECTANCE

21.4.1 METHODS

This study was conducted near Lubbock, Texas, on an Acuff sandy clay loam (fine-loamy, mixed, superactive, thermic, Aridic Paleustoll) from 2007 to 2009 and was reported in Nusz.[47] Drip tape was placed in the center of every other furrow at a depth of 12 and water flowed at a rate of 1 L min^{-1} at 0.08 MPa. AFD 5065 B2F cotton was planted in mid-May and harvested in late October. The experimental design was a randomized complete block design, one-way factorial with three replications or blocks. Blocks consisted of 40, 1 m rows that were 180 m long. Each block was divided into five, eight-row plots that were randomly assigned to the five N-fertilized treatments. However, for the purposes of this chapter's emphasis on energy, we only address the zero-N, soil test-based N management, and reflectance-based N management treatments. Each eight-row plot has its own irrigation and fertilizer injection station. The N fertilizer requirement of 134 kg N ha^{-1} was based on a 1400 kg lint ha^{-1} yield. The requirement was modified based on the amount of nitrate-N contained in the spring soil samples (0–60 cm) and estimated amount of N in the irrigation water (22 kg N ha^{-1}). After the credits were subtracted from the requirement, the predicted N rate (71 kg N ha^{-1}) was determined (Table 21.3). Nitrogen (UAN) fertilizer was injected into the drip system 5 days a week, between late June (early square) and early August (mid-bloom). In the reflectance-based strategy treatment, the N injection rate was initially set to the 50% of the soil test treatment. Every week, canopy reflectance measurements were made with Crop Circle ACS-210 (Holland Scientific Inc., Lincoln, NE) and GreenSeeker (NTech Industries, Ukiah, CA) spectroradiometers at 1 m above the canopy on one row per plot.

Normalized difference vegetative index (NDVI) was calculated by the equation:

$$\text{NDVI} = \frac{(\text{reflectance}_{\text{NIR}} - \text{reflectance}_{\text{visible}})}{(\text{reflectance}_{\text{NIR}} + \text{reflectance}_{\text{visible}})}$$

TABLE 21.3
CottonSeed and Energy Yields as Affected by Reflectance-Based N Fertilizer Management with Subsurface Drip Irrigation, Lubbock, Texas, 2007–2009

Nitrogen Treatment	Nitrogen Applied ($kg\ ha^{-1}$)	Seed Yield ($kg\ ha^{-1}$)	Energy from N Fertilizer Production ($MJ\ ha^{-1}$)	Gross Energy ($MJ\ ha^{-1}$)	Net Energy ($MJ\ ha^{-1}$)
Soil test based	71	2,676a	4,903	35,603a	30,700b
Reflectance based	49	2,790a	3,388	37,130a	33,742a
Zero	0	2,003b	0	27,452b	27,452c
LSD		158		2,149	2,135

Numbers in a column followed with the same letter are not significantly different from each other ($p = 0.05$).

The remote-sensing-based N rate was calculated by[20]

1. Starting with an N fertilizer injection rate at first square of 50% of soil test-based rate.
2. If $NDVI_{reflectance\text{-}based}$ was statistically $< NDVI_{soil\ test\text{-}based}$, then the N fertilizer injection rate was increased to match the soil test-based N injection rate.

Hand harvesting of lint and seed were harvested from 8 m of row at three DGPS-referenced points in each 180 m long plot in October of each year. The hand samples were ginned and the unique percentage turnout of lint and seed for each DGPS point was calculated. In the absence of fat and digestible nutrient data, gross energy value of cottonseed was calculated from relationships between seed yield and gross energy in the center-pivot case study for N-fertilized and zero-N plots.

21.4.2 Results

Cottonseed yields were much greater in the drip irrigation study (case study 2) than those observed in case study 1 (Table 21.3). Zero-N plot yields were very high with an average total N uptake of 87 kg N ha^{-1} (data not shown).[48] Averaged across the 3 years of the study, N fertilizer application resulted in increased seed yields (Table 21.3). Reflectance-based N management and soil test-based management resulted in a 39% and 33% "delta yields," respectively, above the zero-N seed yield of 2003 kg ha^{-1}. When compared with the soil test strategy, the reflectance-based approach recommended 31% less N. This is in contrast to the first case study and suggests greater potential for saving N fertilizer with SSNM of cotton based on canopy reflectance compared to grid soil sampling and variable-rate N maps. The lower N usage and greater seed yields and delta seed yields resulted in a positive energy return to N fertilizer compared to the zero-N treatment. Notably, the site-specific, reflectance-based approach had significantly greater net energy return than the soil test-based N management (Table 21.3).

21.5 CASE STUDY 3: VARIABLE-RATE N BASED ON CROP REFLECTANCE

21.5.1 METHODS

Reflectance sensors like those described in case study 2 were used to control variable-rate N applications to over 100 corn fields in Missouri. These fields are part of a demonstration program conducted by the University of Missouri. When possible, these demonstrations included multiple (3–15) replicates of two N rate strategies: a constant N rate chosen by the producer and a variable-rate N application controlled in real time by crop reflectance sensors. Both Crop Circle (ACS-210) and GreenSeeker (red light model) sensors were used in these demonstrations. From 2004 to 2008, there were 55 fields with side-by-side comparisons between constant- and variable-rate N management. We will present the story of one of those fields in this case study.

The study field was located in Audrain County, Missouri, in 2007. All practices were carried out by the cooperating producer. Corn was planted on 24 April at a rate of 75,000 seeds ha^{-1}. A high-N reference area measuring 10 m × 18 m was installed on 10 May by hand-spreading ammonium nitrate at a rate of 240 kg N ha^{-1}. No preplant or early-season N fertilizer was applied to the rest of the field. Irrigation was applied through a center-pivot system as needed.

Constant and variable-rate treatments were applied on 13 June, when corn was at the V8 growth stage (about thigh high). A Rogator sprayer equipped with drop nozzles and a 25 m boom was used to apply UAN solution between corn rows. Two GreenSeeker sensors were mounted on a custom-made boom on the front of the Rogator. Custom software averaged the values from the two sensors each second (about 10 values per sensor per second) and converted this average to an N rate using an equation similar to those published by Scharf and Lory[28] and Schmidt et al.[29] This equation requires a value measured from the high-N reference area; therefore, we measured the red/near-infrared ratio of the high-N reference area first, and then used this value in calculating N rates in the variable-rate demonstration areas using the equation:

$$\text{N rate (kg N ha}^{-1}) = 280 \times \frac{\text{red}_{\text{sample}}/\text{NIR}_{\text{sample}}}{\text{red}_{\text{reference}}/\text{NIR}_{\text{reference}}} - 224$$

where

the $\text{red}_{\text{sample}}$ and $\text{NIR}_{\text{sample}}$ were the reflectance values at the demonstration site

$\text{red}_{\text{reference}}$ and $\text{NIR}_{\text{reference}}$ were the reflectance values in the well-fertilized controls

The actual rates of N fertilizer applied to the fields were developed after discussions with the collaborating producer. After this discussion, the minimum and maximum rates of 60 and 180 kg N ha^{-1} for the sensor-based N treatment were selected. In the constant N rate strip, 112 kg N ha^{-1} was applied. Nitrogen rates applied to this field are shown in Figure 21.4. The average N rate based on sensor measurements was 30 kg N ha^{-1} lower than the rate chosen by the producer.

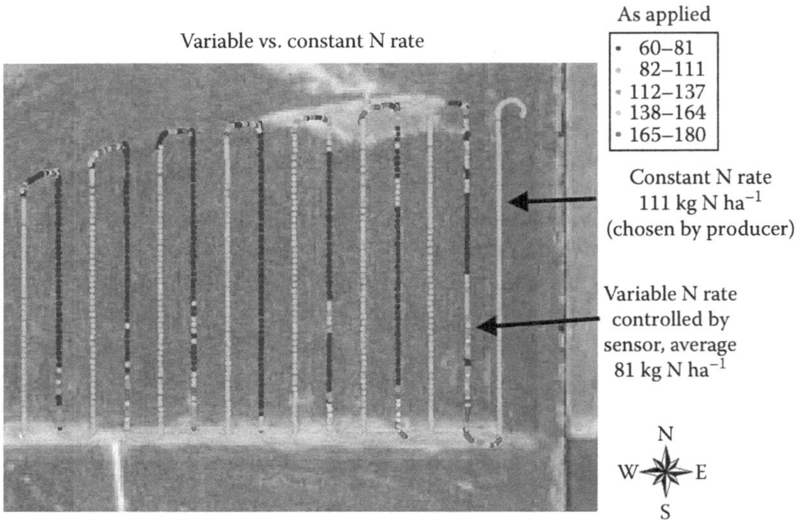

FIGURE 21.4 Nitrogen fertilizer rates applied at corn growth stage V8 in case study 3. Light grey strips are the constant N rate chosen by the producer. Strips with various shades are based on crop reflectance measured by sensors mounted on the front of the N applicator. Use of sensors in this field reduced N use by 30 kg N ha^{-1}. The photo on which the application data are overlaid is a stock USDA photo (NAIP) and not from the year of the case study.

21.5.2 Results

The lower N rates applied with sensor-based SSNM did not result in any apparent deficiency in an aerial photograph acquired 7 weeks after N application (Figure 21.5), nor was yield negatively affected (Figure 21.6).

Notably, the energy balance of this field was improved by using crop sensors to guide N rates (Table 21.4). This field was chosen because it best represents the

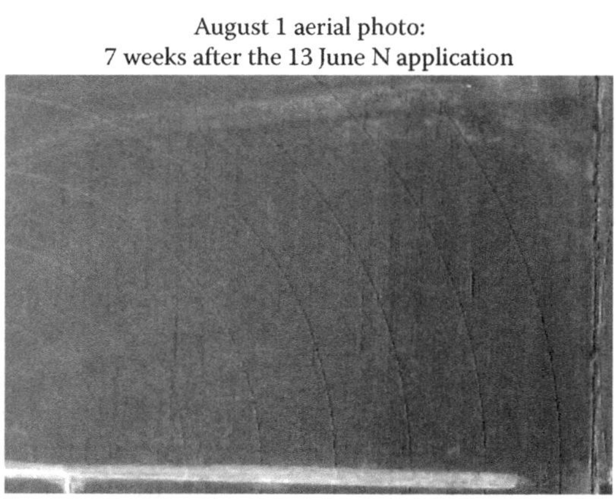

FIGURE 21.5 No evidence of N stress is seen in this August 1, 2007 aerial photo of the case study area, providing evidence that the lower N rates recommended by the sensors were adequate to fully supply crop needs for N.

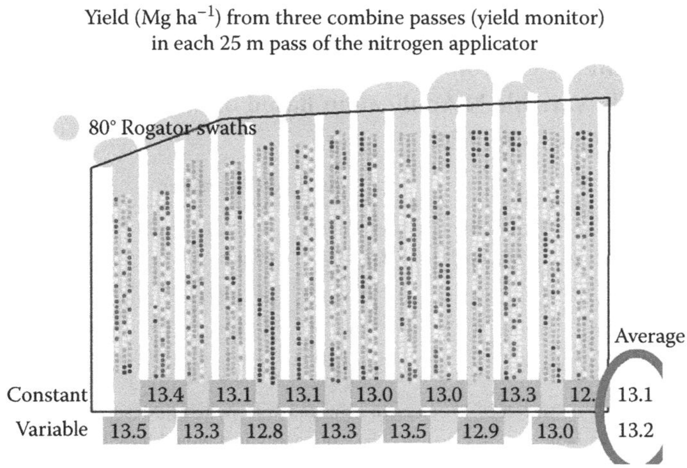

FIGURE 21.6 Yields were high in both treatments. Nitrogen rates supplied by the sensor-based variable-rate N treatment were sufficient to produce yields as high as, or higher than, the N rate chosen by the producer while reducing total N use by 30 kg N ha^{-1}.

TABLE 21.4
Energy Outcome for Spatially Variable N Application Based on Reflectance Sensors in Case Study 3

	Value for Parameter			
Nitrogen Strategy	**Grain Yield (Mg ha^{-1})**	**Feed Energy (GJ ha^{-1})**	**N Rate (kg ha^{-1})**	**N Production Energy (GJ ha^{-1})**
Constant	13.1	214	111	−7.3
Sensor (variable)	13.2	215	81	−5.3
Difference	0.1	1	−30	2

Energy for the production of N is shown as negative to indicate consumption of energy. Sensor-based variable-rate N saved 2 GJ ha^{-1} of energy that would have been used to produce the additional N used in the producer's normal N rate while maintaining or increasing the feed energy output in the corn grain produced.

average energy outcome of the 55 fields for which we have replicated comparisons of a constant N rate (chosen by the producer) with variable N rates (guided by sensor measurements in real time) (Table 21.5). Thus, in our experience, an outcome of using sensor technology to guide N rates is to improve the energy balance of the system. It is apparent in Tables 21.4 and 21.5 that feed energy values used for corn grain result in system energy outputs that far outweigh energy inputs as N fertilizer. However, this energy output is in a very different form than the hydrocarbon energy input (as methane) used in N fertilizer production. Comparing hydrocarbon energy inputs to hydrocarbon energy outputs (as ethanol fuel) is in many ways a

TABLE 21.5
Average Energy Outcome for 55 Fields with Demonstrations of Spatially Variable N Application Based on Reflectance Sensors (Similar to Case Study 3)

	Value for Parameter			
Nitrogen Strategy	Yield ($Mg\ ha^{-1}$)	Feed Energy ($GJ\ ha^{-1}$)	N Rate ($kg\ ha^{-1}$)	N Production Energy ($GJ\ ha^{-1}$)
Constant	9.8	160	130	−9
Sensor (variable)	9.9	162	116	−8
Difference	0.1	2	−14	1

Energy for the production of N is shown as negative to indicate consumption of energy. Sensor-based variable-rate N saved 1 GJ ha^{-1} of energy that would have been used to produce the additional N used in the producer's normal N rate while increasing the feed energy output in the corn grain produced by 2 GJ ha^{-1}. Net energy gain to sensor-based N management is 3 GJ ha^{-1}; this value is the same as for the field in case study 3, which was chosen because it best represented the energy outcome from the entire group of demonstration fields.

more appropriate analysis. Using average values cited by references from Shapouri et al.,[40] for energy inputs and outputs in ethanol production from corn grain, we calculated that in-season application of a constant N rate in our 55 demonstration fields increased net energy output by 13% (relative to state-average values representing mainly preplant N application) (Figure 21.7).

Variable-rate N fertilization based on crop sensors increased net energy output by 29%, again relative to state average values for yield and N rate. This shows the importance of efficient N management to the energy balance of ethanol, and the potential for spatially variable N management to increase N efficiency and energy output.

21.5.3 Use of GIS

Although GIS is not, strictly speaking, required to implement this SSNM approach, it was needed to help communicate the results with the producers and provide an opportunity for the producers to override treatments (Figure 21.4). A secondary benefit of the technology was that it could be used as a training tool where the producers could compare their knowledge with the predications (Figure 21.6).

21.6 CASE STUDY 4: CORN REFLECTANCE AND MANAGEMENT ZONES

21.6.1 Background

The amount of N needed by crops varies within fields[11,12] and is most often attributed to soil and landscape properties that affect soil N supply (i.e., mineralization) and

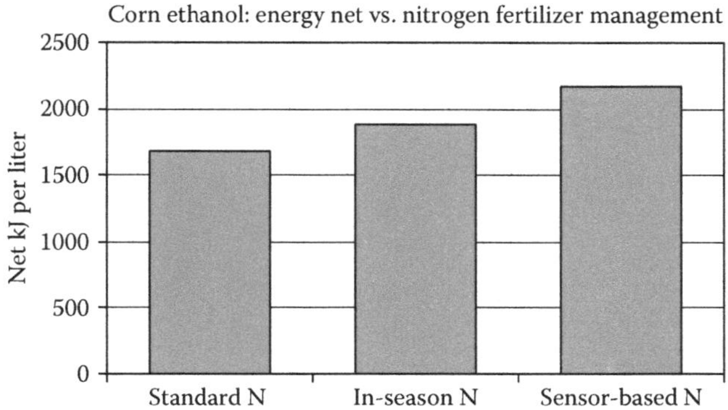

FIGURE 21.7 Net energy gain to corn ethanol production as a function of N fertilizer management strategy. Net energy gain for standard N management is taken as the average of 10 (widely varying) estimates presented in Table 1 of Shapouri et al.[40]. Nitrogen use per unit of corn grain for standard N management was taken from Table 3 of Shapouri et al.,[40] then converted to energy required to produce the N to grow the corn to produce a liter of ethanol. Average nitrogen use and corn grain yield for 55 demonstration fields in Missouri were used to calculate N energy savings per liter ethanol for in-season and sensor-based N management. These calculated savings were added to the net energy estimate for standard N management. "In-season N" is the constant N rate chosen by the producer in our sensor N demonstration fields. Improved N management, and specifically N management that accounts for spatially variable N needs, can substantially improve net energy gains in corn ethanol production.

soil water supply.[19,48] However, the variability in nutrients need can be further exacerbated by historic and current management practices.[49] The following swine (*Sus scrofa* L.) manure management case study demonstrates how management zones and in-season crop reflectance can be integrated. In this case study, management zone maps were created to represent three unique levels of slurry manure application. These maps were then used in concert with in-season corn canopy reflectance sensing to target SSNM. The goal of this field-scale project was to determine if N fertilizer inputs could be reduced and optimal yields maintained when using this variable-rate strategy as compared to a uniform N application.

21.6.2 Methods

A 49 ha Missouri field located near a large swine production facility is uniquely managed during the growing season with lagoon effluent applications through center-pivot irrigation systems. Figure 21.8A provides an aerial view of the operation with two overlapping center-pivot systems.

The boundary of the case study field is shown in white on this same figure. Soil mapping indicates five unique soils (primarily Vertic Epiaqulfs and Vertic Albaqualfs), with topography varying from 0% to 9% slope. The field sits in the landscape adjacent to continuous deciduous forest, which blocks the center-pivots from completing a full 360° circle (Figure 21.8A).

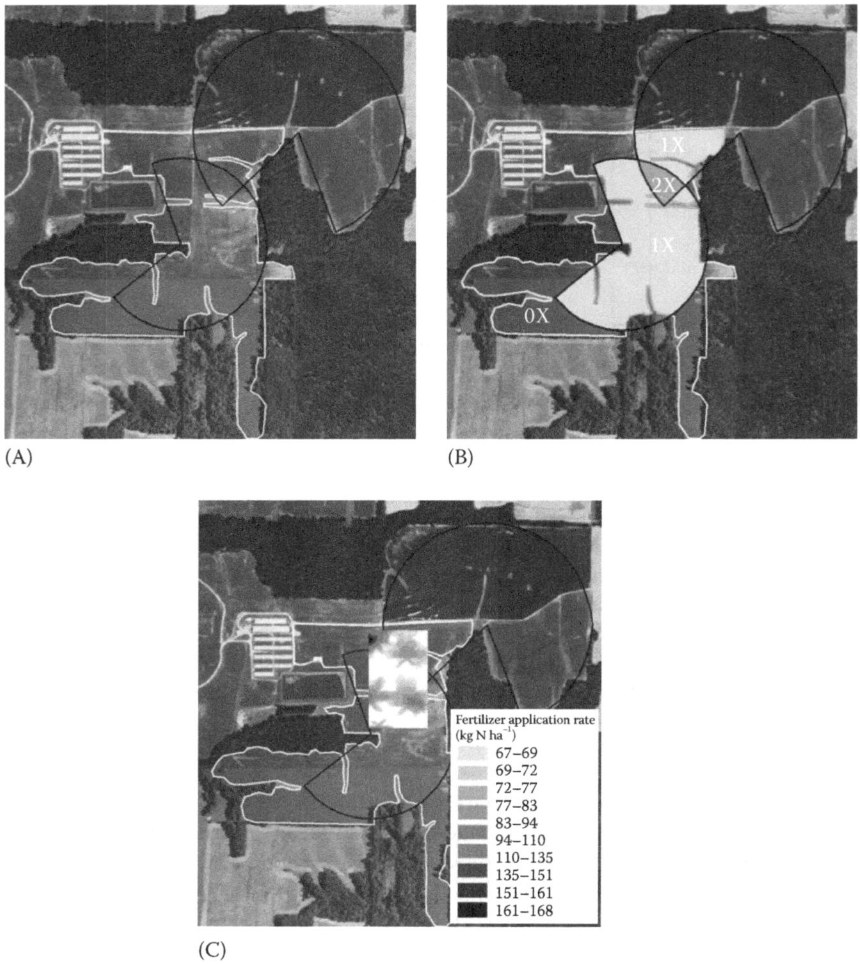

FIGURE 21.8 A case study is provided showing how GIS tools were used on a Missouri corn field to integrate manure management zones with reflectance sensing to do variable-rate N applications. Panel A shows the case study field boundary (white line) along with coverage of two partial center-pivot systems. Panel B shows the slurry management zones of the case study field, with some field area receiving no-slurry (0X, dark grey), some areas receiving slurry from one center-pivot (1X, white), and some receiving slurry from both center-pivots (2X, medium grey). Panel C provides the 2006 variable-rate N map that was obtained on a portion of the case study field using the management zones (panel B) and canopy reflectance sensors.

Swine lagoon effluent is applied through the center-pivots twice during the growing season. The primary purpose of the center-pivots is not for water irrigation, but to apply the effluent to cropland. The first manure application occurs during early corn vegetative growth stages (V3–V5). The second typically is planned during the mid- to late-season vegetative growth stages (V12–V16). Historic nutrient content testing and monitored slurry rates have shown that an average of 45 kg of inorganic N ha^{-1}

was applied with each center-pivot application. Thus as shown in Figure 21.8B, some areas of the field receive no manure (in dark grey as 0X), some receive slurry from either one of the two center-pivots (in white as 1X), and a small portion of the field receives slurry from both center-pivots (in medium grey as 2X). Respectively, these three areas receive through manure applications approximately 0, 90, and 180 kg N ha^{-1} during the growing season.

Additional N fertilization as fluid UAN was sidedressed between V8 and V10. The applicator was equipped with crop-canopy reflectance sensing and a variable-rate controller that with each pass traversed thirty-two 0.76 m spaced corn rows. Details for sensor mounting and operating procedures are similar to that described previously.[30] The timing of this in-season N fertilization was between the two center-pivot lagoon effluent applications. It was presumed the crop had taken up N from the first slurry application, and that crop canopy sensing would detect differences as the boundary between no-slurry and slurry was crossed. Since the second slurry application was planned after the canopy-sensed N fertilization, a credit of up to 45 kg N ha^{-1} (1X areas) or up to 90 kg N ha^{-1} (2X areas) was built into the application algorithm, but only for rates called for greater than 67 kg N ha^{-1}. This minimum rate of 67 kg N ha^{-1} was built into the algorithm to ensure that an adequate amount of N would be available to corn late in the growing season.

A study was conducted on a portion of this case study field in 2006 and 2007. The study area is represented by the rectangle shown in Figure 21.8C. Within this area, uniform (151 kg N ha^{-1}) and canopy-based variable-rate N applications were compared. Treatments were applied in randomized paired strips, oriented north to south, within this area. Within the paired N strips, N rates (recorded from as-applied maps) and grain yield (obtained from combine yield-monitoring maps) were extracted using GIS tools. Based on this information, N response relationships were determined. Nitrogen applications and yield response determined from this study area were presumed representative for the whole field and were used to calculate field-level mass and energy differences between uniform and variable-rate N management systems.

Generalized GIS steps for this analysis using ArcGIS software included (1) the addition of field boundaries, N treatment strips, yield strips, and slurry zones as shape files over the raster aerial image of the case study field; (2) the use of GIS-based tools to calculate the size of the treatment areas; (3) the extraction of yields for the different N treatment strips using the tool "Spatial Analyst/Extraction/Extract by Mask"; and (4) the use of an Excel spreadsheet to calculate N responses and energy efficiency (Table 21.6).

21.6.3 Results

Using the strips of senor-based variable-rate N from 2006, a variable-rate map was generated using ordinary kriging interpolation methods, and this is shown overlaid on the field aerial photo in Figure 21.8C. The most notable feature is the relative increase in N fertilizer rate in the northwest corner, where slurry was not applied. Much of the test area under center-pivot only received 67–83 kg N ha^{-1}, regardless of whether it was in the 1X or 2X slurry zones. Based on the experimental protocol, a minimum of 67 kg N ha^{-1} was applied to all areas.

TABLE 21.6

Nitrogen Application Rates and Corn Yield Response Are Shown in Energy Metrics for Both Uniform and Variable-Rate N Management Systems on a 49 ha Missouri Corn Field over 2 Years

	No-Slurry Zone		1× Slurry Zone		2× Slurry Zone		Area-Weighted Average or Total		Field-Level Difference
	Uniform	Variable	Uniform	Variable	Uniform	Variable	Uniform	Variable	(Uniform—Variable)
Fraction of field	0.46	0.46	0.51	0.51	0.03	0.03			
Area of field (ha)	22.36	22.36	24.77	24.77	1.49	1.49			
				2006					
Nitrogen fertilizer									
Average rate									
Mass (kg N ha^{-1})	151	136	151	72	151	68	151	101	50
Energy (GJ ha^{-1})	9.8	8.8	9.8	4.7	9.8	4.4	9.8	6.6	3.2
Field N usage									
Mass (kg)	3376	3041	3740	1783	225	101	7342	4926	2416
Energy (GJ)	219	198	243	116	15	7	477	320	157
Crop									
Average yield									
Mass (Mg ha^{-1})	9.07	8.89	9.13	9.43	9.46	9.67	9	9	0
Energy (GJ ha^{-1})	148	145	149	154	154	158	149	150	−1.2

Field yield									
Mass (Mg)	203	199	226	234	14	14	443	447	–4
Energy (GJ)	3306	3240	3686	3807	230	235	7222	7282	–61
Net energy (GJ)	3086	3042	3443	3691	215	228	6745	6962	–218
				2007					
Nitrogen fertilizer									
Average rate									
Mass (kg N ha^{-1})	151	140	151	71	151	78	151	103	48
Energy (GJ ha^{-1})	9.8	9.1	9.8	4.6	9.8	5.1	9.8	6.7	3.1
Field N usage									
Mass (kg)	3376	3130	3740	1759	225	116	7342	5005	2336
Energy (GJ)	219	203	243	114	15	8	477	325	152
Crop									
Average yield									
Mass (Mg ha^{-1})	5.75	6.36	7.36	7.60	7.68	8.08	7	7	0
Energy (GJ ha^{-1})	94	104	120	124	125	132	109	116	–6.6
Field yield									
Mass (Mg)	129	142	182	188	11	12	322	343	–20
Energy (GJ)	2096	2318	2972	3069	187	196	5254	5583	–329
Net energy (GJ)	1876	2115	2728	2954	172	189	4777	5257	–481

Portions of the field received different amounts of swine slurry through a center-pivot irrigation system (see Figure 21.8) and are the basis of the slurry zones shown here. Conversion values used: 65 MJ kg^{-1} N; 16.3 GJ Mg^{-1} grain.

The N fertilizer applied and grain harvested of the two N management systems were compared on both a mass and energy basis for the whole field (Table 21.6). The N amounts shown do not account for N in the slurry, but only account for differences in N fertilizer. While there was a slight reduction in N fertilizer used in the no-slurry zone using the variable system, the greatest reduction in N fertilizer came in the zones receiving slurry. For these zones, an average of 79 kg ha^{-1} less N was used with the SSNM system. Significantly, yield was equal or slightly higher with the variable-rate system. While these differences were not statistically tested, the trend observed in both the years was real for this field. When the estimated amount of N that was applied with slurry is combined with the fertilizer N, the total N input for the uniform N system was 241 and 331 kg N ha^{-1} for the 1X and 2X zones, respectively. Typical corn N rates in this region would not exceed 200 kg N ha^{-1}. We suspect the slightly lower yields with the uniform system may have been the result of enhanced early-season vegetative growth from excess N, resulting in accelerated soil–water use early in the growing season, and subsequent greater water stress during grain set and grain fill late in the season.

The combination of less N fertilizer used and greater harvested yield with the variable-rate N system produced an average energy benefit over the uniform system of 7.1 GJ ha^{-1} $year^{-1}$. For this 49 ha field, that translated into an average of 350 GJ $year^{-1}$ energy savings using the variable-rate system. In hindsight, the variable-rate algorithm probably should have been adjusted so that N credit from the second slurry would have been increased. Had this adjustment been made, without a loss in yield potential, an additional benefit of 1.8 GJ ha^{-1} or 79 GJ $year^{-1}$ for the field would have been realized.

In this case study targeting N fertilization to account for both known management differences (by slurry manure zones), as well as less-quantified soil/landscape differences (by canopy sensing), proved to be an effective strategy for decreasing energy inputs and increasing crop energy produced. Such site-specific management and assessment would be impossible without GIS mapping and tools.

21.7 CASE STUDY 5: CORN N RATES BASED ON AERIAL PHOTO

21.7.1 Background

Case studies 2–4 utilize crop reflectance sensors to diagnose N status of corn or cotton, based on the principle that as N need increases, reflectance of visible light increases (and reflectance of near-infrared light often decreases). The same principle can be used to translate information from aerial photographs into N rate decisions.[50,51] The limitation with aerial photographs is that they need to be either acquired after full-canopy development,[51] or acquired at ultra-high resolution so that soil background can be filtered out.[50] Both of these options present substantial logistical difficulties in corn, especially with N application after full canopy, when the corn is tall.

However, in fields with center-pivot irrigation and fertilizer injection pumps, applying N fertilizer after full-canopy development is not a limitation. This situation presents an ideal opportunity to use aerial photographs to guide N rate decisions. We have worked with a small number of producers to try this approach. Our first trial field is presented here.

21.7.2 Methods

Using an approach for photograph interpretation similar to Scharf and Lory,[50] but based on unpublished full-canopy (growth stages V10–V16) aerial photographs, green values were translated into N rate recommendations. Details for this calculation are provided below.

This approach relies on having a high-N reference area to compensate for the effects of growth stage, hybrid, and photographic procedures on the measured greenness of the corn. The producer created a field map with the area under the center-pivot defined as a separate zone in the field. He then applied anhydrous ammonia at his normal rate (220 kg N ha^{-1}) outside of the center-pivot zone, but at half that rate under the center-pivot (shown as dark grey in Figure 21.9), knowing that he could easily supplement N by injecting UAN solution into the center-pivot water.

The area outside of the center-pivot thus acted as the high-N reference area. The areas north and south of the center-pivot point were managed as two separate fields and were planted to different hybrids, so they were analyzed separately, each with its own high-N reference area.

A digital aerial photograph of the study field was acquired on 13 June when the corn was approximately waist high (growth stage V11) (Figure 21.10).

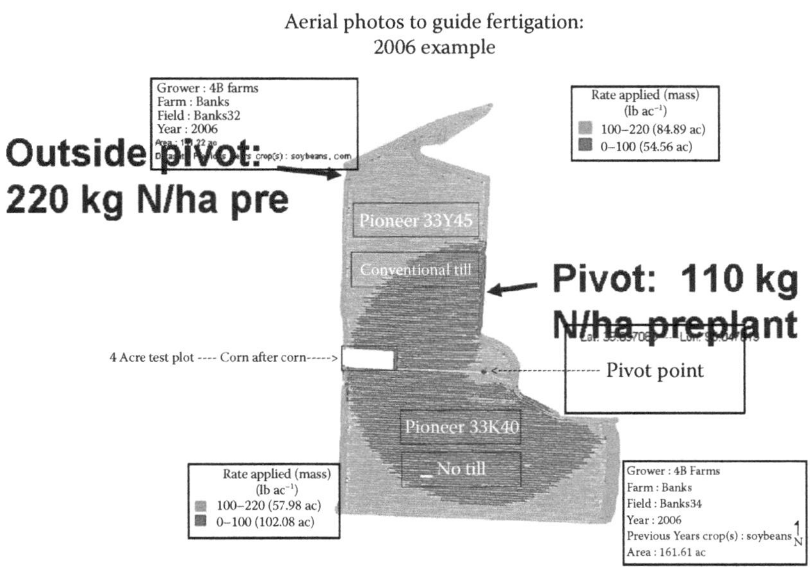

FIGURE 21.9 Preplant N applications to the case study field. Areas outside of the irrigation center-pivot's reach (light grey) received the producer's normal N rate of 220 kg N ha^{-1}. Areas under the center-pivot received half that rate (dark grey), with the plan to supply additional N in the irrigation water at rates suggested by analysis of an aerial photograph. This analysis requires comparing the area to be fertilized with a high-N reference area of the same hybrid. The areas north and south of the center-pivot point were planted to different hybrids, so they were analyzed independently, with the light grey area outside of the center-pivot's reach serving as the high-N reference area for each half of the field.

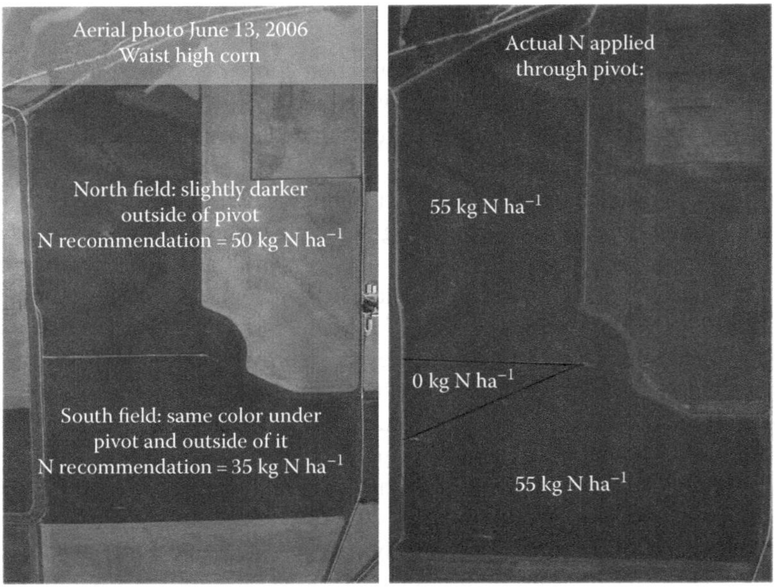

FIGURE 21.10 An aerial photo acquired on June 13, 2006 showed little evidence of N stress in the areas of the field that had received half of the producer's normal N rate. Image analysis gave relative green values of 1.0 in the south field and 1.05 in the north field, translating into N rate recommendations of 35 and 50 kg N ha^{-1}, respectively. A relative green value of 1.0 means that average green value is the same for the high-N and low-N areas, and they are indistinguishable from each other. Our calibration data suggested that even when a low-N area shows no visible N stress at this stage, it may sometimes need N. For simplicity's sake, the producer applied 55 kg N ha^{-1} over the entire field with his first irrigation, except in a small wedge of the south field where he applied no N in the irrigation water to allow us to estimate yield response.

No irrigation water had been applied. The photograph was georeferenced, field zone boundaries overlaid (under the center-pivot vs. outside the center-pivot), and average green value was measured for seven areas: north high-N zone, two north low-N zones (dark grey in Figure 21.9, divided radially), south high-N zone, and three south low-N zones (again divided radially). Relative green value was calculated for each of the five low-N zones by dividing their average green value by the average green value of the corresponding high-N zone.

21.7.3 Results

Average green value was nearly identical for all four south zones, giving relative green values of 1.0 for all of the south low-N zones. Our calibration data suggested that even when the low- and high-N zones have similar colors, an additional 35 kg N ha^{-1} is needed to optimum profits in the low-N areas. In the north half of the field, N stress was also not immediately apparent in the area receiving the low-N rate (Figure 21.10). However, image analysis revealed that the high-N

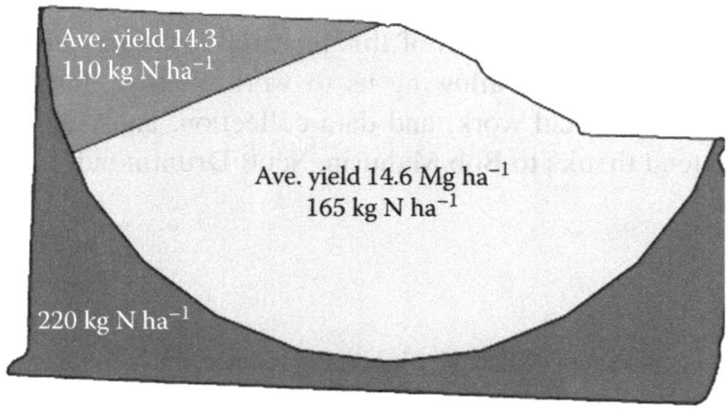

FIGURE 21.11 Yield outcome of reduced (light grey) vs. zero supplemental N in irrigation (medium grey) N rates for this case study. The very small difference in yield shows that the 165 kg N ha^{-1} N rate was at or above the economically optimal N rate for this field, and that no economic penalty resulted from reducing N rates relative to the producer's normal 220 kg N ha^{-1} N rate. This use of imagery/GIS saved 640 GJ of energy that would have been used to produce the N fertilizer that was saved.

area outside the center-pivot was slightly darker green than the two areas under the center-pivot (relative green ≈ 1.05), resulting in a suggested N rate of 50 kg N ha^{-1}. Because the variable N rates were similar for the different zones, the producer opted to apply a constant rate of 55 kg N ha^{-1} to the entire field except the zero-N control area. The purpose of the no-N control area was to assess the N responsiveness of the site.

GIS analysis of the south field showed a slightly lower yield in the unfertilized control area than area where 55 kg N ha^{-1} was applied (Figure 21.11). However, this yield enhancement was not sufficient to pay for the additional N. This confirms that the optimal N rate for the south field was at or below the rate that it received. By analogy, the same is likely true for the north field. Thus, a substantial amount of N, energy, and money was saved in this field with minimal or no cost in terms of lost yield. For the 64 ha area under the center-pivot irrigation system, calculations suggest that relative to the producer's normal practice of applying 220 kg N ha^{-1} to the whole field before planting, SSNM reduced the total N applied by 9910 kg N. This reduced the amount of energy invested in the field by 640 GJ, reduced the production costs by $3310, and reduced the amount of CO_2 released into the atmosphere by 39.2 Mg CO_2.

21.8 CONCLUSIONS

Net energy return to N fertilizer with SSNM can be greater than with conventional, soil-test-based regional, blanket-N applications in cotton and corn. The SSNM approaches tested included grid soil sampling, management zone strategies, aerial photography, and canopy reflectance. Improved energy return to N fertilizer with SSNM was usually due to N savings without a reduction in yield.

ACKNOWLEDGMENTS

We thank the producer cooperators of this research for their interest, cooperation, comments, and generosity in allowing us to work on their farms. For help with equipment preparation, field work, and data collection, analysis, and development of figures we extend thanks to Bob Mahurin, Scott Drummond, Larry Mueller, and Vicky Hubbard.

REFERENCES

1. Borlaug, N.E. and Dowswell, C.R. Mobilising science and technology to get agriculture moving in Africa. *Dev. Policy Rev.* 13, 115, 1995.
2. Borlaug, N.E. The green revolution: Its origins and contributions to world agriculture. *J. Bioresource Sci.* 4, 11022, 2003.
3. Gellings, C.W. and Parmenter, K.E. Energy efficiency in fertilizer production and use. Efficient use and conservation of energy. In C.W. Gellings and K. Blok (eds.). *Encyclopedia of Life Support Systems (EOLSS)*. Eolss Publishers, Oxford, U.K., 2004.
4. The Fertilizer Institute. Supply & demand, energy drive global fertilizer prices, 2009. http://www.tfi.org/publications/pricespaper.pdf
5. Hood, C.F. and Kidder, G. Fertilizers and energy. Fact Sheet EES-58, November 1992. University of Florida, Florida Cooperative Extension Service, Gainesville, FL, 1992.
6. Hülsbergen, K.-J., Feil, B., Biermann, S., Rathke, G.-W., Kalk, W.-D., and Diepenbrock, W.A. method of energy balancing in crop production and its application in a long-term fertilizer trial. *Agric. Ecosyst. Environ.* 86, 303, 2001.
7. Kuesters, J. and Lammel, J. Investigations of the energy efficiency of the production of winter wheat and sugar beet in Europe. *Eur. J. Agron.* 11, 35, 1999.
8. Eichner, M.J. Nitrous oxide emissions from fertilized soils: Summary of available data. *J. Environ. Qual.* 19, 272, 1990.
9. Bronson, K.F., Mosier, A.R., and Bishnoi, S.R. Nitrous oxide emissions in irrigated corn as affected by nitrification inhibitors. *Soil Sci. Soc. Am. J.* 56, 161, 1992.
10. Adviento-Borbe, M.A.A., Haddix, M.L., Binder, D.L., Walters, D.T., and Dobermann, A. Soil greenhouse gas fluxes and global warming potential in four high-yielding maize systems. *Global Change Biol.* 13, 1972, 2007.
11. Mamo, M., Malzer, G.L., Mulla, D.J., Huggins, D.R., and Strock, J. Spatial and temporal variation in economically optimum nitrogen rate for corn. *Agron. J.* 95, 958, 2003.
12. Scharf, P.C., Kitchen, N.R., Sudduth, K.A., Davis, J.G., Hubbard, V.C., and Lory, J.A. Field-scale variability in economically-optimal N fertilizer rate for corn. *Agron. J.* 97, 452, 2005.
13. Raun, W.R. and Johnson, G.V. Improving N use efficiency for cereal production. *Agron. J.* 91, 357, 1999.
14. Cassman, K.G., Dobermann, A.D., and Walters, D.T. Agroecosystems, N-use efficiency, and N management. *AMBIO* 31, 132, 2002.
15. Oritiz-Monasterio, J.I., Sayre, K.D., Rajaram, S., and McMahon, M. Genetic progress in wheat yield and nitrogen use efficiency under four nitrogen rates. *Crop Sci.* 37, 898, 1997.
16. Castleberry, R.M., Crum, C.W., and Krull, C.F. Genetic improvement of U.S. maize cultivars under varying fertility and climatic environments. *Crop Sci.* 24, 33, 1984.
17. USDA-Economic Research Service. Fertilizer trade summary, 2009. Available at http://www.ers.usda.gov/Data/FertilizerTrade/summary.htm
18. Murrell, S. Fertilizer nitrogen BMPs for corn in the North Central Region. *Better Crops Plant Food.* 90, 16, 2004.

19. Kitchen, N.R., Goulding, K.W.T., and Shanahan, J.F. Proven practices and innovative technologies for on-farm crop nitrogen management. In J.L. Hatfield and R.F. Follett (eds.). *Nitrogen in the Environment*, Chap. 15. Elsevier Science, Amsterdam, the Netherlands, 2008, pp. 483–517.
20. Yabaji, R., Nusz, J.W., Bronson, K.F., Malapati, A., Booker, J.D., Nichols, R.L., and Thompson, T.L. Nitrogen management for subsurface drip irrigated cotton: Ammonium thiosulfate, timing, and canopy reflectance. *Soil Sci. Soc. Am. J.* 73, 589, 2009.
21. Chang, J., Clay, D.E., Carlson, C.G., Reese, C.L., Clay, S.A., and Ellsbury, M.M. Defining yield goals and management zones to minimize yield and nitrogen and phosphorus fertilizer recommendation errors. *Agron. J.* 96, 825, 2004.
22. Miao, Y., Mulla, D.J., Batchelor, W.D., Paz, J.O., Robert, P.C., and Wiebers, M. Evaluating management zone optimal nitrogen rates with a crop growth model. *Agron. J.* 98, 545, 2006.
23. Koch, B., Khosla, R., Frasier, W.M., Westfall, D.G., and Inman, D. Economic feasibility of variable-rate nitrogen application utilizing site-specific management zones. *Agron. J.* 96, 1572, 2004.
24. Bronson, K.F., Booker, J.D., Bordovsky, J.P., Keeling, J.W., Wheeler, T.A., Boman, R.K., Parajulee, M.N., Segarra, E., and Nichols, R.L. Site-specific irrigation and nitrogen management for cotton production in the Southern High Plains. *Agron. J.* 98, 212, 2006.
25. Scharf, P.C., Kitchen, N.R., Sudduth, K.A., and Davis, J.G. Spatially variable corn yield is a weak predictor of optimal nitrogen rate. *Soil Sci. Soc. Am. J.* 70, 2154, 2006.
26. Raun, W.R., Solie, J.B., Johnson, G.V., Stone, M.L., Mullen, R.W., Freeman, K.W., Thomason, W.E., and Lukina E.V. Improving nitrogen use efficiency in cereal production with optical sensing and variable rate application. *Agron. J.* 94, 815, 2002.
27. Flowers, M., Weisz, R., Heiniger, R., Osmond, D., and Crozier, C. In-season optimization and site-specific nitrogen management for soft red winter wheat. *Agron. J.* 96, 124, 2004.
28. Scharf, P.C. and Lory, J.A. Calibrating reflectance measurements to predict optimal sidedress nitrogen rate for corn. *Agron. J.* 101, 615, 2009.
29. Schmidt, J.P., Dellinger, A.E., and Beegle, D.B. Nitrogen recommendations for corn: An on-the-go sensor compared with current recommendation methods. *Agron. J.* 101, 916, 2009.
30. Kitchen, N.R., Sudduth, K.A., Drummond, S.T., Scharf, P.C., Palm, H.L., Roberts, D.F., and Vories, E.D. Ground-based canopy reflectance sensing for variable-rate nitrogen corn fertilization. *Agron. J.* 102, 71, 2010.
31. Harmel, R.D., Kenimer, A.L., Searcy, S.W., and Torbert, H.A. Runoff water quality impact of variable rate sidedress nitrogen application. *Prec. Agric.* 5, 1657, 2004.
32. Hong, N., Scharf, P.C., Davis, J.G., Kitchen, N.R., and Sudduth, K.A. Economically optimal nitrogen rate reduces soil residual nitrate. *J. Environ. Qual.* 36, 354, 2007.
33. Roberts, D.F., Kitchen, N.R., Scharf, P.C., and Sudduth, K.A. Will variable-rate nitrogen fertilization using corn canopy reflectance sensing deliver environmental benefits? *Agron. J.* 102, 85, 2010.
34. Inman, D., Khosla, R., Westfall, D.G., and Reich, R. Nitrogen uptake across site specific management zones in irrigated corn production systems. *Agron. J.* 97, 169, 2005.
35. Liska, A.J., Yang, H.S., Bremer, V.R., Klopfenstein, T.J., Walters, D.T., Erickson, G.E., and Cassman, K.G. Improvements in life cycle energy efficiency and greenhouse gas emissions of corn-ethanol. *J. Ind. Ecol.* 13, 58, 2008.
36. Nelson, R.G., Hellwinckel, C.M., Brandt, C.C., West, T.O., De La Torre Ugarte, D.G., and Marland, G. Energy use and carbon dioxide emissions from cropland production in the United States, 1990–2004. *J. Environ. Qual.* 38, 418, 2009.
37. Tilman, D., Hill, J., and Lehman, C. Carbon-negative biofuels from low-input high-diversity grassland biomass. *Science* 314, 1598, 2006.

38. Verma, S.B., Dobermann, A., Cassman, K.G., Walters, D.T., Knops, J.M., Arkebauer, A.E., Suyker, T.J. et al. Annual carbon dioxide exchange in irrigated and rainfed maize-based agroecosystems. *Agric. Forest Meteorol.* 131, 77, 2005.
39. Hülsbergen, K.-J., Feil, B., and Diepenbrock, W. Rates of nitrogen application to achieve maximum energy efficiency for various crops: Results of a long-term experiment. *Field Crops Res.* 77, 61, 2002.
40. Shapouri, H., Duffield, J.A., and Wang, M. The energy balance of corn ethanol: An update/USDA Agricultural Economic Report No. 813, 2002.
41. Lorenz, D. and Morris, D. How much energy does it take to make a gallon of ethanol? Revised and updated. Institute for Local Self-reliance, Washington, DC, 1995.
42. Shapouri, H., Duffield, J.A., and Graboski, M.S. Estimating the net energy balance of corn ethanol. USDA-ERS, AER-721, 1995.
43. Adamsen, F.J., Bigelow, D.S., and Scott, G.R. Automated methods for ammonium, nitrate, and nitrite in 2M KCl-phenylmercuric acetate extracts of soil. *Commun. Soil Sci. Plant Anal.* 16, 883, 1985.
44. Zhang, H., Raun, B., Hattey, J., Johnson, G., and Basta, B. OSU soil test interpretations. Publication no. F-2225, Oklahoma Cooperative Extension Service, Oklahoma State University, Stillwater, OK, 1998.
45. National Research Council. *Nutrient Requirements of Beef Cattle*, 6th revised edn. National Academy Press, Washington, DC, 1984.
46. Rashid, U., Anwar, F., and Knothe, G. Evaluation of biodiesel obtained from cottonseed oil. *Fuel Process. Technol.* 90, 1157, 2009.
47. Nusz, J.W. Remote sensing to improve nitrogen management in subsurface drip irrigation cotton. MS thesis, Texas Tech University, Lubbock, Texas, 2009.
48. Shahandeh, H., Wright, A.L., Hons, F.M., and Lascano, R.J. Spatial and temporal variation of soil nitrogen parameters related to soil texture and corn yield. *Agron. J.* 97, 772, 2005.
49. Clay, D.E., Kitchen, N.R., Gregg, C.G., Kleinjan, J., and Chang, J. Using historical management to reduce sampling errors. In F.J. Pierce and D.E. Clay (eds.). *GIS Applications in Agriculture Series*. CRC Press, Boca Raton, FL, 2007, pp. 49–64.
50. Scharf, P.C. and Lory, J.A. Calibrating corn color from aerial photographs to predict sidedress nitrogen need. *Agron. J.* 94, 397, 2002.
51. Sripada, R.P., Heiniger, R.W., White, J.G., and Weisz, R. Aerial color infrared photography for determining late-season nitrogen requirements in corn. *Agron. J.* 97, 1443, 2005.
52. Alam, M.S., Alam, M.R., and Islam, K.K. Energy flow in agriculture: Bangladesh. *Am. J. Environ. Sci.* 1, 213, 2005.

22 Geographic Information and the Management of Animal Manure

D.A. Crouse and J.L. Havlin

CONTENTS

22.1 EXECUTIVE SUMMARY

The production of animals for meat is a significant source of income in many parts of the United States. However, animal production is not uniformly distributed across the United States. In fact, production is concentrated in specific regions and states. Swine production is common in Illinois, Iowa, Minnesota, and North Carolina. Poultry production, including turkeys or broiler chickens, greatly contributes to the economies of Alabama, Arkansas, Georgia, Minnesota, Missouri, and North Carolina. Even within these states, production is highly concentrated in local areas surrounding feed mills. This intensive production in small regions leads to the excess accumulation of plant nutrients, resulting from repeated manure applications, which results in environmental contamination issues. This chapter discusses the use of geographical information system (GIS) tools to improve the distribution of manure nutrients at the national, local, and field scales as a means to minimize environmental contamination.

22.2 INTRODUCTION

All animal manures contain nutrients that are essential to plant life. For centuries, animal wastes, such as poultry litter as well as swine and cattle manure, have been land applied as a fertilizer to primarily address the nitrogen (N), phosphorus (P), and potassium (K) needs of agronomic crops. Unfortunately, the N:P:K ratio in manures rarely matches the N:P:K ratio needed by the receiving plant. Traditional manure application rates are based on supplying sufficient N to meet crop needs. This approach considers N the most limiting nutrient for application since (1) it is often needed by plants at the highest rates and (2) N, when transported off-site, has traditionally been considered the nutrient most likely to cause negative environmental impacts. These N-based application rates result in P being applied at rates greater than crop needs, leading to a buildup of soil P. The problems associated with excessive phosphorous in the soil are well known.[7] Sediment-bound P can be deposited in surface waters following erosion. These P-laden sediments then provide aquatic algae, a readily available source of nutrition. Algae population expands resulting in a tremendous volume of organic biomass produced and deposited in the aquatic ecosystem. As algae die, the aerobic decomposition of their biomass consumes oxygen, leading to hypoxic conditions and die-off of fish and shellfish.[7] The problem is not only environmental, but also economic as fisheries and tourism industries are negatively impacted.

Phosphorous is not the only concern. Other elements, such as Cu and Zn, are added to animal diets in mineral forms. These nutrients improve animal health or feed-to-meat conversion ratios. Unfortunately, the animals do not retain all of the minerals, and as a result significant quantities of both Cu and Zn exist in many animal types of manure. Since Cu and Zn nutrients are needed at very low rates, relative to N, they are often overapplied. Although these elements have no significant impact on surface or subsurface water quality, the elevated concentrations in soil can result in plant toxicities.

Properly managing manure is challenging due to current agricultural practices. Just to list a few, there are issues related to manure storage, timing of applications, and maintenance of sufficient cropland to receive the manure nutrients generated on-farm. Not only are the issues numerous, but the scale at which these challenges exist also varies. For example, animal production tends to be clustered in specific regions of the country, specific regions of an individual state, and even, specific regions of a single county. Much of this is driven by the economics of moving feed to the animals and moving the animals to market. Technological innovation and shifts to larger, more specialized operations have led to increases in productivity, reduced production costs, and lower prices.[13]

Traditionally, land managers, regulatory personnel, and other decision makers have used maps to understand how "things change in space." Early mapping techniques required extensive surveying, a highly skilled cartographer, and many hours of map development just to record, describe, and illustrate simple geospatial patterns. Information technologies available today have greatly improved our understanding of spatial variability. Highly accurate location and navigation is achieved now with relatively inexpensive global positioning equipment. The computational burden of interpretation and visualization has been greatly reduced with faster computers and the use of geographic information databases. With regard to implementing management decisions at the field scale, there are variable-rate application technologies that can place plant nutrients where needed while avoiding areas where nutrients are not needed. This chapter explores the use of a variety of geospatial technologies that can improve manure management decisions at the national, local, and field scales.

22.3 NATIONAL SCALE

Due to economics, animal production tends to be regionalized. For example, of the 66.8 millions hogs on farms as of September 2009, nearly 62% of those animals were located in Iowa, North Carolina, Minnesota, and Illinois.[26] Likewise, Alabama, Georgia, Arkansas, and North Carolina accounted for 49.8% of all broiler chickens produced in the United States in 2008.[26] Turkey production in 2009 followed similar trends, with 52.8% of all birds being produced in Minnesota, North Carolina, Arkansas, and Missouri.[26] Cattle production is not nearly as concentrated; however, 33.5% of January 1, 2009, cattle inventory were located in Texas, Nebraska, Kansas, and Oklahoma.[26]

As previously mentioned, in North Carolina and other major animal-producing states, the amount of manure nutrients produced in manure often exceeds crop requirements, especially for P, Cu, and Zn.[2] Using geographic information reported by the U.S. Department of Agriculture, Maguire et al.[16] developed a nationwide P balance map. As discussed below, the balance considered acreages and yields for the crops that most commonly receive applications of animal manure. The study calculated, at the county level, the average mass of P that would be removed per hectare of cropped land (Figure 22.1). It calculated the mass of manure-P as a function of harvested crop area (Figure 22.2). The authors then determined whether there was regional evidence of a P surplus or deficit (Figure 22.3).

The study found that manure-P (kg ha^{-1}) tended to be highest in the counties in states having significant production of animals (i.e., North Carolina, Georgia, Alabama, etc.). There are many reasons for this concentration of manure-P. Perhaps the most obvious reason is the number of animals produced per unit available crop land. For example, Iowa and North Carolina rank first and second in swine production in the number of animal unit's produced.[26] However, North Carolina has many more counties with manure-P in excess of 15 kg ha^{-1}. This is due to the additional animal species produced. North Carolina is second in the production of turkeys and

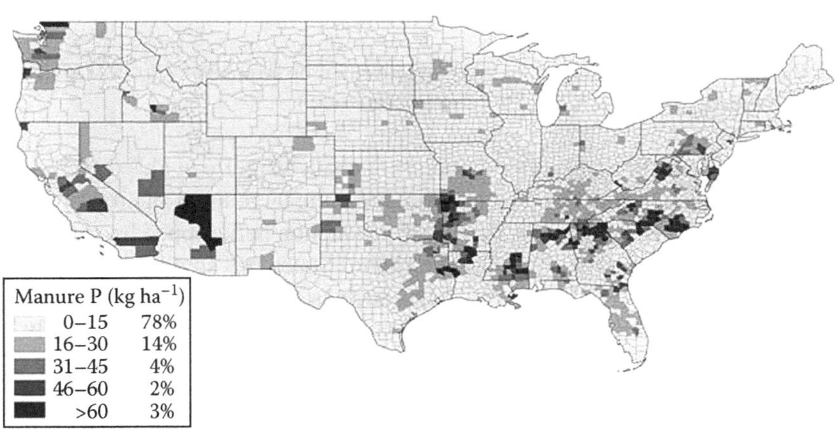

FIGURE 22.1 Cumulative excreted livestock manure-P (2002). Presented as kg P per hectare of crop harvested area, averaged across the census years 1987, 1992, 1997, and 2002. (From Maguire, R.O. et al., *J. Environ. Qual.*, 36, 1235, 2007. With permission.)

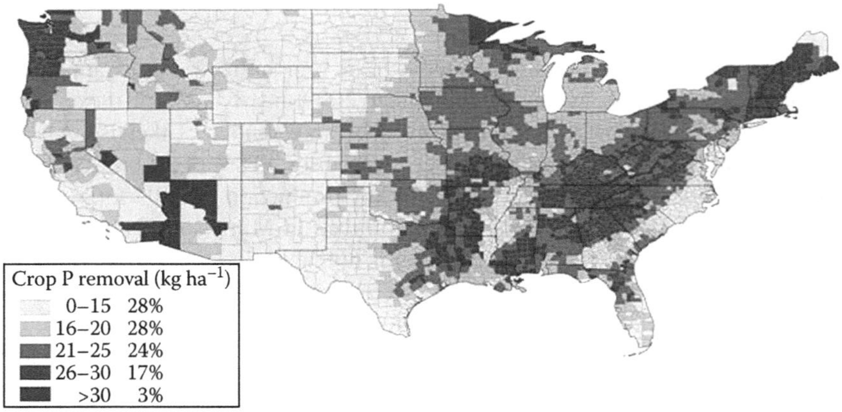

FIGURE 22.2 Phosphorus, in units of kg ha^{-1}, removed in harvested portion of crop, averaged across the census years 1987, 1992, 1997, and 2002. (From Maguire, R.O. et al., *J. Environ. Qual.*, 36, 1235, 2007. With permission.)

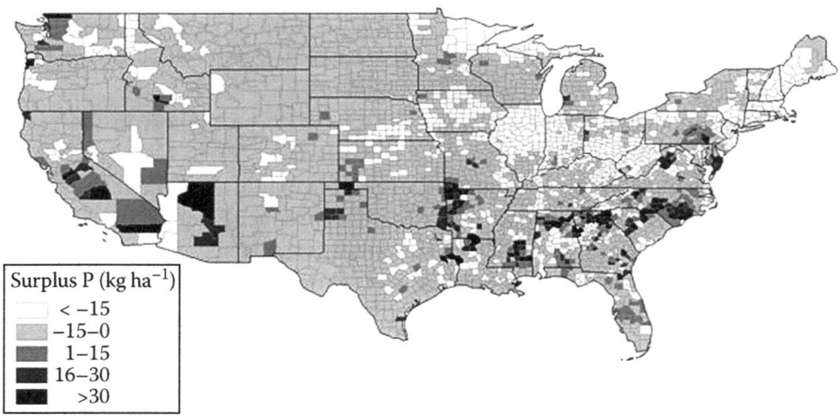

FIGURE 22.3 Manure-P surplus or deficit relative to crop P removal. (From Maguire, R.O. et al., *J. Environ. Qual.*, 36, 1235, 2007. With permission.)

fourth in the production of broiler chickens. Second, Maguire et al.[16] reported when manure-P production is divided by cropped area, the large areas of crop production in the Midwest translate into low manure-P production per cropped hectare. For the most part, the southeastern region of the United States has small, highly dissected farms due to other predominant land usage, such as forest and urban land. This low crop area means even small amounts of manure-P can translate into relatively high manure-P per cropped hectare.

As previously mentioned, the second part of the Maguire et al.[16] nutrient balance addressed what the authors termed assimilative capacity. Using crop yields and harvested areas from the USDA National Agricultural Statistics Service,[26] and typical crop P removal rates, the authors calculated a crop P removal rate per hectare. According to the study, 28% of all U.S. counties had <15 kg P ha^{-1}crop P removal in the harvested portion of the crop, and more than half of all counties had P removal of <20 kg P ha^{-1} (Figure 22.2). The authors reported that in geographic regions the greatest crop P removal tended to be areas with pasture, mostly in the eastern and northwestern parts of the United States (Figure 22.2).

In addition, 92% of all U.S. counties produced ≤30 kg manure-P ha^{-1}, whereas 97% of U.S. counties had the ability to assimilate as much as 30 kg manure-P ha^{-1}. Although that sounds positive, in many cases, the counties producing high quantities of manure-P are not the same counties that can assimilate this amount of manure-P. The end result is either a P deficit or surplus. The study revealed that 90% of all U.S. counties in fact have a P deficit (Figure 22.3). The other 10% had a surplus, with 3% of the counties exhibiting a surplus >30 kg manure-P ha^{-1} of cropped land. One of the assumptions for a nutrient balance calculated using the methodology of Maguire et al.[16] is that the manure nutrients are applied to all available crop land within any given county. However, as the authors noted, that may not be the case with surpluses of P possibly existing on certain farms or fields, due to uneven spatial generation of manure and the expense of transport. This leads to the next geographic analysis that can be performed at the local, within-county scale.

22.4 LOCAL SCALE

Within states and even counties, animal production tends to be clustered. One obvious factor determining the location of farms is the proximity to animal feed. With swine and poultry production in North Carolina, the state annually has a grain deficit, meaning grain consumption by swine and poultry exceeds statewide grain production. As a result, grain is regularly imported into the North Carolina. The vast majority of the grain comes from the Midwest, arriving by rail, and as a result feed mills are adjacent to railroads. This in turn creates a geographically fixed point from which the farm locations radiate. For example, of the 4148 swine farms registered in North Carolina, more than half are within a 50 km radius of one of the main feed mills. The result is a dense population of farms in a centralized geographic region (Figure 22.4).

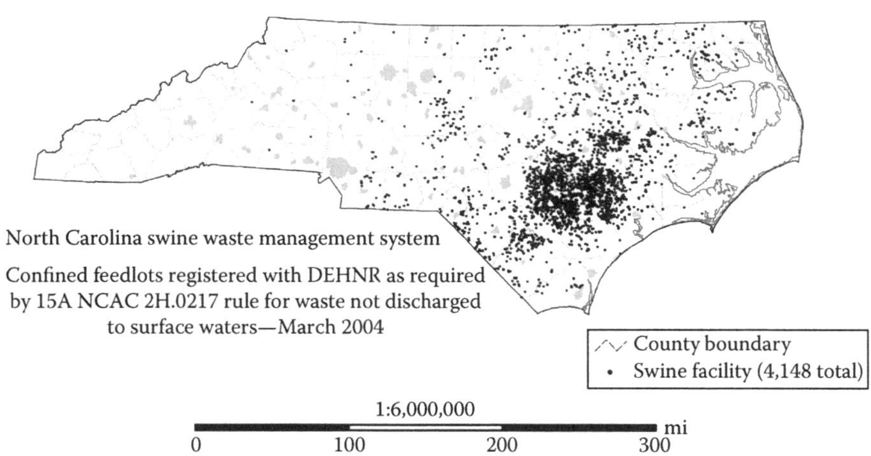

FIGURE 22.4 Distribution of farms in North Carolina producing more than 250 head of swine per year.

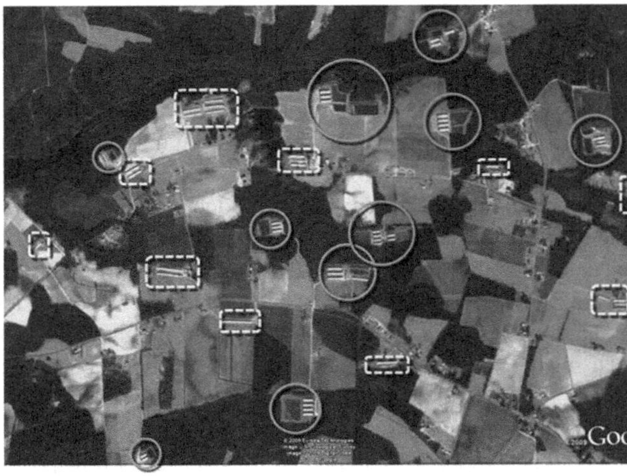

FIGURE 22.5 Clustering of swine and poultry farms in eastern North Carolina. Poultry farms are identified in the figure with the squares and swine farms with circles.

In addition to the clustering of swine farms, poultry production facilities tend to be centrally located around the same feed mills, or neighboring feed mills advantageously located close to the rail transportation. As a result, areas with intensive swine production will often have intensive poultry production (Figure 22.5). In fact, it is not uncommon to have swine and poultry barns on the same farm. These areas of intensive animal production will often have manure nutrients available for application that far exceed P, Cu, and Zn needs of local crops. Continued application of these nutrients has resulted in unsustainable accumulations. One possibility for addressing the local scale issue is to develop markets to move manure nutrients out of local communities. Ultimately it comes down to economics related to how far can you move manure before transportation costs become prohibitive? GISs are ideally suited to assist in these kinds of spatial/economical analyses.

22.5 FIELD SCALE

The improvement of soil and crop productivity with applications of manure is well documented.[8,12,14] These and many other studies document increased soil organic matter (OM), N, P, and many other nutrients with continued manure application.[22] Although few studies have compared variable-to-uniform waste application, recent studies have demonstrated increased crop productivity with variable application of manure. Eghball et al.[9] compared uniform manure application to similar rates applied only to less productive areas based on soil organic C content. These results showed that manure application to less productive areas increased corn grain yield compared to commercial fertilizer, and that variable manure application improved less productive areas within the field.

As discussed earlier, continued application of animal wastes especially based on crop N requirements substantially increases residual soil P and risk of P transport to surface and ground waters.[21,20] Quantifying the spatial distribution of soil P in animal waste amended fields may be useful in directing future manure applications to those field areas with lower soil P levels. A site-specific approach to describe the spatial distribution of soil P would minimize potential environmental impact of manure use and extend the use of fields for future manure application.[15]

Even though variable manure application may not consistently increase grain yield compared with uniform application, Wittry and Mallarino[27] demonstrated that variable liquid swine manure application based on soil test P minimized P application to high soil test P areas and reduced within-field soil P variability. Reducing overapplication of manure especially in areas of high soil test P, which is highly correlated to soluble P transport to surface and ground waters, will require an accurate assessment of the spatial distribution in soil test P as well as the capability to variably apply subsequent manure applications.

Traditionally, the spatial variability of soil properties is obtained by grid or zone sampling, which is labor and time intensive.[17] Advances in remote sensing and related technologies provide a rapid and efficient platform for mapping soil properties. These technologies provide a soil mapping tool for spatial management of animal waste.

A number of investigations have been done with multi- and hyperspectral reflectance sensors to characterize soil properties. In the early 1970s, soil scientists began to investigate use of multispectral remote sensing for differentiating surface soil properties.[6,19] However, compared to hyperspectral images, multispectral imagery is problematic where either soil property prediction accuracy is low or the prediction equation has limited transferability between geographical regions.[3,10] Therefore, hyperspectral remote sensing has been increasingly used in mapping soil properties.[5,11]

Hyperspectral sensor platforms vary from laboratory, field, to airborne and/or satellite-based sensors. Laboratory methods involve collection of soil samples from the field, measuring soil reflectance, then correlating reflectance to soil property values (e.g., soil OM, total N, etc.). Clearly, laboratory-based hyperspectral data, especially if corrected for soil moisture, are more highly correlated to measured soil properties than spectral data collected with either field or airborne/satellite platforms. While satellite hyperspectral imagery has been available since 2000, few attempts have been made using satellite hyperspectral imagery for mapping soil properties.[10]

Although remote sensing of soil OM and N is well established, spectral measures of soil P have not proved as reliable. Several authors used NIR reflectance and reported poor correlation between reflectance of total or Mehlich-3 P.[5,18,24] Recently, Bogrekci and Lee[4] demonstrated a strong relationship between total soil P and NIR absorbance spectra; however, soil samples required sieving and removal of soil moisture. Laboratory-based measures of reflectance of soil properties may be used for calibration purposes, but will not be included in practical application of remote sensing to quantify the distribution of soil properties in the field.

Recent use of airborne hyperspectral remote sensing of soil properties has been documented.[1,23,25] In a recent study, soil samples were collected from several watersheds and analyzed for soil OM, total N, and total P; their corresponding hyperspectral reflectance was measured in the laboratory, field, and using satellite (Hyperion) images.[28] The hyperspectral reflectance data were related to soil OM, total N, and

TABLE 22.1
Relationships between Laboratory, Field, and Satellite Hyperspectral Imagery and Selected Soil Properties

Scale	Soil Properties	Validation (R^2)
Laboratory	OM (%)	0.76
Field		0.74
Satellite		0.74
Laboratory	Total N (%)	0.75
Field		0.79
Satellite		0.72
Laboratory	Total P (mg kg^{-1})	0.67
Field		0.60
Satellite		0.66

Source: Zheng, B. Using satellite hyperspectral imagery to map soil organic matter, total nitrogen, and total phosphorus. MS Thesis, Department of Earth Sciences, Indiana University, Bloomington, IN, 2008.

total P concentration through partial least squares regressions (Table 22.1). These results show slightly decreased model performance when shifting from laboratory-measured spectra to satellite image spectra. Regardless of the spectral data source, the models for estimating SOM and total N consistently outperformed those for estimating total P, although the correlations were as good or better than previous efforts.[4,18] Although 30 m× 30 m resolution obtained with the Hyperion spectral sensor was cited as a limitation to quantifying soil properties for use in a field scale, using 900 m^2 resolution should still provide reliable estimates of soil properties, especially since most grid-based sampling is done on a 0.8–1.2 ha scale.[17]

22.6 CONCLUSIONS

As shown in this chapter, manure management provides challenges at a variety of scales. Nationally, there is the concern of county-wide imbalances between crop nutrient needs and availability of manure nutrients. Regionally, areas of the country are already accumulating more nutrients than can be used on all of the available crop land. The challenge is how do we redistribute the nutrients? Or, perhaps, are there changes that can be made in cropping systems that will result in the utilization of more of the available nutrients. It may be as simple as changing from grain production to forage production, where large quantities of biomass can be removed and transported off the farm, out of the county. Perhaps forestland could be better utilized to receive a portion of the local manure production. Manures could be a key source of nutrients for cellulose-based biofuel production. Ultimately, GISs will have a role in solving these national nutrient imbalances.

Locally, the same problems persist, but this time, nutrients accumulate not only in one or more counties but in isolated areas of individual counties. The challenge is to economically distribute manure nutrients recognizing the farm footprints are already established. Can value-added products be developed that will redistribute manure nutrients out of the rural community and into the urban landscape market? In North Carolina, for example, organic lawn care businesses utilizing manure nutrient sources are expanding throughout the major municipalities. How will green energy policy and the conversion of manure to methane affect nutrient distribution within counties? Again, these are all questions with spatial components which make them ideally suited for geographic information analysis and modeling.

At the field scale, we know spatial distribution of soil OM, total N, and total P can be accurately mapped with airborne or satellite hyperspectral imagery. How will this technological breakthrough affect local decision making of land managers, agronomists, and farmers? We know they can, but will these maps be used to distribute animal waste to less productive areas (low soil OM) and/or reduce waste application rates to areas high in soil P? How long before these less productive areas become productive? How long before they reach a maximum P-, Cu-, or Zn-holding capacity? What do we do then? Again, all these questions have spatial components, and this time, temporal components. That makes them ideal candidates for further analysis using the tools of the modern mapper, working hard to determine "how things change in space."

REFERENCES

1. Bajwa, S.G. and Tian, L.F. Soil fertility characterization in agricultural fields using hyperspectral remote sensing. *Trans. ASAE* 48, 2399, 2005.
2. Barker, J.C. and Zublena, J.P. Livestock manure nutrient assessment in North Carolina. In *Proceedings of the Seventh International Symposium on Agricultural and Food Processing Wastes*, Chicago, IL. ASAE, St. Joesph, MI, 1995, pp. 98–106.
3. Ben-Dor, E. Quantitative remote sensing of soil properties. *Adv. Agron.* 75, 173, 2002.
4. Bogrekci, I. and Lee, W.S. Effects of soil moisture content on absorbance spectra of sandy soils in sensing phosphorus concentrations using UV–VIS–NIR spectroscopy. *Trans. ASABE* 49, 1175, 2006.
5. Chang, C.W., Laird, D.A., Mausbach, M.J., and Hurburgh, C.R. Near-infrared reflectance spectroscopy—Principal components regression analyses of soil properties. *Soil Sci. Soc. Am. J.* 6, 480, 2001.
6. Chen, F., Kissel, D.E., West, L.T., and Adkins, W. Field-scale mapping of surface soil organic carbon using remotely sensed imagery. *Soil Sci. Soc. Am. J.* 64, 746, 2000.
7. Daniel, T.C., Sharpley, A.N., and Lemunyon, J.L. Agricultural phosphorus and eutrophication: A symposium overview. *J. Environ. Qual.* 27, 251, 1998.
8. Dormaar, J.F., Lindwall, C.W., and Kozub, C.G. Effectiveness of manure and commercial fertilizer in restoring productivity of an artificially eroded Dark Brown Chernozemic soil under dryland conditions. *Can. J. Soil Sci.* 68, 669, 1988.
9. Eghball, B., Bauer, C.J., Shapiro, C.A., and Schepers, J.S. Site-specific application effects on corn yield and nitrogen status. *Proceedings of the Animals Residuals Management Conference*, November 12–14, 2000, Kansas City, MO, 2001.
10. Ge, Y., Thomasson, J.A., and Sui, R. Remote sensing of soil properties in precision agriculture: A review. *ASABE Annual International Meeting No. 061176*, Portland, OR, 2006.
11. He, Y., Huang, M., Garcia, A., Hernandez, A., and Song, H. Prediction of soil macronutrients content using near-infrared spectroscopy. *Comput. Electron. Agric.* 58, 144, 2007.

12. Hornick, S.B. and Parr, J.F. Restoring the productivity of marginal soils with organic amendments. *Am. J. Altern. Agric.* 2, 64, 1987.
13. Key, N., McBride, W.D., and Ribaudo, M. Changes in manure management in the hog sector. *American Agricultural Economics Association, 2008 Annual Meeting*, Orlando, FL, July 27–29, 2008.
14. Larney, F.J. and Janzen, H.H. A simulated erosion approach to assess rates of cattle manure and phosphorus fertilization for restoring productivity of eroded soils. *Agric. Ecosyst. Environ.* 65, 113, 1997.
15. Mallarino, A.P. and Schepers, J.S. Role of precision agriculture in phosphorus management practices. In T. Sims and A.N. Sharpley, eds. *Phosphorus: Agriculture and the Environment*. ASA, CSSA, SSA, Madison, WI, 2005, pp. 881–908.
16. Maguire, R.O., Crouse, D.A., and Hodges, S.C. Diet modification to reduce phosphorus surpluses: A mass balance approach. *J. Environ. Qual.* 36, 1235, 2007.
17. Mallarino, A.P., Beegle, D.B., and Joern, B.C. *Soil Sampling Methods for Phosphorus—Spatial Concerns*. SERA-17 Position Paper, 2006. http://www.sera17.ext.vt.edu/Documents/Sampling-P-Spatial%20Concerns.pdf
18. Moron, A. and Cozzolino, D. Measurement of phosphorus in soils by near infrared reflectance spectroscopy: Effect of reference method on calibration. *Commun. Soil Sci. Plant Anal.* 38, 1965, 2007.
19. Ray, S.S., Singh, J.P., Das, G., and Panigrahy, S. Use of high resolution remote sensing data for generating site-specific soil management plan. *XX ISPRS Congress*, Commission 7, Istanbul, Turkey. *The International Archives of the Photogrammetry, Remote Sensing and Spatial Information Sciences*, Vol. 127, 2004.
20. Sharpley, A.N., Daniel, T.C., Sims, J.T., Lemunyon, J., Stevens, R., and Parry, R. Agricultural phosphorus and eutrophication. *USDA Agricultural Research Service Special Publication No. 149,* 1999.
21. Sims, J.T. Agricultural and environmental issues in the management of poultry wastes: recent innovations and long-term challenges. In J.E. Rechcigl and H.C. MacKinnon, eds. *Agricultural Uses of By-Products and Wastes*. American Chemical Society, Washington, DC, 1997, pp. 72–90.
22. Sommerfeldt, T.G., Chang, C., and Entz, T. Long-term annual manure applications increase soil organic matter and nitrogen, and decrease carbon to nitrogen ratio. *Soil Sci. Sci. Am. J.* 52, 1668, 1988.
23. Stevens, A., Van Wesemael, B., Vandenschrick, G., Toure, S., and Tychon, B. Detection of carbon stock change in agricultural soils using spectroscopic techniques. *Soil Sci. Soc. Am. J.* 70, 844, 2006.
24. Udelhoven, T., Emmerling, C., and Jarmer, T. Quantitative analysis of soil chemical properties with diffuse reflectance spectrometry and partial least-square regression: A feasibility study. *Plant Soil* 25, 319, 2003.
25. Uno, Y., Prasher, S.O. Patel, R.M., Strachan, I.B., Pattey, E., and Karimi, Y. Development of field-scale soil organic matter content estimation models in Eastern Canada using airborne hyperspectral imagery. *Can. Biosyst. Eng.* 45, 1.9, 2005.
26. USDA–NASS. United States Department of Agriculture–National Agricultural Statistics Service. http://www.nass.usda.gov/Census_of_Agriculture/index.asp, verified December 2009.
27. Wittry, D.J. and Mallarino, A.P. Use of variable-rate technology for agronomic and environmental phosphorus-based liquid swine manure management. In P.C. Robert et al. eds. *Sixth International Conference on Site-Specific Management for Agricultural Systems*. Proceedings of the CD-ROM, July 14–17, Minneapolis, MN. ASA, CSSA, and SSSA, Madison, WI, 2002.
28. Zheng, B. Using satellite hyperspectral imagery to map soil organic matter, total nitrogen, and total phosphorus. MS Thesis, Department of Earth Sciences, Indiana University, Bloomington, IN, 2008.

23 Spatial Ramifications of Crop Selection: Water Quality and Biomass Energy

M.P. Russelle, D.W. Kelley, A.S. Birr, and D.G. Tiffany

CONTENTS

23.1 EXECUTIVE SUMMARY

The use of GIS in concert with simple or complex simulation modeling provides an unparalleled way to generate new data and to help a variety of audiences understand spatial patterns of data. From improved understanding, policy incentives can be crafted to reduce adverse environmental impacts of agricultural production at lower costs than would be necessary otherwise. In this chapter, two case studies demonstrate how GIS and modeling can be used to understand how crop selection and soils interact to effect environmental outcomes across an agricultural landscape.

We addressed the needs of two distinctly different audiences: (1) a public drinking water supplier faced with increasing nitrate in a ground water source and (2) a variety of stakeholders involved with planning a new biomass conversion facility to produce renewable fuels from grain or cellulosic feedstock. In both cases, the GIS output documents the benefits of the perennial legume alfalfa (*Medicago sativa* L.) in particular landscape areas, and provides a mechanism to compare alfalfa with corn (*Zea mays* L.) and soybean (*Glycine max* [L.] Merr.).

23.2 INTRODUCTION

Farmers usually know that particular crops perform better on some soils or landscape positions than on others. This knowledge accrues over years of observation, either directly or by previous generations who farmed that land. Although there are management approaches that can reduce such variation (e.g., fertilization, artificial drainage, irrigation, etc.), the fundamental basis for the variation remains—the soils and their characteristics, and how these interact with weather.[1]

However, crop yields affect less obvious outcomes, such as net energy production of various biofuel crops or nitrate leaching losses to shallow drinking water aquifers. Here we present two case studies as examples of ways GIS can be used to help an audience gain new understanding of problems it faces and to visualize how to achieve better solutions by growing different crops in the landscape. Diversifying cropping systems may increase management effort, equipment needs, and other costs, but alternative crops can offer significant advantages. The primary alternative crop we consider here is alfalfa, a legume that requires no fertilizer N, fits well into rotations with other crops, and provides a wide array of environmental services, in addition to high dry matter and protein yields.[2]

In the first case study, we developed maps of estimated nitrate leaching losses into a rural water drinking supply. These maps were used for targeting the "leakiest" fields for improved fertilizer management of corn or better crop rotation. In the second study, we analyzed the yield and net energy production of different biomass crops in a prospective fuelshed near a town interested in producing alternative transportation fuels.

23.3 CASE STUDY 1: ESTIMATING NONPOINT NITRATE FLUX INTO A SHALLOW AQUIFER

In many regions, drinking water is obtained from shallow aquifers that are subject to contamination from inputs on or near the soil surface. Nitrate impairment of drinking water aquifers has been related to agricultural practices (fertilizer and manure application), residential sources (leaking septic systems), industrial activities (leaking fertilizer storages or spills), and geologic sources. The risk of contaminant movement through the soil depends in large part on soil characteristics, including depth, slope, landscape position, texture, and density. These characteristics define how much water infiltrates, how much is held against gravity, how quickly it moves, and how long it will remain in the plant root zone. The risk of nitrate loss by leaching also depends on the management of inputs that affect the concentration of the potential contaminant in the soil solution (e.g., nitrogen [N] fertilizer or animal manure) and the soil water balance (i.e., rootzone water storage capacity, effective precipitation, and crop water use).

The probability of N loss generally increases with the intensity of agricultural production, which usually is related to greater water and N inputs and shorter crop rotations that include only annual crops, such as corn and soybean. Within a particular crop rotation, nitrate leaching loss can be minimized with a combination of optimum N rate and source, and of timing and method of fertilizer application;[3,4] but further reductions can be achieved only with altered cropping systems.[5,6] As these authors and others have shown, alfalfa can be particularly effective in reducing nitrate leaching.

The nitrate concentration of public drinking water is currently limited by the U.S. EPA to 10 mg nitrate–N L^{-1} (water.epa.gov/drink/contaminants/basicinformation/nitrate.cfm). As nitrate concentration in a public water supply approach this limit, the water supplier must institute one or more means to reduce the concentration. This can be done by dilution with water from less contaminated sources (e.g., from deeper wells or surface water) or by removal using reverse osmosis, ion exchange, or, rarely, distillation. These tactics all require energy and produce waste materials. In the case of reverse osmosis at one treatment facility, 10%–20% of the treated water volume is discharged to a local stream at nitrate concentrations five to ten times that of the influent water. In addition, the discharge contains other salts that were concentrated during reverse osmosis. The advisability and sustainability of this permitted activity could be questioned.

An alternate, preventative approach is to reduce the amount of contaminants transported into the wellhead protection area (WPA). The WPA is defined by the area of land that contributes recharge water to the aquifer (see www.health.state.mn.us/divs/eh/water/swp/whp/fs/swpadfs.html for a more complete description). Most wellfield operators have dealt with known point sources, but it is less common and more difficult to address the nonpoint sources from agricultural fields and feedlots, which are privately owned and managed. In this case study, we were part of a team that provided ideas to a rural water supplier that wanted to reduce nitrate flux into the aquifer from nonpoint sources.

23.3.1 Methods

The subject of investigation was a drinking water supply management area (DWSMA) for the Holland wellfield near Pipestone, Minnesota. This DWSMA encompasses about 8800 ha with 30 soil series and 49 soil mapping units (Figure 23.1). The aquifer is outwash sand and gravel deposited in a glacial meltwater channel that formed in clayey deposits. Loamy surficial deposits cover most of the underlying materials (Figure 23.2). Water table depth usually is within 6 m of the soil surface, and low-lying soils near the creek often are saturated or flooded during late autumn through spring. Due to their drainage condition, most of these low-lying soils are either not farmed or are used for pasture.

Base maps were generated with (1) a soils layer, delineated using the U.S. Department of Agriculture-Natural Resource Conservation Service (USDA-NRCS)

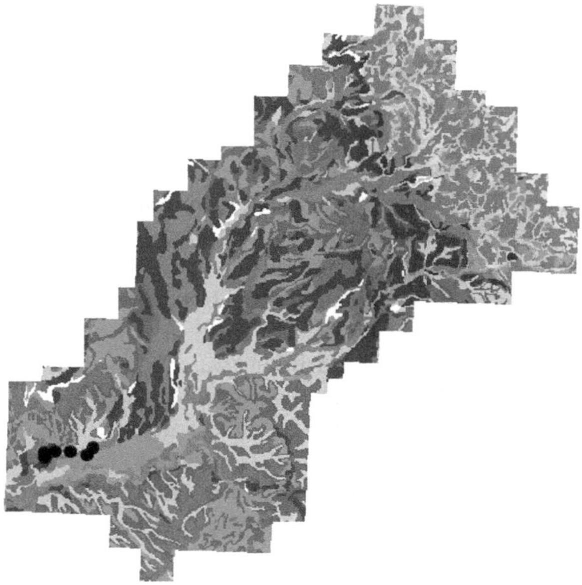

FIGURE 23.1 Soil units in the Holland, Minnesota, DWSMA. The DWSMA is comprised of 30 soil series and 49 mapping units, but the legend is omitted for clarity. Dark circles indicate the wells at the wellhead.

Soil Survey Geographic (SSURGO) database for Pipestone county, Minnesota (www.soils.usda.gov/survey/geography/ssurgo/); (2) field boundaries estimated using 1991–1992 1 m USGS Digital Orthophoto Quad (DOQ) for the Holland (NE, NW, SE, SW) and Pipestone North (NE, NW, SE, SW) quadrants, Pipestone county, Minnesota (deli.dnr.state.mn.us/); (3) a subset of cultivated areas generated using Farm Service Agency information (confidential); (4) roads for Pipestone county, Minnesota (deli.dnr.state.mn.us/); and ancillary information.

The Groundwater Loading Effects of Agricultural Management Systems (GLEAMS) modeling tool was used to provide one-dimensional, field-scale, continuous flow estimates to evaluate the effects of agricultural management systems on chemical and nutrient movement within and through the rootzone.[7] This software incorporates four distinct subroutines to predict hydrology, nutrients, erosion, and pesticide dynamics. The model considers chemical interactions, soil characteristics, weather, and management to arrive at selected output for soil, water, nutrient, or pesticide transport. Model input requirements include weather data (daily rainfall, temperature, solar radiation, and wind speed), soil characteristics, pesticide information, fertilizer and tillage data, and crop-specific information. Soil physical and chemical parameters are described for up to five soil genetic horizons, which are further distributed into seven distinct computational soil layers. GLEAMS output includes daily, monthly, or annual values for runoff, percolation volumes, sediment transport, pesticide mass and concentration, and plant nutrient mass and concentration.

The GLEAMS model utilizes the field capacity concept to simulate percolation of water through the computational soil layers. To estimate evapotranspiration

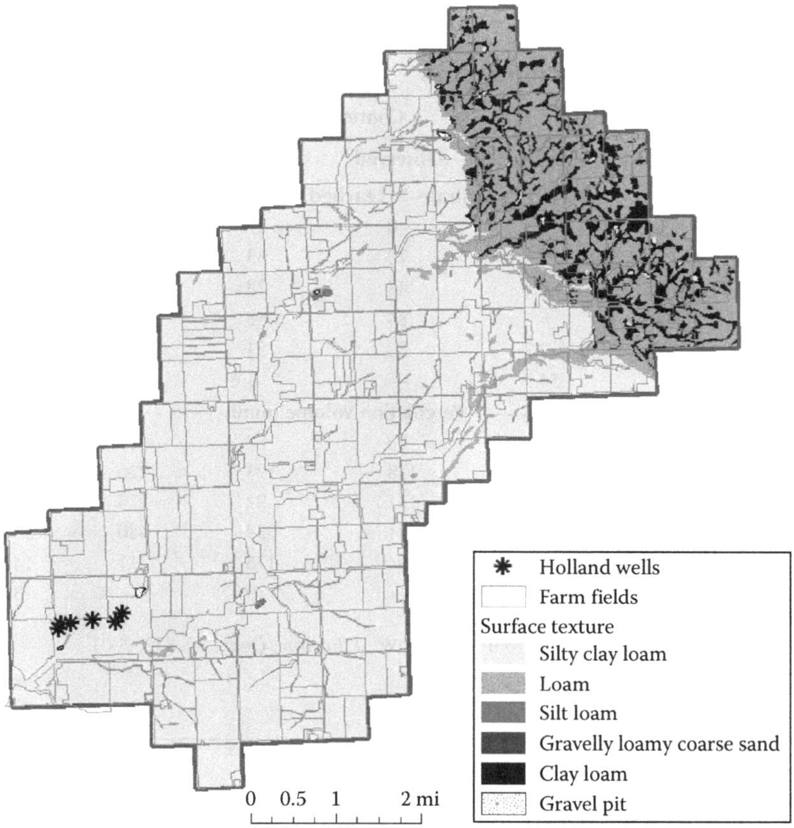

FIGURE 23.2 Generalized soil texture classes in the Holland, Minnesota, DWSMA.

rates and soil water content, model algorithms account for seasonal changes in leaf area index. Water percolation is calculated on a daily basis and is assumed to be zero unless the soil water stored above a given layer exceeds field capacity. Any nitrate available for leaching is transported deeper in the profile only when water moves between layers. In other words, the model ignores diffusive movement in unsaturated conditions. Nitrate leaching loss is assumed only when nitrate is present in the soil solution in the lowest computational layer and when deep percolation occurs out of this layer. We set the maximum depth of rooting to 1.5 m in this simulation for all crops, as information for deeper layers is not available for all soils included here.

23.3.1.1 Model Validation

Because of the number of soil types and the lack of observed data for each modeling case study, model calibration was impractical for each combination of soil, management, and crop. To substantiate the utility of GLEAMS to accurately predict nitrate leaching, we validated the model using observed field data from

TABLE 23.1
GLEAMS Model Validation Results

	Alfalfa		Continuous Corn		Corn–Soybean[a]	
Year	Observed	Simulated	Observed	Simulated	Observed	Simulated
	Nitrate–N Leached (kg N ha^{-1})					
1990	0	0	0	27	0	27
1991	1	5	70	63	81	55
1992	2	0	48	28	32	28
1993	4	6	84	88	67	52
Mean	1.8	2.7	50	51	46	40
% Error	—	54.7	—	1.6	—	(10)
	Percolation Volume (mm)					
1990	n/a[b]	0	20	80	19	80
1991	n/a	21	178	193	220	161
1992	n/a	0	131	92	124	92
1993	n/a	29	441	339	480	238
Mean	—	13	194	176	211	143
% Error	—	—	—	(9)	—	(32)

Source: Observed data obtained from Chung, S.W., *J. Environ. Qual.*, 30, 822, 2001.
[a] Corn grown in 1990 and 1992.
[b] n/a = not available.

the Southwest Research and Outreach Center of the University of Minnesota in Lamberton, Minnesota.[8] Climate data from the station were combined with site-specific field and crop management information to test the effectiveness of the GLEAMS model to predict nitrate losses, leachate percolation volumes, and crop yields for alfalfa, continuous corn, and corn–soybean in rotation (Table 23.1). The main parameter that was modified to achieve acceptable agreement between observed and predicted results was NRCS runoff curve number (CN2), which was set at 75 and 78 for alfalfa and corn, respectively[8] (see Ref. 9 for an in-depth evaluation of CN values).

Validation results were generally within 10% of the mean of observed data, with two exceptions. First, the small average difference between predicted and observed nitrate leaching under alfalfa was large in proportion to the small leaching values, but this error would not be biologically significant. Second, deep percolation under a corn–soybean rotation was underpredicted by 32%. Although this is a large difference for the cropping system, predicted percolation volumes were considerably larger for annual crops than alfalfa, as one would expect given the different temporal and total water demands of these crops. There were no significant differences between the other observed and predicted means (paired *t*-test, $p > 0.05$). For our purposes, we concluded that the model was adequate for predicting percolation volumes below these three crops.

23.3.1.2 Model Application

Following validation, the model was applied to each soil type using the following parameter files for each:

1. Nutrient: current standard management practices for each crop (fertility, timing, field operations) based on a survey of farm management practices[10]
2. Hydrology: predominant soil parameters, mean monthly maximum and minimum air temperatures, planting and harvest dates, irrigation parameters, daily precipitation summaries
3. Erosion: field parameters (slope, contour, CN, % cover, soil erodibility factor)

The effective rooting depth, texture of each soil horizon, effective saturated conductivity of each horizon, soil evaporation parameter based on surface soil texture, and saturated conductivity of each horizon were identified for the typifying pedon of the dominant soil textural class for each polygon. These selections were based upon the SSURGO attribute database and the Official Soil Series Descriptions (ortho.ftw.nrcs.usda.gov/osd/osd.html). Soil textural parameters such as bulk density, porosity, field capacity, permanent wilting point, and hydrologic soil group were also considered.

Mean monthly maximum and minimum air temperatures (°C), solar radiation (MJ m^{-2}), wind speed (km day^{-1}), and dew point temperatures (°C) were input from the GLEAMS climate database, along with the elevation and latitude of Pipestone. Daily data for the 10-year weather records (1989–1998) and daily precipitation and air temperature data for input to the GLEAMS model were obtained from the Minnesota Climatological Working Group Historical Records site (climate.umn.edu/doc/historical.htm) for Pipestone. Daily solar radiation, wind movement, and dew point temperatures were obtained from the GLEAMS model climate database.

Fertility for the corn production simulations consisted of applying urea with a nitrification inhibitor and incorporating to 15 cm 1 week before planting. The rate of N application varied with the crop rotation, and we included two rates for the continuous corn simulation to discern the potential impact of excessive fertilizer rates on nitrate leaching. No N fertilizer was applied to soybean or alfalfa, and we assumed the alfalfa was an established crop. Other essential nutrients were assumed to be optimal.

23.3.1.3 GIS Application

A soil map from the SSURGO database for Pipestone county was used to extract soil characteristics at a mapping scale of 1:24,000. Topology and attribute data were obtained from the USDA-NRCS Soil Data Mart site as SSURGO data in ArcView Shapefile format. These data are at a level of mapping designed for use by landowners and by township and county natural resource planners and managers.

The GLEAMS model output was extended to the individual polygons of the GIS coverages using ArcMap to produce the final map products, which depicted areas of high, moderate, and low nitrate leaching risks.

23.3.2 Results

Predicted nitrate leaching was much lower under alfalfa than under the corn and soybean rotation or continuous corn with higher fertilizer N rates (Figure 23.3). It is evident from the maps that nitrate leaching was substantially higher when continuous corn received higher, rather than lower, fertilizer N rates, especially on the "leakier" soils. The model predicted similar yields for both of these N rates, as supported by current recommendations.[11]

A few fields in the DWSMA are irrigated, which reduces yield loss risks due to drought and often increases the crop yield potential. We ran GLEAMS for the entire DWSMA for irrigated situations, based on the "checkbook" method.[12] Predicted nitrate leaching increased under irrigation for all cropping and N rates (data not shown). This is a recognized risk of irrigation because the soil profile is maintained with more plant-available water and, therefore, has less remaining storage capacity for natural precipitation.

Water quality is improved most rapidly by altering land management within the 10-year time-of-travel area, rather than in other areas of the DWSMA, unless surface water provides a significant input to ground water.[13] The 10-year time-of-travel area usually is delineated using dye tracing and hydrologic modeling during wellhead protection development. In order to help the water utility focus incentives for

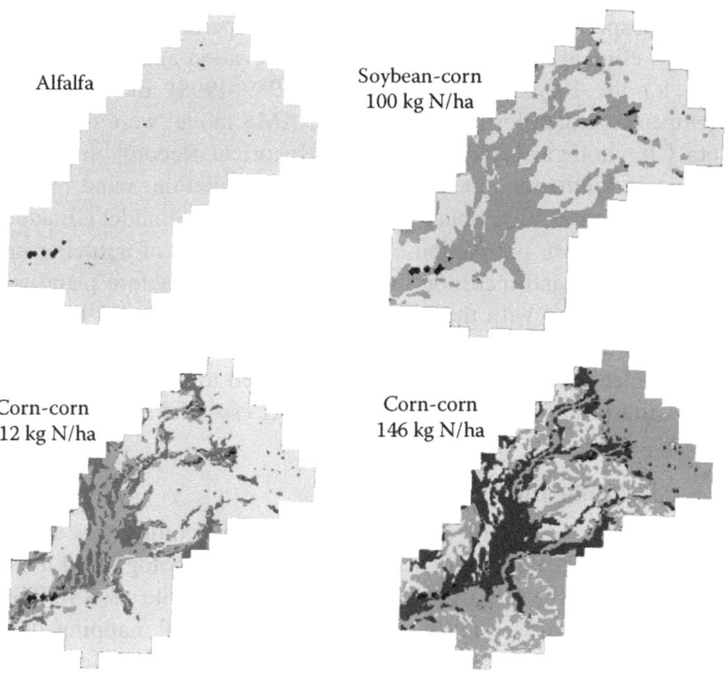

FIGURE 23.3 Effect of crop species and fertilizer N addition on estimated nitrate leaching in the Holland, Minnesota, DWSMA. The grayscale increments, increasing from light gray to black, are <2, 2–4, 4–9, 9–18, and >18 kg NO_3–N ha^{-1}; black circles indicate the wells at the wellhead.

changing cropping practices on these fields, we generated additional maps showing the relative decrease in nitrate leaching for areas within the 10-year time of travel (Figure 23.4). Some fields were characterized into a single category, whereas others were not. The water utility should focus on the fields in which the largest reductions could occur with improved fertilizer or manure management or rotation to a short-lived perennial, such as alfalfa. Conversion of the nearest of these fields will provide the fastest water quality improvement.

However, it is clear from Figure 23.4 that altering land management based on proximity alone is as likely to produce unsatisfactory reductions in nitrate leaching as basing the decision solely on either soil series (Figure 23.1) or soil texture (Figure 23.2). Other functional soil, crop, and management characteristics must be considered. As a case in point, the quarter section of land (about 65 ha) immediately north of the wellhead had been planted to native perennial grasses and forbs a few years before our team began this project. The water suppliers reasoned correctly that this conversion from annuals to perennials would reduce nitrate leaching. As is evident in

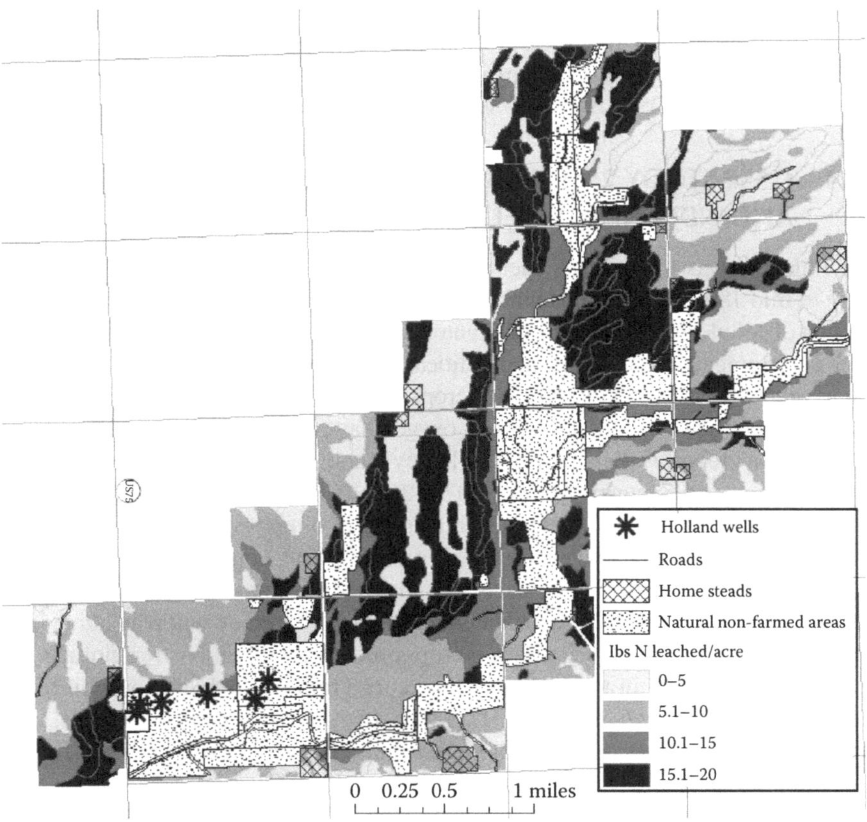

FIGURE 23.4 Reduction in nitrate–N loading estimated in GLEAMS by changing from continuous corn with annual applications of 180 kg N ha^{-1} to a well-adapted perennial crop that does not require N fertilizer.

Figure 23.4, however, nitrate leaching under corn was probably relatively small from these soils compared to other areas of the DWSMA.

Management of private land may be altered by encouraging voluntary changes (e.g., appealing to the manager's sense of public responsibility and providing research results that demonstrate good economic returns for a practice), supporting desired changes with payments (such as leases or per area payments for contracts) or in-kind contributions (seed, soil sampling, etc.), or regulation (such as penalties for excessive manure or fertilizer application rates). The option(s) selected will depend on many conditions we do not address here, such as legal authority, sociological considerations, and access to funds to incentivize improvements.

Sophistication and precision are sacrificed for the visualization of spatial patterns in many cases where GIS is employed, particularly when a rigorous process-based model is not used. But even when process-based models are used, a valid question is whether modeled predictions are reliable and accurate. One can use lookup tables, equations based on simplifying assumptions, or simulation models to estimate the amount of nitrate leaching that may occur. Although results from GLEAMS may not have been accurate for the annual crops, based on the dataset used in our validation process (described above), it is reasonable to expect that more reliable estimates will be provided by models that include crop management and weather variables along with soil characteristics. For the purposes of this project, the combination of simulation modeling and GIS was used to visualize nitrate leaching risks.

23.4 CASE STUDY 2: ESTIMATED BIOMASS FEEDSTOCK YIELD AND NET ENERGY DELIVERY IN A FUELSHED

Ethyl alcohol was the original fuel used by the internal combustion engines in the early to mid-1800s, and both Henry Ford and Charles Kettering promoted the development of cellulosic ethanol for use in automobiles.[14] However, this effort was stopped by the convergence of economic and political conditions offering cheap gasoline from crude oil.[14] And now, again, biomass crops as feedstocks for biofuels are catching fire—metaphorically, at least. Fuel and oxygen have been added by the DOE-USDA "Billion Ton" report,[15] the U.S. Energy Policy Act of 2005 (P.L. 109-58), the Energy Independence and Security Act of 2007 (P.L. 110-140), President Bush's "Biofuels Initiative," and President Obama's establishment of a top-level Biofuels Interagency Working Group, while review articles, editorials, and commentaries have added heat.[16,17] The extent to which herbaceous and woody biomass can provide cellulosic feedstock to replace fossil fuels is contentious, but the fact they will play a role is not.

Currently, only two principal grain crop–product combinations for biofuels have been commercialized in the United States (corn grain ethanol and soybean biodiesel), but sorghum (*Sorghum bicolor* [L.] Moench), wheat (*Triticum aestivum* L.), sugarcane (*Saccharum officinarum* L.) bagasse, and brewery waste are used in some of the 182 facilities operating in early 2010.[18] Of these, 21 were in Minnesota, with 3 using biomass as an additional source of heat and power.[19] Minnesota also hosted three biodiesel production facilities.[19]

Cellulosic plant materials may be converted with fermentation, pyrolysis, or other processes, but decisions about *what* crop to grow for cellulosic biomass will be guided by yield, cost of production, product requirements,[20] the potential to produce valuable coproducts (fats, oils, proteins), agricultural infrastructure, and agronomic capability. In the case of alfalfa, stems can be used as the biomass feedstock and leaves can provide additional income as a high-protein livestock feed.[21] Crop selection should also maximize ancillary environmental benefits and net energy production,[22,23] and alfalfa also fits these criteria.[24]

It is reasonable to expect the U.S. cellulosic biomass industry will be based on a variety of plant species and that facilities will be sited near feedstock production areas because of high transportation costs for this lower-density material. Because crop yields vary on different soils, highest efficiencies and lowest risk can be achieved by growing the selected crop on productive soils nearest the facility. Specific environmental goals (e.g., reduced nitrate leaching or lower risk of soil erosion) may be achieved most reliably by growing the appropriate crop on certain soils or landscape positions. To demonstrate this fuelshed planning approach, we analyzed a hypothetical, 80 km diameter area around Madelia, Minnesota, where a consortium of interests is developing plans for bio-based economic growth.[25,26]

23.4.1 Methods

The subject area comprised about 420,000 ha of cultivated land. A base map was generated with a soils layer, delineated using the SSURGO database, and coverages of roads, municipal areas, and rivers were added. Corn grain and stover (assumed to be 1:1 on a dry matter basis), soybean, and smooth brome grass (*Bromus inermis* L.)-alfalfa yields were provided in the SSURGO database by soil series and county, based on a simple model driven by location and soil series (J. Floren, 1992, internal document, USDA-NRCS). This model was developed with data from field-level crop yields and extended to soils with similar characteristics. In this model, yield potential declines according to a climatology factor that combined growing degree days and growing season precipitation.[27] The model estimated yields for a smooth brome–alfalfa mixture because that was a common practice at the time (latter half of the twentieth century). Currently, it is more common for alfalfa to be grown as a pure stand when it is grown for hay or haylage. In most fields, yields from pure alfalfa stands likely would be similar to the mixture, except on poorly drained soils where the alfalfa would not persist.

This simple approach to predicting yields was appropriate for our purpose and helped us develop maps quickly. Alternatively, the Web Soil Survey tool (websoilsurvey.nrcs.usda.gov) now includes a National Commodity Crop Productivity Index and/or estimated yields for several primary crops in some states. As demonstrated in the first case study, one also could select more detailed crop growth models that are validated for the crop species of interest.

We translated yield into higher heating value (HHV) energy content[28] as delivered to the facility and reduced this by subtracting the energy used in typical crop production inputs in the fuelshed (Table 23.2). Typical and desirable crop production practices were outlined along with best estimates of fuel requirements and amounts of products used. These inputs were assigned HHV for materials utilized (fuels, fertilizers, lime, seed, and pesticides).[29–34] Some of these inputs varied with

TABLE 23.2

Inputs for Production of Corn Grain and Stover, Soybean, and Alfalfa (see Excel Workbook in Appendix)

		Requirement by Crop			
Operation or Material Applied	**Input**	**Corn**	**Corn Stover**[b]	**Soybean**	**Alfalfa**[c]
Tillage and Field Operations					
Dry fertilizer application (Urea + P + K)	Diesel fuel (L ha^{-1})	1.4		1.4	0.7
Lime application	Diesel fuel (L ha^{-1})				**1.8**
Field cultivator	Diesel fuel (L ha^{-1})	3.0		3.0	**0.8**
Roller harrow	Diesel fuel (L ha^{-1})				**0.8**
Planting	Diesel fuel (L ha^{-1})	3.2		3.2	**0.8**
Herbicide application	Diesel fuel (L ha^{-1})	0.9		1.4	**0.2**
Insecticide application	Diesel fuel (L ha^{-1})				0.9
Cultivation	Diesel fuel (L ha^{-1})	4.1		1.6	
Combine grain	Diesel fuel (L ha^{-1})	18.0		18.0	
Stalk raking	Diesel fuel (L ha^{-1})		4.7		
Baling: Large round bales	Diesel fuel (L ha^{-1})		7.2		
Swathing	Diesel fuel (L ha^{-1})				7.8
Raking	Diesel fuel (L ha^{-1})				9.4
Baling: Large square bales	Diesel fuel (L ha^{-1})				6.6
Stalk shredding	Diesel fuel (L ha^{-1})	6.9			
Chisel plow	Diesel fuel (L ha^{-1})			5.6	
Combined disk and V ripper	Diesel fuel (L ha^{-1})	13.7			
Moldboard plow	Diesel fuel (L ha^{-1})				**12.0**
Pickup use (supplies, repairs, etc.)	Gasoline fuel (L ha^{-1})	9.3		9.3	9.3

Seed, Fertilizer and Chemicals Applied					
Seed	Seed (kg ha^{-1})	19.9		87.4	**4.2**
Limestone	Aglime (kg ha^{-1})				**840.0**
Nitrogen[a]	N (kg ha^{-1})	112.0	*11.0*		
Phosphate[a]	P_2O_5 (kg ha^{-1})	*56.0*	*25.3*	*48.8*	*55.9*
Potash[a]	K_2O (kg ha^{-1})	*76.8*	*43.0*	*56.0*	*224.0*
Herbicides	Product (kg ha^{-1})	2.23		1.2	**0.56**
Insecticides	Product (kg ha^{-1})	0.08			0.28
Post Harvest					
Transport of grain or biomass from field to farm	Diesel fuel (L [Mg-km]$^{-1}$)	0.0281	0.0281	0.0281	0.0281
Drying of wet grain	Propane (L Mg^{-1} per % water)	2.97			
Drying of wet grain	Electricity (kJ Mg^{-1} per % water)	1.41			
Transport of feedstock from farm to facility	Diesel fuel (L [Mg-km]$^{-1}$)	0.0178	0.0178	0.0178	0.0178

[a] For purposes of illustration, yield-dependent inputs (italics) are shown for assumed static yields of 11.3, 4.8, 3.4, and 9.0 Mg ha^{-1} for corn grain, corn stover, soybean, and alfalfa, respectively.

[b] Corn grain inputs also apply to stover, with additional inputs required for nutrient replacement and stover collection and transport.

[c] One-time inputs for alfalfa are amortized over the 4-year life of the stand (bold-face).

yield (e.g., diesel required to haul biomass), whereas others were constant per unit area (e.g., diesel use for tillage). See the Appendix for the Excel spreadsheets showing these calculations.

Two harvests per year were assumed for alfalfa biomass.[21] Minnesota guidelines were used for N fertilizer rate assuming corn followed soybean and the price ratio of N to corn grain was 0.15.[11] We assumed replacement of phosphorus (P) and potassium (K) based on the amount of removed biomass of all crops and that lime would be required to prevent soil pH decline during alfalfa production. In most states, P and K recommendations are made on the basis of soil test results rather than on replacement. Furthermore, the availability of added P and K varies with soil chemistry. Therefore, better site-specific estimates could be made by taking these considerations into account. Here, we assumed that soil test levels were optimum and used the University of Wisconsin recommendations to maintain those levels.[35]

Corn grown after alfalfa needs less fertilizer N and insecticide than corn grown after soybean, so these energy savings were assigned to alfalfa. We included only the variable energy inputs rather than the embedded energy in the farm equipment and trucks, which already are part of the food, feed, and fiber production systems. Because end products from each crop varies with technology, we did not include processing energy beyond that required to produce and deliver the crop at a moisture content suitable for storage. We assumed that corn grain and stover were harvested in separate operations and restricted removal of corn stover to 50% in order to sustain soil organic matter (SOM) levels.[36] This removal rate might not be sustainable in a corn–soybean rotation, however, because SOM declines during bean production. On the other hand, sustainable stover removal rates may be higher in rotation with alfalfa, which improves soil carbon storage.[37]

Mean distance between fields and a farmstead was assumed to be 6.4 km (based on unpublished research in southern Minnesota by the senior author). To account for typical rectilinear road patterns, transportation distance to the processing facility was calculated as twice the square root of 2 multiplied by the vector distance of the field to Madelia. To simplify this calculation, we estimated energy required to transport each crop from concentric rings of 1 mile (1.6 km) radius. Inputs were aggregated and expressed in the relevant units, i.e., per unit area, mass, or distance. After calculating total yield, net energy (delivered), and net energy value for each cultivated hectare, corresponding coverages were produced using ArcGIS. All associated shapefiles and ancillary data layers are provided in the Appendix.

23.4.2 Results

Energy input for corn production is considerably higher than for the legumes, soybean, or alfalfa, mainly because of N fertilizer requirements and the need to artificially dry corn grain most years (Figure 23.5). More than one-third of the energy input to corn receiving 112 kg N ha^{-1} in a corn–soybean rotation is due to the N fertilizer and nearly one-fifth of the input is due to grain drying. The figure for grain drying would vary widely with the weather conditions in autumn and the timing of harvest. Transportation is a small contributor to the net energy balance compared to the energy inputs of fertilizer, but was a larger factor for the cellulosic crops

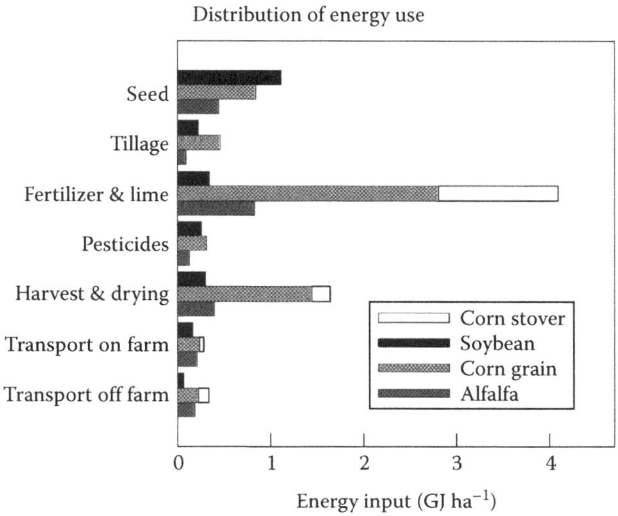

FIGURE 23.5 Energy use in different inputs for corn, soybean, and alfalfa production at assumed yields of 11.3, 3.4, and 9.0 Mg ha^{-1}, respectively.

(about 8% for alfalfa and 5% for the extra energy input to produce corn stover) than for grain (4% for corn and 3% for soybean). Note that the values in Figure 23.5 are static figures based on selected yields.

Unlike analyses that use static figures, such as those above and in most regional and national reports, our use of GIS provided products that showed how the energy required to produce and deliver biomass crops varied with soil type and distance to the biofuel manufacturing facility. Across the fuelshed, net energy yields of soybean ranged from 14 to 58 GJ ha^{-1}, corn grain from 33 to 206 GJ ha^{-1}, and alfalfa from 41 to 226 GJ ha^{-1}. Maps of the entire fuelshed are included in the Appendix, but the variability due to soils can be seen in an example subarea (Figure 23.6). The maps

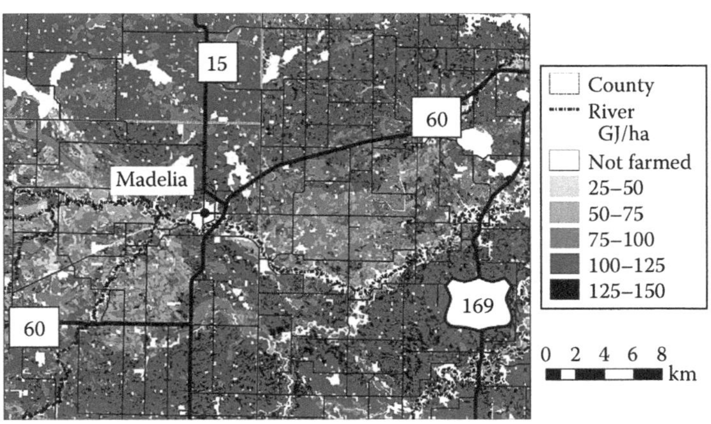

FIGURE 23.6 Spatial variation in net energy yield of corn grown on soils around Madelia, Minnesota.

TABLE 23.3
Examples of Energy Input and Output for Several Biomass Crops, Assuming a Fixed Yield and Distance (24 km) from a Processing Facility in Madelia, Minnesota

Crop (Yield)	Energy Input ($GJ\ ha^{-1}$)	Delivered Energy ($GJ\ ha^{-1}$)	Net Energy ($GJ\ ha^{-1}$)	Output-to-Input Energy Ratio
Soybean (3.36 $Mg\ ha^{-1}$)	5.4	54.3	48.9	9.0
Corn grain (11.3 $Mg\ ha^{-1}$)	15.6	138.0	122.4	7.8
Corn grain + stover (11.3 $Mg\ ha^{-1}$ and 4.8 $Mg\ ha^{-1}$ stover dry matter)	20.1	211.9	191.7	9.5
Alfalfa (13.4 $Mg\ ha^{-1}$)	6.6	192.1	185.4	27.9

illustrate that fields to the North and South of Madelia can be expected to produce higher net energy yields than those to the East and West. Further enlargements of net energy yield can assist in field-level decision making (examples for alfalfa and soybean are included in the Appendix).

Biofuel facility planners can minimize costs and maximize energy production by contracting with farmers on high-producing soils (Figure 23.6). Although the effect of transportation distance is small in terms of energy balance, it directly reduces the net yield of liquid fuel (not calculated here), so selecting farms based on productivity and proximity would help achieve the national goals expressed in several public laws.

The large amount of input energy needed for corn production concerns some authors.[22] In this fuelshed, alfalfa was about three times more efficient in producing energy per unit input than either corn or soybean, although gross and net energy production were similar for alfalfa and corn grain plus 50% stover (Table 23.3). This does not imply that alfalfa should be grown on all land parcels, but rather that its inclusion in crop rotations would improve net energy production in the fuelshed, in addition to providing many environmental benefits (e.g., better water, soil, and air quality, and improved wildlife habitat). Our analysis shows that alfalfa grown in riparian areas west of Madelia yield three to four times more net energy than soybean (Figure 23.7) while contributing to improved water quality.

Requirements for liquid fuels and natural gas vary among the crops, and these energy sources differ in their economic costs.[38] Economic costs of energy inputs were set at \$9.00 GJ^{-1} for liquid fuels, \$6.73 GJ^{-1} for natural gas, and \$22.75 GJ^{-1} for electricity. Using these prices, the net energy delivered per dollar of energy input ranged from 570 to 800 MJ $\$^{-1}$ for soybean, 380 to 1020 MJ $\$^{-1}$ for corn grain, and 790 to 2540 MJ $\$^{-1}$ for alfalfa. Fine-scale maps covering about 1550 ha are included in the Appendix to illustrate how this information might be used at the farm scale. However, these maps were not intended for within-field management decisions. The inherent variability within the soil series polygons (due to inclusions and indistinct borders) and the effects of past field management (erosion, manure application, etc.) limits the utility of these broad-scale maps to larger areas.

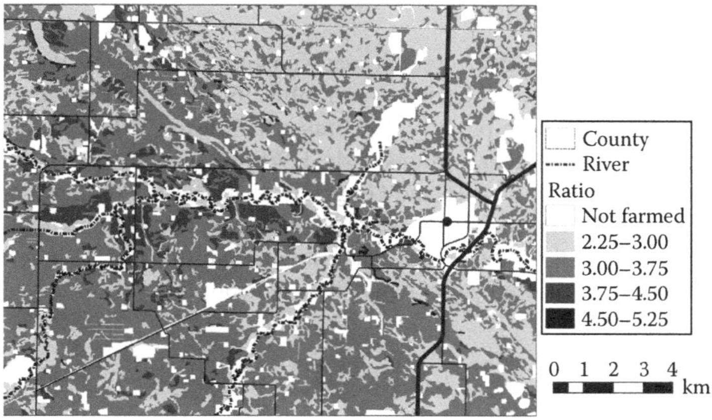

FIGURE 23.7 Relative net energy yield of alfalfa compared to soybean delivered to a hypothetical facility in Madelia, Minnesota.

As stated earlier, other considerations should be taken into account when deciding which biomass crop to select and where to grow it in the landscape. Data layers (digital elevation models, location of shallow aquifers, estimates of greenhouse gas emissions, etc.) can be added to these GIS-based analyses.[39,40]

For the grain crops and eventually for lignocellulosic crops, the land area required per megaliter of ethanol or biodiesel may be calculated from mapped yields,[41] allowing more complete life-cycle assessment of selected scenarios. Predicted net energy yields can be validated by long-term local records of crop yields, when available. Such maps and data layers may also help regional, state, and federal policy makers evaluate how they might encourage particular biomass production systems to achieve policy goals.

23.5 CONCLUSIONS

Soil information can be used to guide decision making when public water supply managers, biomass fuelshed operators, and others are selecting crop species, but only after the information is synthesized to produce interpretable outputs. Soil units are defined by a number of characteristics that do not necessarily relate to the problem of interest. The importance of particular soil characteristics, however, depends on the crop species and management, which are taken into account in simulation models. GIS allows us to aggregate the model output into categories, making the map product easy to interpret. In the DWSMA, specific fields were highlighted to help the managers prioritize their prevention activities and note which fields are not likely contributing to the problem of excess ground water nitrate. In the prospective biomass fuelshed, planners could visualize which areas of the fuelshed could supply feedstock most efficiently, assess how feedstock selection will affect net energy production, and consider areas where additional environmental payments may be available to improve farm profitability.

ACKNOWLEDGMENTS

The wellhead protection area research was supported in part by the Legislative Commission on Minnesota Resources (#L3916016101). In-house funding from USDA-ARS supported both projects. Dr. Prasanna Gowda, Kyle J. Symoniak, and Kyle Kvien assisted with other portions of the DWSMA mapping. We appreciate the editorial reviews by Ryan Maher and Ginger Walker. Results of these two studies have been published previously only as abstracts of technical presentations. Mention of specific commercial products was made for the purposes of completeness and does not imply endorsement by the organizations involved in this research.

APPENDIX

Primary Folder	Contents
Holland_DWSMA	Contains each of the following folders and files required for data generation and mapping as described in this chapter.
Subfolder	
Color figures	Color versions of Figures 23.1 through 23.4.
Holland_DWSMA_ GISfiles	
Subfolder/file	
DOQ_color	Color digital orthoquads of the DWSMA area.
DWSMA	Arc shapefiles for the DWSMA.
Excel_files	MS Excel workbooks containing the GLEAMS model output for the different crop–soil management scenarios used in the study.
Fields	Arc shapefiles for field boundaries, natural areas, and nonfarmed areas.
GLEAMS	Simulation program files, including the modified version we used in the research, and the required input files, including the 10 year weather file (Pip10pcp.dat), crop-specific management parameter input files, by crop, soil type, and irrigation (present or absent), and in the case of corn, N fertilizer timing and rate. These files may be read and modified using a text editor, such as Notepad. The .par files that end in "e" include erosion parameters, "h" are hydrology parameters, and "n" are nutrient parameters. Output files also are included for each scenario. In order to understand these files, users must have some familiarity with the file structure required and produced by GLEAMS; documentation is provided within the GLEAMS_Program folder.

mn117_pipestone_mn	Arc shapefiles for the NRCS digital soil survey SSURGO soils for Pipestone COUNTY, Minnesota.
N_Leaching	Arc shapefiles for nitrate–N leaching losses as predicted by GLEAMS. The three shapefile groups are 1_N_leach (<2 lb/ac), 4_N_Leach (2–4 lb/ac), and 8_N_Leach (8–16 lb/ac).
Soildata	Arc shapefiles for soil polygons and for soil attributes joined to GLEAMS output for corn and alfalfa and the nitrate–N reductions achieved by converting from corn to alfalfa.
Support	Arc shapefiles for Minnesota counties, DWSMA wells, roads, and the area.

HollandDWSMA_crop_modeling.mxd
The Arc map document file that contains data frames for:

1. Holland Wellfield Landcover—location of Pipestone COUNTY, Minnesota; the DWSMA; farm fields; unfarmed areas; roads; wells.
2. Holland Landcover Imagery—Farm Services Administration Color Orthophotos 2003–2004; the DWSMA; roads; wells.
3. Holland Wellfield Soils—NRCS digital soil survey (SSURGO soils) for Pipestone COUNTY, Minnesota; the DWSMA; farm fields; wells.
4. N Loading Reductions with Management—SSURGO soils joined to GLEAMS model output; the DWSMA; farm fields; unfarmed areas; roads; wells.

Madelia_fuelshed	Contains each of the following folders and files required for data generation and mapping as described in this chapter.
Subfolder/file	
Color figures	Color versions of Figures 23.5 through 23.7, and additional figures of the entire fuelshed, by crop, of (1) net energy at the farm (without transport to Madelia), (2) net energy with transport delivered to Madelia, and (3) total energy delivered per dollar cost of energy invested.
Net energy calculations.xlsx	Annotated Excel workbook with separate worksheets for each crop to calculate energy input, energy output, and net energy production at a defined yield, moisture content, and hauling distance. Typical U.S. units are given for most operations.

Madelia_fuelshed_GISfiles

All GIS files are in the following coordinate system:
UTM
Zone 15
Nad83
Meters

Alfalfa_shapefiles

Layer: alfalfa_energy2010.shp

Overview: The spatial boundaries of this layer are comprised of SSURGO soil map units, Minnesota 1990s census of the land use boundaries, and 1 mile (1.6 km) concentric circles surrounding the city of Madelia. The following attributes are the resulting energy calculations associated with the alfalfa yields derived from the SSURGO data. Example values are presented in parentheses.

Field	Description
AREA	Polygon area in m^2 (1,127.64550420000).
PERIMETER	Polygon perimeter m (280.32763887900).
MALFYLDLU_	Polygon id (2).
MUSYM	SSURGO map unit symbol (386).
MUKEY	SSURGO map unit key (400,598).
MUSYMNAME	SSURGO map unit symbol name (OKOBOJI M).
BROMALFHUN	Bromegrass–alfalfa yield units normalized by acres (Tons).
BROMALFHYL	Bromegrass–alfalfa yield (3.00000000000).
ENERGYHAY	Harvested energy content of alfalfa in kcal/ac = BROMALFHYL × 3,551,020 (10,653,060.00000000000).
LUSE_CODE	Minnesota 1990s census of the land use code (21 = cultivated land).
DISTANCE	Concentric circle distance in m (40,225.00000000000).
ALFOUTENG	Output energy from alfalfa in kcal/ac = 12,481,200 BTU/ton/ac × BROMALFAHYL × 0.252 (9,435,787.19999999000).
ALFIN1ENG	Input energy associated with field operations and seed in kcal/ac = 1,198,466 BTU/ac × 0.252 (302,013.43200000000).
ALFIN2ENG	Input energy associated with nutrients and transportation in kcal/ac = 195,276 BTU/ton/ac × BROMALFHYL × 0.252 (147,628.65599999900).
ALFIN3ENG	Input energy associated with transportation to the plant in kcal/ac = 2,842 BTU/ton/mile × BROMALFHYL × (DISTANCE/1609 m/mile) (53,713.80000000000).
ALFNETENG	Net energy after accounting for transport energy associated with transport to the plant in kcal/ac = (ALFOUTENG + ALFOUT2ENG) − (ALFIN1ENG + ALFIN2ENG + ALFIN3ENG) (9,771,334.52399999000).

ALFNETENG2 Net energy without transport energy associated with transport to the plant in kcal/ac = (ALFOUTENG + ALFOUT2ENG) − (ALFIN1ENG + ALFIN2ENG) (13,142,617.48000000000).

ALFYLDMET BROMALFHYL expressed in Mg/ha = BROMALFHYL × 2.24 (6.72000000000).

ALFNETMET ALFNETENG expressed in GJ/ha = ALFNETENG × 1.0341396 × 10^{-5} (101.04923976116).

ALFNETMET2 ALFNETENG2 expressed in GJ/ha = ALFNETENG2 × 1.0341396 × 10^{-5} (135.91301183720).

ALFNETMETJ Total net energy for each concentric circle in TJ = (ALFNETMET × (AREA/10,000))/1,000 (0.01139477209).

ALFOUT2ENG Energy associated with nitrogen fertilizer benefit to next two corn crops and insecticide benefit to next corn crop in kcal/ac = 3,328,981 BTU/ac × 0.252 (838,903.21200000000).

ALFIN1DOL Input costs associated with field operations and seed in $/ac = $14.52810000000.

ALFIN2DOL Input costs associated with nutrients and transportation in $/ac = $2.5567 × BROMALFHYL (8.57010000000).

ALFIN3DOL Input costs associated with transportation to the plant in $/ac = $0.0270 × BROMALFHYL × (DISTANCE/1,609 m/mile) (2.02500000000).

ALFINDOL Input costs after accounting for transport energy associated with transport to the plant in $/ac = (ALFIN1DOL + ALFIN2DOL + ALFIN3DOL) (25.12320000000).

ALFINDOLM ALFINDOL expressed as $/ha = ALFINDOL × 2.47 (62.05430400000).

ALFMJDOL Output energy delivered per value total energy input in MJ/$ = (ALFOUTENG × 0.01033448)/ALFINDOLM (1,571.42934199465).

ALFOUTMETJ Total harvested energy for each concentric circle in TJ = (ALFOUTMEN × (AREA/10,000))/1,000 (0.01100347597).

ALFINDOLMT Total value of input costs for each concentric circle in $ = (ALFINDOLM × (AREA/10,000))/1,000 (0.00699752569).

ALFOUTMET ALFOUTENG expressed as GJ/ha = ALFOUTENG × 1.0341396 × 10^{-5} (97.57921200693).

Corn_shapefiles

Layer: corn_energy2010.shp

Overview: The spatial boundaries of this layer are comprised of SSURGO soil map units, Minnesota 1990s census of the land use boundaries, and 1 mile (1.6 km) concentric circles surrounding the city of Madelia. The following attributes are the resulting energy calculations associated with the corn yields derived from the SSURGO data. Example values are presented in parentheses.

Field	**Description**
AREA	Polygon area in m^2 (1127.64550420000).
PERIMETER	Polygon perimeter m (280.32763887900).
MCORYLDLU_	Polygon id (2).
MUSYM	SSURGO map unit symbol (386).
MUKEY	SSURGO map unit key (400598).
MUSYMNAME	SSURGO map unit symbol name (OKOBOJI M).
CORNUNIT	Corn yield units normalized by acres (bu).
CORNYLD	Corn grain yield (136.00000000000).
ENERGYCORN	Harvested energy content of corn grain + stover in kcal/ac = CORNYLD × 190,996.8 (25,975,565.00000000000).
ENERGYCRNG	Harvested energy content of corn grain in kcal/ac = ENERGYCORN – ENERGYCRNS (12,841,854.00000000000).
ENERGYCRNS	Harvested energy content of corn stover in kcal/ac = CORNYLD × 96,571.4 (13,133,714.00000000000).
LUSE_CODE	Minnesota 1990s census of the land use code (21 = cultivated land).
DISTANCE	Concentric circle distance in m (40,225.00000000000).
CGSOUTENG	Output energy from corn grain and stover in kcal/ac = 494,612 BTU/bu/ac × CORNYLD × 0.252 (16,951,342.46400000000).
CGSIN1ENG	Input energy for corn grain and stover associated with field operations, seed, fertilizer, pesticides in kcal/ac = 4,409,879 BTU/ac × 0.252 (1,111,289.50799999000).
CGSIN2ENG	Input energy for corn grain and stover associated with drying, nutrients, and transportation to the farm in kcal/ac = 15,926 BTU/bu/ac × CORNYLD × 0.252 (545,815.87199999900).
CGSIN3ENG	Input energy for corn grain and stover associated with transportation to the plant in kcal/ac = 147 BTU/bu/mile/ac × CORNYLD × (DISTANCE/1609 m/mile) (856.80000000000).
CGSNETENG	Net energy for corn grain and stover after accounting for transport energy associated with transport to the plant in kcal/ac = (CGSOUTENG) – (CGSIN1ENG + CGSIN2ENG + CGSIN3ENG) (15,293,380.28400000000).
CGOUTENG	Output energy from corn grain in kcal/ac = 327,644 BTU/bu/ac × CORNYLD × 0.252 (11,229,015.16800000000).
CGIN1ENG	Input energy for corn grain associated with field operations, seed, fertilizer, pesticides in kcal/ac = 4,223,189 BTU/ac × 0.252 (1,064,243.62800000000).

CGIN2ENG	Input energy for corn grain associated with drying, nutrients, and transportation to the farm in kcal/acre = 8664 BTU/bu/acre × CORNYLD × 0.252 (296,932.60800000000).
CGIN3ENG	Input energy for corn grain associated with transportation to the plant in kcal/acre = 80 BTU/bu/mile/acre × CORNYLD × (DISTANCE/1609 m/mile) (68,544.00000000000).
CGNETENG	Net energy for corn grain after accounting for transport energy associated with transport to the plant in kcal/acre = (CGOUTENG) − (CGIN1ENG + CGIN2ENG + CGIN3ENG) (9,799,294.93200000000).
CGSNETENG2	Net energy for corn grain and stover without accounting for transport energy associated with transport to the plant in kcal/acre = (CGSOUTENG) − (CGSIN1ENG + CGSIN2ENG) (17,842,035.96000000000).
CGNETENG2	Net energy for corn grain without accounting for transport energy associated with transport to the plant in kcal/acre = (CGOUTENG) − (CGIN1ENG + CGIN2ENG) (11,844,721.98000000000).
CSNETENG	Net energy for corn stover after accounting for transport energy associated with transport to the plant in kcal/acre = CGSNETENG − CGNETENG (5,494,085.35200000000).
CSNETENG2	Net energy for corn stover without accounting for transport energy associated with transport to the plant in kcal/acre = CGSNETENG2 − CGNETENG2 (5,997,313.98000000000).
CORNYLDMET	CORNYLD expressed in Mg/ha = CORNYLD × 6.273 × 10^{-2} (8.53128000000).
CGSNETMET	CGSNETENG expressed in GJ/ha = CGSNETENG × 1.0341396 × 10^{-5} (158.15490169544).
CGSNETMET2	CGSNETENG2 expressed in GJ/ha = CGSNETENG2 × 1.0341396 × 10^{-5} (184.51155930860).
CGNETMET	CGNETENG expressed in GJ/ha = CGNETENG × 1.0341396 × 10^{-5} (101.33838941261).
CGNETMET2	CGSNETENG2 expressed in GJ/ha = CGSNETENG2 × 1.0341396 × 10^{-5} (122.49096050508).
CSNETMET	CSNETENG expressed in GJ/ha = CSNETENG × 1.0341396 × 10^{-5} (56.81651228283).
CSNETMET2	CSNETENG2 expressed in GJ/ha = CSNETENG2 × 1.0341396 × 10^{-5} (62.02059880352).
CGSNETMETJ	Total net energy for each concentric circle in TJ = (CGSNETMET × (AREA/10,000))/1,000 (0.01783426639).
CGNETMETJ	Total net energy for each concentric circle in TJ = (CGNETMET × (AREA/10,000))/1,000 (0.01142737792).

CSNETMETJ	Total net energy for each concentric circle in TJ = (CSNETMET × (AREA/10,000))/1,000 (0.00640688846).
CGSIN1DOL	Input costs associated with seed, nitrogen fertilizer, and pesticides in \$/acre = 44.61910000000.
CGSIN2DOL	Input costs associated with nutrients and transportation in \$/acre = \$0.1471 × CORNYLD (20.00560000000).
CGSIN3DOL	Input costs associated with transportation to the plant in \$/acre = \$0.0014 × CORNYLD × (DISTANCE/1,609 m/mile) (4.76000000000).
CGSINDOL	Input costs after accounting for transport energy associated with transport to the plant in \$/acre = (CGSIN1DOL + CGSIN2DOL + CGSIN3DOL) (69.38470000000).
CGIN1DOL	Input costs associated with seed, nitrogen fertilizer, and pesticides in \$/acre = 42.84550000000.
CGIN2DOL	Input costs associated with nutrients and transportation in \$/acre = \$0.1471 × CORNYLD (11.12480000000).
CGIN3DOL	Input costs associated with transportation to the plant in \$/acre = \$0.0014 × CORNYLD × (DISTANCE/1,609 m/mile) (2.72000000000).
CGINDOL	Input costs after accounting for transport energy associated with transport to the plant in \$/acre = (CGIN1DOL + CGIN2DOL + CGIN3DOL) (56.69030000000).
CSINDOL	Input costs after accounting for transport energy associated with transport to the plant in \$/acre = (CGSINDOL – CGINDOL) (12.69440000000).
CGINDOLM	CGINDOL expressed as \$/ha = CGINDOL × 2.47 (140.02504100000).
CGSINDOLM	CGSINDOL expressed as \$/ha = CGSINDOL × 2.47 (171.38020900000).
CSINDOLM	CSINDOL expressed as \$/ha = CSINDOL × 2.47 (31.35516800000).
CGSMJDOL	Output energy delivered per value total energy input in MJ/\$ = (CGSOUTENG × 0.01033448)/CGSINDOLM (1,022.19101429243).
CGMJDOL	Output energy delivered per value total energy input in MJ/\$ = (CGOUTENG × 0.01033448)/CGINDOLM (828.75199924700).
CGSOUTMETJ	Total harvested energy for each concentric circle in TJ = (CGSOUTMET × AREA/10,000))/1,000 (0.01976768716).
CGOUTMETJ	Total harvested energy for each concentric circle in TJ = (CGOUTMET × AREA/10,000))/1,000 (0.01309463598).
CGSINDOLMT	Total value of input costs for each concentric circle in \$ = (CGSINDOLM × (AREA/10,000))/1,000 (0.01932561222).

CGINDOLMT	Total value of input costs for each concentric circle in \$ = (CGINDOLM × (AREA/10,000))/1,000 (0.01578986080).
CGSOUTMET	CGSOUTENG expressed as GJ/ha = CGSOUTENG × 1.0341396 × 10^{-5} (175.30054515184).
CGOUTMET	CGOUTENG expressed as GJ/ha = CGOUTENG × 1.0341396 × 10^{-5} (116.12369254229).
CSOUTMET	CSOUTENG expressed as GJ/ha = CSOUTENG × 1.0341396 × 10^{-5} (59.17685260955).

Soybean_shapefiles

Layer: soybean_energy2010.shp

Overview: The spatial boundaries of this layer are comprised of SSURGO soil map units, Minnesota 1990s census of the land use boundaries, and 1 mile (1.6 km) concentric circles surrounding the city of Madelia. The following attributes are the resulting energy calculations associated with the soybean yields derived from the SSURGO data. Example values are presented in parentheses.

Field	**Description**
AREA	Polygon area in m^2 (1,127.64550420000).
PERIMETER	Polygon perimeter m (280.32763887900).
MSOYYLDLU_	Polygon id (2).
MUSYM	SSURGO map unit symbol (386).
MUKEY	SSURGO map unit key (400,598).
MUSYMNAME	SSURGO map unit symbol name (OKOBOJI M).
SOYUNIT	Soybean yield units normalized by acres (bu).
SOYYLD	Soybean grain yield (38.00000000000).
ENERGYSOY	Harvested energy of soybeans in kcal/acre = SOYYLD × 130,204.1 (4,947,756.00000000000).
LUSE_CODE	Minnesota 1990s census of the land use code (21 = cultivated land).
DISTANCE	Concentric circle distance in m (40,225.00000000000).
SBOUTENG	Output energy from soybean grain in kcal/acre = 458,473 BTU/bu/acre × SOYYLD × 0.252 (4,390,337.44799999000).
SBIN1ENG	Input energy for soybean grain associated with field operations, seed, and pesticides in kcal/acre = 1,941,560 BTU/acre × 0.252 (489,273.12000000000).
SBIN2ENG	Input energy for soybean grain associated with drying, nutrients, and transportation to the farm in kcal/acre = 6,539 BTU/bu/acre × SOYYLD × 0.252 (62,617.46400000000).
SBIN3ENG	Input energy for soybean grain associated with transportation to the plant in kcal/acre = 83 BTU/bu/mile/acre × SOYYLD × (DISTANCE/1609 m/mile) (19,870.20000000000).

SBNETENG	Net energy for soybean grain after accounting for transport energy associated with transport to the plant in kcal/acre = (SBOUTENG) − (SBIN1ENG + SBIN2ENG + SBIN3ENG) (3,818,576.66399999000).
SBNETENG2	Net energy for soybean grain without accounting for transport energy associated with transport to the plant in kcal/acre = (SBOUTENG) − (SBIN1ENG + SBIN2ENG) (4,605,517.47599999000).
SOYYLDMET	SOYYLD expressed in Mg/ha = SOYYLD × 6.7211 × 10^{-2} (2.55401800000).
SBNETMET	SBNETENG expressed in GJ/ha = SBNETENG × 1.0341396 × 10^{-5} (39.48941343878).
SBNETMET2	SBNETENG2 expressed in GJ/ha = SBNETENG2 × 1.0341396 × 10^{-5} (47.62748000424).
SBNETMETJ	Total net energy for each concentric circle in TJ = (SBNETMET × (AREA/10,000))/1,000 (0.00445300595).
SBIN1DOL	Input costs associated with field operations, seed, and pesticides in $/acre = 26.83770000000.
SBIN2DOL	Input costs associated with nutrients and transportation in $/acre = $0.0973 × SOYYLD (3.69740000000).
SBIN3DOL	Input costs associated with transportation to the plant in $/acre = $0.0008 × SOYYLD × (DISTANCE/1609 m/mile) (0.76000000000).
SBINDOL	Input costs after accounting for transport energy associated with transport to the plant in $/acre = (SBIN1DOL + SBIN2DOL + SBIN3DOL) (31.29510000000).
SBINDOLM	SBINDOL expressed as $/ha = SBINDOL × 2.47 (77.29889700000).
SBMJDOL	Output energy delivered per value total energy input in MJ/$ = (SBOUTENG × 0.01033448)/SBINDOLM (586.96639034328).
SBOUTMETJ	Total harvested energy for each concentric circle in TJ = (SBOUTMET × AREA/10,000))/1,000 (0.00511976071).
SBINDOLMT	Total value of input costs for each concentric circle in $ = (SBINDOLM × (AREA/10,000))/1,000 (0.00871657537).
SBOUTMET	SBOUTENG expressed as GJ/ha = SBOUTENG × 1.0341396 × 10^{-5} (45.40221812340).

Net_energy_ratio_shapefiles

Layer: ratio_energy2010.shp

Overview: The spatial boundaries of this layer are comprised of SSURGO soil map units, Minnesota 1990s census of the land use boundaries, and 1 mile (1.6 km) concentric circles surrounding the city of Madelia. Example values are presented in parentheses.

Field	Description
AREA	Polygon area in m^2 (1,127.64550420000).
PERIMETER	Polygon perimeter m (280.32763887900).
DISTANCE	Distance in m (40,225.00000000000).
ALFCGSGJ	Ratio of net energy from alfalfa to net energy from corn grain and stover = ALFNETMENT/CGSNETMET (0.639).
ALFCGGJ	Ratio of net energy from alfalfa to net energy from corn grain = ALFNETMENT/CGNETMET (0.997).
ALFSBGJ	Ratio of net energy from alfalfa to net energy from soybean grain = ALFNETMENT/SBNETMET (2.559).
ALFCGSMD	Ratio of alfalfa energy output delivered per value energy input to corn grain and stover energy output delivered per value energy input (1.537).
ALFCGMD	Ratio of alfalfa energy output delivered per value energy input to corn grain energy output delivered per value energy input (1.896).
ALFSBMD	Ratio of alfalfa energy output delivered per value energy input to soybean grain and energy output delivered per value energy input (2.677).

Madelia_base_shapefiles

Layer	Description
madelhwy2 (route: coverage)	Major roads and highways within the Madelia fuelshed.
madeliabfcty.shp	Minnesota county boundaries within the Madelia fuelshed.
madeliabfwat.shp	Eight-digit HUC boundaries within the Madelia fuelshed.

REFERENCES

1. Kaspar, T.C., Pulido, D.J., Fenton, T.E., Colvin, T.S., Karlen, D.L., Jaynes, D.B., and Meek, D.W. Relationship of corn and soybean yield to soil and terrain properties. *Agron. J.* 96, 700, 2004.
2. Putnam, D., Russelle, M., Orloff, S., Kuhn, J., Fitzhugh, L., Godfrey, L., Kiess, A., and Long, R. *Alfalfa, Wildlife and the Environment: The Importance and Benefits of Alfalfa in the 21st Century*. California Alfalfa and Forage Association, Novato, CA, 2001. Document available at alfalfa.ucdavis.edu/-files/pdf/Alf_Wild_Env_BrochureFINAL.pdf
3. Randall, G.W. and Sawyer, J.E. Nitrogen application timing, forms, and additives, In *Final Report: Gulf Hypoxia and Local Water Quality Concerns Workshop*. September 26–28, 2005. Ames, IA. American Society of Agricultural and Biological Engineers, St. Joseph, MI, 2008, p. 73. Document available at asae.frymulti.com/azdez.asp?JID=6&AID=24245&CID=ghlw2008&T=2

4. Sawyer, J.E. and Randall, G.W. Nitrogen rates, In *Final Report: Gulf Hypoxia and Local Water Quality Concerns Workshop*. September 26–28, 2005. Ames, IA. American Society of Agricultural and Biological Engineers, St. Joseph, MI, 2008, pp. 59–71. Document available at asae.frymulti.com/azdez.asp?JID=6&AID=24244&CID=ghlw2008&T=2
5. Bergström, L. Nitrate leaching and drainage from annual and perennial crops in tile-drained plots and lysimeters. *J. Environ. Qual.* 16, 11, 1987.
6. Burkart, M.R. and James, D.E. Agricultural-nitrogen contributions to hypoxia in the Gulf of Mexico. *J. Environ. Qual.* 28, 850, 1999.
7. Knisel, W.G., ed. *GLEAMS Manual, Version 2.10*. UGA-CPES-BAED Publication 5. University of Georgia, Coastal Plain Experimental Station, Tifton, GA, 1993. Documentation for the current version (v. 3) available at www.tifton.uga.edu/sewrl/Gleams/gleams.htm
8. Chung, S.W., Gassman, P.W., Huggins, D.R., and Randall, G.W. EPIC tile flow and nitrate loss predictions for three Minnesota cropping systems. *J. Environ. Qual.* 30, 822, 2001.
9. Van Mullem, J.A., Woodward, D.E., Hawkins, R.H., and Hjelmfelt, A.T. Runoff curve number method: Beyond the handbook, In *Hydrologic Modeling for the 21st Century. Second Federal Interagency Hydrologic Modeling Conference*, Las Vegas, July 28 to August 1, 2002.
10. Bruening, D. *Pipestone County Holland Wellfield Survey*. Minnesota Department of Agriculture, 2002. Document available at www.mda.state.mn.us/sitecore/content/Global/MDADocs/protecting/soilprotection/fanmapholland.aspx
11. Rehm, G., Randall, G., Lamb, J., and Eliason, R. *Fertilizing Corn in Minnesota*. University of Minnesota Extension, Publication FO-03790, 2006. Document available at www.extension.umn.edu/distribution/cropsystems/DC3790.html
12. Wright, J. *Irrigation Scheduling: Checkbook Method*. University of Minnesota Extension FO-01322, 2002. Document and worksheets available at mn4h.com/distribution/cropsystems/components/DC1322_01.html
13. Trojan, M.D. and Stockinger, J.M. Estimating surface water contribution to ground water recharge in the Verdi Aquifer, Southwest Minnesota, *Environmental Bulletin*, Minnesota Pollution Control Agency, 2004. Abstract available at www.pca.state.mn.us/index.php/view-document.html?gid=11454
14. Kovarick, B., Henry Ford, Charles Kettering and the fuel of the future, Automotive History Review. 32, 7, 1998. Document available at www.radford.edu/wkovarik/papers/fuel.html
15. Perlack, R.D., Wright, L.L., Turhollow, A.F., Graham, R.L., Stokes, R.J., and Erbach, D.C. *Biomass as Feedstock for a Bioenergy and Bioproducts Industry: The Technical Feasibility of a Billion-Ton Annual Supply*. Oak Ridge National Laboratory, Oak Ridge, TN, 2005. Document available at feedstockreview.ornl.gov/pdf/billion_ton_vision.pdf
16. Melillo, J.M., Reilly, J.M., Kicklighter, D.W., Gurgel, A.C., Cronin, T.W., Paltsev, S., Felzer, B.S., Wang, X., Sokolov, A.P., and Schlosser, C.A. Indirect emissions from biofuels: How important? *Science* 326, 1397, 2009.
17. Patzek, T.W. Can the Earth deliver the biomass-for-fuel we demand? In: Pimentel, D. ed. *Biofuels, Solar and Wind as Renewable Energy Systems*. Springer-Science+Business Media B.V. vol. 19, 2008.
18. Renewable Fuels Association. Biorefinery locations. Accessed March 28, 2010. Online at www.ethanolrfa.org/biorefinery-locations/
19. Minnesota Pollution Control Agency. *Alternative Fuels*, 2008. Online resource at www.pca.state.mn.us/energy/fuels/

20. U.S. Department of Energy. *Alternative & Advanced Fuels.* Alternative Fuels & Advanced Vehicles Data Center, 2009. Online resource at www.afdc.energy.gov/afdc/fuels/index.html
21. Lamb, J.F.S., Sheaffer, C.C., and Samac, D.A. Population density and harvest maturity on leaf and stem yield in alfalfa. *Agron. J.* 95, 635, 2003.
22. Hill, J., Nelson, E., Tilman, D., Polasky, S., and Tiffany, D. Environmental, economic, and energetic costs and benefits of biodiesel and ethanol biofuels. *Proceedings of the National Academy of Sciences.* 103, 11206, 2006.
23. Wang, M. *Ethanol—The Complete Energy Life Cycle.* Argonne National Laboratory, 2007. Document available at www.transportation.anl.gov/pdfs/TA/345.pdf
24. Russelle, M., Jung, H., and Lamb, J. Alfalfa biomass: Energy benefits of alfalfa vs. corn. *Minnesota Crop eNews.* University of Minnesota Extension, 2007. Document available at www.extension.umn.edu/cropEnews/2007/pdf/07MNCN16.pdf
25. Russelle, M.P., Birr, A.S., and Tiffany, D.G. Estimated net energy yields in a biomass fuelshed [abstract]. ASA-CSSA-SSSA Annual Meeting Abstracts. CD-ROM. Paper No. 167–172, 2006.
26. Meschke, L., ed. *Madelia Bio-Based Eco-Industrial Assessment.* Rural Advantage, Fairmont, MN, 2007. Document available at ruraladvantage.org/pdf/MadeliaReport.pdf
27. Anderson, J.L., Robert, P.C., and Rust, R.H. *Productivity Factors and Crop Equivalency Ratings for Soils of Minnesota.* Minnesota Extension Service. AG-BU-2199-F, 1992. See update at www.mngeo.state.mn.us/chouse/cer.html
28. Kim, S. and Dale, B.E. Cumulative energy and global warming impact from the production of biomass and biobased products. *J. Ind. Ecol.* 7,147, 2004.
29. AURI Fuels Initiative. *Agricultural Renewable Solid Fuels Data,* Agricultural Utilization Research Institute. (not dated). Document available at www.auri.org/research/fuels.pdf
30. Mudahar, M.S. and Hignett, T.P. *Energy and Fertilizer: Policy Implications and Options for Developing Countries.* International Fertilizer Development Center, Muscle Shoals, AL. Technical Bulletin, T-20, 1982.
31. Graboski, M.S. *Fossil Energy Use in the Manufacture of Ethanol.* National Corn Growers Association, 2002. Document available at www.oregon.gov/ENERGY/RENEW/Biomass/docs/FORUM/FossilEnergyUse.pdf
32. Wilke, B. *Energy Costs for Corn Drying and Cooling,* University of Minnesota Extension Service and University of Wisconsin-Extension, 2004. Document available at www.extension.unn.edu/cropenews/2008/pdfs/08MNCN27.pdf
33. Hanna, M. *Fuel Required for Field Operations.* Iowa State University Extension Service, Ames, IA, 2005. Document available at www.extension.iastate.edu/agdm/crops/html/a3-27.html
34. Lazarus, W. and Selley, R. *Farm Machinery Economic Cost Estimates for Late 2005,* University of Minnesota Extension Service, 2005. Updated document available at www.extension.umn.edu/distribution/businessmanagement/DF6696.pdf
35. Laboski, C.A., Peters, J.B., and Bundy, L.G. *Nutrient Application Guidelines for Field, Vegetable, and Fruit Crops in Wisconsin,* University of Wisconsin Extension, Madison, WI. A2809, 2006.
36. Wilhelm, W.W., Johnson, J.M.F., Karlen, D.L., and Lightle, D.T. Corn stover to sustain soil organic carbon further constrains biomass supply. *Agron. J.* 99, 1665, 2007.
37. Angers, D.A. Changes in soil aggregation and organic carbon under corn and alfalfa. *Soil Sci. Soc. Am. J.* 56, 1244, 1992.
38. Cleveland, C.J. Net energy from the extraction of oil and gas in the United States. *Energy.* 30, 769, 2005.

39. Burkart, M.R., James, D.E., Liebman, M.Z., and Herndl, C. Impacts of integrated crop-livestock systems on nitrogen dynamics and soil erosion in western Iowa watersheds. *J. Geophys. Res.-Biogeosci.* 110, G1, G01009, 10.1029/2004JG000008, 2005.
40. Scheffran, J. and BenDor, T. Bioenergy and land use: A spatial-agent dynamic model of energy crop production in Illinois. *Int. J. Environ. Pollut.* 39, 4, 2009.
41. Hammerschlag, R. Ethanol's energy return on investment: A survey of the literature 1990-present. *Environ. Sci. Technol.* 40, 1744, 2006.

24 Estimating Soil Productivity and Energy Efficiency Using the USDA Web Soil Survey, Soil Productivity Index Calculator, and Biofuel Energy Systems Simulator

Kurtis D. Reitsma, R. Kyle Heimerl, and Thomas E. Schumacher

CONTENTS

24.1 EXECUTIVE SUMMARY

Soils have varying production capacities for a specific plant or sequence of plants under defined management strategies. The production capacity or "productivity" can be quantified as a mathematical function of a soil's ability to sufficiently sustain plant growth and development. The result of this function is a productivity index

(PI) value that can be used to estimate crop yield and develop management strategies. This chapter demonstrates a simulation of how erosion can impact productivity, profitability, and energy efficiency for a typical corn production field in South Dakota. Energy and productivity values are calculated in an MS Excel workbook using data sourced from an online soil survey and a biofuel systems model. These data are used with spatial soils data to demonstrate the spatial variation of these parameters within a field.

24.2 INTRODUCTION

The production capacity of a soil depends on rooting depth, topsoil thickness, available water capacity (AWC), plant nutrient storage, surface runoff, soil tilth, and soil organic carbon (SOC).[1] The soil PI is calculated as a function of these parameters that can be translated into biomass and energy yield and production estimates for individual soils or aggregated for field-scale estimates. Several models have been developed for calculating soil PI values. The model presented in this chapter was developed by Pierce et al.,[2] which assesses the effects of erosion on soil productivity.

Water, wind, and gravity assisted by tillage are erosive forces that move soil from upper to lower landscape positions. Tillage loosens and exposes soil to erosive forces in addition to direct translocation of soil resulting in tillage erosion.[3,4] In eroded landscapes, managers may attempt to maintain or improve productivity by increasing material inputs, which increases energy inputs. Increasing material inputs may fail or only partially recover productivity as damage to many soil properties (available water holding capacity (WHC), rootzone depth, and soil structure) is not easily repaired. Together, increased inputs and decreased production capacity reduce energy efficiencies of crop production systems in eroded landscapes.

24.2.1 Data Sources Used in Exercise

This chapter expands on the use of soils data available from the Web Soil Survey (WSS ver. 2.2.6, 2009)[5] and the Biofuel Energy Systems Simulator (BESS ver. 2008.3.1)[6] presented in Chapters 2 and 5, respectively. Soils data sourced from WSS is entered into an MS Excel® (Microsoft® Office Excel™)[7] workbook; "Soil PI Calculator" (SPIC)[8] that uses methods developed by Pierce et al.[2] for assessing long-term impact of erosion on soil productivity. This exercise will use a 96 ac (39 ha) field located in Lake County, South Dakota, United States. Initial PI values (T_0) are calculated and compared to PI values after erosion has removed the top layer of the soil (T_1). PI values are translated into grain yields, used to build scenarios for the Biofuel Energy Systems Simulator (BESS). The BESS provides estimates of crop production energy inputs, efficiency, and net production, which demonstrate the energy impacts of erosion on a biofuels production system.

Spatial data retrieved from WSS is joined to results summarized in the Soil Productivity Index Calculator (SPIC), demonstrating the spatial variation of long-term erosion impacts on biomass and energy production. For the purpose of this chapter, an assessment was developed using the following simplifications:

- Soil conditions at T_0 (time 0) represent optimal productivity of each soil map unit (MU).
- The first defined soil layer is completely removed by erosion at T_1 (time 1) for soils likely to erode.
- A dryland production and moldboard plow tillage system was used between T_0 and T_1.
- Material inputs were not applied at a variable rate and do not change between T_0 and T_1.
- Material inputs are optimal for plant growth at T_0 and T_1.
- Initial (T_0) proven corn yield for the field was 160 bu ac^{-1}.

These simplifications assume a worst-case scenario for erosion but provide conservative crop and energy productivity impact estimates. Due to the robust nature of the WSS, an Internet connection of 100 Mb s^{-1} or greater is recommended. Minimum system recommendations include IBM-PC Pentium® 4 or higher processor, with 1 Gb RAM, operating system, Microsoft Windows (Windows 7, Vista, or XP Service Pack 2), installed software, MS Excel™ ver. 2003 or higher, ESRI® ArcGIS™ (ver. 9.3).[9] All units used in the PI workbook and discussed in this chapter are consistent with those used in the WSS or otherwise noted. Currency, units are U.S. dollars ($US).

24.2.2 Productivity Index

A PI uses soil characteristics the soils production capacity as a single value. The PI model demonstrated in this chapter can be used with site-collected or soil survey data to predict the long-term effect of erosion on soil productivity.[10,11] The general premise of this method is that productivity declines with increasing erosion. Erosion has been shown to affect physical and chemical characteristics of soil including particle size, bulk density, WHC, aggregate stability, and pH, resulting in reduced yield in corn grain.[12,13]

The PI model modified by Pierce et al.[2] derives values from the sum of sufficiency rating values of 0–1 (0 = no productivity and 1 = optimal productivity) for AWC, soil pH, bulk density, and a root distribution weighting factor. Sufficiency ratings can be found in individual soil worksheets of the SPIC. This model assumes that plant nutrient availability, climate, and management are optimal, and plant differences are constant. The resulting PI value represents conditions in the soil rootzone, which impact yield and productivity.

Based on Pierce et al.,[2] the mathematics of the PI model is expressed as

$$\mathrm{PI} = \sum_{i=1}^{r} (\mathrm{A}_i \times \mathrm{C}_i \times \mathrm{D}_i \times \mathrm{WF})$$

where

PI is the productivity index
r is the number of soil horizons
A is the sufficiency of AWC
C is the sufficiency of bulk density
D is the sufficiency of pH
WF is the weighting factor

24.3 USING THE SPIC

A blank copy of the SPIC MS Excel™ workbook (Ch24_PI_Model.xls) has been included on the CD accompanying this book. The SPIC allows calculation of PI values for up to 10 soils using the algorithm described in Section 24.2.1, but only 9 will be used in this exercise. PI values are calculated using initial conditions (T_0) for each soil and compared to PI values assuming erosion of the top layer of soil (T_1). Users may force the SPIC to remove any depth of soil by dividing or aggregating soils data. For example, to simulate erosion of the top 2 in. of a soil that has a 6 in. top layer, designate layer 1 with a lower depth of 2 in. and layer 2 with a lower depth of 6 in. Soil parameters will be the same for both layers.

Users indicate if a soil is likely to erode; soils indicated not likely to erode result in PI values at T_1 equal to PI values at T_0. The default value is "No," changing field "Likely to Erode" to "Yes" in the SPIC simulates erosion. In this exercise, erosion likelihood is based on landscape position and slope reported by WSS. Soils occurring at toe-slope and foot-slope positions or with slopes less than 2% are assumed unlikely to erode.

Yield (bu ac^{-1}) and crop value (\$US bu^{-1}) estimates are user defined and entered on the "Summary" worksheet of the SPIC. The yield value is the yield at T_0 and may be based on proven field average yield, county average yield, or other yield data. Yield at T_0 is assumed to be 160 bu ac^{-1}. Crop value may be based on contract, average annual, or current offering prices. In this exercise, corn value is based on a selling price of \$3.50 US bu^{-1} (\$154 Mg^{-1}).

All fields shaded green in the SPIC are user populated. Individual soils data are entered on worksheets "SOIL_1....SOIL_10" while field values and individual soil values from BESS are entered on the **"Summary"** worksheet. Results are summarized in two worksheets (**Summary & Summary_GIS**). The **"Summary"** worksheet provides results by soil MU, BESS yield values, and an overall field production and energy summary. The **"Summary_GIS"** worksheet provides results by soil mapping unit, which can be joined to spatial data for mapping and spatial analysis. In this exercise, these data will be joined to an ESRI® shapefile of soil mapping units that occur in the area of interest (AOI), sourced from WSS.

24.3.1 Important Tips for Using the PI Workbook

- All "green" shaded fields are populated by the user; leave these fields blank if there is no data.
- Use abbreviations provided on the worksheet(s) for soil textures.
- Area values may be obtained from WSS or results from spatial analysis of actual site-collected spatial data.
- Maintain a blank copy of the workbook, saving populated workbooks under another name. The data CD that accompanies this book has a blank workbook (Ch_24_PI_Model.xls) and one populated with data used in this excercise (Ch_24_PI_Model_Key.xls).
- Data cleared using the "Clear" command cannot be recovered if the workbook is saved.

24.3.2 Retrieving Data from Web Soil Survey and Populating SPIC

Before proceeding, copy the entire "*X*:\Chapter_24" (*X*:\ is the CD/DVD drive) folder to the root directory of your hard drive (C:\Chapter_24). Important: Failure to save example data to the proper location will complicate following the procedures described below. WSS datasets are delivered via e-mail, cued by accessing an Internet data server at an address provided in an e-mail sent to the user. The WSS uses an Internet map server as an interface to aggregate tabular data, and display user-defined map layers, Figure 24.1 illustrates the WSS environment.

We recommend calibrating your screen to the map scale to achieve maximum functionality of the interface. Screen calibration is performed as follows:

1. Open your web browser and navigate to the WSS (http://websoilsurvey.nrcs.usda.gov/) and click "Start Web Soil Survey."
2. Click the Scale button; a dialog box opens with instructions for screen calibration.
3. Placing a ruler on the screen, adjust the bar to 1 in. and click OK.

The PI model found on the accompanying CD requires several soil characteristics in order to complete the algorithm (Table 24.1).

Begin data retrieval from WSS by selecting the AOI. In this example, the field of interest (AOI) is located in Lake County, South Dakota, section 26, township 107N, range 53W, of the Fifth Principle Meridian. Using what was learned in Section 5.3, gather the necessary information to populate fields in the

FIGURE 24.1 Illustration of WSS environment.

TABLE 24.1
Parameters Required for the SPIC

Parameter	Worksheet Variable	Unit
MU name[a]	MU name	—
Number of horizons[a]	Number of horizons	—
MU symbol[a]	MU symbol	—
Horizon depth (bottom)[a]	Depth (in.)	Inches
AWC[a]	PAWHC	Inches/inch
Bulk density[a]	BD	g cm^{-3}
Soil pH[a]	pH	—
Hydrolic conductivity[a]	Ksat	Micrometer per second ($\mu m\ s^{-1}$)
Percent clay[a]	Clay%	Percent (%)
Percent passing #200 sieve[a]	Passing 200	Percent (%)
Maximum ideal yield[b]	MAX Ideal YD	Bushel per acre (bu ac^{-1})
Crop[c]	Crop	—
Value of crop[c]	Value (bu)	U.S. dollars $bushel^{-1}$ (\$ bu^{-1})
Area of soil MU[a]	Area (square feet)	Square feet (ft^2)

[a] Values obtained from WSS.
[b] User-defined value based on county average or proven yields.
[c] User-defined values.

SPIC. Area (acres) results may be slightly different depending on the accuracy and similarity of AOI delineation.

1. Under Quick Navigation, select PLSS (Section, Township, Range) and select "South Dakota" from the State: combo box.
2. Select "Fifth Principle" from the Principle Meridian combo box.
3. Enter "26," "107," "53" in the Section, Township, and Range input boxes, respectively.
4. Select the North and West radio buttons for Township and Range, respectively.
5. Click View. The map window should zoom to a scale of ≈1:10,000, showing an aerial image of section 26. The AOI for this example will be the east ½, northwest ¼, section 26, township 107N, range 53W. Note that in this example the field appears light brown, indicating that the field as shown may be a harvested small grain field.
6. Select the field as the AOI by clicking the Select AOI by Polygon tool. Begin the polygon by clicking a mark in the northeast corner of the northwest quarter of section 26. Continue the polygon by marking the northwest corner. Continue along the west border of the field marking vertices to trace the outline of the irregular shape of the west border until reaching the southwest corner, click a mark. Continue to the southeast corner, close the polygon by double clicking a mark in the southeast corner of the field. Figure 24.2 shows selection of the AOI.

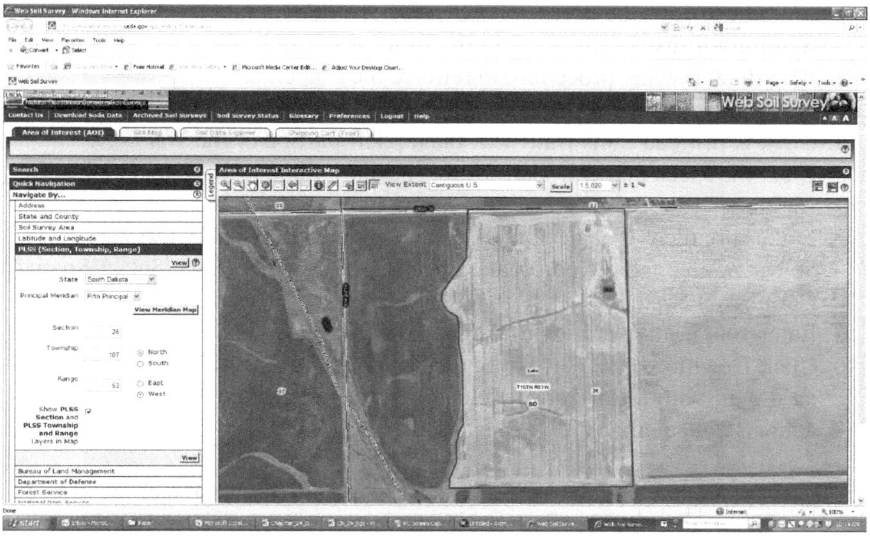

FIGURE 24.2 AOI selection in WSS.

7. The AOI selection can be re-delineated by clicking the "Clear AOI" button, and beginning the process over.
8. Click on the "Soil Map" tab on the top left corner of the WSS. The map will now show polygons of soil MUs occuring in the AOI, there should be nine total soil MUs (see Figure 24.3).

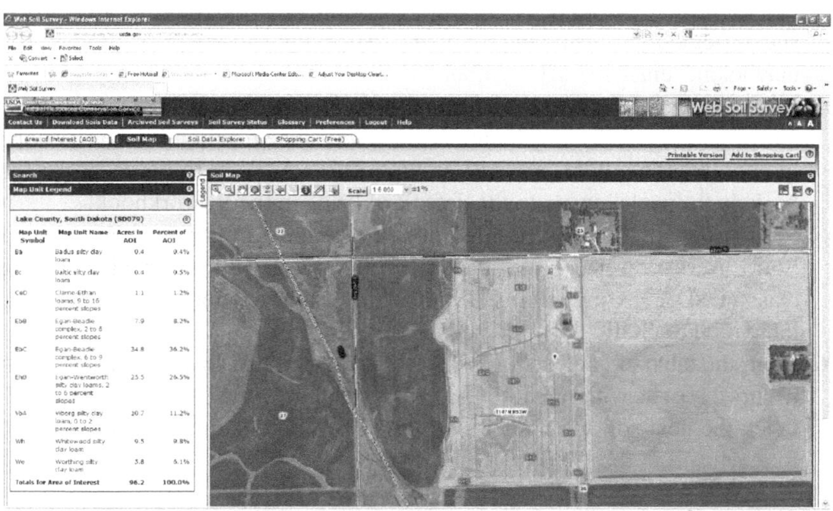

FIGURE 24.3 AOI with soil MUs shown in the WSS environment.

The WSS provides the option of retreiving spatial and tabular data for the AOI. This exercise will use spatial data in the form of a ESRI® shapefile in later procedures. Use the following procedure to retreive the shapefile:

1. Click Download Soils Data near the top left portion of the WSS.
2. Clear the check marks from the boxes next to "Tabular Data" and "Template Database."
3. Make sure there is a check mark in the box next to "Spatial Data."
4. Select "Geographic Coordinate System (NAD-83)" below "Select Spatial Coordinant System."
5. Enter your e-mail address in the input box next to "Email Address."
6. Click Download, an alert will appear noting your position in que and that an e-mail with a link for downloading the data will be sent to the e-mail address indicated.

Other data provided by the WSS is used to populate the MS Excel™ workbook to calculate PIs for individual soils considering the impact of erosion. For the purpose of this exercise, soils with slopes greater than 0%–2% will be assumed to likely erode. Estimated PI values, yields, and production for indivdual soils and the field in the AOI are calculated by the SPIC using the following procedure:

1. Navigate to "C:\Chapter_24\PI_Workbook" and open the SPIC workbook (Ch24_PI_Model.xls). Note: this workbook is included on the CD included with this book and should have been copied to your computer.
2. Upon opening the SPIC, you will note that there are worksheets for up to 10 soils, this exercise will use 9. You will also note that some cells are shaded green, this indicates fields that are user populated. For the most part, data for these fields will be obtained from WSS and BESS.
3. Click on the "SOIL_1" tab in the PI workbook to activate the first worksheet. Begin by entering the soil name (MU name), in this case "Badus silty clay loam," enter the MU symbol, "Ba."
4. In the WSS, click on the MU name to view the "Map Unit Description." Note that Badus silty clay loam occurs at toe-slope positions and has a slope of 0%–2%. Therefore, we will assume it is not likely to erode, leave the "Likely to Erode" value to the default of "No" in the PI workbook.
5. Scroll down to "Typical Profile," enter the number of horizons (Number of Soil Layers), lower depth of each horizon (Depth [in]), and USDA texture class (USDA texture) using USDA texture abbreviations provided in the table on the "SOIL_1" worksheet of the SPIC.
6. From the table of contents in WSS, it is estimated that Badus silty clay loam occurs on 0.4 ac within the AOI. Enter "0.4" in the "Area (Acres)" in the SPIC.
7. Click the next soil worksheet tab in the PI workbook and repeat steps 3–6 for each soil in the WSS table of contents.

The "Soil Reports" section of the WSS provides detailed information about each soil MU by horizon. Chemical and physical properties of each soil MU are required to estimate PI values and yield for each soil. Use the following procedure to populate the remaining parameters in the SPIC.

1. In WSS, click the "Soil Data Explorer" tab, click the "Soil Reports" tab.
2. Select "Soil Chemical Properties" from the table of contents.
3. Clear the check mark next to "Include Minor Soils."
4. Click "View Soil Report."
5. Scroll down to view the report, enter the representative pH value for each horizon of each soil MU. Representative values will need to be calculated by dividing the difference of the minimum and maximum values by 2 and adding it to the minimum value. For example, horizon 1 (0–14 in.) of Badus silty clay loam has a range of pH values of 7.4 – 8.4; 8.4 – 7.4 = 1.0, 1.0/2 = 0.5, 7.4 + 0.5 = 7.9. Therefore, the representative value would be 7.9. Repeat this procedure until all pH values for all horizons are entered. Note that more than one soil may appear for complex soil MUs, select properties for the dominant component of the soil MU.
6. Repeat steps 1–5 until all horizons for all soils are entered.

The SPIC requires physical properties as well. These are also retreived from WSS by using the following procedure:

1. Select "Soil Physical Properties" from the table of contents in WSS.
2. Select "Physical Soil Properties," clear the check-mark in the box next to "Include Minor Soils," click "View Soil Report."
3. Enter representative values for all parameters except for "Passing 200," some may need to be calculated as shown in step 5 of the previous procedure.
4. Select "Engineering Properties," clear the check-mark in the box next to "Include Minor Soils," click "View Soil Report," record "Passing 200" values for each horizon of all soils.
5. Refer to Table 24.2 for corresponding WSS and SPIC parameters.
6. After populating fields for all soils, go to the "Summary" worksheet in the SPIC. The "Field Production & Profitability Summary" should resemble Table 24.3.

TABLE 24.2
Description of Corresponding WSS and SPIC Soils Parameters

Description	WSS	SPIC Workbook
Water holding capacity[a]	AWC	PAWHC
Bulk density[a]	Moist bulk density	BD
Hydraulic conductivity[a]	Saturated hydraulic conductivity	Ksat
Percent clay[a]	Clay	Clay%
Percent passing #200 sieve[b]	Percentage passing sieve number—(200)	Passing 200

[a] Data source from Soil Physical Properties—Physical Soil Properties.
[b] Data source from Soil Physical Properties—Engineering Properties.

TABLE 24.3
Field Production and Profitability Summary

Crop	Corn
Crop value (\$US bu^{-1})	\$3.50
Total area (ac)	96
Field average yield (bu ac^{-1})	160
Average PI, T_0	0.79
Average PI, T_1	0.75
Average yield, T_0 (bu ac^{-1})	160
Average yield, T_1 (bu ac^{-1})	151
Average yield difference, $T_1 - T_0$ (bu ac^{-1})	−9
Average partial profit change, $T_1 - T_0$ (\$US ac^{-1})	−\$29.90
Total production, T_0 (bu)	17,376
Total production, T_1 (bu)	15,909
Total production change, $T_1 - T_0$ (bu)	−1,467
Net annual partial profit change, $T_1 - T_0$ (\$US)	−\$5,135.75

24.4 USING THE BESS MODEL TO CALCULATE ENERGY GAINS OR LOSSES

Chapter 2 demonstrates the calculation of net energy gains (or losses), comparing nutrient management at two landscape positions using the BESS ver. 2008.3.1.[6] The BESS model calculates the energy efficiency, greenhouse gas emission, and natural resource requirements of a corn to ethanol biofuel production system. This exercise will focus on the crop production component of the model assuming material inputs and management remains constant between T_0 and T_1.

Modified representative management (input) parameters of the U.S. Midwest provided in the BESS model (2-U.S. Midwest Average—UNL.stg) are used to build T_0 and T_1 scenarios for each soil MU. BESS uses these scenarios to calculate a biofuel production life cycle analysis for each individual soil MU. Results for T_0 and T_1 can be compared using BESS or the SPIC.

Begin by downloading and installing the BESS model using steps 1–3 in Section 2.4. Additional information is available in the BESS model user's guide[6] available at http://www.bess.unl.edu/.

BESS model scenarios (setting files) can be built from scratch or by modifying scenarios provided by the BESS model authors. This exercise will use modified management inputs provided in the BESS model (2-U.S. Midwest Average—UNL.stg) and yields for each soil MU calculated by the SPIC for T_0 (YLD_BESS_T0) and T_1 (YLD_BESS_T1). Create a model scenario file for T_0 and T_1 for each soil MU based on this scenario assuming a dryland, moldboard production system with no manure, potassium, lime, or insecticides applied. BESS model scenario files for each soil MU at T_0 and T_1 are provided on the CD that accompanies this book (C:\Chapter_24\BESS_Scenarios\). Users are encouraged to create their own scenarios to observe how results change with tillage system, material inputs, and yield (Figure 24.4). Create

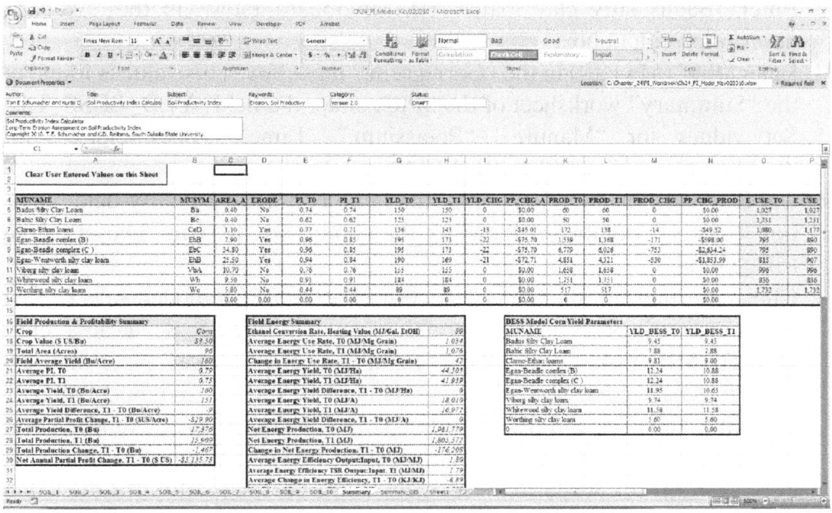

FIGURE 24.4 Completed summary worksheet of SPIC shown in MS Excel™ (ver. 2007) environment.

BESS model scenario files and calculate energy balance and efficiency values using the following procedure:

1. Start the BESS model, click "Start the model" to activate the "Input: Operation settings" section. Figure 24.5 shows a completed input scenario.
2. Click "Open a scenario," select "-U.S. Midwest Average—UNL.stg" from the list.

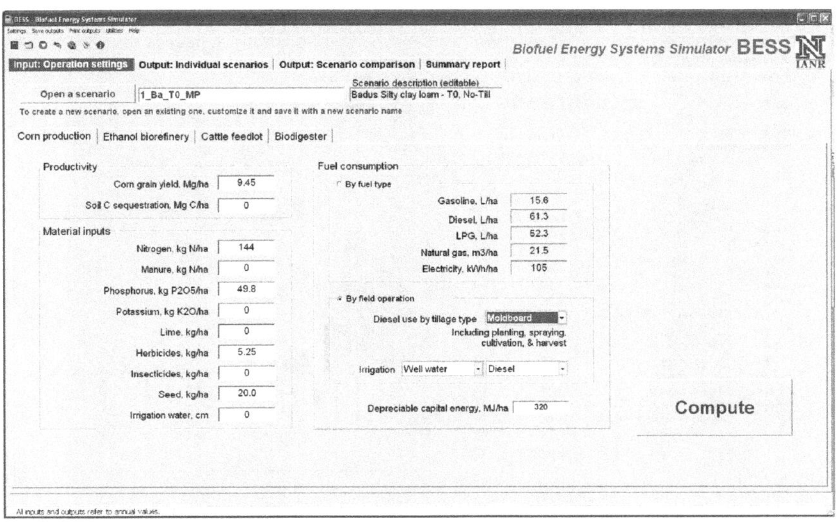

FIGURE 24.5 Input operation settings tab of the BESS model showing input parameters for Badus silty clay loam, T_0, moldboard plow.

3. Enter “Badus silty clay loam—T0, Moldboard Plow” in the “Scenario description” input box.
4. Enter “9.5” (Mg ha^{-1}) in the “Corn grain,” input box. This value is found on the “Summary” worksheet of the SPIC, under the field “YLD_BESS_T0.”
5. Set values for “Manure,” “Potassium,” “Lime,” “Insecticides,” and “Irrigation water” to “0” in the BESS model.
6. Under “Fuel Consumption” mark the radio button next to “By field operation,” select “Moldboard Plow” in the combo box next to “Diesel use by tillage type.”
7. Click “Settings” in the upper left corner of the BESS model, select “Save operation settings…” Navigate to “C:\Chapter_24\BESS_Scenarios,” enter “Ba_T0_MP” in the “File name” input box, click “Save.” Note that scenario files have been created, names include the soil MU symbol (Ba), indicate time step (T_0), and tillage system (MP), separated by “_” (underscore). Avoid the use of spaces in file naming, substitute underscores for spaces. To avoid overwriting scenario files, create a subfolder or modify the file name.
8. Click “Compute,” an output summary appears summarizing crop production results (Figure 24.6).
9. Click the “Output Individual Scenarios” tab, record the “Energy Use Rate” (1027 MJ Mg^{-1}) from BESS, in the “E_USE_T0” field of the “Summary” worksheet in the SPIC.
10. Click on the “LC analysis” tab in BESS (Figure 24.7), enter the “Net energy yield” and “Net energy ratio output:input” values from BESS in the “E_YLD_T0” and “E_EFCY_T0” fields, respectively for Badus silty clay loam in the “Summary” worksheet of the SPIC.

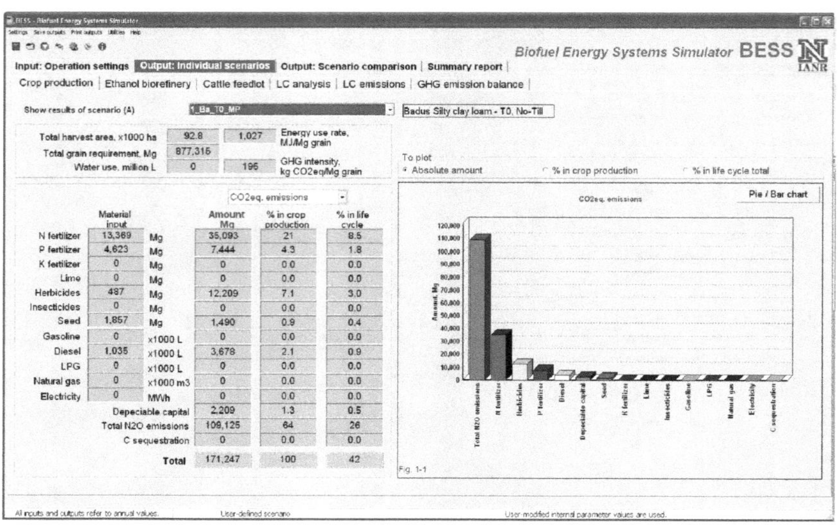

FIGURE 24.6 BESS model crop production output summary sheet showing Badus silty clay loam, T_0, moldboard plow simulation results.

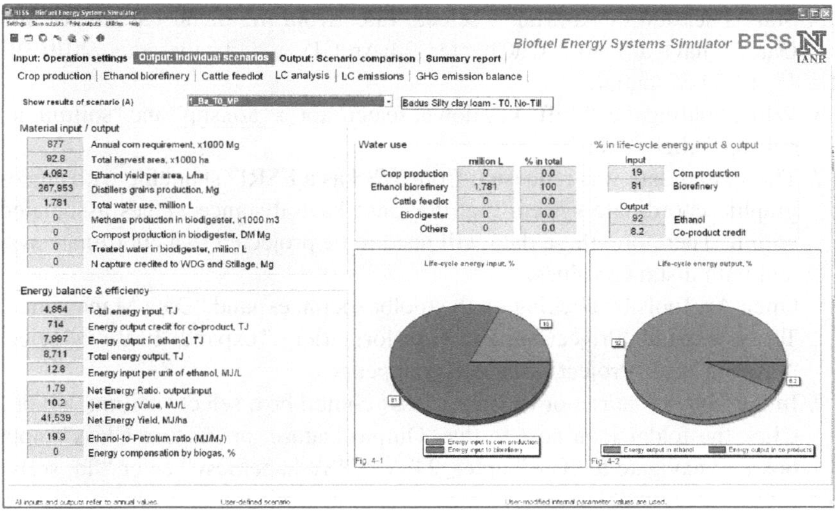

FIGURE 24.7 BESS model life cycle analysis (LC analysis) summary report for simulation of Badus silty clay loam, T0, moldboard plow.

11. Repeat the above procedure to calculate values for "E_YLD_T1" and "E_EFCY_1," editing the "Scenario description" and "Corn grain (dry matter), Mg ha^{-1}" values, save scenario file as "BA_T1_MP."
12. Repeat the above procedure for all soil MUs at T_0 and T_1.

24.5 USING ARCGIS™ TO CREATE FIELD-SCALE MAPS

Up to this point, the impacts of erosion on crop yield, profitability, energy production, and energy efficiency have been demonstrated. Field-scale erosion can be reduced to acceptable levels by implementing precision conservation efforts. Essentially, precision conservation implements conservation practices specifically designed for individual areas in a field. A geographic information system is a tool that can be used to spatially depict the impact of erosion on productivity and energy efficiency; targeting areas in the field where conservation efforts may have the greatest impact. A field-scale map of productivity or energy parameters calculated in this exercise can be created based on soil MU using the following procedure:

1. Open the e-mail received from WebSoilSurveyReports@ftc.usda.gov, click the link to the USDA-WSS ftp server, click "Save" when prompted by the "File Download" dialog box.
2. Navigate to "C:\Chapter_24\ArcGIS\wss_data", click "Save."
3. Click "Start" in MS Windows, select "My Computer," navigate to "C:\Chapter_24\ArcGIS\wss_data," extract compressed file using a typical file compression software, extracting all files and subfolders to "C:\Chapter_24\ArcGIS\wss_data."
4. After extraction, notice that a new folder has been created.

5. Start a session of ArcMap™, select "File" from the menu bar, click "Add Data," navigate to "C:\Chapter_24\ArcGIS\wss_datawss_aoi_2010-01-19_12-54-32spatial."
6. While holding the "Shift" key down, select "aoi_a_aoi.shp" and "soilmu_a_aoi.shp," click "Add."
7. The spatial data was requested from WSS as a ESRI® shapefile with a geographic coordinate system that does not have distance values associated with it. Therefore, these data will need to be projected to a coordinate system with distance values.
8. Open ArcTools by clicking on the toolbox icon, expand "Data Management Tools," expand "Projections and Transformations," expand "Feature," select "Project," the "Project," dialog box appears.
9. In the "Input Dataset or Feature Class" combo box, select "soilmu_a_aoi." Click the folder icon next to the "Output Dataset or Feature Class" input box and navigate to "C:\Chapter_24\ArcGIS\Shapefiles\," enter "fld_soils" in the name input box, click "Save."
10. Click the icon next to the "Output Coordinate System" input box, the "Coordinate System" dialog box appears.
11. Click "Select" in the "Coordinate System" dialog box, select "Projected Coordinate Systems," select "Continental," select "North America," select "USA Contiguous Albers Equal Area Conic.prj," click "Add."
12. Click "Modify," choose "Foot" in the "Linear Unit:, Name" combo box, click "Ok."
13. Click "Ok" to set the coordinate system, click "Ok" to create a projected shapefile stored in the defined location.
14. Repeat the procedure above to project the "aoi_a_aoi" theme, naming it "fld_bnd."
15. After projecting both shapefiles, click the "Page" icon in the upper left corner to create a new map, click "No" when asked if you want to save the current map.
16. Add both projected shapefiles to the new map, by clicking "Add Data," navigating to "C:\Chapter_24\ArcGIS\Shapefiles\," hold the shift key down, select "fld_bnd.shp"; "fld_soils.shp," click "Add."

The next part of this exercise involves joining data from the SPIC to spatial data in ArcMap™. ArcMap™ is a fully functional relational database allowing users to directly join spatial data with tabular data that can be used as attributes for spatial map features. ArcMap™ is fully compatible with Microsoft applications such as Excel™, Access™, and SQL™. Before joining and symbolizing soils data, we will create symbology for the field boundary (fld_bnd) using the following procedure:

1. Click the "Add Data" icon, navigate to "C:\Chapter_24\PI_Workbook\," select the SPIC MS Excel™ workbook (Ch24_PI_Model.xls), select the "Summary_GIS" worksheet (Summary_GIS$), click "Add." IMPORTANT; If using MS Excel ver. 2007, SPIC must be saved as a MS Excel 97–2003 workbook. Note that the ArcMap™ table of contents appears different as the

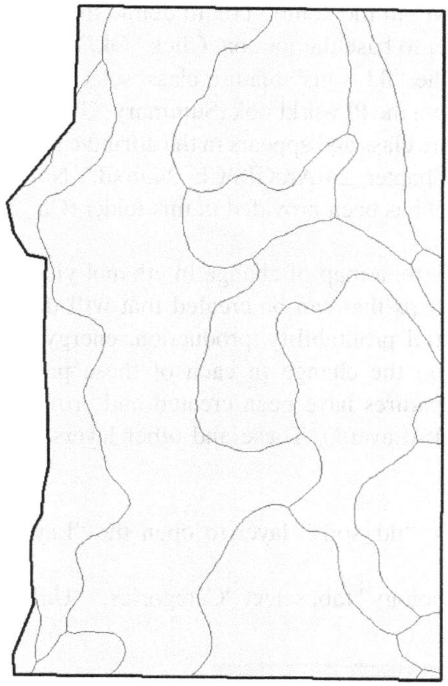

FIGURE 24.8 Map rendering of field boundary (fld_bnd.shp) and soil MU (fld_soils.shp) map feature classes.

"Source" tab is now activated. When the table of contents is in "Source" mode, paths to data are shown in headings above feature classes. Click "Display" to return to "Display" mode, which is more compact.

2. Double click the "fld_bnd" feature class to activate the "Layer Properties" dialog box.
3. Click "Symbology," click "Symbol," select "Hollow" from the symbol selector, increase outline size to "2," click "Ok," click "Ok." A map of the field with soil MUs shown in all one color should appear (Figure 24.8).

A join or relationship connects one dataset to another using a common field. Fields do not need to have the same name but must have a unique identifier to base the join or relationship on. Join data from the SPIC (Summary_GIS$) to the spatial dataset (fld_soils) using the following procedure:

1. Double click the "fld_soils" feature class to activate the "Layer Properties" dialog box, click "Joins and Relates," click "Add" to activate the "Join Data" dialog box.
2. Select "MUSYM" in the combo box to define the field in "fld_soils" that the join will be based on.
3. Select "Summary_GIS$" from the combo box in step 2 of the "Join Data" dialog box.

4. Select "MUSYM" in the combo box to define the field in the "Summary_GIS$" worksheet to base the join on. Click "Ok," click "Ok."
5. Right click on the "fld_soils" feature class, select "Open Attribute Table," note that data from the PI workbook (Summary_GIS$) has been joined to the "fld_soils" feature class and appears in the attribute table. Save the ArcMap™ project as "C:\Chapter_24\ArcGIS\Ch_24.mxd." Notice that a completed ArcMap™ project has been provided in this folder (Ch_24_key.mxd).

In this next procedure, a map of change in ethanol yield is created. There are a number of different maps that can be created that will depict the spatial variability of crop yield, partial profitability, production, energy yield, energy efficiency, energy production, and the change in each of these parameters due to erosion. Some example map features have been created and provided as ESRI® layer files (C:\Chapter_24\ArcGIS\Layers\). These and other layers can be created using the following procedure.

1. Double click the "fld_soils" layer to open the "Layer Properties" dialog box.
2. Click the "Symbology" tab, select "Categories," "Unique Values."

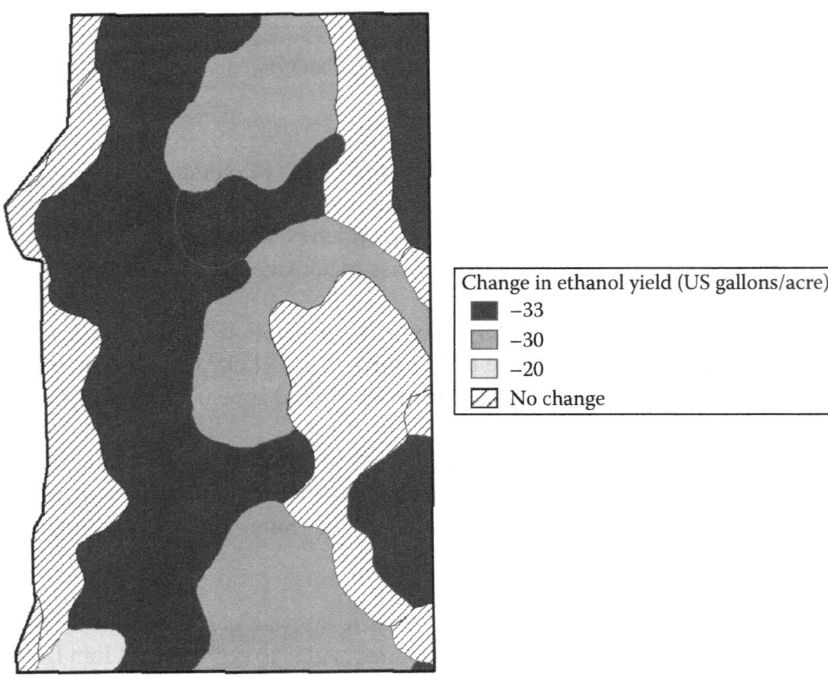

FIGURE 24.9 Map rendering of field boundary (fld_bnd.shp) and soil MU (fld_soils.shp ↔ Ch24_PI_Model.xls, GIS_Summary$, ETOH_YLD_CHG) symbolized to depict spatial variation of change in ethanol yield (U.S. gallons? ac^{-1} due to erosion).

3. Select "ETOH_YLD_CHG" in the "Value Field" combo box, click "Add All Values."
4. Unique values are selected as we are working with very little data. Click in the "Label" field for each value and round values to the nearest gallon for labeling, set "0" to "No Change." Edit the "<Heading>" label in the same manner, changing it to "US Gallons/acre."
5. Uncheck, "all other values."
6. Select an appropriate color ramp for the symbology, click "Apply." Map provided in the example ArcGIS™ project (Ch_24_key.mxd) is symbolized using gray-scale and crosshatching for "No Change" (Figure 24.9).
7. Click the "General" tab, enter "Change in Ethanol Yield," click "Apply," click "Ok."
8. This procedure can be used to map other data generated by the SPIC. Note that other methods for defining symbology may need to be used.

24.6 CONCLUSION

This simple scenario selects one 97 ac (240 ha) field with complex slopes as an example to demonstrate how erosion can impact production and energy aspects of a cropping system. The analysis assumes worst-case erosion scenario as SPIC calculates a PI value at T_1 based on removing the top layer of soil; assumed to be the first soil layer. Results from the energy analysis are conservative estimates as material inputs (fertilizer, manure, etc.) are kept constant between T_0 and T_1. However, these results demonstrate that the use of corn for biofuel production adds another facet to the importance of sustainable crop production.

Loss of top soil is predicted to lead to an overall field production loss of nearly 1500 bu annually; equating to annual partial profit loss of over $5000 (Table 24.3). The BESS predicts that energy use rate increase of 42 MJ Mg^{-1} as yield declines (Table 24.4). Overall field average yield declined by 9 bu ac^{-1} equating to a loss of 23 gal EtOH ac^{-1} annually. On an annual basis, simulation results suggest that erosion would increase crop production energy use by 42 MJ Mg^{-1} and with an overall field loss of 176,208 MJ equating to a loss of ≈2,200 gallons of ethanol (80 MJ gal $EtOH^{-1}$).[6]

Mapping results of this simulation provides an indication of spatial variability and expanse across the landscape. The map in Figure 24.8 depicts spatial variability and expanse of the change in ethanol yield across the study field. Areas that are most impacted by erosion show the greatest decline in ethanol yield making them priority areas for considering investment in conservation practices, precisely placing and designing practices to realize the greatest return.

This chapter focuses on erosion but is only one facet of sustainable agricultural production. Tillage system, and SOC, residue, water, and pest management are among other considerations for designing a sustainable cropping system. As the biofuels industry evolves and demand for biofuels increases, energy efficiency of crop production is likely to become more important.

TABLE 24.4
Field Energy Summary

Ethanol conversion rate, heating value (MJ gal $EtOH^{-1}$)	80
Average energy use rate, T_0 (MJ Mg $Grain^{-1}$)	1,034
Average energy use rate, T_1 (MJ Mg $Grain^{-1}$)	1,076
Change in energy use rate, $T_1 - T_0$ (MJ Mg $Grain^{-1}$)	42
Average energy yield, T_0 (MJ ha^{-1})	44,503
Average energy yield, T_1 (MJ ha^{-1})	41,939
Average energy yield difference, $T_1 - T_0$ (MJ ha^{-1})	–2,564
Average energy yield, T_0 (MJ ac^{-1})	18,010
Average energy yield, T_1 (MJ ac^{-1})	16,972
Average energy yield difference, $T_1 - T_0$ (MJ ac^{-1})	–1,037
Net energy production, T_0 (MJ)	1,981,779
Net energy production, T_1 (MJ)	1,805,572
Change in net energy production, $T_1 - T_0$ (MJ)	–176,208
Average energy efficiency output:input, T_0 (MJ MJ^{-1})	1.80
Average energy efficiency output:input, T_1 (MJ MJ^{-1})	1.79
Average change in energy efficiency, $T_1 - T_0$ (kJ kJ^{-1})	–8.89
Net ethanol production, T_0 (gal EtOH)	24,772
Net ethanol production, T_1 (gal EtOH)	22,570
Change in net ethanol production, $T_1 - T_0$ (gal EtOH)	–2,203
Change in ethanol yield, $T_1 - T_0$ (gal EtOH ac^{-1})	–23

ACKNOWLEDGMENTS

Partial funding was provided by: South Dakota Agricultural Experiment Station; South Dakota Corn Utilization Council; SunGrant Initiative—North Central Regional Center; Sustainable Agricultural Research and Education (SARE); United States Department of Agriculture, Conservation Innovation Grant Program; South Dakota Soybean Council; South Dakota Wheat Council; and South Dakota 2010 Initiative.

REFERENCES

1. McCormack, D.E., Young, K.K., and Kimberlin, L.W. Current criteria for determining soil loss tolerance. American Society of Agronomy, Special Publication No. 45. *Soil Sci. Soc. Am.*, Madison, WI, 1982.
2. Pierce, F.J., Larson, W.E., Dowdy, R.H., and Graham, W.A.P. Productivity of soils: Assessing long-term changes due to erosion. *J. Soil Water Conserv.* 38, 39–44, 1983.
3. Lindstrom, M.J., Nelson, W.W., Schumacher, T.E., and Lemme, G.D. Soil movement by tillage as affected by slope. *Soil Till. Res.* 17, 255–264, 1990.
4. Govers, G., Lobb, D., and Quine, T.A. Tillage erosion and translocation: Emergence of a new paradigm in soil erosion research. *Soil Till. Res.* 51, 167–174, 1999.
5. Soil Survey Staff, Natural Resources Conservation Service, United States Department of Agriculture. Web Soil Survey (WSS). Available online at http://websoilsurvey.nrcs.usda.gov/, accessed 02/03/2010.

6. Liska, A.J., Yang, H.S., Bremer, V., Walters, D.T., Erickson, G., Klopfenstein, T., Kenney, D., Tracy, P., Koelsch, R., and Cassman, K.G. *BESS: Biofuel Energy Systems Simulator. Life Cycle Energy and Emissions Analysis Model for Corn–Ethanol Biofuel.* vers. 2008.3.1. www.bess.unl.edu. University of Nebraska, Lincoln, NE, 2009.
7. Microsoft® Office Excel™, Microsoft Office Enterprise 2007, Copyright ©2006, Microsoft Corporation, Redmond, WA.
8. Schumacher, T.E. and Reitsma, K.D. *Soil Productivity Index Calculator*. Microsoft® Office, Excel™ workbook. South Dakota State University, 2010.
9. ESRI® ArcMap™ 9.3, ESRI® ArcGIS™ 9.3, Copyright ©1999–2008, ESRI Inc, Redlands, CA.
10. Lindstrom, M.J., Schumacher, T.E., Jones, A.J., and Gantzer, C. Productivity index model comparison for selected soil in North Central United States. *J. Soil Water Conserv*. 47, 491–494, 1992.
11. Schumacher, T.E., Lindstrom, M.J., Schumacher, J.A., and Lemme, G.D. Modeling spatial variation in productivity due to tillage and water erosion. *Soil Till. Res*. 51, 331–339, 1999.
12. Frye, W.W., Ebelhar, S.A., Murdock, L.W., and Blevins, R.L. Soil erosion effects on properties and productivity of two Kentucky soils. *Soil Sci. Soc. Am. J*. 46, 1051–1055, 1982.
13. Gollany, H.T., Schumacher, T.E., Lindstrom, M.J., Evenson, P.D., and Lemme, G.D. Topsoil depth and desurfacing effects on properties and productivity of a Typic Argiustoll. *Soil Sci. Soc. Am. J*. 56, 220–225, 1992.

Index

D

E

H

I

K

L

M

R

S

T

U

Y

Z

9781420092707
T - #0043 - 221019 - C0 - 234/156/25 [27] - CB - 9781420092707